Living Rivers: Trends and Challenges in Science and Management

Developments in Hydrobiology 187

Series editor
K. Martens

Living Rivers: Trends and Challenges in Science and Management

Edited by

R.S.E.W. Leuven[1,5], A.M.J. Ragas[1,5], A.J.M. Smits[2,3,5] & G. van der Velde[4,5]

[1] *Department of Environmental Science, Institute for Wetland and Water Research, Radboud University Nijmegen, Nijmegen, The Netherlands*
[2] *Centre for Water and Society, Radboud University Nijmegen, Nijmegen, The Netherlands*
[3] *Erasmus Centre for Sustainability and Management, Erasmus University Rotterdam, The Netherlands*
[4] *Department of Animal Ecology and Ecophysiology, Institute for Wetland and Water Research, Radboud University Nijmegen, Nijmegen, The Netherlands*
[5] *Members of the Netherlands Centre for River Studies (NCR), Delft, The Netherlands*

Reprinted from Hydrobiologia, Volume 565 (2006)

Springer

Library of Congress Cataloging-in-Publication Data

A C.I.P. Catalogue record for this book is available from the Library of Congress.

ISBN 1-4020-4780-0

Published by Springer,
P.O. Box 17, 3300 AA Dordrecht, The Netherlands

Cover illustration: View in upstream direction of the river Allier (France) just south of the city of Moulins (30 July 2002; © G.W. Geerling)

Printed on acid-free paper

Printed in the Netherlands

TABLE OF CONTENTS

Hydrobiologia (2006) 565:ix–x
R.S.E.W. Leuven, A.M.J. Ragas, A.J.M. Smits & G. van der Velde (eds), Living Rivers: Trends and Challenges in Science and Management
DOI 10.1007/s10750-005-1901-y

Preface

This volume of Hydrobiologia and Developments in Hydrobiology is dedicated to Prof. Dr. Pieter Hendrik Nienhuis. Piet Nienhuis worked for almost 40 years in all aspects of aquatic ecology and environmental sciences. He produced important contributions to ecological research in rivers, estuaries and coastal zones and he was a key player in national and international scientific advisory boards. On October 31, 2003 the Radboud University Nijmegen organised a symposium in honour of Piet Nienhuis, on the occasion of his retirement. The proceedings of this symposium, together with a number of complementary invited papers, are presented in this volume.

The book starts with a paper that describes the curriculum vitae and scientific career of Piet Nienhuis and his contributions to estuarine ecology and environmental science. Subsequently, the following papers are organised into three sections: (1) Functioning of river systems; (2) Rehabilitation of riverine ecosystems; (3) Challenges and obstacles to sustainable management. The issue ends with a synthesis paper, which gives an overview of the achievements, structured along the three section themes and addresses trends and challenges in river science and management.

Sustainable river management is one of the leading principles in the European Water Framework Directive. The evidence, however, to underpin the full scope of 'sustainability' is rather scanty. In this book a truly environmental sciences approach is demonstrated, implying an integrative perspective on trends and challenges in river science and management. The three pillars underneath sustainable water management, ecology, economy and sociology, are elaborated by experts in their fields. A number of papers integrate the present knowledge on 'living rivers'. We included papers about the structure, functioning and management of the rivers Allier, Meuse, Rhine, Sava and Tagliamento in Europe and the river Illinois in the USA. Sustainable river management asks for unorthodox rehabilitation programmes and ecosystem based transboundary river basin management.

The organisation of the symposium and the publication of this special issue were financially supported by the Bargerveen Foundation, water board Waterschap Brabantse Delta, European InterregIIIb 'Freude am Fluss project', and Netherlands Institute of Ecology and Radboud University Nijmegen (i.e. Faculty of Science, Centre of Water and Society, Centre for Wetland Ecology, Institute for Wetland and Water Research, University Centre for Environmental Sciences and Sustainable Development, Department of Experimental Plant Ecology, Department of Aquatic Ecology and Environmental Biology, Department of Environmental Science and Department of Animal Ecology and Ecophysiology).

We like to thank 67 reviewers from 10 European countries and from the United States of America, for their thorough work, which certainly improved the standard of this volume: B. Aarts M.Sc. (Stichting Staring Advies, Zelhem), Dr. C. Ahn (George Mason University, Fairfax), Prof. Dr. C. Amoros (University Claude Bernard, Lyon, Villeurbanne), Dr. J.M. Armitage (Stockholm University, Stockholm), Prof. Ir. E. van Beek (WL|Delft Hydraulics, Delft), Prof. Dr. E.P.H. Best (U.S. Army Engineer Research & Development Center, Vicksburg), Dr. A.T. Blowers (Open University, Milton Keynes), Dr. G. Bornette (University Claude Bernard, Lyon, Villeurbanne), Dr. F.W.B. van den Brink (Province of Limburg, Maastricht), Dr. A.D. Buijse (RIZA – Institute for Inland Water Management and Waste Water Treatment, Lelystad), W. Bosman M.Sc. (RAVON – Society for Reptiles, Amphibians and Fish Surveys, Nijmegen), Dr. H. Coops (RIZA – Institute for Inland Water Management and Waste Water Treatment, Lelystad), Dr. I. Cousins (Stockholm University, Stockholm), Dr. S. Declerck (Catholic University, Leuven), Dr. J.T.A. Dick (Queens

University Belfast, Belfast), N. Douben M.Sc. (UNESCO-IHE Institute for Water Education, Delft), Dr. I. Durance (Cardiff University, Cardiff; University of Rouen, Rouen), Dr. D.J. Gilvear (University of Stirling, Stirling), Dr. P. Goethals (Ghent University, Ghent), Dr. R. Gulati (NIOO-KNAW – Centre for Limnology, Nieuwersluis), Prof. Dr. Ir. A.J. Hendriks (Radboud University Nijmegen, Nijmegen), Dr. G. Huang (University of Regina, Regina), Prof. Dr. C. den Hartog (emeritus Radboud University Nijmegen, Nijmegen), Dr. D. Hering (University of Duisburg-Essen, Essen), Prof. Dr. V. de Jonge (University of Hull, Hull; University of Groningen, Groningen), Dr. M.M. van Katwijk (Radboud University Nijmegen, Nijmegen), Prof. Dr. V. Kesminas (Vilnius University, Vilnius), Dr. M. Knoester (Ministry of Public Works, Traffic and Water Management, Middelburg), Dr. T. Koellner (ETH – Swiss Federal Institute of Technology, Zürich), Dr. Ir. L. Kooistra (Wageningen University and Research – Alterra, Wageningen), Dr. E. Krywko (Catholic University of Lublin, Lublin), Dr. J.J. de Leeuw (Netherlands Institute for Fisheries Research, Wageningen), Dr. H.J.R. Lenders (Radboud University Nijmegen, Nijmegen), Prof. Dr. Ir. W. van Leussen (University of Twente, Enschede), Prof. Dr. H.J. MacIsaac (University of Windsor, Windsor), Dr. S. Meijerink (Radboud University Nijmegen, Nijmegen), Ir. K.S. Meijer (WL|Delft Hydraulics / Delft University of Technology, Delft), Dr. H. Middelkoop (Utrecht University, Utrecht), Prof. Dr. B. Mitchell (Waterloo University, Waterloo), Dr. N. Mount (University of London, London), Prof. Dr. J. O'Keeffe (UNESCO-IHE Institute for Water Education, Delft), Dr. H. Olde Venterink (ETH – Geobotanical Institute, Zürich), Prof. Dr. S. Ormerod (Cardiff University, Cardiff), Prof. Dr. L. O'Toole (University of Twente, Enschede), C. van Overdijk M.Sc. (Institute for Great Lakes Research, Windsor), Prof. Dr. N. de Pauw (Ghent University, Ghent), Dr. D.T. Patten (Montana State University, Bozeman), Dr. M. Pintar (University of Ljubljana, Ljubljana), Dr. N.L. Poff (Colorado State University, Fort Collins), Prof. Dr. M. Reich (University of Hannover, Hannover), Dr. K. Reise (Alfred Wegener Institute for Polar and Marine Research, Sylt), Dr. L. Sandin (Swedish University of Agricultural Sciences, Uppsala), Dr. M. Scheringer (ETH – Swiss Federal Institute of Technology, Zürich), Prof. Dr. R.W. Scholz (Swiss Federal Institute of Technology – ETH, Zürich), Dr. M.F. Smith (Southern Illinois University, Edwardsville), C. Storm M.Sc. (Ministry of Public Works, Traffic and Water Management, Rotterdam), Dr. H. Strijbosch (Radboud University Nijmegen, Nijmegen), Dr. J.C. Stromberg (Arizona State University, Tempe), Dr. R.M. Teeuw (University of Portsmouth, Portsmouth), Dr. K. Tockner (EAWAG/ETH – Swiss Federal Institute of Technology, Duebendorf), Dr. A. bij de Vaate (RIZA – Institute for Inland Water Management and Waste Water Treatment, Lelystad), Dr. J. van der Velden (Ministry of Public Works, Traffic and Water Management, Rotterdam), Ir. J.M. Verhallen M.Sc. (Wageningen University and Research, Wageningen), Dr. Ir. J.E. Vermaat (Free University, Amsterdam), Dr. C. Wolter (Leibniz-Institute of Freshwater Ecology and Inland Fisheries, Berlin), Dr. J.J.G. Zwolsman (Kiwa Water Research, Nieuwegein),

Last, but not least, we wish to thank R.M.M. Delmee and N.J. Ruitenbeek-Mohr (Department of Environmental Science, Radboud University Nijmegen) for secretarial support during the organisation of the symposium and editing of the book.

R.S.E.W. Leuven
A.M.J. Ragas
A.J.M. Smits
G. van der Velde

Hydrobiologia (2006) 565:1–18
R.S.E.W. Leuven, A.M.J. Ragas, A.J.M. Smits & G. van der Velde (eds), Living Rivers: Trends and Challenges in Science and Management
DOI 10.1007/s10750-005-1902-x

Pieter Hendrik Nienhuis: aquatic ecologist and environmental scientist

R.S.E.W. Leuven[1,*], P.J. van den Heuvel[2], M.M. van Katwijk[1], P.M.J. Herman[1,3], G. van der Velde[4] & A.M.J. Ragas[1]

[1]*Department of Environmental Science, Institute for Wetland and Water Research, Radboud University Nijmegen, P.O. Box 9010, 6500 GL Nijmegen, The Netherlands*
[2]*University Centre for Environmental Sciences and Sustainable Development, Radboud University Nijmegen, P.O. Box 9010, 6500 GL Nijmegen, The Netherlands*
[3]*Netherlands Institute of Ecology (NIOO), Royal Academy of Sciences and Arts (KNAW), Korringaweg 7, 4401 NT Yerseke, The Netherlands*
[4]*Department of Animal Ecology and Ecophysiology, Institute for Wetland and Water Research, Radboud University Nijmegen, P.O. Box 9010, 6500 GL Nijmegen, The Netherlands*
*(*Author for correspondence: E-mail: r.leuven@science.ru.nl)*

Key words: coastal zone management, estuarine ecology, environmental sciences, interdisciplinary research, river basin management, river ecology

Abstract

Prof. Dr. Pieter Hendrik (Piet) Nienhuis worked for almost 40 years in all aspects of aquatic ecology and environmental science and retired on 31 October 2003. He can be characterised as a distinguished scientist, shaped in an applied estuarine and aquatic research ambience of the former Delta Institute for Hydrobiological Research (DIHO) in Yerseke in the Netherlands. His appointment as a full professor at the Radboud University Nijmegen offered him a challenging step from monodisciplinarity in ecology, via multidisciplinarity in the application of ecological knowledge in river science to interdisciplinarity in environmental science and management. This paper describes his education, teaching activities, research, scientific publications, science management, and significance for various scientific disciplines. He made important contributions to biosystematics of angiosperms and algae, the ecology of seagrasses, nutrient cycling and eutrophication in estuarine ecosystems, and the integrated modelling of the ecological functioning of estuaries. Subsequently, he paid much attention to environmental problems in river basins, ecological rehabilitation and sustainable development. His work influenced the view of ecologists, aquatic scientists and water managers in the Netherlands as well as abroad, in particular regarding the drawbacks of compartmentalization of the estuaries and the importance of connectivity and morphodynamics in river systems. In hindsight, it appears as a logical line that he gradually moved from estuarine ecological research that became increasingly driven by societal and environmental problems to the field of environmental science and management.

Introduction

The special issue 'Living rivers: trends and challenges in science and management' of Hydrobiologia is dedicated to Prof. Dr. Pieter Hendrik (Piet) Nienhuis (Fig. 1), on the occasion of his 65th birthday and retirement (Leuven et al., 2006). He worked for almost 40 years in all aspects of aquatic ecology and environmental science and was one of the key players in the Dutch scientific arena and international scientific boards for ecology and management of rivers, their estuaries and

Figure 1. Prof. Dr. Pieter Hendrik Nienhuis.

coastal zones. On 31 October 2003 the Radboud University Nijmegen organised a symposium in honour of Piet Nienhuis, including his valedictory lecture (Nienhuis, 2003, 2006). The special issue presents the proceedings of this symposium, together with a number of complementary invited papers. The present paper briefly describes his education and scientific career. It particularly focuses on his major contributions to estuarine ecology and environmental science and to the education of undergraduates and PhD students.

Education and family circle

Piet Nienhuis was born on 29 October 1938 in Groningen, a city in the north-eastern part of the Netherlands. His love for biology was already triggered at primary school, where an enthusiastic schoolmaster took the class on excursions and taught the pupils to identify wild flowers. He became an active member of a youth society for nature, and he prepared a detailed herbarium of the flora around Groningen and a geological collection of stones from glacial deposits in the province of Drenthe. He developed a broad interest in geology and palaeontology, but he decided to choose biology and never regretted that choice.

He passed high school (Christelijk Lyceum: HBS-B) in Amsterdam. With his solid Calvinistic background he struggled with problems of creation and evolution, and it was a matter of course that he started his study in biology at the Free University of Amsterdam. He carried out several doctorate studies that comprised both theoretical and practical work and he specialized in plant ecology, plant physiology, animal ecology and biogeography. He worked on the biosystematics and ecology of the Sheep's sorrel (*Rumex acetosella* s.l.), in garden experiments and quantitative field studies, demonstrating the phenotypic plasticity of this complex of plant species. He studied the feeding ecology of birds of prey, i.e. three sympatric living Harrier species (*Circus* species) at the island of Terschelling, during the breeding season. The experimental life cycle studies of freshwater algae of the wetland area Botshol have eventually been decisive for his career as an aquatic ecologist. In 1965 he received his 'doctoral' graduation in Biology (equivalent to Master of Science).

During his study in Amsterdam he met his wife Arine Snaterse and they married in 1964. Piet and Arine have three children, Martine (1969), Pieter (1970) and Arjan (1976). Over the period 1966–1994 the family lived in Goes, a city that is situated in the delta of the rivers Rhine, Meuse and Scheldt. In 1994 they moved to the town of Zaltbommel (and later on to the village Rossum) in the scenic Rhine-Meuse river district in the central part of the Netherlands.

A professional career affected by flooding disasters

As an undergraduate student Piet Nienhuis was already employed as teacher biology at a college for elementary school-masters 'Christelijke Kweekschool voor Onderwijzers' in The Hague (1962–1963) and he also was appointed as student-assistant for plant taxonomy and plant ecology at the Free University of Amsterdam (1963–1965).

In October 1965 he took service as biologist (aquatic ecologist, phycologist) at the Department of Aquatic Botany of the former Delta Institute for Hydrobiological Research (DIHO) of the Dutch Royal Academy of Sciences and Arts in Yerseke (now Centre for Estuarine and Marine Ecology of the Netherlands Institute of Ecology; NIOO-CEME). The flooding disaster in the Rhine-Meuse-Scheldt estuary (south-western part of the Netherlands) of 1953, with more than 1835 human casualties and uncountable damage to human goods and chattels (Nienhuis, 2006), had led to the foundation of this research institute in 1957. The DIHO was founded to analyse the ecological consequences of the execution of the Delta Works, i.e. the construction of a system of high-tech dams to close the mouths of the estuaries of the rivers Rhine and Meuse, in order to prevent future damage by storm surges from the North Sea. Piet Nienhuis became head of the Department Aquatic Botany and he was known in that group as "Piet macrofiet" (Piet macrophyte). Over the periods 1976–1994 and 1994–1998 he was appointed as research group leader of the Centre for Estuarine and Marine Ecology (NIOO-CEME) in Yerseke and the Centre for Limnology of the Netherlands Institute of Ecology (NIOO-CL) in Nieuwersluis, respectively.

Over the period 1988–2001 he was appointed as visiting professor in tropical marine ecology and management at the Free University of Brussels (VUB) in Belgium. From 1988 until 1994 he was also connected to the University of Nijmegen (now Radboud University Nijmegen), as part-time professor Estuarine Ecology (Nienhuis, 1988). In 1994 he became full professor environmental science, head of the Department of Environmental Science of the Faculty of Science and chairman of the University Centre for Environmental Sciences and Sustainable Development (UCM-DO) of the Radboud University Nijmegen. He joined a motivated and ambitious group of young scientists, executing mainly environmental education and education-sustaining research on generic topics, such as sustainability indicators, environmental utility space and environmental quality standards (Anonymous, 1992). The Department of Environmental Science was founded in 1991 and had three main tasks: (1) to co-ordinate the Undergraduate School of Environmental Science (MSc school); (2) to perform scientific research; (3) to provide services to society on environmental science. The education of students predominantly occupied his time. However, a considerable effort was also invested in environmental research focusing on the interaction between the natural environment and the human society, both based on funding by the university and by external sources. Already in 1995 part of the work at the department was focused on the ecological basis of sustainable river management (Nienhuis, 1995). This approach gained momentum owing to the narrow escape from extremely high floods in the rivers Rhine and Meuse in 1993 and 1995. In 1995 there was a real danger that the river dikes along the rivers Rhine and Meuse should be overtopped or should breach. As a precautionary measure approximately 250,000 citizens, including the family Nienhuis, were evacuated from the threatened area. The threat of floods triggered the development of a new water management strategy in the Netherlands (Water Management in the 21st century; Room for the Rivers) and an academic focus on sustainable river basin management (e.g. Van Stokkom et al., 2005; Wiering & Arts, 2006). From 1995 onwards research activities of Piet Nienhuis and his co-workers were concentrating on river basin oriented research (Nienhuis, 1995; Nienhuis et al., 1999).

In total Piet Nienhuis co-authored about 233 scientific publications, among which are 123 international journal papers, books and book chapters (Appendix), and 110 professional and other publications. He was invited speaker and organiser of roughly 50 symposia and seminars in foreign countries. After his retirement he continued publishing activities. For instance, at present he is writing a new book on the ecological history of the lowland rivers in the Netherlands. His motto is "publish or perish".

Significance for estuarine ecology

His PhD thesis was devoted to the biosystematics and ecology of *Rhizoclonium riparium*, a complex of filamentous green algae, and a number of other estuarine algae occurring under the same environmental conditions (Nienhuis, 1975). He showed in field studies and laboratory experiments that *Rhizoclonium riparium* is in fact a holeuryhaline green alga, demonstrating an enormous phenotypic plasticity, occurring both under extreme saltwater conditions as well as in eutrophic freshwater stretches of rivers. Comparable strategies could be demonstrated for other estuarine algal species. He defended his thesis at the State University of Groningen in 1975; his promotor was Prof. Dr. C. van den Hoek.

Over the years 1975–1994 he acquired major research grants (in total more than 10 million Euros) from the European Commission and the Netherlands Organisation for Scientific Research (NWO), the Dutch Royal Academy of Sciences and Arts, the Dutch ministries of Transport and Public Works (Rijkswaterstaat) and of the Housing, Physical Planning and Environment. These grants were allotted to the Netherlands Institute of Ecology. As research group leader of this institute Piet Nienhuis was responsible for the scientific management of a number of projects, e.g. on the ecology of lake Grevelingen (ZOWEC), the Oosterschelde (BALANS I and II, and EOS), suspension feeders (FAR) and eutrophication and primary producers (BEON-EUTRO). He coordinated research programmes and carried out research projects on sustainable management of estuaries and coastal zones in the Netherlands and several other western European countries, Central Europe, USA, Indonesia and Mauritania (West Africa). Most of these projects were carried out in cooperation with staff members of the Dutch ministry of Transport and Public Works. He also was project leader and co-ordinator of a large research contract on eutrophication and marine macrophytes (EUMAC project) in the Fourth Framework Programme Environment and Climate of the European Commission (DGXII) and over the period 1993–1996 he organised several EUMAC-workshops in Paris, Thessaloniki, Venice and Sête.

He was an early adopter of systems ecology, inspired by the brothers Odum and their colleagues in the USA (Nienhuis, 2006). Already in the 1970's, he enthused a group of colleagues at the Netherlands Institute of Ecology in Yerseke, WL | Delft Hydraulics, the Netherlands Institute for Sea Research (NIOZ) and the Dutch ministry of Transport and Public Works to join forces in systems-oriented research of the major carbon and nutrient fluxes in the estuaries under transition. Although he was not a modeller himself, he attached great importance to ecological simulation modelling right from the start. The first big project in a series was focused on lake Grevelingen, a former estuary that was transformed into a saline lake in the early 1970's. In a comprehensive list of studies, many of which were bundled in a 1984 issue of Netherlands Journal of Sea Research, Nienhuis and colleagues investigated primary production, nutrient exchanges (with special emphasis on benthic processes) and benthic consumption set in a hydrodynamically sound context (Lambeck et al., 1984). His own original scientific contributions already at that time focused on eelgrass and macrophytes, and some of his best-cited papers examine biomass production and consumption processes of eelgrass in lake Grevelingen (Nienhuis & de Bree, 1977, 1980).

With the shift of the major Delta works to the Oosterschelde, the BALANS projects in the 1980's extended the system-oriented approach to this estuary in its original state. In the second half of the 1980's, the ecology of the Oosterschelde (EOS) project evaluated the changes in the estuarine system caused by the construction of a storm surge barrier. Many results of these studies have been bundled in a special issue of Hydrobiologia (Nienhuis & Smaal, 1994), which still is a reference publication for ecological processes and changes in this estuary. The Oosterschelde studies were multidisciplinary, encompassing approaches to geomorphology, hydrodynamics, dynamics of salt marshes, primary production by phytoplankton and macrophytes, zooplankton, zoobenthos and mussel fisheries. Again, mathematical modelling was used as the unifying and co-ordinating vehicle to streamline the diverse approaches. Nienhuis' own contributions remained focused on macrophytes, and his major role as inspirator and co-ordinator of a large and diverse research project remains somewhat hidden in his publication output. In so doing, however, he has given ample opportunity to many young scientists to profile themselves and to start their career in estuarine ecology.

During the 1980's, he continued in-depth studies of eelgrass ecology in lake Grevelingen, where at that time a spectacular blooming of *Zostera marina* was observed, and he extended his research to tropical areas. He took part of the Snellius II expedition and his first publications on seagrasses in Indonesia appeared, and later he participated in studies of the Banc d'Arguin in Mauritania. At the European level, he started to co-ordinate several projects on macrophytes (in the context of eutrophication) and on the ecology of benthic filter feeders.

In the early 1990s this diversification in study systems, together with an increased focus on seagrasses and their fascinating ecology, became the leading theme of his estuarine ecological research. The increase in his publication output on seagrasses strangely coincided with dramatic changes in the population density of seagrass in lake Grevelingen and other estuaries in the south-western part of the Netherlands. Figure 2 illustrates how environmental problems have always been a source of inspiration to direct his research. After having studied mainly the effects of engineering works on the functioning of estuarine ecosystems, eutrophication became a major leading theme in the European projects he co-ordinated. Eutrophication was also suspected to be a hidden cause behind the disappearance of eelgrass in the south-western part of Netherlands, although the relation could never be proven and it appeared that changes in salinity (caused by engineering works!) were probably more important (Nienhuis, 2006). Figure 3 visualises the changes in content of his seagrass publications over the period 1974–1997.

He also made a contribution to invasion biology as he discovered the Japanese brown alga *Sargassum muticum* in the Netherlands. Because fishermen suffered from contact dermatitis caused by this alga he published also a paper in a medical journal (Van de Willigen et al., 1988). He was very proud on his chapters in a book in Dutch entitled 'De Nederlandse Delta' in fact a bundle of scientific popularized papers (Duursma et al., 1982). This voluminous book on the ecology of the delta and the influence of the delta works was widely spread and was even available on the table of his barber for reading by the waiting clients. The same procedure was followed for a series of papers, earlier published in the Dutch journal 'Natuur & Techniek', on the transition of the lake Grevelingen from an estuary to a saltwater lake (Nienhuis, 1985). These books present the work of the Delta Institute for Hydrobiological Research for a wide audience in the Netherlands, leading to a better awareness of the impact of hydraulic works on the ecosystem and an integration of these views in present day management.

Contributions to environmental science

Owing to his teaching commitment as full professor, from 1995 onwards his disciplinary orientation was predominantly focused on environmental science. Nienhuis and his co-workers choose to concentrate on river basin-oriented research, thus exploiting both the already existing research focal points within the Radboud University Nijmegen and the geographical position of the city of Nijmegen in the Rhine-Meuse river district (Nienhuis, 1995; Nienhuis et al., 1999). His research

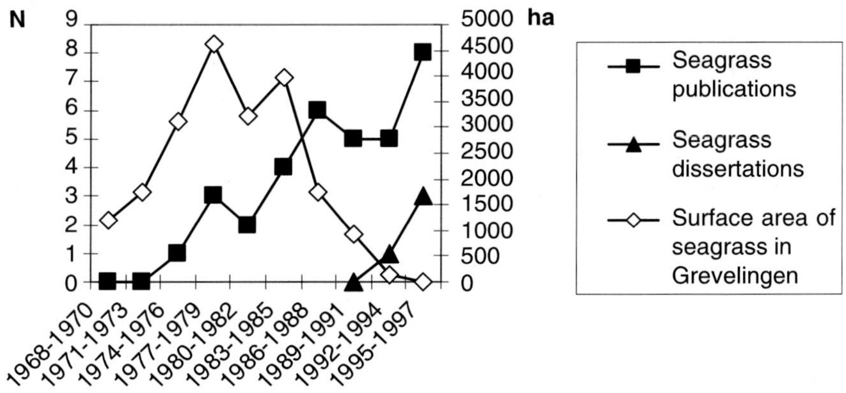

Figure 2. Effects of enclosure dams on surface area of seagrass in lake Grevelingen triggered much scientific research of Nienhuis and co-workers, cumulating in 35 publications and 4 dissertations on seagrass in 1997.

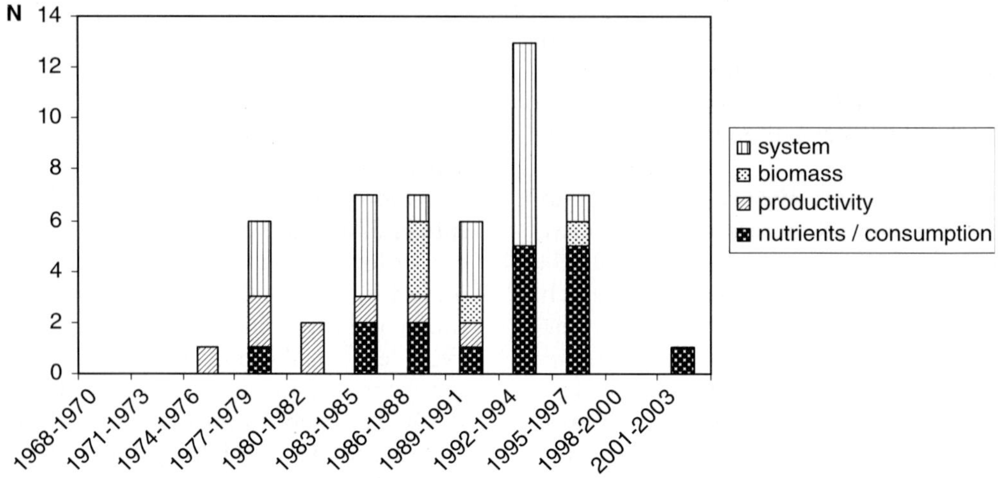

Figure 3. Changing content of seagrass publications of Nienhuis and co-workers (N: number of publications).

programme focused on ecological quality (i.e. water- and sediment quality and habitat quality) and ecological rehabilitation of river basins. Nienhuis also introduced his tradition of tropical expertise in river mouths and coastal areas into the research programme of Department of Environmental Science. The lessons learnt in European river basins were applied in developing countries. Starting from different types of human impact in the societal domain, which influenced river ecosystem health, problems were formulated in the domain of the natural sciences, to be studied in a river ecosystem model, including validation and uncertainty analysis. The results of these model exercises were used to sustain policy and management options, again in the domain of human society. So, his research programme resulted in (1) integration of physical, chemical and biological (mainly ecological and toxicological) knowledge in validated (computer) models that describe and predict the development and rehabilitation of disturbed river ecosystems (river ecosystem model), and (2) tools for management and policy purposes that balance and integrate environmental (ecological) quality of river systems with other functions (contributions to river ecosystem health). An external assessment committee of research quality stated that his research programme had "an important scientific output and a potential for future expansion and funding" (Anonymous, 2000). The research programme as a whole remained very relevant both for the theoretical advancement of science and for policy, and this was reflected in the high relevance score. The committee was pleased to read that the Radboud University Nijmegen did make a commitment that it will continue to support this programme after his retirement, so that river research will be kept as a central topic in environmental science in the Netherlands.

Figure 4a shows the trends that can be observed in the environmental science publications of Nienhuis during his active career (1967–2004). His move from Yerseke and Nieuwersluis to Nijmegen in 1995 clearly triggered a change in his main research topic: from the ecology of estuaries towards the sustainable management of river basins. The change from the monodisciplinarity of ecology to the interdisciplinarity of environmental science seems quite abrupt, but already during his period in Yerseke and Nieuwersluis he developed a strong interest in management issues and he published a considerable number of papers about integrated assessments of human impacts on estuaries. Over the period 1967–1995, the nature of his publications changed from descriptive ecological papers on the status of estuarine systems to more prospective papers presenting management alternatives to conserve and restore these valuable ecosystems (Fig. 4b). His move to Nijmegen matches well with his evolving interest for conservation and management issues.

Although river basins became the main object of research in Nijmegen, Nienhuis' affinity for estuaries never ceased to exist. He was a strong

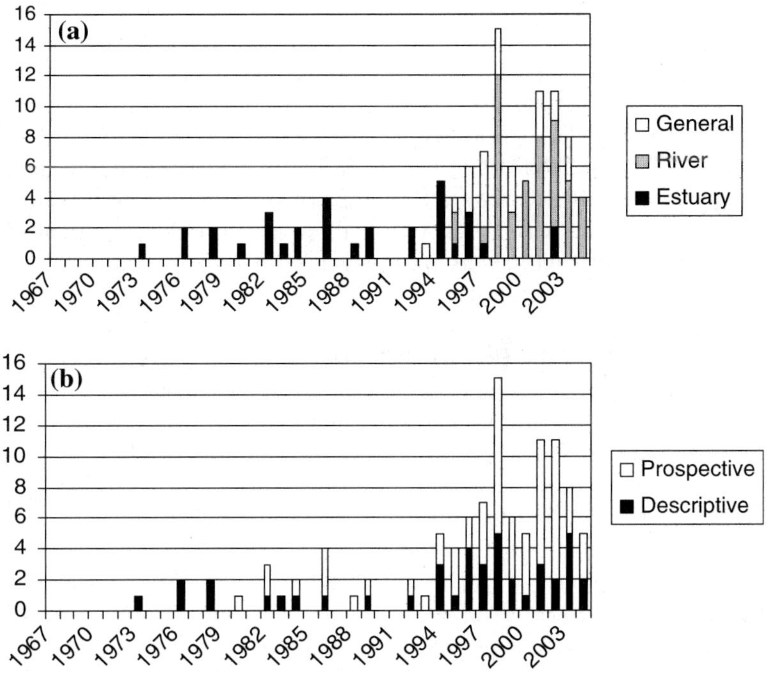

Figure 4. Characterization of Nienhuis' publications in environmental science. Panel a: publications dealing with (1) estuaries, (2) river basins and (3) general environmental science. Panel b: development in (1) prospective and (2) descriptive publications.

proponent of the river continuum concept (Vannote et al., 1980) which considers the estuary to be an integral part of the river basin. His key publications in this period concentrated on the importance of ecological concepts for rehabilitation and sustainable management of river systems (Nienhuis & Leuven, 1998a, 1998b; Nienhuis & Leuven, 2001; Nienhuis et al., 2002). Nienhuis stressed that a good understanding of human impacts on river basins requires integral consideration of the entire environmental impact chain and the river continuum from source to estuary (Nienhuis & Leuven, 1998b). Although Nienhuis' main topic was river basins, he also contributed significantly to the development of the environmental science in general (Leroy & Nienhuis, 1997; Van den Heuvel et al., 1997); and concepts such as sustainability (Sprengers et al., 1996; Nienhuis & Leuven, 1997; Nienhuis, 1998) and biodiversity (Aarts & Nienhuis, 1999; Lenders et al., 2001). Noteworthy is a joint paper with his colleague Leroy of the Nijmegen School of Management in which they accurately characterize the development of environmental science in the Netherlands from holistic discipline to serial specialism (Leroy & Nienhuis, 1997).

In the course of the years Piet Nienhuis succeeded in integrating his research programme in a number of national and international scientific networks, guaranteeing long-term viability of the Department of Environmental Science. Important networks are the Netherlands Centre for River Studies (NCR), the Netherlands Centre for Nature Studies (NCN) and the research school Socio-Economic and Natural Sciences of the Environment (SENSE). The NCR is a good example of a fruitful collaboration of major developers and users of expertise on river science and management (Leuven et al., 2003). The NCN is an alliance of the Radboud University Nijmegen, the Bargerveen Foundation, the Dutch Centre for Field Ornithology (SOVON), the Institute of Reptiles, Amphibians and Fish Research in the Netherlands (RAVON) and the Institute of Flora en Fauna Research (VOFF). This alliance is active in the field of nature conservation and policy. The SENSE Research School is a joint venture of environmental research institutes of eight Dutch universities. This national PhD school is accredited by the Dutch Royal Academy of Sciences. The scientific mission of SENSE is to develop and promote an integrated understanding of environmental change in terms of the mechanisms that cause it and the consequences that result from it. To fulfil this mission, the combined programmes of research and education within SENSE are aimed at the development and further improvement of scientific concepts and methods that are required for an effective disciplinary and multidisciplinary understanding of environmental change, so as to promote effective environmental management. Piet Nienhuis was member of the General Board and manager of the core programme Environmental change and ecosystem dynamics of SENSE.

Participation of the Department of Environmental Science in the above-mentioned networks gave rise to substantial external funding and offered new opportunities for acquisition of large multidisciplinary and interdisciplinary research projects. For instance, the near flooding disasters of 1993 and 1995 in the Rhine-Meuse river district, triggered research on sustainable river basin management and new approaches for river management (e.g. combining the reduction of flooding risks with the conservation and rehabilitation of biodiversity values and ecosystem functions). As a member of the supervisory board of the NCR (1998–2003), Piet Nienhuis played an important role in the acquisition and the scientific advisory board of the IRMA-SPONGE project (1999–2002), a large research project on sustainable flood risk management in the Rhine and Meuse river basins financed by the European Commission (Klijn et al., 2004). The NCN acquired several integrated research projects, concerning the effects of various harmful factors (e.g. acidification, eutrophication and desiccation), ecological rehabilitation and management measures on biodiversity and the functioning of species and ecological communities (Nienhuis, 2006).

University Centre for Environmental Sciences and Sustainable Development

Over the period 1994–2003 Piet Nienhuis was the first official chairman of the Centre for Environmental Sciences and Sustainable Development (UCM-DO) at the Radboud University Nijmegen. This interdisciplinary centre was already found in 1993 by his predecessor Dr. D.J.W. Schoof and colleagues from several faculties. Originally, the aim of the centre was to coordinate interdisciplinary environmental education and to stimulate students of environmental sciences to look critically at the environment and environmental issues from different scientific angles. The centre could offer a comprehensive curriculum of widely varying courses and field-work on environmental sciences. From its inception the centre also played a coordinating role in environmental research at the university. It served as intermediary for awarding environmental research to the best-suited department or research group.

Although Nienhuis was a genuine natural scientist he was intrigued by the concept of interdisciplinary sciences. The interdisciplinary cooperation was sometimes frustrated by the restructuring of environmental education and faculties, but Nienhuis always found ways to reach his goals or to find even better ways to cooperate. A good example is the initiation of the "Middle Waal project", a project in which students of several disciplines (i.e. natural sciences, social sciences and law) are stimulated to work together in research projects concerning the spatial and socio-economic developments within the river district between Nijmegen and Tiel.

When the Radboud University Nijmegen became signatory to the so-called Copernicus Charter for Sustainable Development at Universities, a covenant in which European universities pledge to promote sustainable development at their institutions, the implementation of the charter was entrusted to the UCM-DO. This expanded the centre's original objectives and tasks to include promotion and the staging of a host of activities specifically related to sustainable development, e.g. the development of several courses, the organisation of the 'year of sustainability' with several conferences and colloquia, disciplinary reviews on science for sustainable development (Van Hengstum, 2001) and the involvement of visual artists to deepen the discussion on the role and responsibilities of scientists to reach a sustainable world (Dankelman et al., 2003).

Cooperation with the Wageningen University and Research Centre was intensified. Together with the Centre for Geo-information Science of this University and the Institute for Environmental

Studies of the Free University of Amsterdam, the Geo-information for integrating personal learning environments by web and mobile information and communication technology systems (GIPSY) project was drafted and approved by the Dutch higher education and research partnership organisation for network services and information and communication technology (SURF education). This project aimed at the enhancement of GIS knowledge at the three universities and strongly stimulated the use of GIS tools in research and education at the Radboud University.

Another task of the UCM-DO was to generate discussion on environmental issues. For this purpose several conferences on environmental issues, such as the use of pesticides, water management, environmental management and sustainable science, were organised in cooperation with other faculties of the university (e.g. Bauland et al., 1995; UCM, 1997; Van den Heuvel et al., 1997; De Jong et al., 1999).

In 1999 he also initiated the Victor Westhoff lectures. The aim of these yearly lectures is to stimulate the debate on nature development and conservation in the Netherlands and abroad and to offer a counterbalance to the dominant concern for economic growth and the unshakable trust in technological advancement. The lectures are published in a book series and have become widely known (e.g. McNeely, 2001).

Meanwhile, another university centre had been set up for education and research related to integrated water management and river basin management: the Centre for Water and Society (CWS). Piet Nienhuis also became the first chairman of this centre. The CWS coordinated its activities closely with UCM-DO and expanded rapidly. Keys to its success were, among others, its international orientation, its close collaboration with universities, organisations, and authorities at home and abroad, and its innovative and practical approach of looking for solutions that are not only scientifically sound but also socially acceptable.

Editor and science agent

Piet Nienhuis was a fair referee for many research proposals of scientific programmes and projects of the European Commission DGXII, the European Science Foundation, and the Netherlands Organisation for Scientific Research (NWO) and the Dutch Technology Foundation (STW). His subtle and constructive review activities were also highly appreciated by the editorial boards of scientific journals. He regularly reviewed manuscripts for more than twenty five international journals, among which Aquaculture, Aquatic Botany, Aquatic Ecology, Estuaries, Journal of experimental Marine Biology and Ecology, Marine Ecology Progress Series and Oikos. He has been appointed as editor of several journals and at present he still is an active member of the editorial boards of River Research and Applications, Web Ecology and last but not least Hydrobiologia.

Writing and editing books was one of his passions. He edited several books of the Kluwer series Developments in Hydrobiology about coastal sea research in Mauritania (1993), ecological functioning of the estuary Oosterschelde (1993) and ecological restoration of aquatic and semi-aquatic ecosystems (2002). He also (co)-edited a volume on intertidal and littoral ecosystems of the Elsevier series Ecosystems of the world (1991), a volume of the renowned Springer series Ecological Studies on effects of eutrophication on marine benthic vegetation (1996) and two books on new concepts and approaches in river management in the Aquatic Sciences series of Backhuys Publishers (1998, 2000). Due to his commitment to higher education he also was editor of several Dutch textbooks on ecology and nature management and development.

Travelling around the world and participating in meetings abroad, field trips and expeditions to tropical ecosystems gave him much pleasure in his scientific life. During his professional career he visited symposia, workshops and seminars in more than 25 countries of all continents. He also visited several marine biological research institutes, e.g. the 'Station Biologique' in Roscoff (1969), Woods Hole Oceanographic Institute (1981) and Virginia Institute of Marine Science in Gloucester Point (1981). He organised the seagrass research of the Snellius II expedition to Indonesia (1983–1984) and co-organised an international expedition to the Banc d'Arguin in Mauritania in West Africa (1987–1988).

During his entire professional career he was an active member of national and international learned societies, such as the Netherlands Society for Aquatic Ecology (NVAE), Netherlands Flemish Ecological Society (NEVECOL; secretary over the period 1989–1994), European Ecological Federation (EEF; president of the executive board of the EEF over the period 1995–1999), International Ecological Society (INTECOL), British Phycological Society, Netherlands Botanical Society (KNBV), Netherlands Institute of Biologists (NIBI) and Dutch Society of Environmental Scientists (VVM). He co-organised numerous scientific meetings, workshops or other events of these societies.

Piet Nienhuis was a sought-after advisor and consultant for ecology, coastal, estuarine and river management, nature management and environmental sciences and fulfilled several positions in national and international science management. He chaired the Commission Tropical Marine Biology (WOTRO) and the board of the Biological Research section Aquatic Ecology of the Netherlands Organisation for Scientific Research (BION-NWO) and was a board member of the NWO Foundation for Marine Research. He also was a member of the advisory board for research in Indonesia of the Royal Netherlands Academy of Sciences, the National Coordination Group Policy-Oriented Ecological Research in Dutch Coastal Waters (BEON), the Belgian–Dutch board of the International Scheldt Faculty and the IGBP-LOICZ Core Office Implementation Commission. Over the period 1989–1993 he was chairman of the BRIDGE-COST 48 programme on European Co-operation in the field of Scientific and Technical Research and organized workshops and seminars on the use of primary marine biomass such as macro-algae, e.g. in La Rochelle (1987), Thessaloniki (1987), Brussels (1991), St. Malo (1991), Sardinia (1991) and Vienna (1991). As vice-president of the Scientific Council of the International Centre of Ecology of the Polish Academy of Sciences (1998–2003) he was involved in the initiation of this institute in Warsaw.

Separated from his professional career, he developed broad experience in Dutch national governing boards and executive councils, being chairman or board member of a considerable number of public and private organisations, viz. school boards, ecclesiastical councils, public health organisations, organisations for water management and nature conservation in the Netherlands (e.g. chairman of Bargerveen Foundation, Secretary Board of Zeeland Landscape Foundation and board member of Gelderland Landscape Foundation). His publications, lectures and advises strongly influenced the view of aquatic scientists and water managers in the Netherlands as well as abroad, in particular regarding the effects of human activities on rivers and their estuaries and the opportunities and constraints for habilitation of aquatic ecosystems.

Inspiring lecturer, supervisor and colleague

During his entire professional career Piet Nienhuis was frequently invited for lectures at many universities and research institutes in the Netherlands and abroad. For instance, in 1981 he gave several lectures in seminars at universities and institutes in the USA (e.g. the University of Durham, Woods Hole Oceanographic Institute and Virginia Institute of Marine Science in Gloucester Point) and over the period 1986–1994 he organised several seminars for Indonesian students at universities in Jakarta and Sulawesi. As visiting professor at the Free University of Brussels (1988–2001) he lectured on coral reef ecology, a course in the programme Fundamental and Applied Marine Ecology (FAME) for MSc students from tropical countries.

As extraordinary professor at Radboud University Nijmegen (1988–1994), he organised advanced courses for undergraduates that were focussed on estuarine ecology and coastal zone management. As full professor of environmental science at Radboud University Nijmegen (1994–2003) he was responsible for environmental education and supervision of many Dutch and foreign undergraduates and graduates (MSc and PhD students). He fulfilled management functions within the educational institutes of the university (director of the Undergraduate School for Environmental Science and chairman of the Examination Committee Environmental Science and the Board of Social Environmental Science) and was board member of the Division Interuniversity Coordination Environmental Science of the Society for University Cooperation in the Netherlands (ICM-VSNU). The environmental science curriculum at Radboud University Nijmegen already started in 1990 and over the period 1990–1995 it evolved into a full educational programme for advanced undergraduate students in bèta-oriented environmental science. He particularly improved the mutual relation between environmental education and academic research, the international orientation of the curriculum and the incorporation in networks such as the Association of University Departments of Environmental Sciences in Europe (AUDES), Environmental Sciences Strengthened in Europe by Networking, Conferences and Education (ESSENCE) and the European Thematic Network on Education and Training (ETNET Environment–Water). External education-quality audits held in 1996 and 2002 judged the environmental science curriculum as 'good' (Anonymous, 1996, 2002). In comparison with other Dutch universities, the environmental education at Radboud University Nijmegen yielded highest marks for the educational approach (e.g. development of a learnable curriculum), international orientation of the programme, quality of graduates and relations with external institutes.

Piet Nienhuis was an inspiring supervisor of 16 PhD projects (Table 1). The PhD dissertations describe the structure and functioning of seagrass

Table 1. Supervised PhD projects

Title thesis	Author	Year
Growth and production of tropical seagrasses: nutrient dynamics in Indonesian seagrass beds	Erftemeijer, P.L.A.	1993
Intraspecific variability of *Zostera marina L.* in the Southwestern Netherlands	Van Lent, F.	1995
Bivalve grazing, nutrient cycling and phytoplankton dynamics in an estuarine ecosystem	Prins, T.C.	1996
Plant-herbivore interactions between seagrasses and dugongs in a tropical small island ecosystem	De Iongh, H.H.	1996
Distribution, year-round primary production and decomposition of Antarctic macroalgae	Brouwer, P.E.M.	1997
Nutrient dynamics in Indonesian seagrass beds: factors determining conservation and loss of nitrogen and phosphorus	Stapel, J.	1997
Growth and photosynthesis of eukaryotic microalgae in fluctuating light conditions, induced by vertical mixing	Flameling, I.A.	1998
Macroalgal mats in a eutrophic lagoon: dynamics and control mechanisms	Malta, E.-j.	2000
Importance of shallow-water bay biotopes as nurseries for Caribbean reef fishes	Nagelkerken, I.	2000
Uncertainty in environmental quality standards	Ragas, A.M.J.	2000
Post-settlement life cycle migrations of fish in the mangrove-seagrass-coral reef continuum	Cocheret de la Morinière, E.	2002
Environmental rehabilitation of the river landscape in the Netherlands. A blend of five dimensions	Lenders, H.J.R.	2003
Incorporating spatial variability in ecological risk assessment of contaminated river floodplains	Kooistra, L.	2004
Integrating ecological knowledge and legal instruments for biodiversity conservation in river management	De Nooij, R.J.W.	2006
Ecological restoration of raised bogs	Van Duinen, G.A.	2006
Nitrogen conservation in marine macrophyte communities of oligotrophic coastal seas in Indonesia	Vonk, J.A.	2007

and macroalgae dominated ecosystems and the rehabilitation and management of river basins (including their estuaries and coastal zones). He was also frequently asked as opponent or member of PhD refereeing committees at several Dutch universities, the State University Gent (Belgium) and the University of Gothenburg (Sweden). Recently, the PhD Award 2005 of the national research school SENSE has been awarded to his PhD student L. Kooistra. The jury was highly impressed by the innovative character of the research and the interdisciplinary approach for incorporating spatial variability in ecological risk assessment of contaminated river floodplains.

For his PhD students and co-workers he always was a friendly and fatherly man, concerned with the main lines, and stressing the importance of keeping it simple and straight to the point (Anonymous, 2003). He always gave them access to his personal library and scientific network. He knows how to enthuse PhD students to write papers and has the ability to transform the wildest ideas to more realistic dimensions, and turn them into publishable results. His skeptical questions triggered stimulating discussions and finally improved the quality of work. He often accompanied his PhD students and co-workers during field trips. He always said: "environmental scientists can't describe and model ecosystems without personal observations of the real world" and "good desk studies always start with field visits".

Acknowledgements

The authors thank B.G.W. Aarts, H.J.R. Lenders, I. Nagelkerken, R.J.W. de Nooij and A.J.M. Smits for useful information or comments on a draft of this paper. This is CWE-publication 417.

References

Aarts, B. G. W. & P. H. Nienhuis, 1999. Ecological sustainability and biodiversity. International Journal of Sustainable Development and World Ecology 6: 89–102.

Anonymous, 1992. Onderzoekscontouren Vakgroep Milieukunde 1992–1997. Vakgroep Milieukunde, Faculteit Natuurwetenschappen, Katholieke Universiteit Nijmegen (in Dutch).

Anonymous, 1996. Onderwijsvisitatie Milieukunde. VSNU, Utrecht: 1–105 (in Dutch).

Anonymous, 2000. Assessment of Research Quality. Environmental Sciences. VSNU, Utrecht: 1–52.

Anonymous, 2002. Onderwijsvisitatie Milieuwetenschappen. VSNU, Utrecht: 1–85 (in Dutch).

Anonymous, 2003. Denkend aan Piet Nienhuis. Liber amicorum ter Gelegenheid van het Emeritaat van Prof. Dr. P. H. Nienhuis. Uitgave Afdeling Milieukunde/Universitair Centrum voor Milieuwetenschappen en Duurzame Ontwikkeling, Katholieke Universiteit Nijmegen: 1–83 (in Dutch).

Bauland, B., P. J. M. van den Heuvel & T. Verheij-van der Linden (eds), 1995. Integraal Beleid en Beheer voor grote Riviergebieden. Mediagroep Katholieke Universiteit Nijmegen, Nijmegen (in Dutch).

Dankelman, I. E. M., M. de Geus & T. van Slobbe, 2003. De Kunst van Duurzaamheid. Uitgeverij Waarde, Beek-Ubbergen (in Dutch).

De Jong, D. M., H. Reynders, G. Laeijendecker & P. J. M. van den Heuvel, 1999. Ruimte voor Milieu over de Grenzen. Universitair Centrum voor Milieuwetenschappen, Katholieke Universiteit Nijmegen, Nijmegen (in Dutch).

Duursma, E. K., H. Engel & Th. J. M. Martens (eds), 1982. De Nederlandse Delta. Natuur en Techniek, Maastricht (in Dutch).

Klijn, F., P. H. Nienhuis & G. B. M. Pedroli (eds), 2004. Towards sustainable flood risk management in the Rhine and Meuse river basins. River Research and Applications 20: 227–357.

Lambeck, R. H. D., H. J. Lindeboom & P. H. Nienhuis (eds), 1984. Lake Grevelingen: from an estuary to a saline lake. Structure and functioning of an evolving marine ecosystem. Netherlands Journal of Sea Research 18: 179–528.

Lenders, H. J. R., R. S. E. W. Leuven, P. H. Nienhuis, R. J. W. de Nooij & S. A. M. van Rooij, 2001. BIO-SAFE: a method for evaluation of biodiversity values on the basis of political and legal criteria. Landscape and Urban Planning 55: 119–135.

Leroy, P. & P. H. Nienhuis, 1997. Van holisme naar serieel specialisme. In Leroy, P. (ed.), Verbondenheid en Samenhang. Vakgroep Milieu, Natuur en Landschap, Katholieke Universiteit Nijmegen: 87–111 (in Dutch).

Leuven, R. S. E. W., A. G. van Os & P. H. Nienhuis, 2003. Current Themes in Dutch River Research. Proceedings NCR-days 2002. NCR-publication 20-2003. Netherlands Center for River Studies, Delft.

Leuven, R. S. E. W., A. M. J. Ragas, A. J. M. Smits & G. van der Velde (eds), 2006. Living rivers: trends and challenges in science and management. Hydrobiologia 565: 1–367.

McNeely, J. A., 2001. Conservation and the Future: Trends and Options towards the Year 2025. Nijmegen University Press, Nijmegen.

Nienhuis, P. H., 1975. Biosystematics and Ecology of *Rhizoclonium riparium* (Roth) Harv. (Chlorophyceae: Cladophorales) in the Estuarine Area of the Rivers Rhine, Meuse and Scheldt. Ph.D. Thesis, Rijksuniversiteit Groningen, 1–240.

Nienhuis, P. H. (ed.), 1985. Het Grevelingenmeer. Van Estuarium naar Zoutwatermeer. DIHO Yerseke, Natuur en Techniek, Maastricht (in Dutch).

Nienhuis, P. H., 1988. Estuariene Oecologie: Visie op een Milieu onder Druk (Estuarine Ecology). Yerseke, DIHO. Inaugurele rede Katholieke Universiteit Nijmegen: 1–15 (in Dutch).

Nienhuis, P. H., 1995. Rivierkunde: 1995–2000. Onderzoek Vakgroep Milieukunde. Katholieke Universiteit Nijmegen Verslagen Milieukunde 100: 1–25.

Nienhuis, P. H, 1998. Duurzame ontwikkeling en ecologie. In: Het Milieu als Offerplaats. Damon, Best: 99–115.

Nienhuis, P. H., 2003. Water en Waarden. Ecologisch Onderzoek als basis voor Waterbeheer en Natuurbeheer. Afscheidsrede. Katholieke Universiteit Nijmegen, Nijmegen: 1–27 (in Dutch).

Nienhuis, P. H., 2006. Water and values: ecological research as the basis for water management and nature management. Hydrobiologia 565: 261–275.

Nienhuis, P. H. & B. H. H. de Bree, 1977. Production and ecology of eelgrass (*Zostera marina* L.) in the Grevelingen estuary, the Netherlands, before and after the closure. Hydrobiologia 52: 55–66.

Nienhuis, P. H. & B. H. H. de Bree, 1980. Production and growth dynamics of eelgrass (*Zostera marina*) in brackish Lake Grevelingen (The Netherlands). Netherlands Journal of Sea Research 14/1: 102–118.

Nienhuis, P. H. & R. S. E. W. Leuven, 1997. The role of the science of ecology in the sustainable development debate in Europe. Verhandlungen der Gesellschaft für Ökologie 27: 243–251.

Nienhuis, P. H. & R. S. E. W. Leuven, 1998a. The conflict between safety precautions and ecological restoration of the Dutch sections of the rivers Rhine and Meuse. In Gayer, J., K. Scheuerlein & Ö. Starosolszky (eds), Proceedings International Conference on European River Development, IAHR/AIRH, Vituki RT., Budapest: 83–90.

Nienhuis, P. H., & R. S. E. W. Leuven, 1998b. Ecological concepts for the sustainable management of river basins: a review. In Nienhuis, P. H. R. S. E. W. Leuven & A. M. J. Ragas (eds), New Concepts for Sustainable Management of River Basins. Backhuys Publishers, Leiden: 7–34.

Nienhuis, P. H. & R. S. E. W. Leuven, 2001. River restoration and flood protection: controversy or synergism. Hydrobiologia 444: 85–99.

Nienhuis, P. H. & A. C. Smaal (eds), 1994. The Oosterschelde Estuary (The Netherlands): a case-study of a changing ecosystem. Hydrobiologia 282/283: 1–597.

Nienhuis, P. H., A. D. Buise, R. S. E. W. Leuven, A. J. M. Smits, R. J. W. de Nooij & E. M. Samborska, 2002. Ecological rehabilitation of the lowland basin of the river Rhine (NW Europe). Hydrobiologia 478: 53–72.

Nienhuis, P. H., H. J. R. Lenders, R. S. E. W. Leuven, A. M. J. Ragas & H. Strijbosch, 1999. Research Programme 1999–2004, Department of Environmental Science, Faculty of Science, University of Nijmegen. Reports Environmental Studies 180: 1–13.

Sprengers, S. A., P. H. Nienhuis & P. Eliáš (eds), 1996. Report of the Workshop 'Sustainability of Ecosystems: Ecological and Economic Factors' 27–29 March 1995, KNAW, Amsterdam: 1–170.

Universitair Centrum voor Milieuwetenschappen (UCM), 1997. Partner of Counterpart: De Machtspositie van de Milieubeweging., Katholieke Universiteit Nijmegen (UCM/NIVAS)/Stichting Natuur en Milieu, Nijmegen (in Dutch).

Vannote, R. L., G. W. Minshall, K. W. Cummins, J. R. Sedell & C. E. Cushing, 1980. The river continuum concept. Canadian Journal of Fisheries and Aquatic Sciences 37: 130–137.

Van den Heuvel, P. J. M., P. Leroy, R. S. E. W. Leuven & P. H. Nienhuis, 1997. Chemische gewasbescherming dertig jaar na "Zilveren Sluiers en verborgen gevaren". Netherlands Journal for Environmental Science 12: 50–59 (in Dutch with English summary).

Van de Willigen, A. H., J. M. W. Habets, Th. van Joost, E. Stolz & P. H. Nienhuis, 1988. Contact allergy to iodine in Japanese sargassum. Contact Dermatitis 18/4: 250–252.

Van Hengstum, G., 2001. Biologie & Duurzame Ontwikkeling. Vakreview Duurzame Ontwikkeling. Netwerk Duurzaam Hoger Onderwijs/Universitair Centrum Milieuwetenschappen, Nijmegen (in Dutch).

Van Stokkom, H. T. C., A. J. M. Smits & R. S. E. W. Leuven, 2005. Flood defense in the Netherlands: a new era, a new approach. Water International 30: 76–87.

Wiering, M. A. & B. J. M. Arts, 2006. Discursive shifts in Dutch river management: 'deep' institutional change or adaptation strategy? Hydrobiologia 565: 327–338.

Appendix: International journal and book publications of Nienhuis & co-workers

Nienhuis, P. H., 1968. The algal vegetation of a brackish inland water basin in the Netherlands (Gat van Ouwerkerk). Acta Botanica Neerlandia 17: 26–32.

Nienhuis, P. H., 1969. The significance of the substratum for intertidal algal growth on the artificial rocky shore of the Netherlands. Internationale Revue der gesamten Hydrobiologie 54: 207–215.

Sterk, A. A., W. M. van der Leeuw, P. H. Nienhuis & J. Simons, 1969. Biotaxonomic notes on the *Rumex acetosella* complex in the Netherlands. Acta Botanica Neerlandica 18: 597–604.

Nienhuis, P. H., 1970. The benthic algal communities of flats and salt marshes in the Grevelingen, a sea-arm in the South-Western Netherlands. Netherlands Journal Sea Research 5: 20–49.

Nienhuis, P. H. & J. Simons, 1971. *Vaucheria* species and some other algae on a Dutch salt marsh, with ecological notes on their periodicity. Acta Botanica Neerlandica 20: 107–118.

Nienhuis, P. H., 1973. Salt marsh and beach plain as a habitat for benthic algae. Hydrobiological Bullletin 7: 15–24.

Nienhuis, P. H., 1973. The use of permanent sample plots in studying the quantitative ecology of algae in salt marshes. Proceedings eleventh International Seaweed Symposium, University of Tokyo Press: 251–254.

Nienhuis, P. H., 1974. Variability in the life cycle of *Rhizoclonium riparium* (Roth) Harv. (Chlorophyceae: Cladophorales) under Dutch estuarine conditions. Hydrobiological Bulletin 8: 172–178.

Nienhuis, P. H., 1975. Biosystematics and ecology of *Rhizoclonium riparium* (Roth) Harv. (Chlorophyceae: Cladophorales) in the estuarine area of the rivers Rhine, Meuse and Scheldt. Ph.D. Thesis Rijksuniversiteit Groningen: 1–240.

Nienhuis, P. H., 1976. Dynamics of environment and algal vegetation in salt marshes in the S.W. Netherlands. Acta Botanica Neerlandica 25: 120–121.

Nienhuis, P. H. & B. H. H. de Bree, 1977. Production and ecology of eelgrass (*Zostera marina* L.) in the Grevelingen estuary, the Netherlands, before and after the closure. Hydrobiologia 52: 55–66.

Nienhuis, P. H., 1978. Primary production in the Grevelingen. Acta Botanica Neerlandica 27: 94.

Nienhuis, P. H., 1978. An ecosystem study in Lake Grevelingen, a former estuary in the SW Netherlands. Kieler Meeresforschung Sonderheft 4: 247–255.

Nienhuis, P. H., 1978. Dynamics of benthic algal vegetation and environment in Dutch estuarine salt marshes, studied by means of permanent quadrats. Vegetatio 38: 103–112.

Nienhuis, P. H., 1978. Lake Grevelingen: a case study of ecosystem changes in a closed estuary. Hydrobiological Bulletin 12: 246–259.

Nienhuis, P. H. & E. J. van Ierland, 1978. Consumption of eelgrass, *Zostera marina*, by birds and invertebrates during the growing season in Lake Grevelingen (SW Netherlands). Netherlands Journal of Sea Research 12: 180–194.

Nienhuis, P. H., 1979. Ecology and production of eelgrass in the former Grevelingen estuary, an evolving saline lake in the S.W. Netherlands. In R. L. Jefferies & A. J. Davy (eds), Ecological processes in coastal environments. Blackwell Scientific Publishers. Oxford: 650.

Nienhuis, P. H., 1980. The eelgrass (*Zostera marina* L.) subsystem in brackish Lake Grevelingen: production and decomposition of organic matter. Ophelia, Supplement 1: 113–116.

Nienhuis, P. H., 1980. The study of the organic carbon cycle in brackish Lake Grevelingen (SW Netherlands). Hydrobiological Bulletin 13: 75–76.

Nienhuis, P. H., 1980. The role of macrophytes, especially *Zostera marina*, in the organic carbon cycle of Lake Grevelingen. Hydrobiological Bulletin 13: 79–80.

Nienhuis, P. H., 1980. The epilithic algal vegetation in the SW-Netherlands. Nova Hedwigia 33: 1–94.

Nienhuis, P. H. & B. H. H. de Bree, 1980. Production and growth dynamics of eelgrass (*Zostera marina*) in brackish Lake Grevelingen (The Netherlands). Netherlands Journal of Sea Research 14: 102–118.

Nienhuis, P. H., 1981. Distribution of organic matter in living marine organisms. In Duursma, E. K. & R. Dawson (eds), Marine Organic Chemistry. Elsevier, Amsterdam. Elsevier Oceanography Series 31: 31–69.

Nienhuis, P. H. & B. H. H. de Bree, 1981. Morphological variability in *Rhizoclonium riparium* (Roth) Harv. (Chlorophyceae: Cladophorales) under natural and experimental conditions. Proceed. 8th Internat. Seaweed Symp. Bangor, North Wales, 1974. In Fogg, G. E. & W. J. Jones (eds), The Marine Science Laboratory Menai Bridge: 434–442.

Nienhuis, P. H., 1981. Manipulations with the tidal factor: a functional comparative approach for two Dutch estuaries. Estuaries 4: 278.

Nienhuis, P. H., 1982. Attached *Sargassum muticum* found in the SW Netherlands. Aquatic Botany 12: 189–195.

Prud'homme van Reine, W. F. & P. H. Nienhuis, 1982. Occurrence of the brown algae *Sargassum muticum* (Yendo) Fensholt in the Netherlands. Botanica Marina 25: 37–39.

Daemen, E. A . M. J., H. Pankow & P. H. Nienhuis, 1982. The benthic diatom-flora of saline Lake Grevelingen (SW Netherlands). Acta Botanica Neerlandia 31: 153–167.

Nienhuis, P. H., 1982. A simulation model of production, seasonal changes in biomass and distribution of eelgrass (*Zostera marina*) in Lake Grevelingen. Hydrobiological Bulletin 16: 286.

Verhagen, J. H. G. & P. H. Nienhuis, 1983. A simulation model of production of eelgrass (*Zostera marina*), seasonal changes in biomass and its distribution in Lake Grevelingen. Marine Ecology Progress Series 10: 187–195.

Nienhuis, P. H., 1983. Temporal and spatial patterns of eelgrass (*Zostera marina* L.) in a former estuary in the Netherlands, dominated by human activities. Marine Technology Society Journal 17: 69–77.

Nienhuis, P. H. & J. C. Huis in 't Veld, 1984. Grevelingen: from an estuary to a saline lake. Water Science & Technology 16: 27–50.

Bannink, B. A., W. P. A. Broeders, M. Knoester, L. Lijklema & P. H. Nienhuis (eds), 1984. Integration of ecological aspects in coastal engineering projects. Proceedings Symposium Rotterdam 6–10 June 1983. Water Science & Technology 16: 1–829.

Lambeck, R. H. D., H. J. Lindeboom & P. H. Nienhuis (eds), 1984. Lake Grevelingen: from an estuary to a saline lake. Structure and functioning of an evolving marine ecosystem. Netherlands Journal of Sea Research 18: 179–528.

Bannink, B. A., J. H. M. van der Meulen & P. H. Nienhuis, 1984. Lake Grevelingen: from an estuary to a saline lake. Netherlands Journal of Sea Research 18: 179–190.

Nienhuis, P. H. & B. H. H. de Bree, 1984. Carbon fixation and chlorophyll in bottom sediments of brackish Lake Grevelingen, The Netherlands. Netherlands Journal of Sea Research 18: 337–359.

Meulstee, C., P. H. Nienhuis & H. T. C. van Stokkom, 1986. Biomass assessment of estuarine macrophytobenthos using aerial photography. Marine Biology 91: 331–335.

Nienhuis, P. H. & A. M. Groenendijk, 1986. Consumption of eelgrass (*Zostera marina* L.) by birds and invertebrates: an annual budget. Marine Ecology Progress Series 29: 29–35.

Vermaat, J. E., M. J. M. Hootsmans & P. H. Nienhuis, 1987. Seasonal dynamics and above-ground production of a perennial intertidal seagrass population (*Zostera noltii*). Aquatic Botany 28: 287–299.

Critchley, A. T., P. H. Nienhuis & J. M. Verschuure, 1987. Presence and development of populations of the introduced brown alga *Sargassum muticum* in the SW Netherlands. Proceed. XIIth Int. Seaweed Symposium. Hydrobiologia 151/152: 245–255.

Nienhuis, P. H., 1986. Background levels of heavy metals in nine tropical seagrass species in Indonesia. Marine Pollution Bulletin 17: 508–511.

Nienhuis, P. H., 1987. Ecology of salt-marsh algae in the Netherlands. A review. In Huiskes, A. H. L., C. W. P. M. Blom & J. Rozema (eds), Vegetation between land en sea. Junk, Dordrecht: 66–83.

Pellikaan, G. C. & P. H. Nienhuis, 1988. Nutrient uptake and release during growth and decomposition of eelgrass, *Zostera marina* L., and its effects on the nutrient dynamics of Lake Grevelingen. Aquatic Botany 30: 189–214.

Van de Willigen, A. H., J. M. W. Habets, Th. van Joost, E. Stolz & P. H. Nienhuis, 1988. Contact allergy to iodine in Japanese sargassum. Contact Dermatitis 18: 250–252.

Meulstee, C., P. H. Nienhuis & H. T. C. van Stokkom, 1988. Aerial photography for biomass assessment in the intertidal zone. International Journal of Remote Sensing 9: 1859–1867.

Nienhuis, P. H., 1989. Eutrophication of estuaries and brackish lagoons in the SW Netherlands. In J. C. Hooghart & C. W. S. Posthumus (eds), Hydro-ecological relations in the Delta-waters of the SW Netherlands. Proceedings & Information 41: 49–70, Den Haag.

Nienhuis, P. H., J. Coosen & H. Kiswara, 1989. Community structure and biomass distribution of seagrasses and macrofauna in the Flores Sea, Indonesia. Netherlands Journal of Sea Research 23: 197–214.

Scholten H., O. Klepper, P. H. Nienhuis & M. Knoester, 1990. Oosterschelde estuary (SW Netherlands): a self-sustaining ecosystem? Hydrobiologia 195: 201–215.

Pérez-Lloréns, J. L., M. Muchtar, F. X. Niell & P. H. Nienhuis, 1991. Particulate organic carbon, nitrogen and phosphorus content in roots, rhizomes and differently aged leaves of *Zostera noltii* Hornem. in Oosterschelde estuary (Southwestern Netherlands). Botanica Marina 34: 319–322.

Van Lent, F., P. H. Nienhuis & J. M. Verschuure, 1991. Production and biomass of the seagrasses *Zostera noltii* Hornem. and *Cymodocea nodosa* (Ucria) Aschers. at the Banc d'Arguin (Mauritania, NW Africa): a preliminary approach. Aquatic Botany 41: 353–367.

Smaal, A. C., M. Knoester, P. H. Nienhuis & P. M. Meire, 1991. Changes in the Oosterschelde ecosystem induced by the Delta works. In Elliott, M. & J.-P.Ducrotoy (eds), Estuaries and coasts: spatial and temporal intercomparisons. ECSA 19 Symp. Olsen & Olsen, Fredenborg: 375–384.

Nienhuis, P. H. & R. H. D. Lambeck, 1991. Kees Bakker: Hydrobiologist. Hydrobiological Bulletin 25: 5–8.

Mathieson, A. C. & P. H. Nienhuis (eds), 1991. Intertidal and littoral ecosystems. Ecosystems of the world 24. Elsevier, London: 1–564.

Nienhuis, P. H. & A. C. Mathieson, 1991. Introduction. In Mathieson, A. C. & P. H. Nienhuis (eds), Intertidal and littoral ecosystems. Ecosystems of the world 24. Elsevier, London: 1–5.

De Heij, H. & P. H. Nienhuis, 1992. Intraspecific variation in isozyme patterns of phenotypically separated populations of *Zostera marina* L. in the south-west Netherlands. Journal of Experimental Marine Biology and Ecology 161: 1–14.

Smaal, A. C., J. van Buuren, R. Dijkema, W. Iedema, J. Mulder, P. H.Nienhuis, T. Pieters & B. Wetsteyn, 1992. Effects of a large-scale coastal engineering project. In N. Dankers, C. J. Smit & M. Scholl (eds), Proceedings 7th International Waddensea Symposium, 1990. Netherlands Institute for Sea Research Publication Series 20: 285–288.

Nienhuis, P. H., 1992. Ecology of coastal lagoons in the Netherlands (Veerse Meer and Grevelingen). Vie et Milieu 42: 59–72.

Nienhuis, P. H., 1992. Eutrophication, water management and the functioning of Dutch estuaries and coastal lagoons. Estuaries 15: 538–548.

Smaal, A. C. & P. H. Nienhuis, 1992. The Eastern Scheldt (The Netherlands), from an estuary to a tidal bay: a review of responses at the ecosystem level. Netherlands Journal of Sea Research 30: 161–173.

Perez-Llorens, J. L., P. de Visscher, P. H. Nienhuis & F. X. Niell, 1993. Light-dependent uptake, translocation and foliar release of phosphorus by the intertidal seagrass *Zostera noltii* Hornem. Journal of Experimental Marine Biology and Ecology 166: 165–174.

Wolff, W. J., J. van der Land, P. H. Nienhuis & P. A. W. J. de Wilde, 1993. The functioning of the ecosystem of the Banc d'Arguin, Mauritania: a review. Hydrobiologia 258: 211–222.

Wolff, W. J., J. van der Land, P. H. Nienhuis & P. A. W. J. de Wilde (eds), 1993. Ecological studies in the coastal waters of Mauritania. Developm. Hydrobiology 86/ Hydrobiologia 258: 1–222.

Nienhuis, P. H., 1993. Nutrient cycling and foodwebs in Dutch estuaries. Hydrobiologia 265: 15–44.

Nienhuis, P. H., 1993. Structure and functioning of Indonesian seagrass ecosystems, In M. K. Moosa, & H. H. de Iongh (eds), Coastal zone management of small island ecosystems, Proceedings International Seminar 7–10 April 1993, Universitas Pattimura, Ambon Indonesia: 82–86.

Nienhuis, P. H., 1994. Causes of the eelgrass wasting disease: Van der Werff's changing theories. Netherlands Journal of Aquatic Ecology 28: 55–61.

Nienhuis, P. H. & A. C. Smaal (eds), 1994. The Oosterschelde Estuary (The Netherlands): a case-study of a changing ecosystem. Developments Hydrobiology 97 / Hydrobiologia 282/283: 1–597. Kluwer Academic Publishers, Dordrecht.

Nienhuis, P. H. & A. C. Smaal, 1994. The Oosterschelde estuary, a case-study of a changing estuary: an introduction. Hydrobiologia 282/283: 1–14.

Nienhuis, P. H., A. C. Smaal & M. Knoester, 1994. The Oosterschelde estuary: an evaluation of changes at the ecosystem level induced by civil-engineering works. Hydrobiologia 282/283: 575–592.

De Jong, D. J., P. H.Nienhuis & B. J. Kater, 1994. Microphytobenthos in the Oosterschelde estuary (The Netherlands), 1981–1990; consequences of a changed tidal regime. Hydrobiologia 282/283: 183–195.

Nieuwenhuize, J., P. L. A. Erftemeijer, Y. E. M. Maas, M. Verwaal & P. H. Nienhuis, 1994. Pre-treatment artefacts associated with the removal of calcareous epiphytes from seagrass leaves. Aquatic Botany 48: 355–361.

Nienhuis, P. H., 1994. Seaweeds in estuaries and salt marshes. In Lobban C. S. & P. J. Harrison (ed.), The physiological Ecology of Seaweeds. Cambridge University Press, Cambridge: 85–87.

Nienhuis, P. H. & A. C. Smaal, 1994. The Oosterschelde (The Netherlands), an estuarine ecosystem under stress: discrimination between the effects of human-induced and natural stress. In Dyer, K. J. & R. J. Orth (eds), Changes in fluxes in estuaries: implications for science and management. ECSA 22/ERF Symposium 1992 Plymouth; Olsen & Olsen, Fredensborg, Danmark: 109–120.

Sterrenburg, F. A. S., P. L. A. Erftemeijer & P. H. Nienhuis, 1995. Diatoms as epiphytes on seagrasses in South Sulawesi (Indonesia). Comparison with growth on inert substrata. Botanica Marina 38: 1–7.

Nienhuis, P. H. & A. C. Smaal, 1995. The Oosterschelde (The Netherlands), an estuarine ecosystem under stress: discrimination between the effects of human-induced and natural stress. In Dyer, K. R. & R. J. Orth (eds), Changes in fluxes in estuaries: implications from science to management: 109–120.

De Leeuw, J. W., W. I. C. Rijpstra & P. H. Nienhuis, 1995. Free and bound fatty acids hydroxy fatty acids in the living and decomposing eelgrass *Zostera marina* L. Organic Geochemistry 23: 721–728.

Schramm, W. & P. H. Nienhuis (eds), 1996. Marine Benthic Vegetation; recent changes and the effects on Eutrophication, Ecological Studies 123, Springer, Berlin: 1–465.

Nienhuis, P. H., 1996. The North Sea coasts of Denmark, Germany and the Netherlands. In Schramm, W. & P. H. Nienhuis (eds), Marine Benthic Vegetation; recent changes and the effects on eutrophication. Ecological Studies 123, Springer, Berlin: 187–221.

Herman, P. M. J., M. A. Hemminga, P. H. Nienhuis, J. M. Verschuure & E. G. J. Wessel, 1996. Wax and wane of eelgrass *Zostera marina* and water column silicon levels. Marine Ecology Progress Series 144: 303–307.

Nienhuis, P. H., B. H. H. de Bree, P. M. J. Herman, A. M. B. Holland, J. M. Verschuure & E. G. J. Wessel, 1996. Twentyfive years of changes in the distribution and biomass of eelgrass, *Zostera marina*, in Grevelingen lagoon, the Netherlands. Netherlands Journal of Aquatic Ecology 30/2–3: 107–117.

Critchley, A. T., P. R. M. de Visscher & P. H. Nienhuis, 1990. Canopy characteristics of the brown alga *Sargassum muticum* (Fucales, Phaeophyta) in Lake Grevelingen, southwest Netherlands. Hydrobiologia 204/205: 211–217.

De Vries, I., A. C. Smaal, P. H. Nienhuis & J. C. A. Joordens, 1996. Estuarine management strategies and the predictability of ecosystem changes, Journal of Coastal Conservation 2: 139–148.

Leuven, R. S. E. W., M. W. H. Thörig, P. H. Nienhuis & B. J. van de Laar, 1997. Environmental utilisation space: a useful concept underpinning sustainable management of river catchments? In Smith, P. & A. Tenner (eds), Dimensions of sustainability. Published by the International Network of Engineers and Scientists for Global Responsibility, NOMOS Verlagsgesellschaft, Baden-Baden: 361–362.

Nienhuis, P. H. & R. S. E. W. Leuven, 1997. The role of the science of ecology in the sustainable development debate in Europe. Verhandlungen der Gesellschaft für Ökologie 27: 243–251.

Nienhuis, P. H. & R. S. E. W. Leuven, 1998. The conflict between safety precautions and ecological restoration of the Dutch sections of the rivers Rhine and Meuse. In Gayer, J., K. Scheuerlein & Ö. Starosolszky (eds), Proceedings International Conference on European River Development, IAHR/AIRH, Vituki RT., Budapest: 83–90.

Nienhuis, P. H. & R. S. E. W. Leuven, 1998. Ecological concepts for the sustainable management of river basins: a review. In Nienhuis, P. H., R. S. E. W. Leuven & A. M. J. Ragas (eds), New concepts for sustainable management of river basins. Backhuys Publishers, Leiden: 7–34.

Nienhuis, P. H., R. S. E. W. Leuven & A. M. J. Ragas, 1998. General introduction. In Nienhuis, P. H., R. S. E. W. Leuven & A. M. J. Ragas (eds), New concepts for sustainable management of river basins. Backhuys Publishers, Leiden: 1–6.

Nienhuis, P. H., R. S. E. W. Leuven & A. M. J. Ragas, 1998. Discussion and state of the art: new concepts emerging. In Nienhuis, P. H., R. S. E. W. Leuven & A. M. J. Ragas (eds), New concepts for sustainable management of river basins. Backhuys Publishers, Leiden: 353–367.

Nienhuis, P. H., R. S. E. W. Leuven & A. M. J. Ragas (eds), 1998. New concepts for sustainable management of river basins. Backhuys Publishers, Leiden: 1–370.

Nienhuis, P. H., N. J. W. Hofman, M. G. Rietbergen, S. S. H. Ligthart & T. Prus, 1998. Water quality management of the lower reach of the Vistula river in Poland. In Nienhuis, P. H., R. S. E. W. Leuven & A. M. J. Ragas (eds), New concepts for sustainable management of river basins. Backhuys Publishers, Leiden: 321–332.

Lenders H. J. R., R. S. E. W. Leuven, P. H. Nienhuis, K. D. Oostinga & P. J. M. van den Heuvel, 1998. Ecological rehabilitation of floodplains along the middle reach of the river Waal: a prosperous future for fauna target species? In Nienhuis, P. H., R. S. E. W. Leuven & A. M. J. Ragas (eds), New concepts for sustainable management of river basins. Backhuys Publishers, Leiden: 115–130.

Leuven, R. S. E. W., P. H. Nienhuis, J. M. A. Kesseleer & W. Zwart, 1998. Emission of pollutants from secondary building materials in hydraulic engineering structures in Dutch river basins. In: Man and River Systems, The Functioning of River Systems at Basin Scale. Presses Ponts et Chaussées, Paris: 167–169.

Leuven, R. S. E. W., W. A. Zwart, J. M. A. Kesseleer & P. H. Nienhuis, 1998. Emission of metals from minestone in hydraulic engineering structures in Dutch river basins. In: Nienhuis, P. H., R. S. E. W. Leuven & A. M. J. Ragas (eds), New concepts for sustainable management of river basins. Backhuys Publishers, Leiden: 273–282.

Van den Heuvel, P. J. M., N. G. M. ten Brink & P. H. Nienhuis, 1998. Conflicting functions in the Middle-Waal area: a preliminary study. In Nienhuis, P. H., R. S. E. W. Leuven & A. M. J. Ragas (eds), New concepts for sustainable management of river basins. Backhuys Publishers, Leiden: 309–320.

Aarts, B. G. W. & P. H. Nienhuis, 1999. Ecological sustainability and biodiversity. International Journal of Sustainable Development and World Ecology 6: 89–102.

Leuven, R. S. E. W., P. H. Nienhuis, J. M. A. Kesseleer & W. Zwart, 1999. Annual emissions of pollutants from mine stone applications in drainage basins of Dutch rivers. Hydrobiologia 410: 315–323.

Leuven, R. S. E. W., A. J. M. Smits & P. H. Nienhuis, 2000. From integrated approaches to sustainable river basin management In Smits, A. J. M., P. H. Nienhuis & R. S. E. W. Leuven (eds), New approaches to river management. Backhuys Publishers, Leiden: 329–347.

Nienhuis, P. H., J. C. Chojnacki, O. Harms, W. Majewski, W. Parzonka & T. Prus, 2000. Elbe, Odra, and Vistula: reference rivers for the restoration of biodiversity and habitat quality. In Smits, A. J. M., P. H. Nienhuis & R. S. E. W. Leuven (eds), New approaches to river management. Backhuys Publishers, Leiden: 65–84.

Smits, A. J. M., P. H. Nienhuis & R. S. E. W. Leuven (eds), 2000. New approaches to river management. Backhuys Publishers, Leiden: 1–355.

Smits, A. J. M., P. H. Nienhuis & R. S. E. W. Leuven, 2000. New approaches to river management: general introduction. In: Smits, A. J. M., P. H. Nienhuis & R. S. E. W. Leuven (eds), New approaches to river management. Backhuys Publishers, Leiden: 7–14.

Kooistra, L., R. S. E. W. Leuven, R. Wehrens, L. M. C. Buydens & P. H. Nienhuis, 2001. A procedure for incorporating spatial variability in ecological risk assessment of Dutch river floodplains. Environmental Management 28: 359–373.

Lenders, H. J. R., R. S. E. W. Leuven, P. H. Nienhuis, R. J. W. de Nooij & S. A. M. van Rooij, 2001. BIO–SAFE: A method for evaluation of biodiversity values on the basis of political and legal criteria. Landscape and Urban Planning 55: 119–135.

Nienhuis, P. H. & R. S. E. W. Leuven, 2001. River restoration and flood protection: controversy or synergism. Hydrobiologia 444: 85–99.

Kooistra, L., R. Wehrens, L. M. C. Buydens, R. S. E. W. Leuven & P. H. Nienhuis, 2001. Possibilities of soil spectroscopy for the classification of contaminated areas in river floodplains. Journal of Applied Earth Observation and Geoinformation 3: 337–344.

Geerling, G. W., C. B. L. van Gestel, M. Sheehy Skeffington, M. G. C. Schouten, P. H. Nienhuis & R. S. E. W. Leuven, 2002. Blanket bog degradation in river catchments in the West of Ireland. In: Leuven, R. S. E. W., I. Poudevigne, & R. M. Teeuw, (eds), Application of geographic information systems and remote sensing in river studies. Backhuys Publishers, Leiden: 25–40.

Kooistra, L., R. S. E. W. Leuven, R. Wehrens, L. M. C. Buydens, & P. H. Nienhuis, 2002. GIS as a tool for incorporating the spatial variability of pollutants in ecological risk assessments of Dutch river floodplains. In: Leuven, R. S. E. W., I. Poudevigne & R. M. Teeuw (eds), Application of geographic information systems and remote sensing in river studies. Backhuys Publishers, Leiden: 133–146.

Malta, E.-J., J. M. Verschuure & P. H. Nienhuis, 2002. Regulation of spatial and seasonal variation of macroalgal biomass in a brackish, eutrophic lake. Helgoland Marine Research 56: 211–220.

Nagelkerken, I., C. M. Roberts, G. van der Velde, M. Dorenbosch, M. C. van Riel, E. Cocheret de la Morinière & P. H. Nienhuis, 2002. How important are mangroves/seagrass beds for coral reef fish? The nursery hypothesis tested on island scale. Marine Ecology Progress Series 244: 299–305.

Nienhuis, P. H. & R. D. Gulati (eds), 2002. Ecological restoration of aquatic and semi-aquatic ecosystems in the Netherlands (NW Europe). Hydrobiologia 478: 1–233.

Nienhuis, P. H. & R. D. Gulati, 2002. Ecological restoration of aquatic and semi-aquatic ecosystems in the Netherlands: an introduction. Hydrobiologia 478: 1–6.

Nienhuis, P. H., A. D. Buise, R. S. E. W. Leuven, A. J. M. Smits, R. J. W. de Nooij & E. M. Samborska, 2002. Ecological rehabilitation of the lowland basin of the river Rhine (NW Europe). Hydrobiologia 478: 53–72.

Nienhuis, P. H., J. P. Bakker, A. P. Grootjans, R. D. Gulati & V. N. de Jonge, 2002. The state of the art of aquatic and semi-aquatic ecological restoration projects in the Netherlands. Hydrobiologia 478: 219–233.

Poudevigne, I., D. Alard, R. S. E. W. Leuven & P. H. Nienhuis, 2002. A systems approach to river restoration: a case study in the lower Seine valley, France. River Research and Applications 18: 239–247.

Aarts, B. G. W. & P. H. Nienhuis, 2003. Fish zonation and guilds as the basis for assessment of ecological integrity of large rivers. Hydrobiologia 500: 157–178.

Kooistra, L., R. Wehrens, R. S. E. W. Leuven, P. H. Nienhuis & L. M. C.Buydens, 2003. A comparison of methods to relate grass reflectance to soil metal contamination. International Journal of Remote Sensing 24: 4995–5010.

Kooistra, L., R. S. E. W. Leuven, P. H. Nienhuis & R. Wehrens, 2003. Incorporation spatial variability of exposure in ecological risk assessment of polluted floodplains using a combined GIS-Monte Carlo approach. In Leuven, R. S. E. W., A. G. van Os & P. H. Nienhuis (eds), Proceedings NCR-days 2002, current themes in Dutch river research. NCR-publication 20-2003. Netherlands Centre for River Studies, Delft: 148–151.

Leuven, R. S. E. W., A. G. van Os & P. H. Nienhuis (eds), 2003. Current themes in Dutch river research. Proceedings NCR-days 2002. NCR-publication 20-2003. Netherlands Centre for River Studies, Delft: 1–186.

Leuven, R. S. E. W., A. G. van Os & P. H. Nienhuis, 2003. Current themes in Dutch river research: general introduction. In: Leuven, R. S. E. W., A. G. van Os & P. H. Nienhuis (eds), Proceedings NCR-days 2002, current themes in Dutch river research. NCR-publication 20-2003. Netherlands Centre for River Studies, Delft: 10–17.

Nolte, B. T., R. S. E. W. Leuven & P. H. Nienhuis, 2003. Linking sedimentation and ecology in riverine landscapes; influences of climate change and physical reconstruction measures. In Leuven, R. S. E. W., A. G. van Os & P. H. Nienhuis (eds), Proceedings NCR-days 2002, current themes in Dutch river research. NCR-publication 20-2003. Netherlands Centre for River Studies, Delft: 144–147.

Aarts, B. G. W., F. W. B. van den Brink & P. H. Nienhuis, 2004. Habitat loss as the main cause of the slow recovery of fish fauna of regulated large rivers in Europe: the transversal floodplain gradient. River Research and Applications 20: 3–23.

De Nooij, R. J. W., H. J. R. Lenders, R. S. E. W. Leuven, G. de Blust, N. Geilen, B. Goldschmidt, S. Muller, I. Poudevigne & P. H. Nienhuis, 2004. BIO-SAFE: assessing the impacts of physical reconstruction on protected and endangered species. River Research and Applications 20: 299–313.

Klijn, F., P. H. Nienhuis & G. B. M. Pedroli (eds), 2004. Towards sustainable flood risk management in the Rhine and Meuse river basins. River Research and Applications 20: 227–357.

Kooistra, L., E. A. L. Salas, J. G. P. W. Clevers, R. Wehrens, R. S. E. W. Leuven, P. H. Nienhuis & L. M. C. Buydens, 2004. Exploring field vegetation reflectance as an indicator for soil contamination in river floodplains. Environmental Pollution 127: 281–290.

De Nooij, R. J. W., H. Hendriks, R. S. E. W. Leuven, H. J. R. Lenders & P. H. Nienhuis, 2005. Evaluation of floodplain rehabilitation: a comparison of ecological and policy oriented biodiversity assessment. Archiv für Hydrobiologie Supplement 155, Large Rivers 15: 413–424.

Part One
Functioning of River Systems

Hydrobiologia (2006) 565:21–38
R.S.E.W. Leuven, A.M.J. Ragas, A.J.M. Smits & G. van der Velde (eds), Living Rivers: Trends and Challenges in Science and Management
DOI 10.1007/s10750-005-1903-9

BasinBox: a generic multimedia fate model for predicting the fate of chemicals in river catchments

A. Hollander[1,*], M.A.J. Huijbregts[1], A.M.J. Ragas[1] & D. van de Meent[1,2]
[1]*Department of Environmental Science, Institute for Wetland and Water Research, Radboud University Nijmegen, P.O. Box 9010, 6500 GL Nijmegen, The Netherlands*
[2]*National Institute of Public Health and the Environment, Laboratory for Ecological Risk Assessment, P.O. Box 1, 3720 BA Bilthoven, The Netherlands*
*(*Author for correspondence: E-mail: a.hollander@science.ru.nl)*

Key words: multimedia fate model, river catchment, Rhine, risk assessment

Abstract

Multimedia fate models have proven to be very useful tools in chemical risk assessment and management. This paper presents BasinBox, a newly developed steady-state generic multimedia fate model for evaluating risks of new and existing chemicals in river basins. The model concepts, as well as the intermedia processes quantified in the model, are outlined, and an overview of the required input parameters is given. To test the BasinBox model, calculations were carried out for predicting the fate of chemicals in the river Rhine basin. This was done for a set of 3175 hypothetical chemicals and three emission scenarios to air, river water and cropland soils. For each of these hypothetical chemicals and emission scenarios the concentration ratio between the downstream area and the upstream area was calculated for all compartments. From these calculations it appeared that BasinBox predicts significant concentration differences between upstream and downstream areas of the Rhine river basin for certain types of chemicals and emission scenarios. There is a clear trend of increasing chemical concentrations in downstream direction of the river basin. The calculations show that taking into account spatial variability between upstream, midstream and downstream areas of large river basins can be useful in the predictions of environmental concentrations by multimedia fate models.

Introduction

For more than 20 years now, multimedia fate models have been used for the prediction of chemical fate and exposure in the environment. Baughman & Lassiter (1978) and Mackay & Paterson (1981) first introduced this kind of models, the so-called 'fugacity models' or 'box models', in which the fate of chemicals in different environmental media was calculated based on physical–chemical properties of the compound, environmental characteristics and emission data. The first models, e.g. Mackay's unit world model (Mackay & Paterson, 1981; Mackay et al., 1983), were relatively simple in structure and detail, but over time the models became more complex and sophisticated. Different types of compartments were added to the models, i.e. vegetation compartments (Trapp & Matthies, 1996; Severinsen & Jager, 1998; Cousins & Mackay, 2001), and organic film-compartments coating impervious surfaces (Diamond et al., 2001). Nested, dynamic and GIS-based models were developed (Brandes et al., 1996; Woodfine et al., 2001; Suzuki et al., 2004), models with layered air and soil compartments arose (McKone, 1993; Toose et al., 2004), and models for multi-species chemicals were introduced (Fenner et al., 2000; Cahill & Mackay, 2003).

Multimedia fate models have proven to be very useful tools in chemical risk assessment and management. Their multimedia character and relatively low complexity make them particularly useful to evaluate the fate of new and existing chemicals. In the European Union the EUSES model (Vermeire et al., 1997; Vermeire et al., 2005) is currently being used for risk assessment and management purposes, while in the USA, the Total Risk Integrated Methodology (TRIM.FaTE; US-EPA, 2002) is typically applied. Most multimedia fate models used in the risk assessment of chemicals are based on a geo-political parameterization, e.g. per country or province, and originate from a generic environmental approach, as for example described in the EU Technical Guidance Documents (ECB, 2003). For water quality management purposes in the European Union, however, the regulatory concept is shifting more and more towards an environmentally specific approach. This regulatory concept is described in the European Water Framework Directive (EC, 2000), in which the European Union defined ecological and chemical water standards at a catchment scale. This catchment approach could also be very advantageous for the risk assessment policy of new and existing chemicals, since river basins form more clearly defined physical entities than political regions. Except for air-borne transport, transport of chemicals occurs only within the basin, which facilitates the description of transport flows in the model. Moreover, in the case of river basin modelling, floodplain areas, which are being regarded as valuable areas for nature conservation, can be modelled separately. It will therefore be useful to implement the concept of catchment based modelling in the process of risk management of chemicals.

Very few multimedia fate models have been developed using river catchments as the basic environmental unit, since chemical modelling in catchments usually focuses on in-stream water quality only. This, for example, is the case in the Great-ER model (Feijtel et al., 1997) and the LOIS model (Boorman, 2003). Coulibaly et al. (2004) did develop a multimedia catchment model for the Passaic River Watershed in the USA, and Suzuki et al. (2004) described a multimedia model built up from 38,000 river catchments in Japan (G-CIEMS), but both are very site-specific GIS-based models of relatively small basins (up to 200 km^2) that cannot be easily applied to other river catchments.

The goal of this study is to develop a generic steady-state multimedia fate model to evaluate risks of new and existing chemicals in river basins. Here, this model, called BasinBox, is presented. The model concept, as well as the environmental processes considered in the model are outlined, and an overview of the required input parameters is given. Furthermore, the model is applied in a case study on the river Rhine basin area for a set of 3175 hypothetical chemicals, representing the whole range of chemical property combinations, to test whether the catchment approach applied in BasinBox yields valuable insights in the context of multimedia fate modelling. For that purpose, concentration ratios between the upstream area and the downstream area are analysed for various compartments and emission scenarios.

Materials and methods

Model description

In the BasinBox model, the river basin is subdivided into an upstream, midstream and downstream area, following Schumm's (1977) idealized scheme of a river basin consisting of three zones arranged in downstream sequence. Since many environmental parameters and process intensities vary between the different zones of a river basin, this subdivision allows the modeller to incorporate basic spatial variability into the model. The three model areas are interconnected by single-direction river flows and two-direction air flows. Figure 1a gives a schematic representation of the three sequential areas and the connections between these areas.

Each of the upstream, midstream and downstream areas consists of 21 compartments, representing different environmental media. Nine of these compartments belong to the floodplain zone and 11 compartments belong to the catchment zone. One single air compartment covers both the floodplain and the catchment zone. The floodplain zone is defined as the area of the river basin that consists of the river or its main tributaries and the land that is being flooded temporally each year. We chose to make the distinction between the floodplain and the catchment zone since some processes, like sedimentation and groundwater

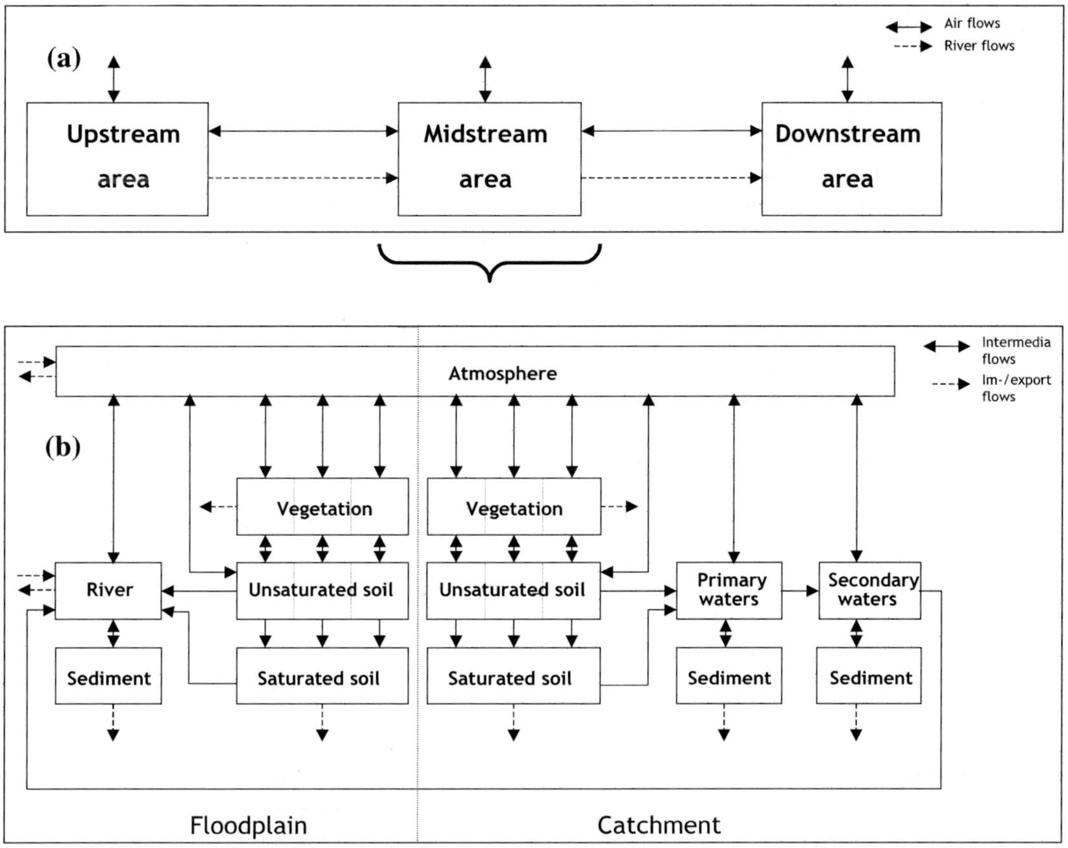

Figure 1. Schematic representation of the BasinBox model. (a) Schematic representation of the three sequential areas (up-, mid- and downstream) and the connections between these areas. (b) Schematic representation of the compartmental construction and the transport routes between the compartments.

flow, proceed differently in these two zones. Moreover, floodplain areas are regarded as valuable habitats for nature conservation and development (Nienhuis et al., 2002; ECNC, 2004; De Nooij et al., 2006). In the floodplain zone, a river and a sediment compartment, three unsaturated soil compartments (pasture-, cropland- and natural soil), saturated soil, and three vegetation compartments (pasture-, cropland-, and natural vegetation) have been distinguished. In the catchment zone one can find compartments for primary waters, secondary/tertiary waters and sediment, three unsaturated soil compartments (pasture-, cropland- and other soil), saturated soil, and three vegetation compartments (pasture-, cropland-, and natural vegetation). A schematic representation of the compartmental construction and the transport routes between the compartments is given in Figure 1b.

The concentrations calculated by BasinBox are affected by emissions, degradation processes and processes that cause chemical mass flows to and from the compartments. Mass balance equations can be written for all compartments, having the following general format:

$$V_i * \frac{\partial C_i}{\partial t} = \text{Emission}_i + \text{Import}_i - \text{Export}_i - \text{Degradation}_i + \text{Advection}_{ij} + \text{Diffusion}_{ij}$$

with V_i being the volume of compartment i (m^3) and C_i being the chemical concentration in that compartment (mol m^{-3}). Emission$_i$, and Degradation$_i$ represent emission to, and degradation from compartment i, respectively. Import$_i$ is the mass flow to compartment i from outside the basin area, while Export$_i$ stands for the mass flow from i out of the basin area. Advection$_{ij}$ and Diffusion$_{ij}$

are the gross advective and diffusive mass flows between compartments i and j within the river basin.

At steady state, the mass flows balance. The set of mass balances is solved using a matrix inversion routine. BasinBox calculates concentrations of chemicals in each of the compartments, using Microsoft Excel® software in combination with the Poptools-extension (CSIRO, 1994).

Since BasinBox is a steady state model, it is assumed that the environmental properties of and emissions to the compartments do not change over time. It is also assumed that equilibrium exists between the different phases within each compartment (e.g. gas, water and solids in soil). Moreover, each compartment is assumed to be completely mixed, which implies that all environmental properties and concentrations are equal throughout a compartment. Exceptions to this are the unsaturated soil compartments, where depth-dependent concentration differences in soil can significantly affect substance flows (e.g. volatilisation, leaching) throughout the soil profile. To overcome this problem, the theoretical principle of exponentially declining soil concentrations with depth of McKone & Bennett (2003), as implemented in multimedia models by Hollander et al. (2004), was introduced in the BasinBox model. This method applies correction factors that account for the deviation of depth-dependent soil concentrations from depth-averaged concentrations.

Model processes

All intermedia mass flows affecting the concentration of a chemical in a compartment (in mol s^{-1}) can be described as the product of a transport coefficient (in $m^3 s^{-1}$) and the concentration (in mol m^{-3}) in the compartment from which the mass flow originates. The transport coefficient is calculated as the product of a mass transfer coefficient (in m s^{-1}) and the interfacial area (in m^2). We distinguish diffusive and advective mass flows and transport coefficients. A diffusive mass flow is treated as a process driven by differences in the chemical potentials in the two media. Advective mass flows proceed by a carrier that physically flows from one compartment to another, e.g. by air or water. The amount of advective mass transport depends on the rate of the carrier flow and the concentration of the compound in the carrier.

Air and water transport flows

In the BasinBox model, air and water are regarded as the main carriers for advective mass flows. Air transport within the modelled river basin as well as into and out of it is dependent on the wind direction and the geometric orientation of the areas in the river basin, e.g. the position of the upstream area with regard to the midstream area, and the position of the midstream area with regard to the downstream area. To calculate the source of imported air, and subsequently, the chemical concentration in the air imported to the area, for each of the possible combinations of wind directions and orientations of the river basin, an air-inflow scenario was formulated for the upstream, midstream and downstream area. This generic calculation method enables the user to enter all possible orientations of a river basin to calculate the source of airflows into and out of the upstream, midstream and downstream area. As constant atmospheric pressure is assumed, the amount of imported air equals the amount of exported air. This amount is estimated based on the residence time of air in the upstream, midstream and downstream area, calculated using the annual average wind speed, the distance across the area in each of the eight compass directions, and the frequencies of wind directions, following the method described by Webster et al. (2004).

Since there is no water flow over the borders of a river catchment, it is possible to accurately construct a water balance, describing all relevant water transport processes within a river basin (Fig. 2). Figure 2a shows the water transport routes between air and soil, air and water, soil and water and soil and vegetation. Water enters the model as precipitation, of which a certain fraction evaporates. A fraction of the water runs off over the soil surface to one of the surface water compartments. Another fraction is discharged as subsurface flow at the mechanical reworking depth, and a third fraction is drained by tube drainage (except in natural soils where no tube drainage is

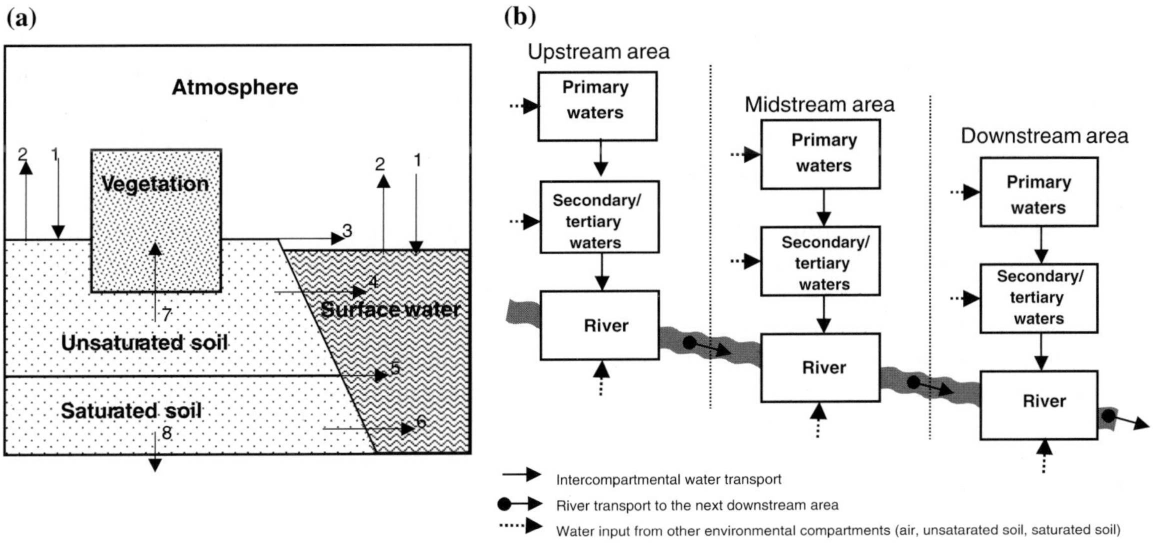

Figure 2. Schematic representation of the water balance elements of BasinBox. (a) Water flows between the air, soil, surface water and vegetation compartments. 1 = rain input (to soil and water), 2 = evaporation, 3 = surface runoff, 4 = subsurface runoff, 5 = tube drainage, 6 = groundwater drainage, 7 = transpiration, 8 = groundwater discharge to deeper aquifers. (b) Water flows between the different surface water compartments.

assumed). The remaining water percolates to the saturated soil compartments. In the catchment zone of the model, it is assumed that a certain amount of groundwater is exported from the saturated soil zone to deeper aquifers and thus exported from the system. The remainder is transported from the saturated soil zone to the surface water compartments. In the floodplain zone all groundwater is assumed to recharge to the surface water compartment, since no groundwater export takes place at a location so close to the drainage base of the system. Figure 2b shows the water transport routes between the different surface water compartments. The primary water compartments (ditches, pools) of the up-, mid- and downstream areas discharge to the secondary and tertiary water compartments (canals, brooks, small rivers) of these areas and those again discharge to their river water compartments. In the river compartments, water is being discharged in a downstream direction. A constant volume of the surface water compartment is assumed; this means that all surplus water is being transported following the routes described in Figure 2b. From the downstream area river compartment, water is exported from the system at the river mouth.

Intermedia chemical transport processes

Air–surface area exchange

Advective air–unsaturated soil and air–water transport occurs by wet and dry deposition. For the calculation of the dry deposition chemical mass flow, the dry deposition velocity of aerosol particles and the fraction of chemicals associated with aerosol in the atmosphere are used (Brandes et al., 1996). An interception fraction for dry aerosol interception by vegetation is introduced. Wet deposition is divided into gas washout and aerosol washout processes, both dependent on rain intensity and the fraction of gas and aerosols in the atmosphere. Gas washout is calculated using the dimensionless air–water partition coefficient (Den Hollander & Van de Meent, 2004), while aerosol washout is calculated based on the aerosol collection efficiency (Mackay, 1991). For aerosol washout an interception fraction for vegetation is inserted. It is assumed that vegetation causes no interception in the case of gas washout.

Diffusive air–water, air–soil and air–vegetation transport occurs by gas absorption and volatilisation. These processes are calculated using the classic two-film resistance model (Schwarzenbach et al., 1993). For the soil–air mass flows a correction factor accounting for the deviation of the concentration at the top of the soil compartment from the average soil concentration (Hollander et al., 2004) is inserted. Diffusive air to vegetation transport vice versa is estimated from the overall mass transfer coefficient at the air–plant interface for natural and agricultural vegetation, following the description of Severinsen & Jager (1998).

Water–sediment exchange and sediment burial
Advective transport between the water and the sediment compartments occurs by sedimentation and resuspension. Gross sedimentation rates are calculated from the settling velocity of suspended particles and the concentration of suspended matter in the water compartments following the method of Brandes et al. (1996). Net sedimentation rates in the up-, mid- and downstream areas were derived from literature (Schwarzenbach et al., 1993; Trapp, 1996; Hofstee & Leenaers, 2002). Resuspension rates equal the difference between the gross sedimentation rates and the net sedimentation rates. Diffusive exchange proceeds by adsorption and desorption processes, based on the partial mass transfer coefficients at the water and the sediment side of the water–sediment interface. In the BasinBox model, the sediment compartments are modelled with a fixed depth. Therefore, a sediment burial flow is introduced, the burial mass transfer coefficient being equal to the product of the net sedimentation rate and the interfacial area.

Unsaturated soil–vegetation exchange
Chemical transport from unsaturated soil to vegetation proceeds by an advective transpiration flow. Severinsen & Jager (1998) described the method used here, using a transpiration stream concentration factor (Briggs et al., 1982). Litter flow causes chemical transport from the vegetation compartments to the unsaturated soil compartments, of which the amount is derived from the growth rate of vegetation, and the harvesting efficiency (Severinsen & Jager, 1998).

Unsaturated soil–saturated soil–surface water exchange
Several advective water-bound soil to water processes are modelled. On agricultural pasture and cropland soils, one water flow occurs at the soil surface (surface runoff), one at the mechanical reworking depth (subsurface flow), and one at the typical depth of drainage tubes (tube drainage). For natural soils, only surface runoff and subsurface flow processes are modelled. The principle of depth-dependent chemical concentrations is used for calculating the amount of chemicals involved in the different processes. For the surface runoff calculation, both the process of solute transport in runoff water and erosion (Asselman, 1997) are modelled.

The fraction of precipitation that does not flow from the unsaturated soil compartments to the surface water compartments percolates to the saturated soil zone. The chemical amount that is transported from the saturated soil zone to deeper aquifers is determined by an export fraction derived from literature (De Wit, 1999), the remainder is recharged to the surface water compartments.

Degradation

All chemical degradation is assumed to obey (pseudo) first order kinetics, following from a degradation rate constant, the volume of the compartment and the concentration. Following the method of Den Hollander & Van de Meent (2004), the degradation rate constant in air is estimated from the fraction of the chemical in air that is associated with aerosol particles, and the OH-radical concentration in air. Values for the degradation rate constants in water, sediments and soils are calculated using the degradation rate constant in the dissolved phase, the bulk degradation rate constant in sediments and in soils respectively, corrected for the actual temperature. The chemical degradation constants in vegetation are assumed to be ten times higher than those in soil (Brandes et al., 1996).

Model parameterization

The input for BasinBox consists of physical–chemical properties of the compound studied, environmental characteristics and emission data, the latter two being user-defined. Required physical–chemical properties of the compound are its molecular weight, octanol–water partition coefficient (K_{ow}), vapour pressure, solubility, melting point, and degradation rate constants for bulk sediment, bulk soil, and the gas phase. The vapour pressure, the solubility and the degradation rates of a chemical are modelled as temperature dependent variables. Using the physical–chemical data the model estimates intermedia partition coefficients. The air-water partition coefficient (K_{aw}) is estimated from the ratio of the vapour pressure and the water solubility of the compound. The air–aerosol partitioning is determined on the basis of the chemical's vapour pressure, according to Junge (1977). The solids–water partition coefficient is calculated based on the relationship with K_{ow} and the organic carbon content of the soil proposed by DiToro et al. (1991). The bioconcentration factor is estimated from the K_{ow} and the fat content of the biota. For vegetation, the concentration ratio between plant tissue and water in thermodynamic equilibrium is estimated from the water and lipid contents of the plant tissue (Severinsen & Jager, 1998).

Values for the environmental parameters and the emission rates are, dependent on the model scenario, to be defined by the user. The environmental parameters that need to be entered in the BasinBox model are listed in Table 1. Emissions in the model can take place to the air compartment, the water compartments and the different soil compartments.

Case study for the river Rhine basin

Model area

To test the BasinBox model, example calculations were carried out for predicting the fate of chemicals in the river Rhine basin. Environmental input parameters for BasinBox were collected for the upstream, midstream, and downstream areas of the Rhine basin (Table 1). The geometrical orientation of the upstream area with regard to the midstream area of the Rhine basin is South, while the orientation of the midstream area with regard to the downstream area is Southeast. Based on this orientation, together with information on the percentages of wind flowing in from the different compass directions, the chemical concentration in the inflowing air in the different areas of the river Rhine basin was calculated (Fig. 3). Figure 3a and Table 1 show the percentage of wind directions occurring in the river Rhine basin. For each of the wind directions, the model estimates the fraction of inflowing air occurring from outside the river basin as well as the fraction of inflowing air taking place from the other areas within the basin. Figure 3b shows the total amounts of inflowing and outflowing air in the different areas of the Rhine basin, calculated using the wind directions of Figure 3a and the residence times of air in the different areas.

In BasinBox it is possible to model periodical inundations of the floodplain zone of the river basin. For the Rhine basin, a yearly inundation of four weeks is assumed in the downstream area of the basin (Hofstee & Leenaers, 2002). During this period gross and net sedimentation rates from the river water to the inundated floodplain soil are assumed to equal those rates from the river water to the sediment compartment. For the upstream and midstream areas of the Rhine basin, no periodical flooding was assumed in this case study.

Set of hypothetical chemicals

For the example calculations, a set of 3175 hypothetical chemicals was used, covering the entire space of plausible chemical partitioning properties and half-lives. The advantage of using this set instead of real chemical data is that even a set of hundreds of real chemicals does not densely cover the space of possible chemical property combinations (Fenner et al., 2005). The set of hypothetical chemicals used here includes all possible combinations of integer values of log K_{aw} from −11 to 2 and log K_{ow} from −1 to 8 with the restriction that $-1 \leq (\log K_{ow} - \log K_{aw}) \leq 15$ (Fenner et al., 2005). For the degradation half-lives, all possible combinations of half-lives in air of 4, 24, 168, 1000, and 8760 h with half-lives in water of 24, 168, 1000, 8760 and 87,600 h were used. The half-life in soil and in sediment was set to twice the half-life in water, in order to limit the chemical properties that were varied to four (Stroebe et al., 2004).

Emission scenarios

With the whole set of hypothetical chemicals, the BasinBox model was run for three emissions, towards air, river water and cropland soils.

Table 1. Environmental input parameters for the BasinBox model in the case study for the river Rhine catchment

Environmental parameter	Unit	Value upstream	Value midstream	Value downstream	Reference
Total area	[m^2]	6.17E+10	6.17E+10	6.17E+10	–
Area fraction river	[–]	1.00E–03	1.62E–03	1.88E–03	a
Area fraction pasture soil floodplain	[–]	1.52E–03	1.29E–03	2.15E–03	a
Area fraction cropland soil floodplain	[–]	4.40E–04	1.38E–03	1.05E–03	a
Area fraction other soil floodplain	[–]	2.05E–03	1.64E–03	5.65E–04	a
Area fraction pasture soil catchment	[–]	3.64E–01	2.92E–01	4.68E–01	a
Area fraction cropland soil catchment	[–]	1.06E–01	3.12E–03	2.30E–01	a
Area fraction other soil catchment	[–]	4.89E–01	3.70E–03	1.23E–01	a
Area fraction primary waters catchment	[–]	2.59E–02	4.72E–04	8.71E–02	a
Area fraction secondary waters catchment	[–]	1.00E–02	1.99E–02	8.71E–02	a
Mixed height air compartment	[m]	1.00E+03	1.00E+03	1.00E+03	b
Depth river compartment	[m]	3.00E+00	3.00E+00	5.00E+00	c
Depth sediment compartments	[m]	3.00E–03	3.00E–03	3.00E–03	b
Depth primary waters compartment	[m]	1.00E+00	1.00E+00	1.00E+00	d
Depth secondary waters compartment	[m]	2.00E+00	2.00E+00	2.00E+00	–
Depth unsaturated soil compartments	[m]	1.00E+00	1.00E+00	1.00E+00	e
Depth saturated soil compartments	[m]	2.00E+00	2.00E+00	2.00E+00	–
Solid phase advection velocity soil	[m s^{-1}]	6.34E–12	6.34E–12	6.34E–12	b
Solid phase turbation coefficient soil	[m s^{-1}]	6.37E–12	6.37E–12	6.37E–12	b
Volume fraction solids soil	[–]	6.00E–01	6.00E–01	6.00E–01	–
Volume fraction water unsaturated soil	[–]	2.00E–01	2.00E–01	2.00E–01	–
Volume fraction air unsaturated soil	[–]	2.00E–01	2.00E–01	2.00E–01	–
Volume fraction water saturated soil	[–]	4.00E–01	4.00E–01	4.00E–01	–
Volume fraction water sediment	[–]	8.00E–01	8.00E–01	8.00E–01	f
Volume fraction solids sediment	[–]	2.00E–01	2.00E–01	2.00E–01	f
Volume fraction water vegetation	[–]	8.00E–01	8.00E–01	8.00E–01	b
Mass fraction lipids vegetation	[–]	1.50E–02	1.50E–02	1.50E–02	b
Leaf area index pasture vegetation	[m^2 m^{-2}]	5.06E+00	5.06E+00	5.06E+00	g
Leaf area index cropland vegetation	[m^2 m^{-2}]	1.71E+00	1.71E+00	1.71E+00	g
Leaf area index natural vegetation	[m^2 m^{-2}]	3.62E+00	3.62E+00	3.62E+00	g
Vegetation cover pasture vegetation	[–]	7.10E–01	7.10E–01	7.10E–01	h
Vegetation cover cropland vegetation	[–]	8.60E–01	8.60E–01	8.60E–01	h
Vegetation cover natural vegetation	[–]	9.00E–01	9.00E–01	9.00E–01	h
Vegetation mass pasture vegetation	[kg m^{-2}]	1.20E+00	1.20E+00	1.20E+00	i
Vegetation mass cropland vegetation	[kg m^{-2}]	1.80E+00	1.80E+00	1.80E+00	i
Vegetation mass other vegetation	[kg m^{-2}]	2.40E+00	2.40E+00	2.40E+00	i
Wet density of vegetation	[kg m^{-3}]	9.00E+02	9.00E+02	9.00E+02	i
Average wind speed	[m s^{-1}]	2.90E+00	2.90E+00	2.90E+00	j
Fraction of wind from direction north	[–]	9.00E–02	9.00E–02	9.00E–02	j
Fraction of wind from direction north-east	[–]	1.00E–01	1.00E–01	1.00E–01	j
Fraction of wind from direction south-east	[–]	1.20E–01	1.20E–01	1.20E–01	j
Fraction of wind from direction south	[–]	1.50E–01	1.50E–01	1.50E–01	j
Fraction of wind from direction south-west	[–]	2.00E–01	2.00E–01	2.00E–01	j
Fraction of wind from direction west	[–]	1.60E–01	1.60E–01	1.60E–01	j
Fraction of wind from direction north-west	[–]	1.00E–01	1.00E–01	1.00E–01	j

Continued on p. 29

Table 1. (Continued)

Environmental parameter	Unit	Value upstream	Value midstream	Value downstream	Reference
Orientation of the up- with regard to the midstream area	[–]	S	S	S	–
Orientation of the mid- with regard to the downstream area	[–]	SE	SE	SE	–
Rain intensity	[m s^{-1}]	1.10E+03	8.23E+02	9.21E+02	k, v, j
Infiltration fraction in floodplain soil	[–]	8.50E–01	9.00E–01	8.50E–01	e
Infiltration fraction in catchment soil	[–]	7.50E–01	8.50E–01	8.50E–01	e
Fraction of rainwater as subsurface flow soil	[–]	3.60E–02	3.60E–02	3.60E–02	–
Fraction of rainwater as tube flow cropland soil	[–]	2.50E–01	2.50E–01	2.50E–01	e
Fraction of rainwater as tube flow pasture soil	[–]	2.50E–01	2.50E–01	2.50E–01	e
Fraction of rainwater exported to deep aquifers floodplain	[–]	0.00E+00	0.00E+00	0.00E+00	–
Fraction of rain water exported to deep aquifers catchment	[–]	1.70E–01	1.70E–01	1.70E–01	l
Fraction of soil water discharging to primary waters	[–]	5.00E–01	5.00E–01	5.00E–01	–
Fraction of soil water discharging to secondary waters	[–]	5.00E–01	5.00E–01	5.00E–01	–
Temperature	[°C]	8.20E+00	9.90E–00	9.70E+00	k, v, j
Specific aerosol surface	[m^2 m^{-3}]	1.50E–04	1.50E–04	1.50E–04	b
Mass fraction organic carbon in suspended matter	[–]	2.00E–01	2.00E–01	2.00E–01	m
Fat content of fresh water fish	[–]	5.00E–02	5.00E–02	5.00E–02	b
Concentration biota in fresh water	[mg l^{-1}]	1.00E+00	1.00E+00	1.00E+00	b
Concentration suspended matter in river water	[mg l^{-1}]	1.50E+01	2.90E+01	3.50E+01	n, w, y
Concentration suspended matter in primary waters	[mg l^{-1}]	5.00E+00	5.00E+00	5.00E+00	–
Concentration suspended matter in secondary waters	[mg l^{-1}]	1.50E+01	2.90E+01	3.50E+01	n, w, y
Mass fraction organic carbon in sediment	[–]	2.00E–02	2.00E–02	2.00E–02	o
Mass fraction organic carbon in unsaturated soil	[–]	2.00E–02	2.00E–02	2.00E–02	b
Mass fraction organic carbon in saturated soil	[–]	6.00E–03	6.00E–03	6.00E–03	o
Deposition velocity of aerosol particles agricultural soil	[m s^{-1}]	5.31E–04	5.31E–04	5.31E–04	p
Deposition velocity of aerosol particles natural soil	[m s^{-1}]	7.30E–03	7.30E–03	7.30E–03	q
Aerosol collection efficiency	[–]	2.00E+05	2.00E+05	2.00E+05	r
Interception of dry aerosol deposition vegetation	[–]	4.40E–01	4.40E–01	4.40E–01	s
Interception of wet aerosol deposition vegetation	[–]	1.40E–01	1.40E–01	1.40E–01	t
Growth rate constant natural vegetation	[s^{-1}]	2.88E–08	2.88E–08	2.88E–08	i
Growth rate constant agricultural vegetation	[s^{-1}]	1.27E–07	1.27E–07	1.27E–07	i
Harvesting efficiency natural vegetation	[–]	0.00E+00	0.00E+00	0.00E+00	b
Harvesting efficiency agricultural vegetation	[–]	5.90E–01	5.90E–01	5.90E–01	b
Water uptake rate natural vegetation	[m s^{-1}]	8.40E–09	8.40E–09	8.40E–09	i
Water uptake rate agricultural vegetation	[m s^{-1}]	9.32E–09	9.32E–09	9.32E–09	i

Continued on p. 30

Table 1. (Continued)

Environmental parameter	Unit	Value upstream	Value midstream	Value downstream	Reference
Settling velocity of suspended particles	[m s^{-1}]	2.89E−05	2.89E−05	2.89E−05	b
Autochthonous production of suspended matter in water	[g m^{-2} y^{-1}]	1.00E+01	1.00E+01	1.00E+01	b
Net sediment accumulation rate in water	[m s^{-1}]	2.36E−11	3.17E−11	4.44E−11	u, x, y
Erosion in floodplain zone	[m s^{-1}]	6.00E−02	1.00E−02	1.00E−04	n
Erosion in catchment zone	[m s^{-1}]	9.00E−02	3.00E−02	1.00E−03	n
Escape rate constant of air to the stratosphere	[s^{-1}]	3.66E−10	3.66E−10	3.66E−10	b
Regional OH-radical concentration	[cm^{-3}]	5.00E+05	5.00E+05	5.00E+05	b
Mechanical reworking depth agricultural soils	[m]	2.00E−01	2.00E−01	2.00E−01	b
Tube drainage depth	[m]	1.00E+00	1.00E+00	1.00E+00	e
Depth of transpiration flow pasture vegetation	[m]	2.00E−01	2.00E−01	2.00E−01	z
Depth of transpiration flow cropland vegetation	[m]	2.00E−01	2.00E−01	2.00E−01	z
Depth of transpiration flow natural vegetation	[m]	8.00E−01	8.00E−01	8.00E−01	z
Population density	[km^{-2}]	1.99E+02	2.34E+02	3.89E+02	a
Emission dose	[kg km^{-2}]	3.17E+00	2.60E+00	8.27E+00	aa

References: a = Nationmaster (2005), b = Den Hollander & Van de Meent (2004), c = McKone (1993), d = CLM (2004), e = Tiktak et al. (2002), f = Paterson & Mackay (1994), g = Scurlock et al. (2001), h = Zeng et al. (2000), i = Severinsen & Jager (1998), j = KNMI (2004), k = MeteoSchweiz (2004), l = De Wit (1999), m = Zhang et al. (2003), n = Asselman (1997), o = McKone et al. (2001), p = McLachlan et al. (2002), q = Horstmann & McLachlan (1998), r = Mackay (1991), s = Chamberlain (1967), t = Scheringer et al., (2000), u = Schwarzenbach et al. (1993), v = DWD (2004), w = Meybeck et al. (2003), x = Trapp (1996), y = Hofstee & Leenaers (2002), z = Jackson (1996), aa = FAO (2001)
S = South, SE = Southeast

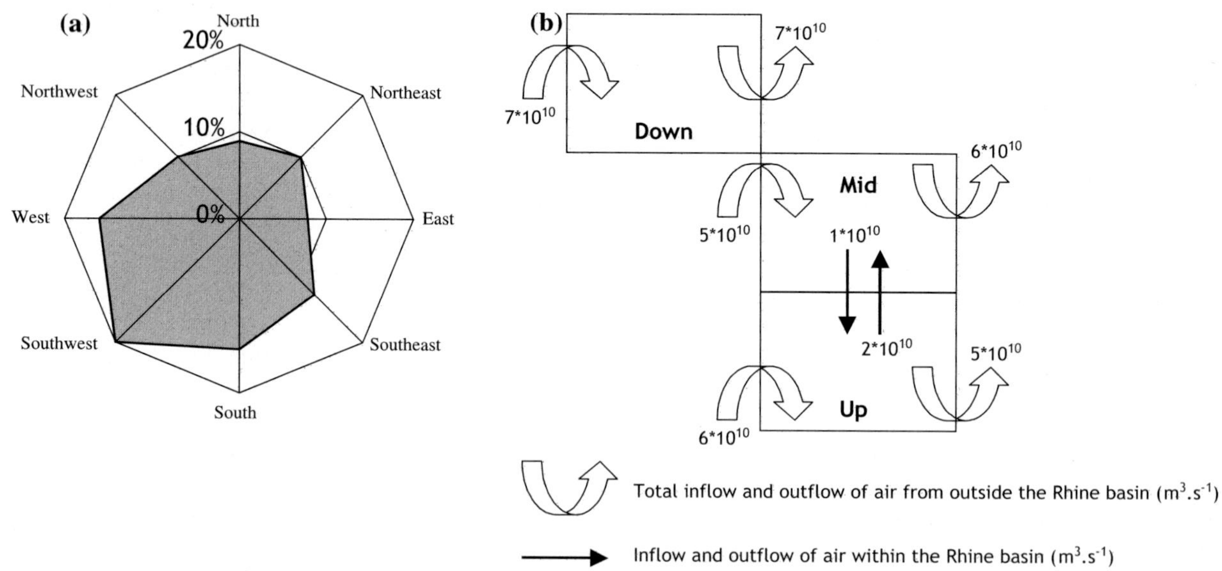

Figure 3. Schematic representation of the calculation of the chemical concentration in the inflowing air in the different areas of the river Rhine basin. (a) The percentage of each of the wind directions occurring in the river Rhine basin (KNMI, 2004). (b) Total amounts of inflowing and outflowing air in the different areas of the Rhine basin.

Emission input towards air and river water was estimated using population density numbers, following Prevedouros et al. (2004). This method assumes that chemicals emitted to air and river water are mainly released from densely populated urban areas. Based on population density numbers in the Rhine basin area (Table 1), air and river water emissions take place in a ratio of 1:1.2:2 between the upstream, midstream and downstream areas (Nationmaster, 2005). Diffuse emissions to cropland soils will typically be pesticides, so input to cropland soils was estimated based on differences in cropland area between the upstream, midstream and downstream areas as well as on differences in the typical pesticide emission dose between the different areas (Table 1). According to this method, the emission ratio between the upstream, midstream and downstream areas in the Rhine basin area is 1:2.6:6.3 for the floodplain soil compartment, and 1:1.9:4.5 for the catchment soil compartment (FAO, 2001; Nationmaster, 2005). Since we were interested in concentration differences between the upstream, midstream and downstream areas, rather than in absolute concentrations, only relative emission rates were required for the calculations.

Concentration ratios

For all hypothetical chemicals, the concentration ratio between the upstream area and the downstream area was calculated for each emission scenario and for all compartment types. These results were analysed in order to select all combinations of chemical properties for which the concentration ratio between the downstream and upstream areas is larger than 100, respectively 10 or smaller than 0.01, respectively 0.1. This was done to select those chemicals for which predicting concentrations within the river Rhine area for the upstream, midstream and downstream area separately can be relevant. For all combinations of emission scenario and compartment type at which concentration ratios >100 or <0.01 occur, series of chemical space plots were created. In these chemical space plots, the concentration ratio between downstream and upstream areas is plotted against K_{ow} and K_{aw} for a certain value of the chemical degradation half-lives in air and water (and soil). Since the calculations were performed for five values of the degradation half-life in water and five values of the degradation half-life in air, for each combination of emission scenario and compartment 25 plots can be made. It appeared that the degradation rate of compounds in air hardly influenced the concentration ratios in our calculations, so plots were created for only one value of the degradation half-life in air. We used the median of the modelled values of the half-life in air (168 h). The plots provide a clear overview of chemical property combinations causing large spatial variation in the predictions of BasinBox.

Results and discussion

Table 2 shows for all combinations of emission scenario and compartment type the percentage of hypothetical chemicals for which the predicted concentrations of BasinBox in the upstream and downstream areas differ more than a factor of 10 and 100, respectively. In more than 95% of all cases, downstream concentrations are higher than upstream concentrations, implying that there is a clear trend of increasing chemical concentrations in downstream direction of the Rhine basin area. This is mainly caused by differences in emissions between the upstream and downstream area. Since the upstream area of the Rhine basin is less densely populated and contains less agricultural soils, emissions to air, river water as well as cropland soils are lower than in the downstream area. Furthermore, water based transport of chemicals, which occurs in downward direction of the river basin, causes variation in chemical concentrations between the upstream and downstream areas. For the river and floodplain compartments, the influence of downward transport can account for up to 60% of the concentration variance between the upstream and downstream areas, particularly when emissions occur to water. For the catchment compartments this downward chemical transport is only of minor influence.

For 13 combinations of emission scenario and compartment type concentration ratios >100 appeared. For all these scenarios, series of chemical space plots were created, consisting of five separate plots for all five modelled values of the degradation half-life in water (and soil). Two series of chemical space plots are given in

Table 2. Percentage of all hypothetical chemicals that show a concentration ratio larger than 10 and 100, per compartment type, per emission scenario

Floodplain compartments

Ratio > 10	Emission	Air	River	River sediment	Pasture soil	Pasture vegetation	Cropland soil	Cropland vegetation	Natural soil	Natural vegetation	Saturated soil
	Air	0.0%	0.0%	0.0%	0.0%	0.0%	0.0%	0.0%	0.0%	0.0%	0.0%
	River	0.0%	0.0%	0.0%	16.5%	0.0%	16.2%	0.0%	8.5%	0.0%	0.5%
	Soil	21.6%	0.0%	0.0%	19.5%	20.4%	0.0%	18.2%	19.5%	20.1%	7.2%

Catchment compartments

Ratio > 10	Emission	Pasture soil	Pasture vegetation	Cropland soil	Cropland vegetation	Natural soil	Natural vegetation	Saturated soil	Primary waters	Primary water sediment	Secondary waters	Secondary water sediment
	Air	0.0%	0.0%	0.0%	0.0%	5.2%	0.0%	0.0%	0.6%	2.7%	0.0%	0.0%
	River	0.0%	0.0%	0.0%	0.0%	10.6%	0.0%	0.0%	0.2%	1.4%	0.0%	0.0%
	Soil	17.6%	28.5%	0.0%	15.1%	62.0%	13.4%	0.5%	6.0%	7.6%	0.0%	0.0%

Floodplain compartments

Ratio > 100	Emission	Air	River	River sediment	Pasture soil	Pasture vegetation	Cropland soil	Cropland vegetation	Natural soil	Natural vegetation	Saturated soil
	Air	0.0%	0.0%	0.0%	0.0%	0.0%	0.0%	0.0%	0.0%	0.0%	0.0%
	River	0.0%	0.0%	0.0%	1.6%	0.0%	1.9%	0.0%	0.0%	0.0%	0.0%
	Soil	2.2%	0.0%	0.0%	1.6%	2.1%	0.0%	2.1%	1.6%	2.1%	0.0%

Catchment compartments

Ratio > 100	Emission	Pasture soil	Pasture vegetation	Cropland soil	Cropland vegetation	Natural soil	Natural vegetation	Saturated soil	Primary waters	Primary water sediment	Secondary waters	Secondary water sediment
	Air	0.0%	0.0%	0.0%	0.0%	0.0%	0.0%	0.0%	0.0%	0.0%	0.0%	0.0%
	River	0.0%	0.0%	0.0%	0.0%	0.0%	0.0%	0.0%	0.0%	0.0%	0.0%	0.0%
	Soil	1.6%	2.6%	0.0%	1.9%	1.1%	1.0%	0.0%	0.0%	0.0%	0.0%	0.0%

Figure 4, while the full set of plots (in colour) can be downloaded from http://www.ru.nl/environmentalscience/research/rivers/basinbox.

River water emission scenario

In case of emissions towards river water, relatively large spatial differences (ratio > 100) are found for about 60 chemical property combinations in cropland soils and pasture soils in the floodplain. Figure 4a shows the chemical space plots for the cropland soil compartment in the floodplain. The main transport routes responsible for the concentration differences are (1) downward transport of chemicals by river water and (2) temporal inundation of downstream floodplain soils and subsequent sedimentation of chemicals onto these soils. Beside that, variations in emission densities between the upstream and downstream areas account for concentration differences. For the natural soil compartments in the floodplain, concentration ratios are somewhat lower, but still larger than 10 for 8.5% of the chemicals. As a result of high concentration ratios in the floodplain soils, also the saturated soil zone in the floodplain shows concentration ratios >10 for some compounds.

In floodplain pasture and cropland soils, chemical property combinations for which the concentration ratio is larger than 100 range between a log K_{aw} of 0 to 2 with a log K_{ow} of 8. These chemicals have a high volatilisation potential from water to air on the one hand, and tend to bind to organic material on the other hand. So, when sedimentation in the floodplain occurs, a large part of these chemicals will settle down to the floodplain soils. Only compounds with a relatively high half-life in water (8760 to 87,600 h) and soil (17,520 to 175,200 h) will show relatively large concentration differences between the upstream and the downstream areas. The degradation half-life of chemicals in air does not largely influence the fate of chemicals. Not many chemicals exist for which the above-mentioned property combinations apply, but octachloro–2-pinene (cas # 25267-15-6; pesticide; PAN, 2005) and 1-iodohexadecane (cas # 544-77-4; pesticide; PAN, 2005) are known compounds in this range. For cropland soils in floodplains, the same applies as for pasture soils, but the range of chemicals for which large concentration ratios occur is somewhat broader. Chemicals with a log K_{aw} of −7, a log K_{ow} of 8 and a water degradation half-life of 8760 h, as well as chemicals with a log K_{aw} of −5, a log K_{ow} of 7 and a water degradation half-life of 24 h show concentration ratios >100. Examples of chemicals that have the above-mentioned chemical property combinations are monomethyl ester (cas # 6983-79-5; pesticide; PAN, 2005) and isodecyl-diphenyl phosphate (cas # 29761-21-5; plasticizer/flame retardant; Chemicalland21, 2005), respectively.

Cropland soil emission scenario

For the emission scenario to cropland soils, 11 compartment types show a concentration ratio between the downstream and upstream areas larger than 100 for a number of hypothetical chemicals. This is the case for the air compartment, all soil and vegetation compartments in the floodplain zone, pasture soils and natural soils in the catchment zone and pasture, cropland and natural vegetation in the catchment zone. The large ratios are mainly caused by differences in emission densities, followed by differences in the volatilisation of chemicals from soil and vegetation to the air. In the saturated soil and in primary waters, concentration differences larger than a factor of 10 occur. These differences are a direct consequence of concentration differences in the soil compartments, which results in different chemical amounts leaching to the saturated soil zone and draining to the surface water compartments.

The series of chemical space plots for the natural soil compartment in the catchment for the soil emission scenario is shown in Figure 4b, but this situation applies approximately for all compartments with emissions to soils and concentration ratios >100. Chemicals with a log K_{aw} ranging from −6 to −3 and a log K_{ow} from 5 or 6–8 show the largest concentration ratios. These compounds tend to bind to organic material on the one hand and are not very volatile on the other hand. Large concentration ratios are only found for chemicals with degradation half-lives in water of 24 or 168 h. This is caused by differences in the soil penetration depth of the chemicals, which is low for compounds with low degradation half-lives in water and soil. As a result, the process of volatilisation becomes relatively important for those

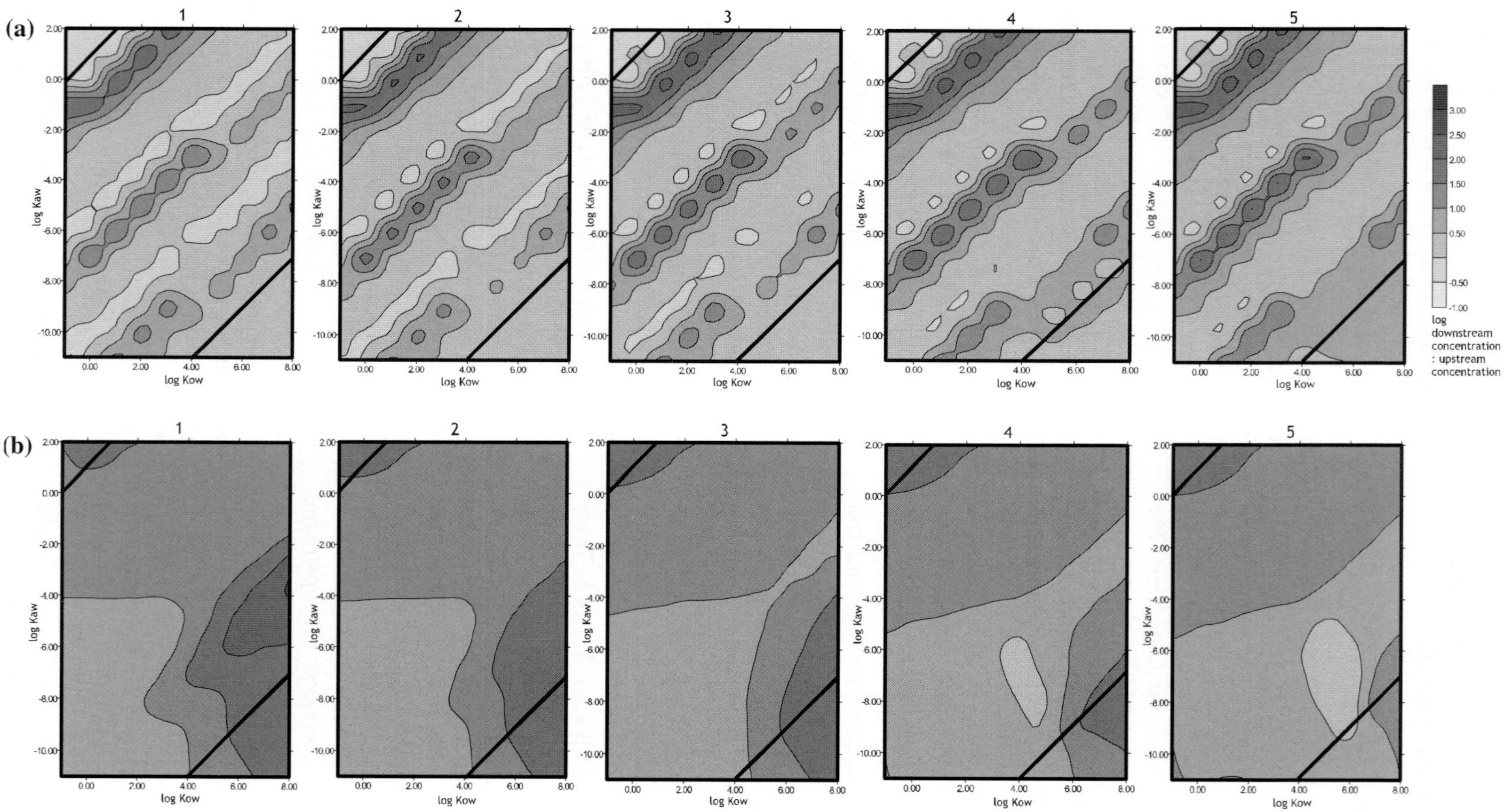

Figure 4. Chemical space plots for cropland soils in the floodplain by emission to river water and a degradation half-life in air of 168 h (a), and for natural soils in the catchment by emission to croplands and a degradation half-life in air of 168 h (b). The numbers 1–5 in (a) and (b) represent the different degradation half-lives in water: 1 = 24 h, 2 = 168 h, 3 = 1000 h, 4 = 8760 h, 5 = 87 600 h. Chemical property combinations represented in the upper left and lower right triangles of the plots are not likely to occur in reality and are not included in the dataset of hypothetical chemicals used in this study.

compounds, so in case of emissions taking place to the cropland soil compartments, the differences in emission densities between the upstream and downstream areas are being strengthened by this volatilisation flow. Therefore, large concentration differences arise in the upstream and downstream areas. In cropland soils their selves, the concentration ratio is not so large, since for those compartments the emission process is much more important than the volatilisation flow. Consequently, the concentration ratio in the cropland soil compartments equals more or less the soil emission ratio. As is the case by emissions to river water, the degradation half-life of chemicals in air does hardly influence their fate. Examples of real chemicals having the property combinations that result in large concentration differences between the upstream and downstream areas are tributyltin adipate (cas # 7437-35-6; pesticide; PAN, 2005), oleic acid (cas # 112-80-1; high production volume chemical, used in consumer products, building materials and pesticides; Scorecard, 2005), dicyclohexyl phtalate (cas # 84-61-7; high production volume chemical, used as plasticizer; Scorecard, 2005), and tetradecanol (cas # 112-72-1; high production volume chemical, used in consumer products, building materials and pesticides; Scorecard, 2005).

Air emission scenario

In case of emissions occurring towards the air compartments, none of the compartment types shows a concentration ratio between the upstream and downstream areas larger than 100, and only a few compartment types show a ratio larger than 10. Due to rapid mixing of chemicals in the atmosphere and because air-based chemical transport takes place both in upstream and in downstream directions, the concentration differences between the downstream area and the upstream area remain relatively low.

Uncertainty

From these calculations it becomes clear, that for certain types of chemicals large concentration differences can exist within one river basin. One remark should be made on the model results for the compounds with low degradation half-lives in air, water and soil and with a high K_{ow} (half-life in water of 24 h, log K_{ow} from 7 to 8). Although these compounds have a low soil penetration depth in reality, predicted soil penetration depths are even considerably lower. The algorithm of the soil penetration depth in BasinBox was primarily designed for compounds with less extreme property combinations, and it is not possible to extrapolate it directly to all chemicals. Therefore, the concept of the soil penetration depth in BasinBox does not apply very well on extreme hydrophobic and rapidly degrading compounds. Since the soil penetration depth largely influences the concentration ratios, especially when emissions take place to the soil, for these compounds the model results are not very reliable. However, for the other compounds the model relations do apply and compounds showing large concentration ratios still exist.

The relevance of the BasinBox model not only depends on relative differences that are found between the upstream, midstream and downstream areas of a river basin, but also on the absolute concentration differences. When concentration differences are large, but absolute chemical concentrations are very low, knowledge about the concentration differences is not very relevant. Since the example calculations were based on relative emissions, the model predicts only relative concentrations. To get an idea about absolute concentrations anyhow, we compared the relative concentrations in cropland soils and in natural soils after emission of chemicals to cropland soils. We assumed that if natural soil concentrations are less than six orders of magnitude lower than cropland soil concentrations (after cropland soil emission), these concentrations might be of serious concern. This appeared to be the case for 89% of the hypothetical chemicals, and for 76% of the chemicals that show concentration ratios >100 between downstream and upstream areas. For 10% of all chemicals the natural soil concentrations are even less than three orders of magnitude lower than cropland concentrations. These numbers indicate that for a large number of chemicals relevant concentrations may be found, in any case in natural soils. For these chemicals also knowledge about concentration variances will be relevant.

Although it is quite difficult to validate this type of generic models with such a large amount of compartments, a validation study has to be carried out to judge whether the spatial variation in predicted concentrations in BasinBox agrees with actual concentration differences throughout a river basin. In a next step in research we will perform a validation study for the BasinBox model using real chemical data in different river catchments, based on real emission scenarios.

Conclusion

BasinBox is a new generic multimedia fate model that predicts environmental concentrations of chemicals in large river basins. It distinguishes an upstream, midstream, and downstream area with different environmental characteristics. Water transport is modelled in a downward direction, and floodplain inundations are taken into account. It appears that BasinBox predicts significant concentration differences between upstream and downstream areas of the river Rhine basin for different types of chemicals and different emission scenarios. There is a clear trend of increasing chemical concentrations in the downstream direction of the river basin. This case study shows that taking into account spatial variability between upstream, midstream and downstream areas of large river basins can be important in the predictions of environmental concentrations by multimedia fate models.

References

Asselman, N. E. M., 1997. Suspended Sediment in the River Rhine. The Impact of Climate Change on Erosion, Transport and Deposition. PhD-thesis, Department of Physical Geography, Utrecht University, Utrecht.

Baughman, G. L. & R. R. Lassiter, 1978. Prediction of environmental pollutant concentration. In Cairns, J., D. L. Dickson & A.W. Maki (eds), Estimating the Hazard of Chemical Substances to Aquatic Life. American Society for Testing Materials (ASTM) 657: 34–54.

Boorman, D. B., 2003. LOIS in-stream water quality modelling. Part 1: catchments and methods. Science of the Total Environment 314: 379–395.

Brandes, L. J., H. den Hollander & D. van de Meent, 1996. SimpleBox 2.0: A Nested Multimedia Fate Model for Evaluating the Environmental Fate of Chemicals. RIVM, Bilthoven.

Briggs, C. G., R. H. Bromilov & A. A. Evans, 1982. Relationships between lipophilicity and root uptake and translocation of non-ionised chemicals by barley. Pesticide Science 13: 495–504.

Cahill, T. M. & D. Mackay, 2003. A high-resolution model for estimating the environmental fate of multi-species chemicals: application to malathion and pentachlorophenol. Chemosphere 53: 571–581.

Chamberlain, A.C., 1967. Transport of lycopodium spores and other small particles to rough surfaces. Proceedings of the Royal Society of London A296: 45–70.

Centrum voor Landbouw en Milieu (CLM), 2004. Werken aan duurzame landbouw en een aantrekkelijk platteland. www.clm.nl (in Dutch).

Chemicalland21, 2005. www.chemicalland21.com.

Commonwealth Scientific and Industrial Research Organization (CSIRO), 1994. CSIRO sustainable ecosystems – Software and resources. http://www.cse.csiro.au/poptools/.

Coulibaly, L., M. E. Labib & R. Hazen, 2004. A GIS-based multimedia watershed model: development and application. Chemosphere 55: 1067–1080.

Cousins, I. T. & D. Mackay, 2001. Strategies for including vegetation compartments in multimedia models. Chemosphere 44: 643–654.

De Nooij, R. J. W., W. C. E. P. Verberk, H. J. R. Lenders, R. S. E. W. Leuven & P. H. Nienhuis, 2006. The importance of hydrodynamics for protected and endangered biodiversity of lowland rivers. Hydrobiologia 565: 153–162.

De Wit, M. J. M., 1999. Nutrient fluxes in the Rhine and Elbe basins. Ph.D. thesis. Department of Physical Geography, Utrecht University, Utrecht.

Den Hollander, H. & D. van de Meent, 2004. SimpleBox 3.0: A Multimedia Fate Model for Evaluating Environmental Behaviour of Chemicals. RIVM, Bilthoven.

Diamond, M. L., D. A. Priemer & N. L. Law, 2001. Developing a multimedia model of chemical dynamics in an urban area. Chemosphere 44: 1655–1667.

DiToro, D. M., C. S. Zarba, D. J. Hansen, W. J. Berry, R. C. Swartz, C. E. Cowan, S. P. Pavlou, H. E. Allen, N. A. Thomas & P. R. Paquin, 1991. Technical basis for establishing sediment quality criteria for nonionic organic-chemicals using equilibrium partitioning. Environmental Toxicology and Chemistry 10: 1541–1583.

Deutscher Wetterdienst (DWD), 2004. www.dwd.de.

EC, 2000. Water Framework Directive. European Commission, Brussels.

ECB, 2003. Technical Guidance Document on Risk Assessment. JRC-Ispra, Italy.

ECNC, 2004. European Centre for Nature Conservation. http://www.ecnc.nl/.

FAO, 2001. Food and Agriculture Organization of the United Nations. http://faostat.fao.org/.

Feijtel, T., G. Boeije, M. Matthies, A. Young, G. Morris, C. Gandolfi, B. Hansen, K. Fox, M. Holt, V. Koch, R. Schroder, G. Cassani, D. Schowanek, J. Rosenblom & H. Niessen, 1997. Development of a geography-referenced regional exposure assessment tool for European rivers – GREAT-ER contribution to GREAT-ER #1. Chemosphere 34: 2351–2373.

Fenner, K., M. Scheringer & K. Hungerbühler, 2000. Persistence of parent compounds and transformation products in a level IV multimedia model. Environmental Science and Technology 34: 3809–3817.

Fenner, K., M. Scheringer, M. MacLeod, M. Matthies, T. McKone, M. Stroebe, A. Beyer, M. Bonnell, A. C. Le Gall, J. Klasmeier, D. Mackay, D. van de Meent, D. Pennington, B. Scharenberg, N. Suzuki & F. Wania, 2005. Comparing estimates of persistence and long-range transport potential among multimedia models. Environmental Science and Technology 39: 1932–1942.

Hofstee, C. & H. Leenaers, 2002. Actief beheer van de waterbodem in landelijk perspectief. TNO-NITG, Utrecht (in Dutch).

Hollander, A., L. K. Hessels, P. de Voogt & D. van de Meent, 2004. Implementation of depth-dependent soil concentrations in multimedia mass balance models. SAR and QSAR in Environmental Research 15: 457–468.

Horstmann, M. & M. S. McLachlan, 1998. Atmospheric deposition of semivolatile organic compounds to two forest canopies. Atmospheric Environment 32: 1799–1809.

Jackson, R., 1996. A global analysis of root distributions for terrestrial biomes. Oecologia 108: 389–411.

Junge, C. E., 1977. Basic considerations about trace constituent in the atmosphere related to the fate of global pollutants. In Suffet, I. H. (ed.), Fate of Pollutants in the Air and Water Environment. Wiley–Interscience: 7–25.

Koninklijk Nederlands Meteorologisch Instituut (KNMI), 2004. Klimaat en klimaatverandering: klimatologische informatie. www.knmi.nl/voorl/weer/ (in Dutch).

Mackay, D., 1991. Multimedia Environmental Models. Lewis, Chelsea.

Mackay, D. & S. Paterson, 1981. Calculating fugacity. Environmental Science and Technology 15: 1006–1014.

Mackay, D., S. Paterson & M. Joy, 1983. Application of fugacity models to the estimation of chemical-distribution and persistence in the environment. ACS Symposium Series 225: 175–196.

McKone, T. E., 1993. CalTOX, A Multimedia Total-exposure Model for Hazardous-wastes Sites. Part 1: Executive Summary. Lawrence Livermore National Laboratory, Livermore.

McKone, T. E. & D. H. Bennett, 2003. Chemical-specific representation of air–soil exchange and soil penetration in regional multimedia models. Environmental Science and Technology 37: 3123–3132.

McKone, T. E., A. B. Bodnar & E. G. Hertwich, 2001. Development and Evaluation of State-specific Landscape Data Sets for Multimedia Source-to-dose Models. School of Public Health, University of California, Berkeley.

McLachlan, M. S., G. Czub & F. Wania, 2002. The influence of vertical sorbed phase transport on the fate of organic chemicals in surface soils. Environmental Science and Technology 36: 4860–4867.

MeteoSchweiz, 2004. www.meteoschweiz.ch.

Meybeck, M., L. Laroche, H. H. Durr & J. P. M. Syvitski, 2003. Global variability of daily total suspended solids and their fluxes in rivers. Global and Planetary Change 39: 65–93.

Nationmaster, 2005. http://www.nationmaster.com.

Nienhuis, P. H., A. D. Buijse, R. S. E. W. Leuven, A. J. M. Smits, R. J. W. de Nooij & E. M. Samborska, 2002. Ecological rehabilitation of the lowland basin of the river Rhine (NW Europe). Hydrobiologia 478: 53–72.

PAN, 2005. Pesticides database. http://www.pesticideinfo.org/List_ChemicalsAlpha.jsp.

Paterson, S. & D. Mackay, 1994. Interpreting chemical partitioning in a soil–plant–air system with a fugacity model. In Trapp, S. & C. McFarlane (eds), Plant Contamination, Modeling and Simulation of Organic Chemical Processes. Lewis Publishers/CRC Press: 191–214.

Prevedouros, K., K. C. Jones & A. J. Sweetman, 2004. European-scale modeling of concentrations and distribution of polybrominated diphenyl ethers in the pentabromodiphenyl ether product. Environmental Science and Technology 38: 5993–6001.

Scheringer, M., F. Wegmann, K. Fenner & K. Hungerbuhler, 2000. Investigation of the cold condensation of persistent organic pollutants with a global multimedia fate model. Environmental Science and Technology 34: 1842–1850.

Schumm, S. A., 1977. The Fluvial System. Wiley–Interscience, New York.

Schwarzenbach, R. P., P. M. Gschwend & D. M. Imboden, 1993. Environmental Organic Chemistry. John Wiley & Sons, New York.

Scorecard, 2005. The pollution information site. www. scorecard.org.

Scurlock, J. M. O., G. P. Asner & S. T. Gower, 2001. Worldwide Historical Estimates of Leaf Area Index, 1932–2000. Oak Ridge National Laboratory, Oak Ridge.

Severinsen, M. & T. Jager, 1998. Modelling the influence of terrestrial vegetation on the environmental fate of xenobiotics. Chemosphere 37: 41–62.

Stroebe, M., M. Scheringer & K. Hungerbühler, 2004. Measures of overall persistence and the temporal remote state. Environmental Science and Technology 38: 5665–5673.

Suzuki, N., K. Murasawa, T. Sakurai, K. Nansai, K. Matsuhashi, Y. Moriguchi, K. Tanabe, O. Nakasugi & M. Morita, 2004. Geo-referenced multimedia environmental fate model (G-CIEMS): model formulation and comparison to the generic model and monitoring approaches. Environmental Science and Technology 38: 5682–5693.

Tiktak, A., D. de Nie, T. van der Linden & R. Kruijne, 2002. Modelling the leaching and drainage of pesticides in the Netherlands: the GeoPEARL model. Agronomie 22: 373–387.

Toose, L., D. G. Woodfine, M. MacLeod, D. Mackay & J. Gouin, 2004. BETR-World: a geographically explicit model of chemical fate: application to transport of alpha-HCH to the Arctic. Environmental Pollution 128: 223–240.

Trapp, S., 1996. Querprofile, WQ-, QW-, WB- und Wu-Regressionen, Einleiterstandorte für den Rhein. Universität Osnabrück, Institut für Umweltsystemforschung, Osnabrück.

Trapp, S. & M. Matthies, 1996. Generic one compartment model for uptake of organic chemicals by foliar vegetation. Environmental Science and Technology 30: 360.

US-EPA., 2002. TRIM.FaTE Technical Support Document. Volume 1: Description of Module. US-Environmental Protection Agency, North Carolina.

Vermeire, T. G., D. T. Jager, B. Bussian, J. Devillers, K. den Haan, B. Hansen, I. Lundberg, H. Niessen, S. Robertson, H. Tyle & P. T. J. van der Zandt, 1997. European Union

System for the Evaluation of Substances (EUSES). Principles and structure. Chemosphere 34: 1823–1836.

Vermeire, T., M. Rikken, L. Attias, P. Boccardi, G. Boeije, D. Brooke, J. de Bruijn, M. Comber, B. Dolan, S. Fischer, G. Heinemeyer, V. Koch, J. Lijzen, B. Müller, R. Murray-Smith & J. Tadeo, 2005. European union system for the evaluation of substances: the second version. Chemosphere 59: 473–485.

Webster, E., D. Mackay, A. Di Guardo, D. Kane & D. Woodfine, 2004. Regional differences in chemical fate model outcome. Chemosphere 55: 1361–1376.

Woodfine, D. G., M. MacLeod, D. Mackay & J. R. Brimacombe, 2001. Development of continental scale multimedia contaminant fate models: integrating GIS. Environmental Science and Pollution Research 8: 164–172.

Zeng, X. B., R. E. Dickinson, A. Walker, M. Shaikh, R. S. DeFries & J. G. Qi, 2000. Derivation and evaluation of global 1-km fractional vegetation cover data for land modeling. Journal of Applied Meteorology 39: 826–839.

Zhang, Q. O., J. C. Crittenden, D. Shonnard & J. R. Mihelcic, 2003. Development and evaluation of an environmental multimedia fate model CHEMGL for the Great Lakes region. Chemosphere 50: 1377–1397.

Hydrobiologia (2006) 565:39–58
R.S.E.W. Leuven, A.M.J. Ragas, A.J.M. Smits & G. van der Velde (eds), Living Rivers: Trends and Challenges in Science and Management
DOI 10.1007/s10750-005-1904-8

Trophic relationships in the Rhine food web during invasion and after establishment of the Ponto-Caspian invader *Dikerogammarus villosus*

M.C. van Riel[1,4], G. van der Velde[1,4,*], S. Rajagopal[1,4], S. Marguillier[2], F. Dehairs[2] & A. bij de Vaate[3,4]
[1]*Department of Animal Ecology and Ecophysiology, Section Aquatic Animal Ecology, Institute for Wetland and Water Research, Radboud University Nijmegen, Toernooiveld 1, 6525 ED Nijmegen, The Netherlands*
[2]*Department of Analytical Chemistry, Vrije Universiteit Brussel, Pleinlaan 2, 1050 Brussels, Belgium*
[3]*Ministry of Transport, Public Works and Water Management, Institute for Inland Water Management and Waste Water Treatment, P.O. Box 17, 8200 AA Lelystad, The Netherlands*
[4]*Member of Netherlands Centre for River Studies, P.O. Box 177, 2600 MH Delft, The Netherlands*
*(*Author for correspondence: E-mail: G.vandervelde@science.ru.nl)*

Key words: macroinvertebrates, invaders, ecotopes, food web, trophic relations, stable isotopes, Rhine

Abstract

The Rhine ecosystem is highly influenced by anthropogenic stresses from pollution, intensive shipping and increased connectivity with other large European rivers. Canalization of the Rhine resulted in a reduction of heterogeneity to two main biotopes: sandy streambeds and riverbanks consisting of groyne stones. Both biotopes are heavily subjected to biological invasions, affecting the rivers food web structure. The Ponto-Caspian amphipods, *Chelicorophium curvispinum* and *Dikerogammarus villosus*, have exerted the highest impact on this food web. The filterfeeding *C. curvispinum* dominated the Rhine food web on the stones in 1998, swamping the stone substrata with mud. However, in 2001 it decreased in numbers, most likely due to top-down regulation caused by increased parasitic and predatory pressure of other more recently invaded Ponto-Caspian species. *D. villosus* showed a fast population increase after its invasion and particularly influenced the macroinvertebrate community on the stones by predaceous omnivory. This species seemed to have maintained its predatory level after its population established. Effects of these mass invaders on the macroinvertebrate community of sandy streambeds in the Rhine are unclear. Here, low densities of macroinvertebrates were observed with the Asiatic clam, *Corbicula fluminea*, as most abundant species. Stable isotope values of food webs from the stones and sand in 2001 were similar. Aquatic macrophytes are nearly absent and the food web is fuelled by phytoplankton and particulate organic matter, originating from riparian vegetation as indicated by similar $\delta^{13}C$ values. Omnivores, filter-, deposit-, and detritus-feeders are the primary and secondary macroinvertebrate consumers and function as keystone species in transferring energy to higher trophic levels. Invaders comprise 90% of the macroinvertebrate numbers, and can be considered ecosystem engineers determining the functional diversity and food web structure of the Rhine by either bottom-up or top-down regulation.

Introduction

The Rhine is an example of a river ecosystem under anthropogenic stress. Dikes reduce its floodplain and the main channel is canalized, reducing its heterogeneity to two main biotopes. The first biotope harbours the river bottom habitats, which are dominated by wave-exposed sand and gravel. The navigation channel itself is disturbed by intensive shipping whirling up the sand sediment, making it unsuitable for zoobenthos. The second biotope is stony substrate found at the riverbanks as riprap or basalt stone groynes (Admiraal et al., 1993). Discharges from potassium mines in France and brown coal mining in Germany salinated the water. Till the 1970s, pollution from municipal and industrial wastewater such as organic loads, heavy metals and pesticides caused a loss of native macrozoobenthos and fish (Van der Velde et al., 1990; Admiraal et al., 1993; Bij de Vaate et al., 2006). When water quality improved by sanitation, recolonization of native macroinvertebrate species and appearance of exotic species was observed in the main channel until the chemical spill of Sandoz in 1986 killed nearly all species living in the Rhine over a stretch of hundreds of kilometres (Den Hartog et al., 1992). Lifting biogeographical barriers by creating canals between large European rivers increased connectivity between the biota of river catchments (Bij de Vaate et al., 2002) and intensified shipping, enabling invasive species to enter the Rhine and to occupy its empty niches. These biological invasions led to interactions between species that had never met before and may have a profound effect on the food web structure of the invaded ecosystem (Dick et al., 1993; Nichols & Hopkins, 1993; Fahnenstiel et al., 1995). Currently, the macroinvertebrate communities of the River Rhine are numerically dominated by exotic species (Van der Velde et al., 2002). Aquatic macrophytes are nearly absent in the Rhine; so particulate organic matter (POM) originating from riparian vegetation and phytoplankton can be considered fuel for the Rhine food web in which macroinvertebrates play a key role transferring energy to higher levels within the food web.

Marguillier et al. (1998) described the Rhine food web between 1996–1998, which harboured successful invaders like the zebra mussel *Dreissena polymorpha* (first observed in 1826), the East-European pikeperch *Sander lucioperca* (1888), the Southern-European freshwater shrimp *Atyaephyra desmarestii* (1915), the Chinese mittencrab *Eriocheir sinensis* (1929), the American crayfish *Orconectes limosus* (1975), the North-American gammarid *Gammarus tigrinus* (1982), the Asiatic clams *Corbicula fluminea* (1988) and *Corbicula fluminalis* (1988), the Ponto-Caspian hydroid *Cordylophora caspia* (1874), and the Ponto-Caspian amphipods *Chelicorophium curvispinum* (1987), *Echinogammarus ischnus* (1989) and *Dikerogammarus villosus* (1995) (Nijssen & De Groot, 1987; Van der Velde et al., 2000, 2002). *C. curvispinum* dominated the Rhine food web, increasing exponentially in numbers since its first colonization in 1987 to densities of hundreds of thousands of specimens per square metre in 1990 (Van den Brink et al., 1991; Rajagopal et al., 1999). This filter-feeder builds housing tubes of mud, changing the stone substrata in the River Rhine to a muddy environment making settling on these substrates difficult for other lithophylic macroinvertebrates. *D. polymorpha* used to be the most dominant species in the Rhine, but was negatively affected by the high densities of *C. curvispinum* (Van den Brink et al., 1991; Jantz, 1996; Van der Velde et al., 1998).

When the Ponto-Caspian invader *D. villosus* entered the Rhine food web in 1995 (Bij de Vaate & Klink, 1995), *C. curvispinum* decreased in numbers (Van der Velde et al., 2000), resulting in a reduction of mud on the stone substrate, leading to an increased diversity of other macroinvertebrates (Haas et al., 2002; Van der Velde et al., 2002). *D. villosus* manifested in the food web as a predator (Van der Velde et al., 2000; Dick et al., 2002), and increased in densities up to 10,000 individuals per square metre of stone substrate. Currently, *D. villosus* and *C. curvispinum* dominate the Rhine food web on this substrate, whereas *C. fluminea* dominates the sandy riverbed in number and biomass.

Since Marguillier et al. (1998) studied the Rhine food web several new invasive Ponto-Caspian species entered the Rhine, such as the triclad flatworm *Dendrocoelum romanodanubiale* (1999), the leech *Caspiobdella fadejewi* (1999), the halacarid *Caspihalacarus hyrcanus* (2000), the isopod *Jaera istri* (1997), the mysids *Hemimysis anomala* (1997) and *Limnomysis benedeni* (1997), and the tubenose goby *Proterorhinus semilunaris* (2002); all of which entered through the Main-Danube Canal opened in 1992 (Bij de Vaate & Swarte, 2001; Van der Velde et al. 2002; Bij de Vaate, 2003). *D. villosus* is assumed to have had the strongest influence on the Rhine food web since its successful invasion. At the time of Marguilliers study, *D. villosus* had just recently invaded the Rhine and possible effects of *D. villosus* had just been noticed in the food web, which was strongly dominated by *C. curvispinum*.

This paper compares the food web of the main channel of the Rhine 1–3 years after the invasion (1996–1998) of *D. villosus*, (Marguillier, 1998) to the food web 6–8 years after its invasion (2001–2003), using macroinvertebrate monitoring data from stone substrates and stable isotope values ($\delta^{13}C$, $\delta^{15}N$). Isotope signatures of organisms are based on actual food assimilation, providing information about the long-term diet of the species (Tieszen et al., 1983; Hobson et al., 1996, 1997) and reveal the direct relationship between the carbon and nitrogen ratios of animals and those of their diets (DeNiro & Epstein, 1978, 1981; Peterson & Fry, 1987). Within a food chain, animals are usually enriched by 1–5‰ (3.4‰ on average) in $\delta^{15}N$ compared to their diet and would allow the determination of trophic position (TP) of the species, whereas $\delta^{13}C$ (1‰ enrichment on average) provides information about the original source of organic matter to the food web (Hobson & Welch, 1992).

By comparing the Rhine food web at different stages of invasion by Ponto-Caspian species we may explain; (1) whether macroinvertebrate communities in the Rhine are changed by invasions (2) the position of successful invaders in the food web, (3) source fuels for the food web and (4) changes to the major trophic pattern of the food web after invasion.

Material and methods

Sampling sites

Food web items were collected from the main distributaries of the Rhine in the Netherlands (river Waal; 5° 48′ E; 51° 51′ N) in the vicinity of Nijmegen during July–September 1996, August 1997 and August 1998. Samples were taken from the main channel of the river Waal from the groyne stone riverbanks, from the sandy streambeds and from cooling water intake sieves of an adjacent electric power plant. Sampling of food web items was repeated in the summers between April 2001 and April 2003. Macroinvertebrate species on the stone substrate and the sandy river bottom were sampled and surveyed monthly at the same site during June–September 1998 and June–September 2001.

Field collection and analysis of macroinvertebrate communities

Groyne stones were sampled at a water depth of 2 m by means of a polyp-grab operated with a hydraulic crane from a ship. Six stones were sampled monthly during June–September 1998 and June–September 2001. Macroinvertebrates were carefully brushed off the stones, collected and preserved in 70% ethanol, sorted by species, and counted in the laboratory using a stereomicroscope. The total surface area of the stones was measured in order to quantify the densities of macroinvertebrate species per square metre substrate.

Macroinvertebrates inhabiting the sandy bottom in between the groynes were sampled using a hydraulic grab (0.475×0.335 m spread, 0.15 m height), which sampled a surface area of 0.159 m^2. Large macroinvertebrates were first sieved out using a coarse sieve (5 mm mesh), and the smaller macroinvertebrates present in the sand samples were collected by subsequently whirling up the sand in water and sieving out the macroinvertebrates. Five sand bottom samples were taken each month.

A microcosm experiment was carried out to investigate the predatory impact of *D. villosus* on *C. curvispinum*. Fifty plastic cups were filled with tap water, provided with five individuals of *C. curvispinum* each and placed in a climate room; 25 cups at 15 °C and 25 cups at 25 °C with a 9/15 h dark/light regime. To 15 cups of each series one individual of *D. villosus* was added. The survival of *C. curvispinum* in absence and presence of *D. villosus* was observed every day. Dead *C. curvispinum* present in the cups with *D. villosus* were replaced. Consumed *C. curvispinum* were counted.

Field collection and laboratory treatments of food web items

Food web items were collected from the shore, water layer, stone substrate and sandy bottom of the river Waal. Samples were sorted by species, washed with distilled water and dried for 48 h at 70 °C after which specimens were ground to a fine powder using a mortar with pestle and liquid nitrogen. Due to low individual weight of some macroinvertebrate species and juvenile fish, a composite tissue sample was prepared by pooling a number of whole specimens. Shells of molluscs were removed before grinding. For adult fish and larger crustaceans, muscle tissue was used because of its slow turnover rate and would provide a history of food assimilation over periods of months and exclude short-term variability (Gearing, 1991).

Samples measured during the summers of 1996 and 1997 were treated as follows (see Marguillier, 1998). For dissolved inorganic carbon ($DIC = \Sigma CO_2 = [CO_{2(aq)}] + [HCO_3^-] + [CO_3^{2-}]$), 50 ml of water was collected in serum vials kept airtight after 1 ml of saturated $HgCl_2$ solution was added to stop bacterial activity. DIC extraction followed procedures by Kroopnick (1974). POM was collected by filtering 1 l of water using Whatman GF/C filters (1.2 μm pore size), dried at 60 °C for 24 h and stored in clean petri dishes until isotopic ratio analysis. The latter was preceded by a hydrochloric acid vapour treatment under vacuum in order to remove calcium carbonate. Zooplankton was collected using a light trap. Silt material fixed on groyne stones of the breakwater was scraped off and macroinvertebrates present in the silt were collected. The $\delta^{13}C$ value for phytoplankton was derived from the $\delta^{13}C$ values of POM and DIC because isolation of phytoplankton from other suspended material was not possible and precluded direct measurement. According to Mook & Tan (1991), the average carbon isotope fractionation value of POM derived from primary production is −23‰ relative to DIC. Addition of this value with the DIC value obtained in this study (−9.65‰) was considered to give a reasonable idea of phytoplankton $\delta^{13}C$ value (−32‰). The $\delta^{15}N$ value (10.56‰) for phytoplankton was estimated by subtracting the value of 3.4‰, set as the difference between two successive trophic levels (Minagawa & Wada, 1984), from the mean $\delta^{15}N$ of all filter-, deposit- and detritus-feeders (13.96‰).

Isotope analysis

Carbon and nitrogen stable isotope compositions were measured with a Carlo Erba NA 1500 elemental analyser coupled online *via* a Finnigan Conflo III interface with a ThermoFinnigan DeltaPlus mass-spectrometer. Carbon and nitrogen isotope ratios are expressed in delta notation ($\delta^{13}C$, $\delta^{15}N$) relative to Vienna PDB and atmospheric nitrogen.

For comparing macroinvertebrate isotope compositions between the food web between 1996 and 1998 and that between 2001 and 2003, baseline variation was corrected for (Cabana & Rasmussen, 1996) by adding 4.39‰ (i.e. the mean decrease in $\delta^{15}N$ values of the primary consumers *D. polymorpha*, *C. fluminea* and *C. fluminalis*) to the mean $\delta^{15}N$ value measured in 2001–2003 for each species.

Trophic levels were estimated according to the method of Vander Zanden & Rasmussen (1999) based on the following overall $\delta^{15}N$–$\delta^{13}C$ relationship:

$$\delta^{15}N_{e,\text{primaryconsumer}} = \frac{6.34}{1 + \exp[9.67 + (0.356 * \delta^{13}C_{m,\text{primaryconsumer}})]} \quad (1)$$

In which $\delta^{15}N_e$ is the estimated $\delta^{15}N$ value for a species, calculated from the measured $\delta^{13}C$ value ($\delta^{13}C_m$) for that species in the food web studied using the overall $\delta^{15}N$–$\delta^{13}C$ relationship (1). The food web specific deviation from the general baseline curve was calculated for every primary consumer in the Rhine food web as follows:

$$\text{residual} = \delta^{15}N_{m,\text{primaryconsumer}} - \delta^{15}N_{e,\text{primaryconsumer}} \quad (2)$$

In which $\delta^{15}N_m$ is the $\delta^{15}N$ value measured for the species in the food web studied. The residual value can be calculated for every primary consumer in the food web studied (2). Subsequently, the mean residual value (U_{residual}) of all primary consumers from the food web can be calculated. The TP of each consumer in the food web can be estimated by combining the general baseline curve (1) with the mean residual value of the Rhine food web:

$$\delta^{15}N_{corrected} = \frac{6.34}{1 + \exp[9.67 + (0.356 * \delta^{13}C_{consumerX})]} + U_{residual} \quad (3)$$

where $\delta^{15}N_{corrected}$ is the food web-corrected baseline $\delta^{15}N$ value. Equation (3) produces an appropriate baseline $\delta^{15}N$ value for each species based on the $\delta^{13}C$ signature of the species, the general $\delta^{15}N$–$\delta^{13}C$ relationship and the food web specific deviation ($U_{residual}$). Finally, the TP of the species in the food web (consumer X) is estimated as follows:

$$TP_{consumer\ x} = ((\delta^{15}N_{consumer\ x} - \delta^{15}N_{corrected})/3.4) + 2 \quad (4)$$

where 3.4 = one trophic level increment (mean enrichment) in $\delta^{15}N$.

Cluster analysis and statistical analysis

Cluster analysis was performed on the stable isotope values for food web items in 1996–1998 (Fig. 4) and 2001–2003 (Fig. 5), using euclidean distances and complete linkage (Statistica 4.5) and was used for interpretation of the food web, not for statistical proof. Differences in $\delta^{15}N$ between different life stages and sexes of *D. villosus* (Fig. 6), and differences in densities of macroinvertebrate species and taxa (Fig. 2) between June–September 1998 and June–September 2001 were tested for each species using a *t*-test (SPSS 11.5) for independent samples.

Results

Changes in the macroinvertebrate community of the Rhine

Macroinvertebrate communities on groyne stones changed during the invasion and population increase of *D. villosus* (Fig. 1). Between June and September 1998, *C. curvispinum* dominated the macroinvertebrate community during the summer with population peaks in June and July (Fig. 1a) and decreased during August and September. Between June and September 2001, population densities of *C. curvispinum* had strongly decreased and peaked in August (Fig. 1b), and other macroinvertebrate species also became abundant. Overall changes in densities of the most dominant macroinvertebrates on stone substrate between both years are shown in Figure 2a. The increase of *D. villosus* ($p = 0.005$) coincides with a significant increase of *D. polymorpha* ($p = 0.005$) and *Ancylus fluviatilis* ($p = 0.02$) and a highly significant decrease of *C. curvispinum* ($p = 0.001$). Populations of *Hypania invalida*, *D. romanodanubiale* ($p = 0.012$) and Gastropoda ($p = 0.003$) other than *A. fluviatilis*, showed a smaller, significant decrease in density. Laboratory experiments confirmed that *D. villosus* could negatively affect populations of *C. curvispinum* (Fig. 8). Survival of *C. curvispinum* (Fig. 8) was higher at a high temperature (25 °C) than at a lower temperature (15 °C). At high temperature survival of *C. curvispinum* was strongly reduced in presence of *D. villosus* due to predation.

On the sandy bottom, *C. fluminea* dominated macroinvertebrate communities during the summer of 1998, with the highest densities observed in July. Oligochaetes and Chironomidae were also relatively abundant, but overall densities were low (Fig. 1c). During 2001, densities of macroinvertebrates were higher than observed in 1998 and biodiversity had increased. *C. fluminea* dominated the community during August and September (Fig. 1d). A comparison of the densities of various taxa in both years shows most taxa had increased in number on the sand bottom (Fig. 2b).

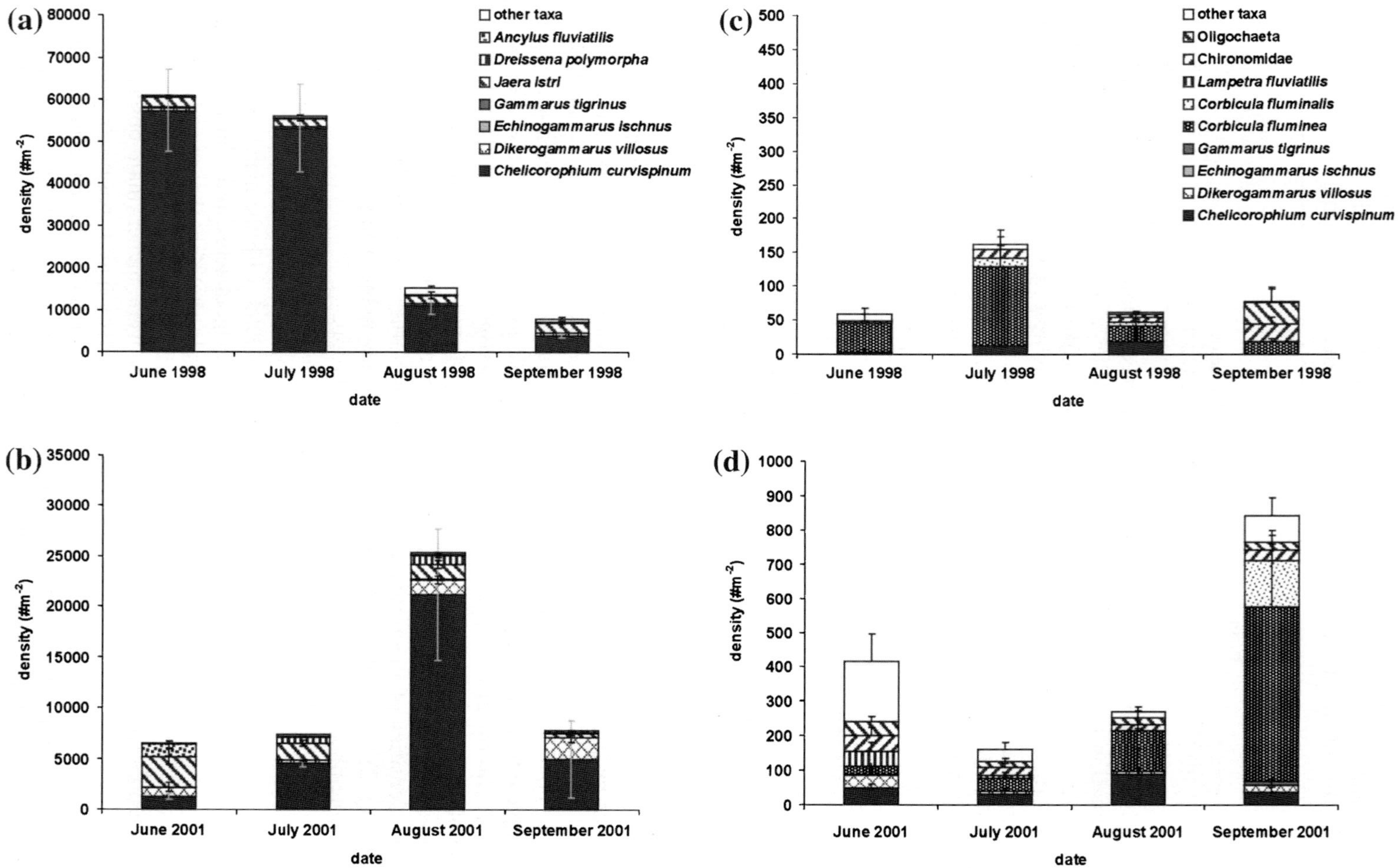

Figure 1. Macroinvertebrate species composition on the stones (a, b) and on the sandy bottom (c, d) in the Waal in June–September 1998 (a, c) and June–September 2001 (b, d).

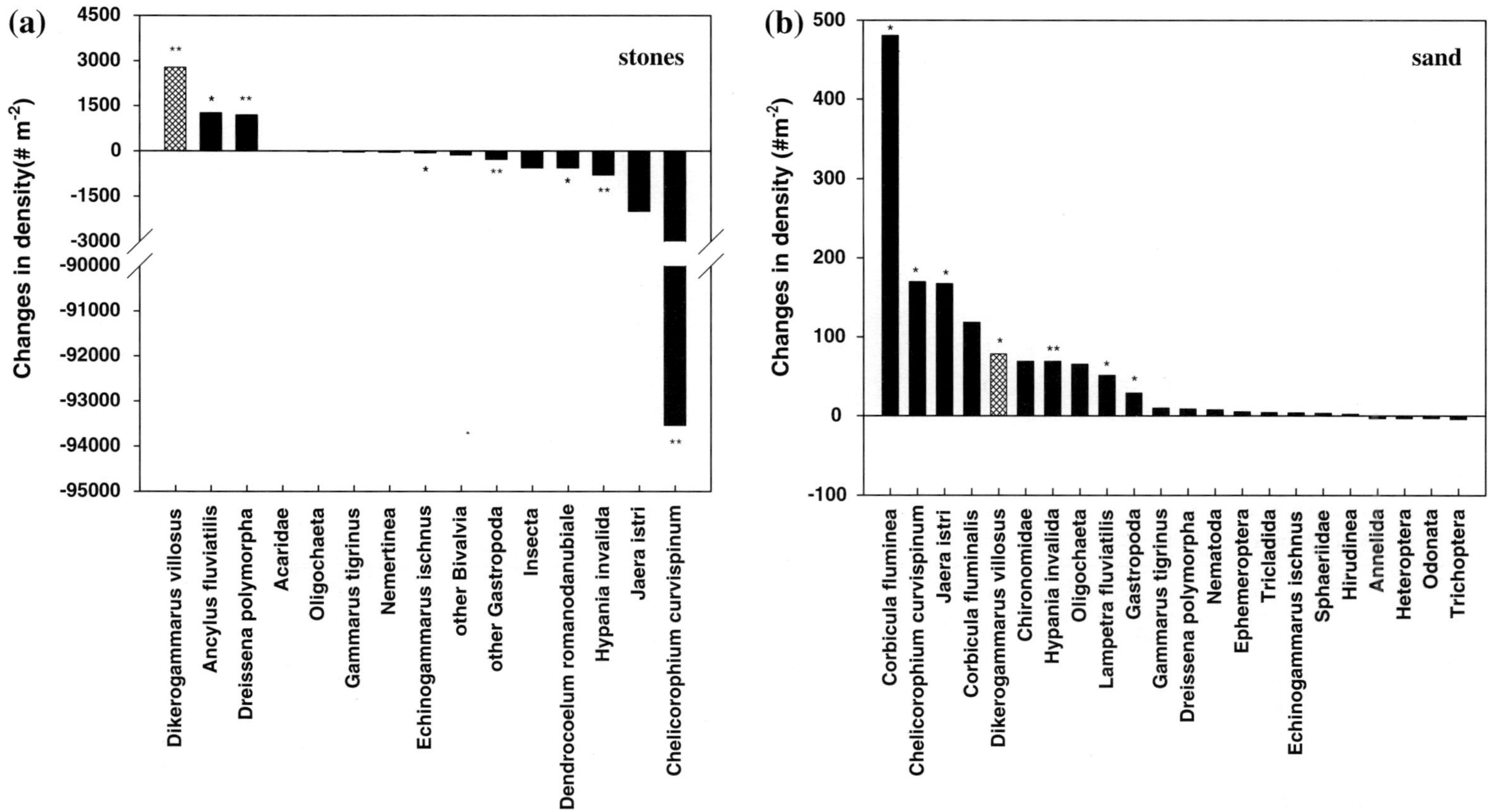

Figure 2. Changes in densities of the total numbers of *D. villosus* (shaded columns) and other macroinvertebrate species and taxa (black columns) on the stone substrate (a) and on the sandy bottom (b) in the Waal in the period 1998–2001. Significance levels: $***p < 0.001$, $**p < 0.01$, $*p < 0.05$.

Table 1. Mean isotopic values ($\delta^{13}C$, $\delta^{15}N$), for items occurring in the Rhine food web in 1996–1998, with corresponding standard deviations. Arranged according to trophic level

	Code	*n*	*n*[b]	Size range (cm)	Means		SD		Range		Trophic level
					$\delta^{15}N$	$\delta^{13}C$	$\delta^{15}N$	$\delta^{13}C$	$\delta^{15}N$	$\delta^{13}C$	
Sander lucioperca	Sl	77	–	3.4–42.6	17.75	−27.14	1.26	2.73	20.67–14.12	−32.44 to −19.85	3.17
Cottus gobio	Cg	6	–	4.9–7.1	17.75	−26.01	0.76	0.96	19.35–17.12	−27.40 to −24.46	3.17
Gymnocephalus cernuus	Gc	44	–	4.0–12.5	17.43	−27.47	0.93	1.17	19.59–15.73	−30.64 to −25.40	3.07
Gobio gobio	Gg	5	–	8.4–11.2	17.30	−25.82	0.34	0.15	19.15–8.57	−54.14 to −13.90	3.04
Platichthys flesus	Pf	3	–	21.8–25.5	17.22	−26.64	0.75	0.87	17.79–16.37	−27.23 to −25.64	3.01
Aspius aspius	Aas	7	–	5.5–8.3	16.57	−26.99	1.39	1.16	18.49–14.47	−28.85 to −25.61	2.82
Perca fluviatilis	Pf	53	–	4.2–18.0	16.54	−28.59	0.81	1.27	18.35–15.06	−30.64 to −24.80	2.81
Scardinius erythrophthalmus	Se	2	–	16.3	16.43	−26.12	0.36	2.29	16.68–16.17	−27.74 to −24.50	2.78
Blicca bjoerkna	Bb	84	–	3.2–31.5	16.42	−27.69	0.99	1.46	20.28–14.15	−32.47 to −25.37	2.78
Anguilla anguilla	Aa	35	–	10.5–68.2	16.35	−28.18	1.35	1.64	18.51–13.49	−31.16 to −24.34	2.76
Osmerus eperlanus	Oe	2	–	6.1–6.5	16.18	−28.37	0.46	0.54	16.50–15.86	−28.75 to −27.99	2.71
Rutilus rutilus	Rr	53	–	3.3–29.0	16.18	−28.09	1.30	1.57	19.42–12.94	−32.02 to −24.62	2.70
*Eriocheir sinensis**	Es	1	–	4.0–5.0	16.16	−28.26	na	na	na	na	2.70
Cobitis taenia	Ct	1	–	5.0	16.01	27.45	na	na	na	na	2.66
Alburnus alburnus	Aal	20	–	4.6–15.0	15.94	−28.11	1.46	1.00	19.13–14.19	−30.28 to −26.29	2.64
Palaemon longirostris	Pl	1[a]	6	na	15.62	−26.73	na	na	na	na	2.54
Atyaephyra desmarestii	Ad	1[a]	5	na	15.56	−30.34	na	na	na	na	2.52
Gasterosteus aculeatus	Ga	5	–	3.4–5.3	15.56	−29.48	3.19	1.92	19.17–9.54	−32.50 to−27.38	2.52
*Orconectes limosus***	Ol	8	–	0.7–6.0 (carapace)	15.58	−26.69	0.75	0.11	na	na	2.43
Leuciscus cephalus	Lc	3	–	4.0–5.2	15.12	−26.29	1.29	0.85	16.55–14.04	−27.06 to −25.39	2.40
Hydropsyche spec.	Hs	3	–	na	15.04	−29.45	na	na	na	na	2.37
Abramis brama	Ab	12	–	9.0–33.0	14.82	−28.12	0.95	1.20	16.28–13.15	−30.46 to −26.89	2.31
Dikerogammarus villosus	Dv	1[a]	>10	na	14.65	−28.17	1.04	1.67	na	na	2.26

Corbicula fluminea	Cfa	1[a]	>10	na	14.48	−29.60	na	na	na	na	2.21
Chironomidae	C	3[a]	±150	na	14.44	−28.41	na	na	na	na	2.19
Ephydatia fluviatilis	Ef	–	–	na	14.37	−28.28	na	na	na	na	2.17
Gammarus tigrinus	Gt	1[a]	>11	na	14.33	−29.52	3.23	1.81	na	na	2.15
Corbicula fluminalis	Cfs	1[a]	>10	na	14.26	−29.33	na	na	na	na	2.14
Bithynia tentaculata	Bt	1[a]	>10	na	14.24	−27.83	na	na	na	na	2.14
Chondrostoma nasus	Cn	1	–	5.7	14.14	−29.30	na	na	na	na	2.11
Cyprinus carpio	Cca	2	–	9.1–9.2	14.13	−24.82	0.46	2.21	14.46–13.80	−26.38 to −23.25	2.10
Chelicorophium curvispinum	Ccu	1[a]	>10	na	14.09	−28.54	na	na	na	na	2.09
Dreissena polymorpha	Dp	1[a]	>10	na	13.78	−29.51	na	na	na	na	2.00
Cordylophora caspia	Cc	1[a]	>10	na	13.61	−28.79	na	na	na	na	1.95
Copepods	Cp	1[a]	>10	na	13.14	−27.85	na	na	na	na	na
Plumatella repens	Pr	1	>10	na	12.67	−31.14	na	na	na	na	1.67
Mud tubes of *C. curvispinum*	T	5	–	na	12.12	−26.69	0.19	0.36	12.00–10.05	−28.62 to −26.69	na
Silt	S	2	–	na	12.08	−26.67	na	na	na	na	na
Lampetra fluviatilis adult (silver colour)	Lfad	3	–	12.0–34.8	11.36	−25.53	0.36	2.57	11.62–10.95	−27.70 to −22.70	1.29
Cladophora glomerata	Cg	1	–	na	11.13	−22.96	na	na	na	na	na
Phytoplankton	Phyt	1	–	na	10.56	−32.00	na	na	na	na	na
POM	POM	1	–	na	10.25	−24.27	0.21	0.23	na	na	na
Lampetra fluviatilis juvenile (brown colour)	Lfj	3	–	9.0–13.0	9.81	−22.53	0.65	1.04	9.81–8.57	−23.65 to −21.60	0.69
Plumatella fungosa	Pf	1	>10	na	9.08	−28.75	na	na	na	na	0.62
DIC	DIC	1	–	na		−9.65	na	na	na	na	na

Items are sorted by their values of d 15N/14N.
[a]Pooled samples.
[b]Size of pooled samples.
na = Not analysed.
* carapace width, ** carapace length

Table 2. Mean isotopic values ($\delta^{13}C$, $\delta^{15}N$), for items occurring in the Rhine food web in 2001–2003, with corresponding standard deviations

	Code	*n*	n^b	Size range (cm)	Means		SD		Range		Trophic level
					$\delta^{15}N$	$\delta^{13}C$	$\delta^{15}N$	$\delta^{13}C$	$\delta^{15}N$	$\delta^{13}C$	
Lampetra fluviatilis adults	Lfa	3	–	13.0–22.5	16.27	−20.20	0.10	0.47	16.35–16.20	−20.54 to −19.86	5.08
Sander lucioperca	SI	28	–	9.5–51.0	15.96	−28.49	1.47	1.60	17.85–11.64	−31.77 to −25.67	3.98
Palaemon longirostris	PI	3	–	5.0–5.5	15.22	−26.75	0.89	0.07	16.02–14.26	−26.82 to −26.69	4.05
Eriocheir sinensis	Es	18	–	4.1–7.5 (carapace)	14.55	−27.03	0.49	0.89	15.85–13.91	−29.14 to −25.57	3.81
Abramis brama	Ab	20	–	10.0–30.0	14.52	−27.54	1.35	2.10	16.11–12.79	−33.77 to −24.66	3.71
Blicca bjoerkna	Bb	22	–	9.5–30.5	14.15	−28.58	1.17	1.61	15.96–11.07	−33.84 to −26.75	3.43
Rutilus rutilus	Rr	14	–	7.0–24.5	13.87	−29.10	0.86	1.72	15.41–11.99	−34.85 to −26.82	3.27
Perca fluviatilis	Pf	22	–	7.5–28.0	13.64	−28.62	1.72	1.98	16.03–10.52	−32.85 to −26.15	3.28
Orconectes limosus	OI	4	–	8.5–10.4	13.56	−26.69	0.28	1.40	13.97–13.37	−28.37 to −24.99	3.57
Porifera gemmula	Pogem	6[a]	500	0.1	13.35	−30.19	1.02	1.10	14.94–12.57	−29.59 to −26.56	na
Ancylus fluviatilis	Af	5[a]	35	na	12.32	−28.79	3.03	1.17	15.11–8.16	−30.19 to −27.34	3.16
Pomphorhynchus spec.	Ps	1[a]	6	na	11.60	−27.67	na	na	na	na	2.83
Dikerogammarus villosus adults > 1.2 cm	Dvbigad	5	10	1.3–2.1	11.27	−26.09	1.00	0.37	12.31–10.61	−28.07 to −26.03	3.03
Porifera	Po	6	–	na	11.06	−28.36	0.51	1.68	11.55–10.08	−30.89 to −26.49	2.56
Hydropsyche bulgaromanorum	Hb	1[a]	15	na	10.95	−29.27	0.07	2.76	11.00–10.89	−31.20 to −27.31	2.39
Dikerogammarus villosus adults	Dvad	9[a]	50–100	1.0–1.3	10.93	−26.33	0.97	0.49	12.63–8.96	−27.92 to −25.64	2.86
Jaera istri	Ji	15[a]	300–500	0.1–0.5	10.88	−26.68	0.83	0.48	14.74–10.03	−27.22 to −26.33	2.78
Chelicorophium curvispinum	Ccu	18[a]	±150	na	10.74	−28.08	0.90	1.12	12.03–8.40	−29.05 to −26.49	2.51
Dikerogammarus villosus mean	Dv	33[a]	10–150	0.4–2.1	10.64	−26.48	1.21	0.49	12.31–8.07	−31.20 to −25.64	2.71
Dikerogammarus villosus juveniles	Dvjuv	22[a]	150	0.4–0.8	10.53	−26.54	1.29	0.51	11.72–8.07	−27.72 to −25.69	2.70
Gammarus tigrinus	Gt	1[a]	30	0.8–1.3	10.48	−26.63	0.65	0.43	11.18–9.32	−27.13 to −25.78	2.68
Echinogammarus ischnus	Ei	3[a]	5	na	10.22	−26.04	0.38	0.48	10.52–9.79	−26.44 to −25.50	2.69
Moss	moss	6	–	na	10.01	−30.82	2.23	5.63	11.53–6.06	−35.15 to −23.25	na
Corbicula fluminalis	Cfs	1[a]	20	na	9.86	−30.37	0.00	0.00	na	na	1.92
Corbicula fluminea	Cfa	1[a]	20	na	9.80	−29.50	0.25	1.16	10.08–9.40	−30.68 to −20.03	2.02
Chironomidae	C	6[a]	±150	na	9.78	−30.64	1.99	1.75	11.63–6.39	32.77 to −28.71	1.87

Dreissena polymorpha	Dp	44[a]	±20	1.0–3.0	9.68	−28.81	0.78	1.73	12.70–8.37	−32.17 to −22.65	2.09
Anodonta anatina	Aa	1	–	10.0	9.61	−29.40	na	na	na	na	1.98
Lampetra fluviatilis juveniles, larvae	Lfj	3[a]	30	1.5–8.5	9.61	−28.11	1.67	3.67	11.49–8.30	−31.70 to −24.37	2.17
Mud tubes of *C. curvispinum*	T	9	–	na	9.50	−24.23	1.08	3.07	12.02–8.43	−31.34 to −21.77	na
Hypania invalida	Hi	1[a]	70	na	9.38	−26.4	0.00	0.00	na	na	2.39
Cordylophora caspia	Cc	1	–	na	9.28	−28.84	0.27	0.22	9.47–9.09	−29.00 to −28.68	1.96
Plant matter	Pm	23	–	na	9.22	−22.77	1.10	2.86	11.68–7.64	−28.88 to −16.56	na
Green algae	Algae	1	–	na	9.22	−16.23	na	na	na	na	na
Bryozoa	Bryo	5	–	na	9.03	−26.70	0.30	2.14	9.44–8.72	−29.59 to −24.54	2.24
Carex acuta	Ca	1	–	na	7.70	−28.54	na	na	na	na	na
Mud	mud	3	–	na	5.88	−18.02	3.36	4.62	9.08–2.39	−22.83 to −13.62	na
Potentilla anserina	Pa	1	–	na	5.75	−28.80	na	na	na	na	na
Cladophora glomerata	Cg	1	–	na	5.57	−21.00	na	na	na	na	na
Clinclidotus spec.	Cs	1	–	na	5.26	−33.84	na	na	na	na	na
Equisetum arvense	Ea	1	–	na	4.95	−28.59	na	na	na	na	na
Lythrum salicaria	Ls	1	–	na	4.83	−27.06	na	na	na	na	na
Festuca arundinacea	Fg	1	–	na	4.62	−30.02	na	na	na	na	na
Plantago major	Pm	1	–	na	4.49	−28.80	na	na	na	na	na
Rubus fruticosus	Rf	1	–	na	4.02	−28.88	na	na	na	na	na
Salix alba	Sa	7	–	na	3.89	−27.81	2.73	1.19	7.64–1.55	−29.94 to −26.50	na
Rorippa sylvestris	Rs	1	–	na	3.86	−28.46	na	na	na	na	na
Eryngium campestre	Ec	1	–	na	3.51	−29.70	na	na	na	na	na
Taraxacum officinale	To	1	–	na	2.60	−29.43	na	na	na	na	na
Cirsium arvense	Ca	1	–	na	2.55	−29.56	na	na	na	na	na
Achillea millefolium	Am	1	–	na	2.18	−29.43	na	na	na	na	na
Calamagrostis epigejos	Ce	1	–	na	1.96	−28.14	na	na	na	na	na
detritus	det	13	–	na	1.86	−27.12	0.28	0.06	2.06–1.66	−27.17 to −27.08	na

[a]Pooled samples.
[b]Size of pooled samples.
na = Not analysed.

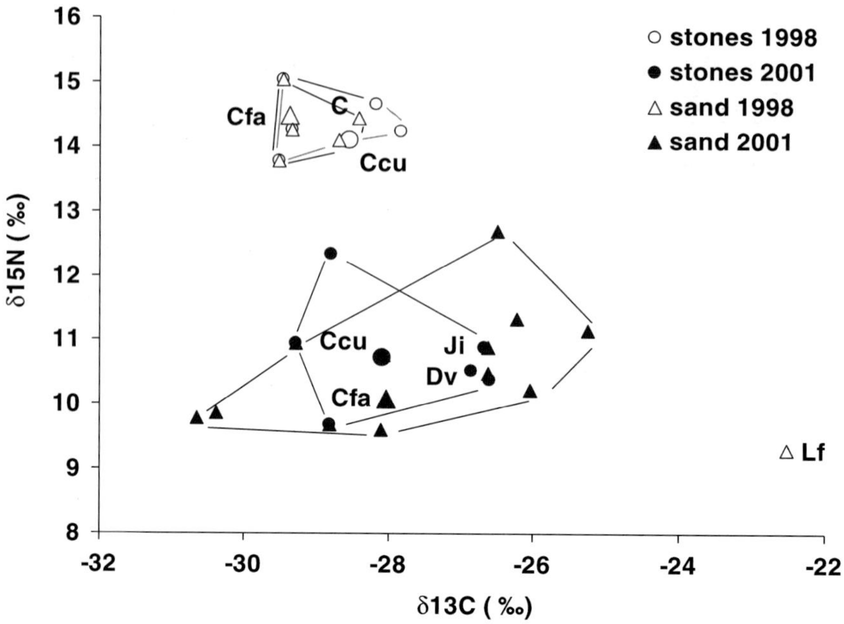

Figure 3. Distributions of $\delta^{13}C$ and $\delta^{15}N$ values in the food web of the two different Rhine biotopes viz. stones and sand in 1998 and 2001. The isotopic values of the dominant species in the biotopes are indicated by enlarged symbols and are coded: Cfa: *Corbicula fluminea*, Ccu: *Chelicorophium curvispinum*, Dv: *Dikerogammarus villosus*, Ji: *Jaera istri* and C: Chironomidae, Lf: *Lampetra fluviatilis* larvae.

Food web studies by stable isotopes

Macroinvertebrate food webs of the different biotopes; groyne stones and sandy bottom

Mean values of $\delta^{13}C$ of the food web on the stones in 1998 were comparable to those observed in 2001 and ranged from −29.51‰ (*D. polymorpha*) to −27.34 i.s.o. −27.83‰ (*A. fluviatilis*) and from −29.76‰ (Bivalvia) to −26.61‰ (Annelida), respectively (Fig. 3). However, mean values of $\delta^{15}N$ of the food web on the sand bottom as well as the stones differed between 1998 and 2001. The mean $\delta^{15}N$ values of the groyne stone food web ranged from 13.78‰ (*D. polymorpha*) to 15.04‰ (Insecta) in 1998 and from 9.68‰ (*D. polymorpha*) to 12.32‰ (*A. fluviatilis*) in 2001. Mean values of $\delta^{13}C$ of species on the sandy bottom food web in 1998 and 2001 ranged from −29.51‰ (*D. polymorpha*) to −22.53‰ (*Lampetra fluviatilis* larvae) and from −30.64‰ (Chironomidae) to −26.04‰ (*E. ischnus*), respectively (Fig. 3), and are comparable to $\delta^{13}C$ values of the food web on stones. The mean $\delta^{15}N$ values of species in the sand food web ranged from 9.31‰ (*L. fluviatilis* larvae) to 15.04‰ (Insecta) in 1998 and from 9.61‰ (*L. fluviatilis* larvae) to 12.70‰ (*D. romanodanubiale*) in 2001. Mean $\delta^{15}N$ values have decreased since 1998 for most species in the food webs of both biotopes.

Total food web of the Rhine

Three trophic levels can be distinguished in the food web between 1996 and 1998 (Fig. 4). The primary level is based on POM (cluster 1) or suspended organic matter and phytoplankton (cluster 2). The second level harboured most species and consisted of primary and secondary consumers (cluster 3), mainly macroinvertebrates inhabiting the stone substrate. The third level consists of top predators (cluster 4). The mean carbon and nitrogen isotope ratios ($\delta^{13}C$, $\delta^{15}N$) of items occurring in the Rhine food web for the period of 1996–1998 ranged from −31.14‰ (*Plumatella repens*) to −22.53‰ (*Lampetra fluviatilis* larvae) for $\delta^{13}C$ and from 9.08‰ (*Plumatella fungosa*) to 17.75‰ (*Sander lucioperca*) for $\delta^{15}N$ (Fig. 4, Table 1).

Two clusters formed the base of the food web, the first cluster consisting of POM (containing both phytoplankton and organic detrital material (Marguillier, 1998)), *Cladophora glomerata* and *L. fluviatilis* larvae, and the second cluster containing

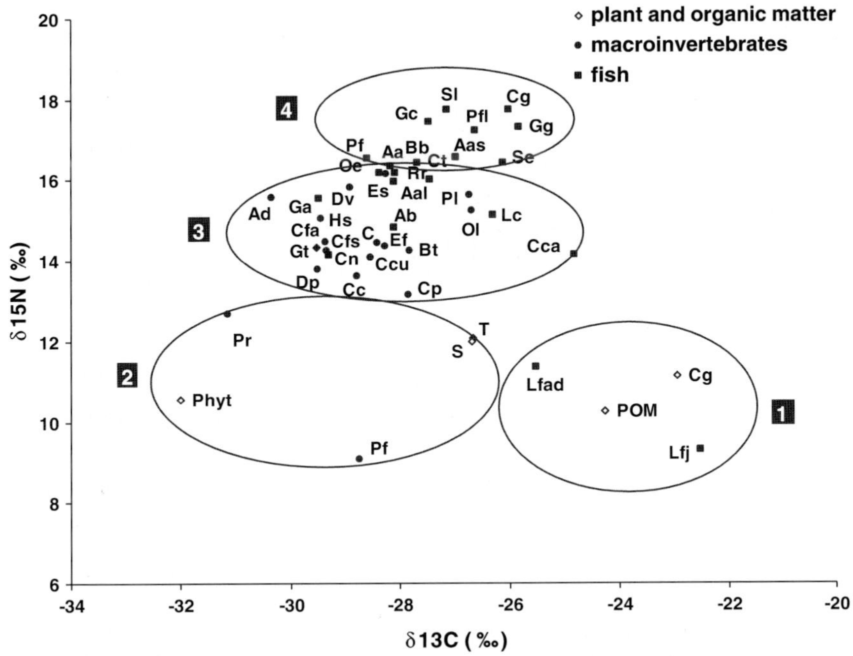

Figure 4. Stable isotopic values of organisms in the River Rhine food web for the period of 1996–1998 according to Marguillier et al. (1998). Mean values are given in the figure, corresponding standard deviations, standard errors of the mean and full names of the abbreviations are shown in Table 1. The circles 1–4 indicate the different clusters of trophic groups (obtained using Statistica 4.5).

phytoplankton, silt, housing tubes of *C. curvispinum* and both *Plumatella* species. The larvae of the river lamprey *L. fluviatilis* showed isotopic values distinct to that of the parasitic adults. Values of $\delta^{13}C$ for muddy housing tubes of *C. curvispinum* are similar to that of silt material. Isotopic carbon composition of DIC from the Waal was −9.65‰, which suggests phytoplankton in the river to have a mean $\delta^{13}C$ value of −32.00‰.

The second level of the food web consisted of primary consumers, filter, deposit and detritus feeders, with $\delta^{15}N$ values ranging from 13.14‰ (copepods) to 14.65‰ (*D. villosus*). The $\delta^{13}C$ values from these primary consumers and omnivores ranged from −24.82‰ *C. carpio* to −31.97 (*A. desmaresti*). These values show considerable depletion in ^{13}C relative to the $\delta^{13}C$ value of POM. *Palaemon longirostris* and *Orconectes limosus* are on average comparable in isotopic level, indicating a similar feeding preference.

Mean isotopic values of zoobenthivorous and/or zooplankton feeding fish like bream (*Abramis brama*), eel (*A. anguilla*) and roach (*Rutilus rutilus*) are comparable to the group of primary and secondary consumers mentioned and are clustered together with secondary consumers as well as with zoobenthivorous fish, functioning as transitive species from the second to the third trophic level of the Rhine food web. Zoobenthivorous fish as *Blicca bjoerkna*, *Gymnocephalus cernuus* and piscivorous fish *Perca fluviatilis* and *S. lucioperca* represent the highest predatory level in the Waal food web.

In the food web between 2001 and 2003, three trophic levels similar to those of the food web between 1996 and 1998 could be distinguished (Fig. 5) extended by a cluster of riparian plant species. The mean carbon and nitrogen isotope ratios ($\delta^{13}C$, $\delta^{15}N$) of items occurring in the Rhine food web for the period of 2001–2003 ranged between −33.84‰ (the moss *Cinclidotus* spec.) and −16.23‰ (green algae) for $\delta^{13}C$ and 1.86‰ (detritus) and 16.27‰ (*L. fluviatilis*, adults) for $\delta^{15}N$ (Fig. 5, Table 2).

Riparian plants and *Cinclidotus* spec. were clustered together (cluster 1), showing the lowest $\delta^{15}N$ values in distance from the aquatic consumer food web. The mean $\delta^{13}C$ values of these plants were similar to those of primary and secondary consumers and ranged from −27.06 (*Lythrum*

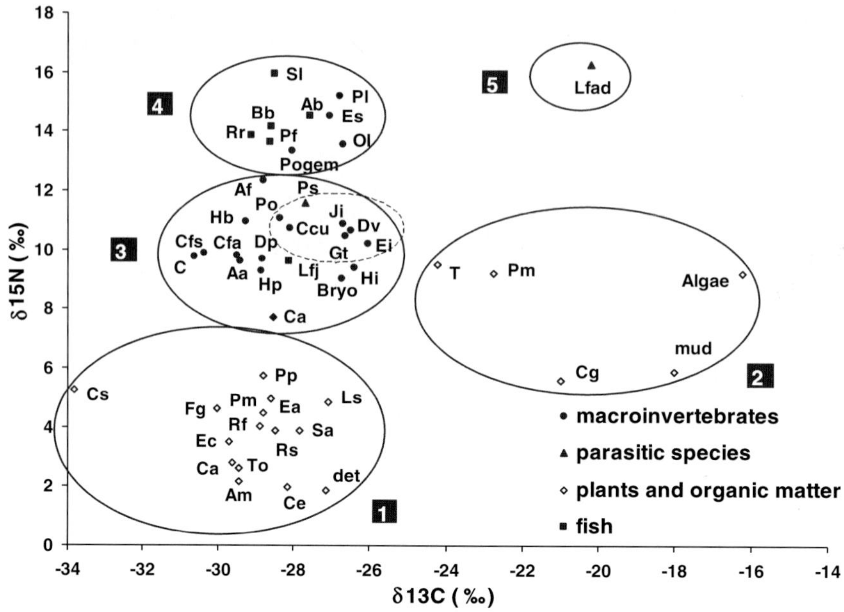

Figure 5. Stable isotopic values of organisms in the River Rhine food web for the period of 2001–2003. Mean values are given in the figure, corresponding standard deviations, standard errors of the mean and full names of the abbreviations are shown in Table 2. The circles 1–5 indicate the different clusters of trophic groups (obtained using Statistica 4.5).

salicaria) to −33.84 (*Cinclidotus* spec.). The second cluster at the first trophic level consisted of green algae, mud, plant matter and housing tubes of *C. curvispinum*. Green algae and decomposing plant matter in the Waal show a scattered pattern of $\delta^{13}C$ values with no clear enrichment in $\delta^{13}C$ compared to detritus and silt.

The $\delta^{13}C$ values from the primary consumers (cluster 3), filter, deposit and detritus feeders and omnivores ranged from −26.04 (*E. ischnus*) to −30.64 (chironomids), showing less depletion in $\delta^{13}C$ relative to detrital organic matter (difference −1.08 to −3.52‰) than in the food web between 1996 and 1998. Isotopic values of the Asian clams *C. fluminalis* and *C. fluminea*, the American gammarid *G. tigrinus*, the Ponto-Caspian amphipods *C. curvispinum*, *E. ischnus* and *D. villosus* and the isopod *Jaera istri* are within the range of other filter, deposit and detritus feeders and omnivores. Among these groups, the amphipods *C. curvispinum* and *D. villosus* have the highest $\delta^{15}N$ value (10.74 ± 0.90‰ and 10.64 ± 1.21‰).

Fish species (cluster 4) show $\delta^{13}C$ values (−29.10 ± 1.72‰ to −24.94 ± 5.05‰) similar to those of most macroinvertebrate species. Their $\delta^{15}N$ values, however, indicate a higher trophic level of feeding, similar to the level of *Palaemon longirostris*, and *Eriocheir sinensis*, all clustered with the highest $\delta^{15}N$ values, and are the top-predators of the food web. *P. longirostris* and *O. limosus* are less comparable in isotopic level in the food web between 2001 and 2003, indicating a less similar feeding preference during this period. Isotopic nitrogen values of the organisms in the food web between 2001 and 2003 were lower than of those of the food web between 1996 and 1998 for almost all species, whereas the isotopic carbon values were similar (Figs. 3–5). Piscivorous fish (cluster 4 and 5) are the top predators in the Rhine food web.

Comparison of the isotope signatures of the most abundant macroinvertebrate species from both food webs after correction of the differences in baseline $\delta^{15}N$ values (Fig. 6) shows comparable isotope values for most macroinvertebrate species for both food webs. Lower mean $\delta^{13}C$ values for *D. villosus* and *G. tigrinus* and a differently related position of *C. curvispinum* to *D. villosus* for the different periods are remarkable. During 1996–1998, isotopic values of *D. villosus* related to those of other macroinvertebrates indicated chironomids and *C. curvispinum* as potential prey items (Figs. 6 and 8) whereas these values were less related in 2001–2003. Standard

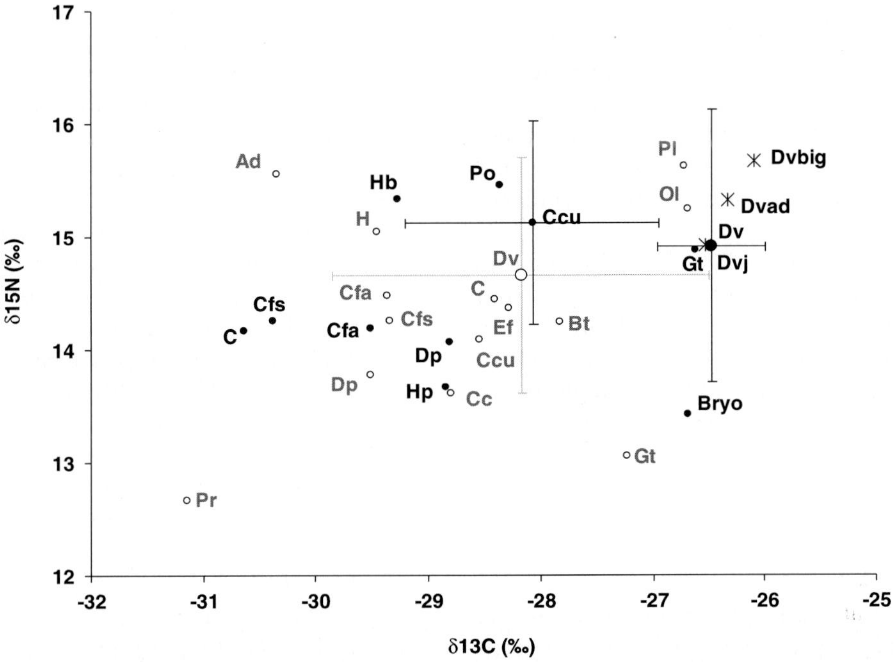

Figure 6. Mean stable isotope values of macroinvertebrate species from both food webs after correction for the nitrogen depletion in the river system after 1998. Standard deviations are shown for the species *C. curvispinum* and *D. villosus*.

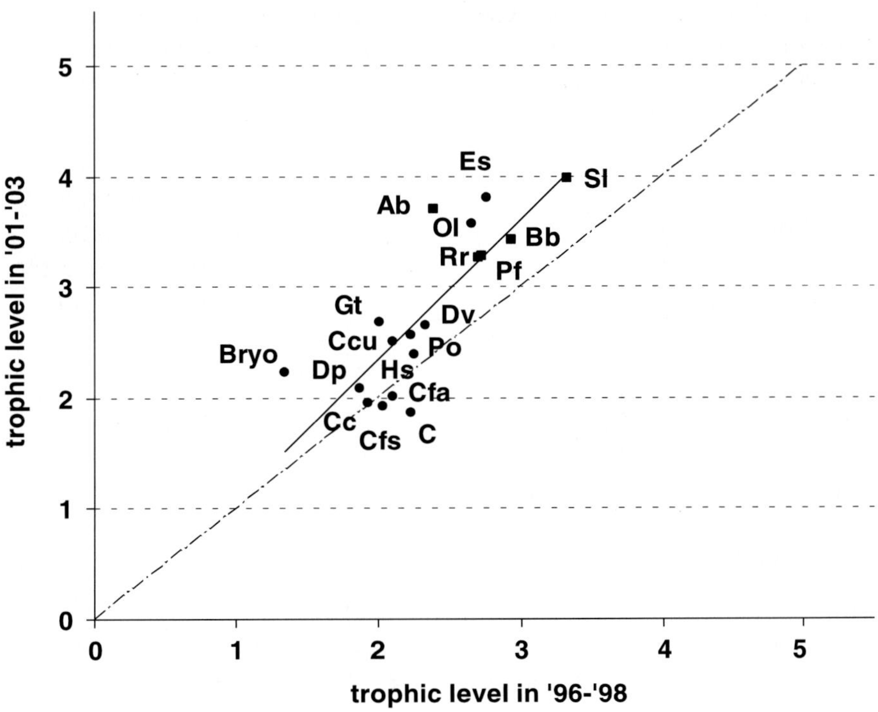

Figure 7. Trophic position of Rhine food web species during invasion and after establishment of *D. villosus*. Codes indicate the species according to the codes given in Table 1 and 2. The trend line compares the TP of the same species between the two different time periods.

deviations of *D. villosus* however still overlap those of *C. curvispinum*, indicating that *D. villosus* with the highest $\delta^{15}N$ values (i.e. bigger adults) would probably still prey on *C. curvispinum*, as is demonstrated in the laboratory experiment (Fig. 8). Differences in $\delta^{15}N$ between different life stages of *D. villosus* were however not significant ($p > 0.05$).

The food web between 2001 and 2003 shows higher trophic levels for most of the primary and secondary consumers (Fig. 7, Tables 1 and 2) than the food web between 1996 and 1998. The TP of *D. villosus* shifted proportionally to changes in the trophic level of most of the species in the food web.

Discussion

The Ponto-Caspian amphipods *C. curvispinum* and *D. villosus* have had a huge impact on the Rhine food web as they represent 80–90% of its macroinvertebrate community in number and biomass. *C. curvispinum* dominated the food web on groyne stones in 1998 and kept its dominance in 2001 next to *D. villosus*, although population densities of *C. curvispinum* had drastically been reduced. Various factors that generate bottom-up or top-down regulation of macroinvertebrate communities could be responsible for this decrease. As abiotic factors on which *C. curvispinum* depends, i.e. temperature and chlorophyll-*a* (Rajagopal et al., 1999), did not inhibit its development as the mean water temperature was somewhat higher in 2001 (13.2 °C in 1998, 13.9 °C in 2001), mean nitrogen concentration decreased from 4.33 g l^{-1} in 1998 to 2.23 g l^{-1} in 2001 and mean chlorophyll-*a* densities increased from 4 μg l^{-1} in 1998 to 10 μg l^{-1} in 2001, its population reduction is not likely caused by low resource availability or by altered abiotic factors. Top-down control by predation by *D. villosus* or by various top-predators (Kelleher et al., 1998) combined with increased predation or parasitic pressure from the triclad *D. romanodanubiale*, the halacarid *Caspihalacarus hyrcanus* and the acanthocephalan *Pomphorhynchus* spec. (Van Riel et al., 2003) could more evidently have caused this decline.

Most of the changes in the higher trophic levels of the Rhine food web between 1998 and 2001 are ascribed to Ponto-Caspian invasions. High numbers of *C. curvispinum* as food item could pave the pathway of invasions of predators and parasites like *E. ischnus*, *D. villosus*, *Jaera istri*, the mentioned predators and parasites as well as the gobies *P. semilunaris* and *Neogobius melanostomus*, causing an invasional meltdown of Ponto-Caspian species in the river Rhine (Ricciardi, 2001; Van der Velde et al., 2002, 2006).

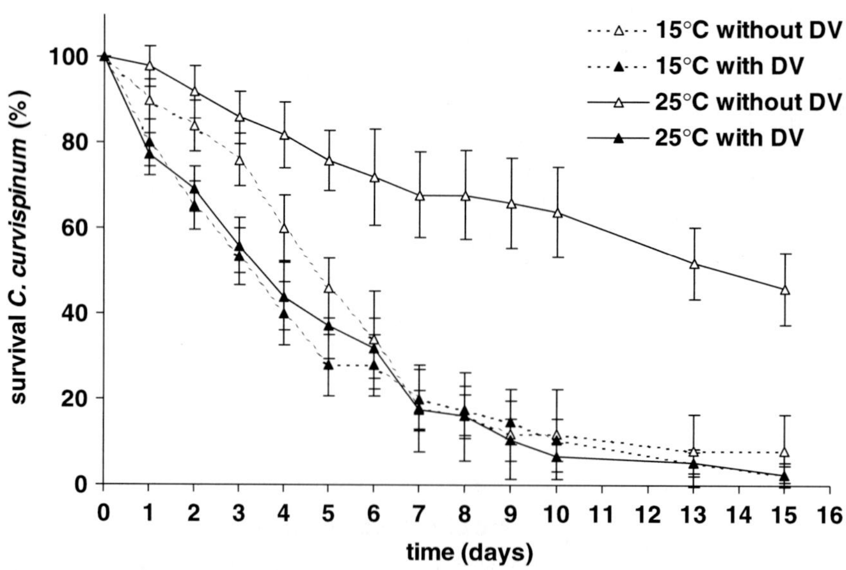

Figure 8. Effect of predation by *D. villosus* (DV) on survival of *C. curvispinum* at two water temperatures in a laboratory experiment. Bars indicate standard error of the mean.

Macroinvertebrate communities on these stone substrata have shown the largest changes in structure due to the subsequent success of various invaders competing for space, refuges and resources. It is assumed that the rapid increase of the omnivorous *D. villosus* that proved to be a strong competitor (Wijnhoven et al., 2003; Van Riel et al., 2004) determined the structure of the macroinvertebrate community on stones in 2001.

On the sand substrate, little has changed since the invasion of *C. curvispinum* and *D. villosus*. The Asiatic clam *C. fluminea* has been dominating the sandy streambeds since 1988 and densities of other macroinvertebrates remained low, probably due to high disturbance of the biotope by waves and shipping whirling up the sediment and to high predation pressure in this open environment. Ponto-Caspian amphipods were present in very low numbers and most individuals found were juveniles. However, the density and diversity of the macroinvertebrate community on the sand substrate increased between 1998 and 2001. Especially, *C. fluminea* and in September also *C. fluminalis* contributed to this increase perhaps related to much higher chlorophyll-*a* values in 2001.

Successful invaders in the Rhine food web are mainly present as primary consumers (e.g. *C. curvispinum, D. polymorpha, J. istri*). The more predaceous omnivorous Ponto-Caspian gammarids such as *D. villosus* are secondary consumers preying on macroinvertebrates as well as scavenging on carcasses and feeding on algae and phytoplankton. *D. villosus* shows high $\delta^{15}N$ values in comparison to other amphipod species, indicating a higher predatory level for this gammarid. Primary consumers invading the Waal did not change their food source years after invasion, but predatory, omnivorous invaders seem to be able to shift their diet, which could be affected by changes in prey species availability as densities of *C. curvispinum*, their most abundant presumed prey species in the period 1996–1998, decreased in the period 1998–2001. The $\delta^{15}N$ values of individual *D. villosus* did not significantly depend on body length and life stage, but adults with body lengths larger than 1.2 cm exceeded the values of the smaller ones, indicating a slightly higher trophic level.

The composition of the food web between 2001 and 2003 is comparable to that of the food web between 1996 and 1998. Input of terrestrial plant material, phytoplankton, benthic algae and POM fuel both food webs. Terrestrial plants seem to fuel the aquatic web, as they show comparable $\delta^{13}C$ values. Input of aquatic macrophytes is of minor influence, as their abundance in the main channel of the Rhine is negligible. The second trophic level is formed by primary consumers, consisting of filter-, detritus- and deposit-feeder and omnivores, of which most are invasive macroinvertebrate species. The most important source of food for this group is probably phytoplankton and suspended organic matter. The zoobenthic-zooplanktivorous and piscivorous fish occupy the highest trophic level of the food web. The isotopic composition of the larvae and adults of *L. fluviatilis* are distinct from other organisms in the Rhine food web and from each other, due to different feeding patterns. Larvae probably feed on benthic algae and microorganisms present in the sand bottom, but as adult become ectoparasitic on fish.

The overall $\delta^{15}N$ values of the Rhine food web decreased in isotopic nitrogen ratio indicating a lower anthropogenic nitrogen input (Cabana & Rasmussen, 1996; Hansson et al., 1997; McClelland et al., 1997) into the river Rhine. From 1998 to 2001, the nitrogen content of the water layer at Lobith, The Netherlands, diminished from 4.33 to 2.23 ± 0.21 g l^{-1}. Carbon isotopic values appeared to be unaltered. Trophic levels in both food webs can be compared after correcting for the variation in baseline $\delta^{15}N$ (Cabana & Rasmussen, 1996; Vander Zanden & Rasmussen, 1999), according to the overall $\delta^{15}N$–$\delta^{13}C$ relationship described by Vander Zanden & Rasmussen (1999). After this correction, the trophic level of most invertebrate species does not seem to have altered much. *D. villosus* continues the high predatory level it showed at the beginning of its invasion, but probably extended its diet as $\delta^{13}C$ shifted to more enriched values, likely to be related to changes in prey species availability.

Summarizing the characteristics of the Rhine food web, roughly three trophic levels can be distinguished. The food web is fuelled by POM, mostly originating from riparian plants, and phytoplankton (Admiraal et al., 1994), which makes the Rhine food web comparable to other large river food webs depending on riparian input (Thorp et al., 1998; Huryn et al., 2002). Macroinvertebrate species are keystone species, transferring food from phytoplankton and POM to higher trophic levels and eventually top-predators. As most of these keystone primary consumers in the Rhine are invaders, mostly crustaceans and bivalves, these invaders could be considered ecosystem engineers affecting functional diversity and food web structure of the community at the expense of insects (Vitousek, 1990; Crooks, 2002; Van der Velde et al., 2006). This engineering becomes more evident as numbers of a high impact invader increase and can result in bottom-up regulation, as *C. curvispinum* showed by swamping the ecosystem with muddy tubes, or top-down regulation by means of predation and competition as is observed for *D. villosus*.

Acknowledgements

The authors would like to thank J. Eygensteyn for stable istotope analysis, F. Cluitmans and B. Kelleher for collecting samples, M.J.E. Orbons, M.G. Versteeg, A. ten Brinke, the staff of the Conrad (RIZA) and the staff of the Schollevaer (OVB) for their assistance and support during fieldwork, and A.D. Buijse, R.S.E.W. Leuven, P.H. Nienhuis, A.M.J. Ragas and anonymous referees for their comments on the manuscript. This project was partly financed by the Institute for Inland Water Management and Waste Water Treatment (RIZA), Lelystad. This is CWE publication no. 398.

References

Admiraal, W., G. van der Velde, H. Smit & W. G. Cazemier, 1993. The rivers Rhine and Meuse in the Netherlands: present state and signs of ecological recovery. Hydrobiologia 265: 97–128.

Admiraal, W., L. Breebaart, G. M. J. Tubbing, B. van Zanten, E. D. de Ruijter van Steveninck & R. Bijkerk, 1994. Seasonal variation in composition and production of planktonic communities in the Lower River Rhine. Freshwater Biology 32: 519–531.

Bij de Vaate, A. & A. Klink, 1995. *Dikerogammarus villosus* Sowinsky (Crustacea: Gammaridae), a new immigrant in the Dutch part of the Lower Rhine. Lauterbornia 20: 51–54.

Bij de Vaate, A. & M. B. A. Swarte, 2001. *Dendrocoelum romanodanubiale* in the Rhine delta: first records from The Netherlands. Lauterbornia 40: 53–56.

Bij de Vaate, A., K. Jazdzewski, H. A. M. Ketelaars, S. Gollasch & G. van der Velde, 2002. Geographical patterns in range extension of Ponto-Caspian macroinvertebrate species in Europe. Canadian Journal of Fisheries and Aquatic Sciences 59: 1159–1174.

Bij de Vaate, A. 2003. Degradation and Recovery of the Freshwater Fauna in the Lower Sections of the Rivers Rhine and Meuse. PhD Thesis, University of Wageningen, Wageningen.

Bij de Vaate, A., R. Breukel & G. van der Velde, 2006. Long-term developments in ecological rehabilitation of the main distributaries in the Rhine Delta: fish and macroinvertebrates. Hydrobiologia 565: 229–242.

Cabana, G. & J. B. Rasmussen, 1996. Comparison of aquatic food chains using nitrogen isotopes. Ecology 93: 10844–10847.

Crooks, J. A., 2002. Characterizing ecosystem-level consequences of biological invasions: the role of ecosystem engineers. Oikos 97: 153–166.

Den Hartog, C., F. W. B. van den Brink & G. van der Velde, 1992. Why was the invasion of *Corophium curvispinum* and *Corbicula* species so successful? Journal of Natural History 26: 1121–1129.

DeNiro, M. J. & S. Epstein, 1978. Influence of the diet on the distribution of the carbon isotopes in animals. Geochimica et Cosmochimica Acta 42: 495–506.

DeNiro, M. J. & S. Epstein, 1981. Influence of the diet on the distribution of the nitrogen isotopes in animals. Geochimica et Cosmochimica Acta 45: 341–351.

Dick, J. T. A., I. Montgomery & R. W. Elwood, 1993. Replacement of the indigenous amphipod *Gammarus duebeni celticus* by the introduced *G. pulex*: differential cannibalism and mutual predation. Journal of Animal Ecology 62: 79–88.

Dick, J. T. A., D. Platvoet & D. W. Kelly, 2002. Predatory impact of the freshwater invader *Dikerogammarus villosus* (Crustacea: Amphipoda). Canadian Journal of Fisheries and Aquatic Sciences 59: 1078–1084.

Fahnenstiel, G. L., T. L. Lang, G. A. Bridgeman, M. J. McCormick & T. F. Nalepa, 1995. Phytoplankton productivity in Saginaw Bay, Lake Huron: effects of zebra mussel (*Dreissena polymorpha*) colonization. Journal of Great Lakes Research 21: 465–475.

Gearing, J. N., 1991. The study of diet and trophic relationships through natural abundance ^{13}C. In Coleman, D. C. & B. Fry, (eds), Carbon Isotope Techniques. Academic Press, San Diego, 201–218.

Haas, G., M. Brunke & B. Streit, 2002. Fast turnover in dominance of exotic species in the Rhine River determines biodiversity and ecosystem function: an affair between amphipods and mussels. In Leppäkoski, E., S. S. Gollasch & S. Olenin (eds), Invasive Aquatic Species of Europe: Distribution, Impacts and Management. Kluwer Academic Publishers, Dordrecht: 426–432.

Hansson, S., J. E. Hobbie, R. Elmgren, U. Larsson, B. Fry & S. Hohansson, 1997. The stable nitrogen ratio as a marker of food-web interactions and fish migration. Ecology 78: 2249–2257.

Hobson, K. A. & H. E. Welch, 1992. Determination of trophic relationships within a high arctic food web using δ^{13}C and δ^{15}N analysis. Marine Ecology Progress Series 84: 9–18.

Hobson, K. A., D. Schell, D. Renouf & E. Noseworthy, 1996. Stable-carbon and nitrogen isotopic fractionation between diet and tissues of captive seals: implications for dietary reconstructions involving marine mammals. Canadian Journal of Fisheries and Aquatic Sciences 53: 528–533.

Hobson, K. A., J. L. Sease, R. L. Merrick & J. F. Piatt, 1997. Investigating trophic relationships of Pinnipeds in Alaska and Washington using stable isotope ratios of nitrogen and carbon. Marine Mammal Sciences 13: 114–132.

Huryn, A. D., R. H. Riley, R. G. Young, C. J. Arbuckle & K. Peacock, 2002. Natural-abundance stable C and N isotopes indicate weak upstream-downstream linkage of food webs in a grassland river. Archiv für Hydrobiologie 153: 177–196.

Jantz B., 1996. Wachstum, Reproduction, Populationsentwicklung und Beeinträchtigung der Zebramuschel (*Dreissena polymorpha*) in einem grossen Fliessgewässer, dem Rhein. PhD Thesis, University of Köln, Köln.

Kelleher, B., P. J. M. Bergers, F. W. B. van den Brink, P. S. Giller, G. van der Velde & A. bij de Vaate, 1998. Effects of exotic amphipod invasions on fish diet in the Lower Rhine. Archiv für Hydrobiologie 143: 363–382.

Kroopnick, P., 1974. The dissolved O_2–CO_2–^{13}C system in the eastern equatorial Pacific. Deep Sea Research 21: 211–227.

Marguillier S., 1998. Stable Isotopes Ratios and Food Web Structure of Aquatic Ecosystems. PhD Thesis, Vrije Universiteit Brussel, Brussels.

Marguillier, S., F. Dehairs, G. van der Velde, B. Kelleher & S. Rajagopal, 1998. Initial results on the trophic relationships based on *Corophium curvispinum* in the Rhine traced by stable isotopes. In Nienhuis, P. H., R. S. W. E. Leuven & A. M. J. Ragas (eds), New Concepts for Sustainable Management of River Basins. Backhuys Publishers, Leiden: 171–177.

McClelland, J. W., I. Valiela & R. H. Michener, 1997. Nitrogen-stable isotope signatures in estuarine food webs: a record of increasing urbanization in coastal watersheds. Limnology and Oceanography 42: 930–937.

Minagawa, M. & E. Wada, 1984. Stepwise enrichment of ^{15}N along food chains: further evidence and the relation between $\delta^{15}N$ and animal age. Geochimica et Cosmochimica Acta 48: 1135–1140.

Mook, W. G. & F. C. Tan, 1991. Stable carbon isotopes in rivers and estuaries. In Degens, E. T., S. Kempe & J. E. Richey (eds), Biochemistry of Major World Rivers. J. Wiley and Sons Ltd, Chicester: 245–264.

Nichols, K. H. & G. J. Hopkins, 1993. Recent changes in Lake Erie (north shore) phytoplankton: cumulative effects of phosphorus loading reductions and the zebra mussel introduction. Journal of Great Lakes Research 19: 637–646.

Nijssen, H. & S. J. De Groot, 1987. De vissen van Nederland. Natuurhistorische bibliotheek 43. Koninklijke Nederlandse Natuurhistorische Vereniging, Hoogwoud (in Dutch).

Peterson, B. J. & B. Fry, 1987. Stable isotopes in ecosystem studies. Annual Review of Ecology and Systematics 18: 293–320.

Rajagopal, S., G. van der Velde, B. G. P. Paffen, F. W. B. van den Brink & A. de bij Vaate, 1999. Life history and reproductive biology of the invasive amphipod *Corophium curvispinum* (Crustacea: Amphipoda) in the Lower Rhine. Archiv für Hydrobiologie 144: 305–325.

Ricciardi, A., 2001. Facilitate interactions among aquatic invaders: is an "invasional meltdown" occurring in the Great Lakes? Canadian Journal of Fisheries and Aquatic Siences 58: 2513–2525.

Thorp, J. H., M. D. Delong, K. Greenwood & A. F. Casper, 1998. Isotopic analysis of three food web theories in constricted and floodplain regions of a large river. Oecologia 117: 551–563.

Tieszen, L., T. W. Boutton, K. G. Tesdahl & N. H. Slade, 1983. Fractionation and turnover of stable carbon isotopes in animal tissues: implications for ^{13}C analysis of diet. Oecologia 57: 32–37.

Van den Brink, F. W. B., G. van der Velde & A. bij de Vaate, 1991. Amphipod invasion on the Rhine. Nature 352: 576.

Van Riel, M. C., G. van der Velde & A. de bij Vaate, 2003. *Pomphorhynchus* spec. (Acanthocephala) uses the invasive amphipod *Chelicorophium curvispinum* (G. O. Sars, (1895) as an intermediate host in the River Rhine. Crustaceana 76: 241–247.

Van Riel, M.C., G. van der Velde & A. bij de Vaate, 2004. Alien amphipod invasions in the river Rhine due to river connectivity: a case of competition and mutual predation. In Douben, N. & A. G. Van Os (eds), Proceedings NCR-days (2003); Dealing With Floods Within Constraints. NCR-publication 24-(2004). Netherlands Centre for River Studies, Delft: 51–53.

Van der Velde, G., G. van Urk, F. W. B. van den Brink, F. Colijn, W. A. Bruggeman & R. S. E. W. Leuven, 1990. Rein Rijnwater, een sleutelfactor in chemisch oecosysteemherstel. In Hekstra, G. P. & F. J. M. van Linden (eds), Flora en Fauna Chemisch Onder Druk. Pudoc, Wageningen: 231–266 .

Van der Velde, G., S. Rajagopal, F. W. B. van den Brink, B. Kelleher, B. G. P. Paffen, A. J. Kempers & A. bij de Vaate, 1998. Ecological impact of an exotic amphipod invasion in the River Rhine. In Nienhuis, P. H., R. S. E. W. Leuven & A. M. J. Ragas (eds), New Concepts for Sustainable Management of River Basins. Backhuys Publishers, Leiden: 159–169.

Van der Velde, G., S. Rajagopal, B. Kelleher, I. B. Muskó & A. bij de Vaate, 2000. Ecological impact of crustacean invaders: general considerations and examples from the Rhine River. Crustacean Issues 12: 3–33.

Van der Velde, G., I. Nagelkerken, S. Rajagopal & A. bij de Vaate, 2002. Invasions by alien species in inland freshwater bodies in Western Europe: The Rhine Delta. In Leppäkoski, E., S. Gollasch & S. Olenin (eds), Invasive Aquatic Species of Europe: Distribution, Impacts and Management. Kluwer Academic Publishers, Dordrecht: 360–372.

Van der Velde G., S. Rajagopal, M. Kuyper-Kollenaar, A. bij de Vaate, D. W. Thieltges & H. J. MacIsaac, 2006. Biological invasions – concepts to understand and predict a global threat. In Bobbink R., B. Beltman, J. T. A. Verhoeven & D. F. Whigham (eds), Wetlands as a natural resource. Volume 2. Wetlands: Functioning, Biodiversity, Conservation and Restoration. Ecological Studies 191. Springer Verlag Dordrecht (in press).

Vander Zanden, M. J. & J. B. Rasmussen, 1999. Primary consumer $\delta^{13}C$ and $\delta^{15}N$ and the trophic position of aquatic consumers. Ecology 80: 1395–1404.

Vitousek, P. M., 1990. Biological invasions and ecosystem processes: towards an integration of population biology and ecosystem studies. Oikos 57: 7–13.

Wijnhoven, S., M. C. van Riel & G. van der Velde, 2003. Exotic and indigenous freshwater gammarid species: physiological tolerance to water temperature in relation to ionic content of the water. Aquatic Ecology 37: 151–158.

Hydrobiologia (2006) 565:59–69 © Springer 2006
R.S.E.W. Leuven, A.M.J. Ragas, A.J.M. Smits & G. van der Velde (eds), Living Rivers: Trends and Challenges in Science and Management
DOI 10.1007/s10750-005-1905-7

Seasonal dependent effects of flooding on plant species survival and zonation: a comparative study of 10 terrestrial grassland species

W.H.J.M. van Eck, J.P.M. Lenssen, H.M. van de Steeg, C.W.P.M. Blom & H. de Kroon*
Department of Experimental Plant Ecology, Institute for Wetland and Water Research, Radboud University Nijmegen, Toernooiveld, 6525 ED Nijmegen, The Netherlands
*(*Author for correspondence: Tel.: +31-24-365-3380; Fax: +31-24-365-2409; E-mail: h.dekroon@science.ru.nl)*

Key words: floodplain, plants, riparian grassland, river Rhine, season, species distribution, summer flood, winter flood

Abstract

Past research has provided compelling evidence that variation in flooding duration is the predominant factor underlying plant species distribution along elevation gradients in river floodplains. The role of seasonal variation in flooding, however, is far from clear. We addressed this seasonal effect for 10 grassland species by testing the hypothesis that all species can survive longer when flooded in winter than when flooded in summer. We carried out an inundation experiment under simulated conditions of summer and winter flooding in the greenhouse. The results showed that all species survived longer under winter floods than under summer floods. However, responses upon flooding were species-specific. All summer flood-tolerant species had high tolerance for winter floods as well, but summer flood sensitive species survived either a little longer, or dramatically longer when flooded under simulated winter conditions. Next, we examined whether winter or summer survival best predicted the lower distribution limits of the species as measured in a natural flooding gradient after an extremely long winter flood. We found a strong significant relationship between the lower distribution limits of species in the field and their tolerance to summer floods, although we measured the lower limits 14 years after the latest major summer flood. In contrast, no such significant relationship existed with species tolerance to winter floods. Some relatively intolerant species occurred at much higher floodplain elevations as was expected from their tolerance to winter inundation in the experiments. This suggests that zonation patterns as created by occasional summer floods may be maintained for a long time, probably due to the limited ability of species to re-colonise lower positions in the floodplain.

Introduction

Flooding is the predominant environmental factor determining plant distribution in river floodplains. It indirectly determines soil composition through erosion and sedimentation (Day et al., 1988; Henry et al., 1996) and directly affects plant growth by reducing oxygen and light availability (Setter et al., 1997). Tolerance to the direct effects of flooding strongly differs among species and these differences are reflected by species zonation along elevation gradients in river floodplains (Lenssen & de Kroon, 2005). Here, the most tolerant species dominate the lower, frequently flooded positions, whereas intolerant species are restricted to the highest elevations of the floodplain (Squires & Van der Valk, 1992; Carter & Grace, 1990; Sand-Jensen & Frost-Christensen, 1999; He et al., 1999; Vervuren et al., 2003; Van Eck et al., 2004). This tight correlation between flooding tolerance and elevational position indicates that tolerance may be an important tool to predict species responses to changes in river flooding regimes as a consequence of global warming, canalization or floodplain excavation.

However, before accurate predictions can be made it may be necessary to gain further flooding reflects the species tolerance to winter flooding, we collected field data in July and August of 2001 at two floodplain locations along the river Waal: Ewijkse Waard (51° 58′ N 5° 45′ E) and Klompenwaard (51° 53′ N 6° 01′ E). In the winter of 2000/2001 flooding was relatively severe. For instance at 11.50 m above sea level, the lower limit of perennial grassland, flooding lasted for 7 weeks, whereas the average winter floods during the preceding 30 years lasted for 5.3 weeks. At this position, summer floods are usually absent (Vervuren et al., 2003); the last flooding events occurred in June and August 1987 and these lasted for 37 and 10 days, respectively.

At both floodplain sites grasslands were present with extensive grazing by cattle and horses. At Ewijkse Waard five transects were placed and at Klompenwaard one, perpendicular to the water line. In each transect vegetation data were recorded by determining the presence or absence of the selected species in sample plots of 0.45×0.45 m that were established on contour lines at 0.10 m elevation intervals along the transect. Each contour line contained 10-sample plots except one transect at Ewijkse Waard that encompassed five sample plots. Distance between the plots on a contour line was at least 2 m. Distance between transects in Ewijkse Waard was at least 50 m. The elevations of the contour lines were determined with a surveyor's level. To allow comparison between the different sites and transects, all elevations mentioned in this paper refer to elevations above sea level as standardized to Lobith gauge station. Contour lines ranged from approximately 10–14 m above sea level.

Statistical analysis

Plant survival under water was analyzed using the SAS procedure LIFEREG (SAS Institute, 2001) with the Weibull model as the baseline function (Vervuren et al., 2003) and block (nested within season), season, species and interactions as covariates. We then calculated ratios of mean deviance changes, which approximately followed the F-distribution (McCullagh & Nelder, 1991) and will hereafter be referred to as Quasi *F*-values. Mean deviances allowed us to test differences between simulated seasons against variation among blocks and to treat species and block as random factors and season as a fixed factor. For each species and seasonal treatment, we subsequently estimated flooding tolerance by computing the median lethal time (LT_{50}; the flooding duration) at which 50% of the plant individuals from a given species had died) on the basis of Weibull equation parameters (Vervuren et al., 2003).

The lower field distribution limits of species along the flooding gradient were expressed as 10th percentile values based on species' frequency in the transects. For these analyses all species were included as far as sufficient field distribution data were available. At least five points are required to compute 10th percentile values (SPSS version 10.1). Since *Medicago falcata*, *Daucus carota* and *Rumex acetosa* were not present or only present in a few sample plots they were excluded from the analysis.

Relations between species' distribution and the experimentally obtained values for LT_{50} in simulated winter and summer floods were determined by correlation analysis (SPSS version 10.1).

Results

All species survived longer in the simulated winter floods than in the simulated summer floods (Fig. 1). However, the magnitude of the effect of the simulated flooding season was strongly species-dependent as indicated by the Species×Season interaction (Table 1). Some species survived only a little longer in the simulated winter flood than in the simulated summer flood (i.e. *Medicago falcata*, *Plantago lanceolata*, *Arrhenatherum elatius* and *Daucus carota*), but for others with relatively low tolerance to summer floods, survival was dramatically increased (i.e. *Rumex acetosa*, *Alopecurus pratensis*, *Rumex thyrsiflorus* and *Festuca rubra*) (Fig. 1). Survival of summer flood-tolerant species (i.e. *Elytrigia repens* and *Rumex crispus*) was also extended under winter floods.

The distributions of the species along the flooding gradient differed widely (Fig. 2). Likewise, the lower limits (10th percentile values) depended on the species and ranged from 11.35 to 13.27 m. For some species the lower limits were situated at relatively high elevations corresponding with relatively low flood durations (e.g. *Arrhenatherum elatius*), whereas others occurred at

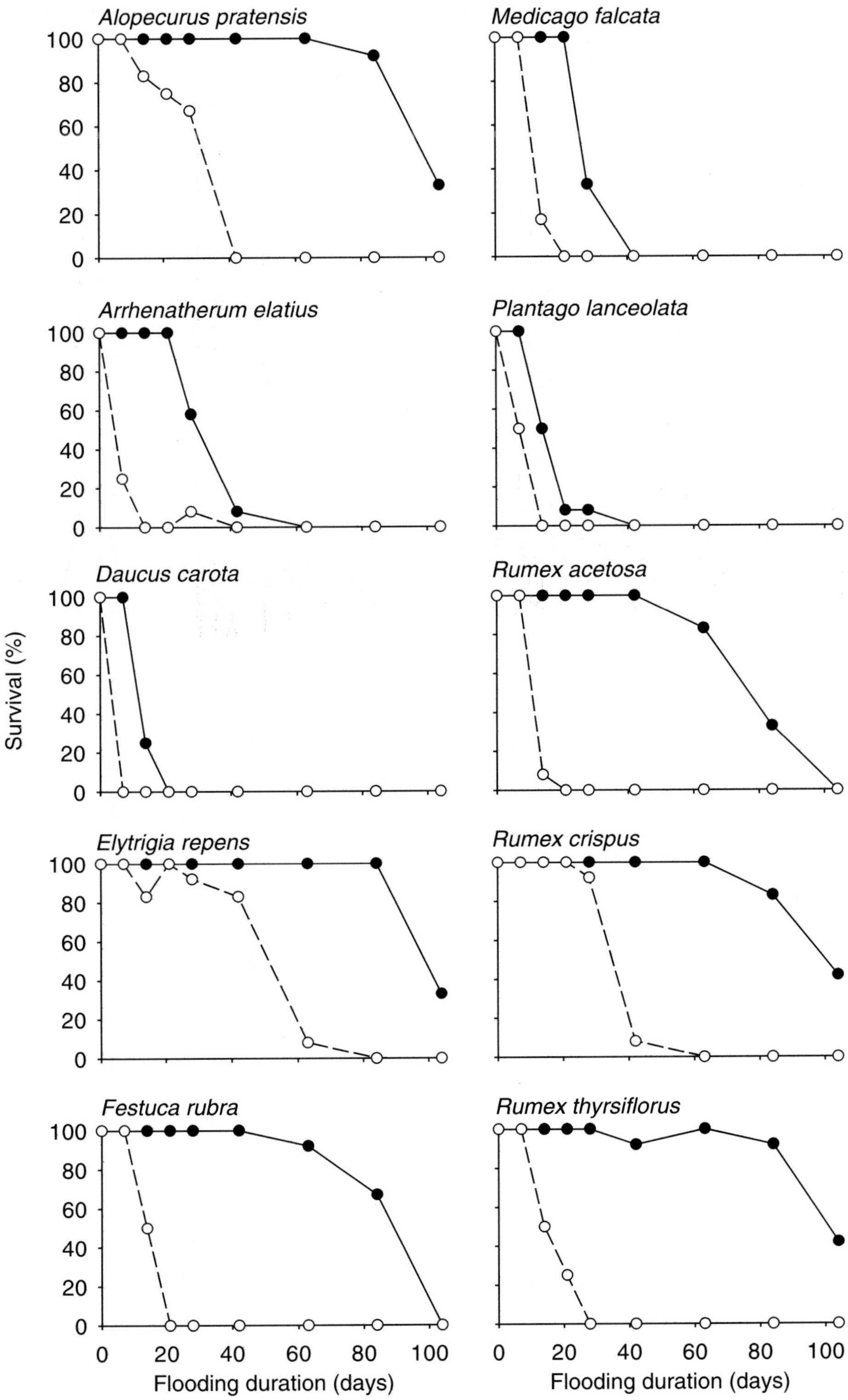

Figure 1. Relationship between flooding duration and survival (based on 12 replicate plants per flooding duration) of 10 river floodplain grassland species in simulated winter (closed symbols) and simulated summer floods (open symbols).

Table 1. Analysis of deviance for survival time under influence of species, simulated flooding season and block (season)

Source of variation	df (effect, error)	Mean deviance	Quasi *F*-value
Block(season)	6, 1895	0.09	0.12
Season	1, 6	0.00	0.00
Species	9, 54	4.27	21.68***
Season×Species	9, 54	1.16	5.90***
Species×Block(season)	54, 1895	0.20	0.25
Residual	1895	0.77	

Mean deviance, Quasi *F*-values and significance levels are presented; df, degrees of freedom (effect, error); ***, $p < 0.001$.

much lower elevations corresponding with longer flood durations (e.g. *Rumex crispus*). *Elytrigia repens* was present at relatively high frequencies along the whole flooding gradient. There was a highly significant negative correlation between species tolerance to summer flooding and the lower distribution limits along the flooding gradient (Fig. 3a) indicating that summer flood intolerant species were restricted to higher elevations in the floodplain (e.g. *Arrhenatherum elatius*) while more summer flood tolerant species (e.g. *Elytrigia repens* and *Rumex crispus*) occurred at lower elevations. Species with intermediate tolerances to summer floods occurred at elevations in the mid-range of the flooding gradient (e.g. *Alopecurus pratensis*).

In contrast, the correlation between species' tolerance to winter floods and their lower distribution limits along the flooding gradient was not significant (Fig. 3b). The lower distribution limits of a number of winter flood tolerant species ranged

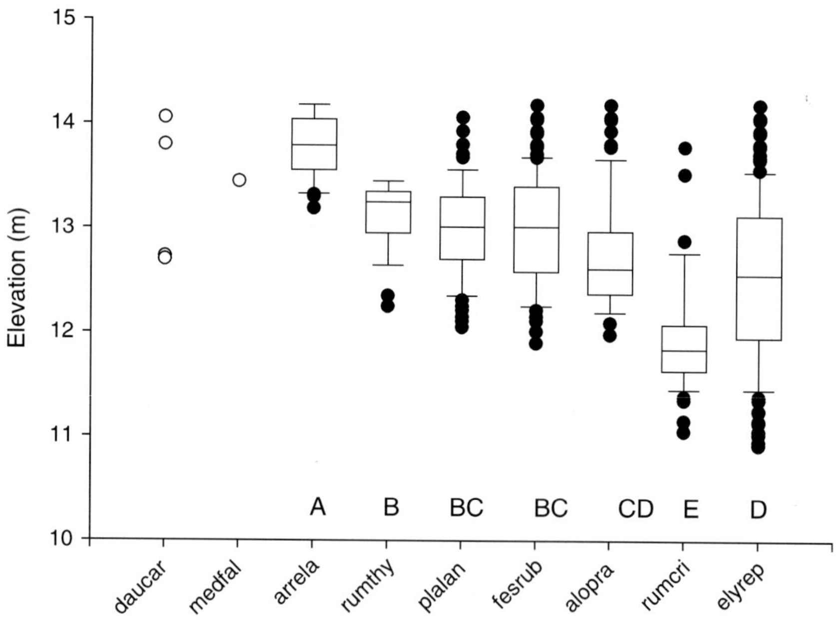

Figure 2. Species distribution along the flooding gradient based on six transects at two floodplain sites along the river Rhine in the Netherlands after the winter flood of 2000/2001. The end of the boxes defines the 25th and 75th percentiles, with a line at the median and error bars defining the 10th and 90th percentiles. Circles define data points beyond the 10th and 90th percentiles. For *Medicago falcata* and *Daucus carota* there were less than the minimum required number of data points (indicated by open circles) to compute a reliable set of percentiles and therefore excluded from analysis. *Rumex acetosa* was not observed within the sample plots along the transects. Species abbreviations: alopra = *Alopecurus pratensis*, arrela = *Arrhenatherum elatius*, daucar = *Daucus carota*, elyrep = *Elytrigia repens*, fesrub = *Festuca rubra*, medfal = *Medicago falcata*, plalan = *Plantago lanceolata*, rumcri = *Rumex crispus* and rumthy = *Rumex thyrsiflorus*. Letters above the species names denote the Tukey-grouping after one-way ANOVA; species with the same letter have the same distribution along the flooding gradient ($p < 0.05$).

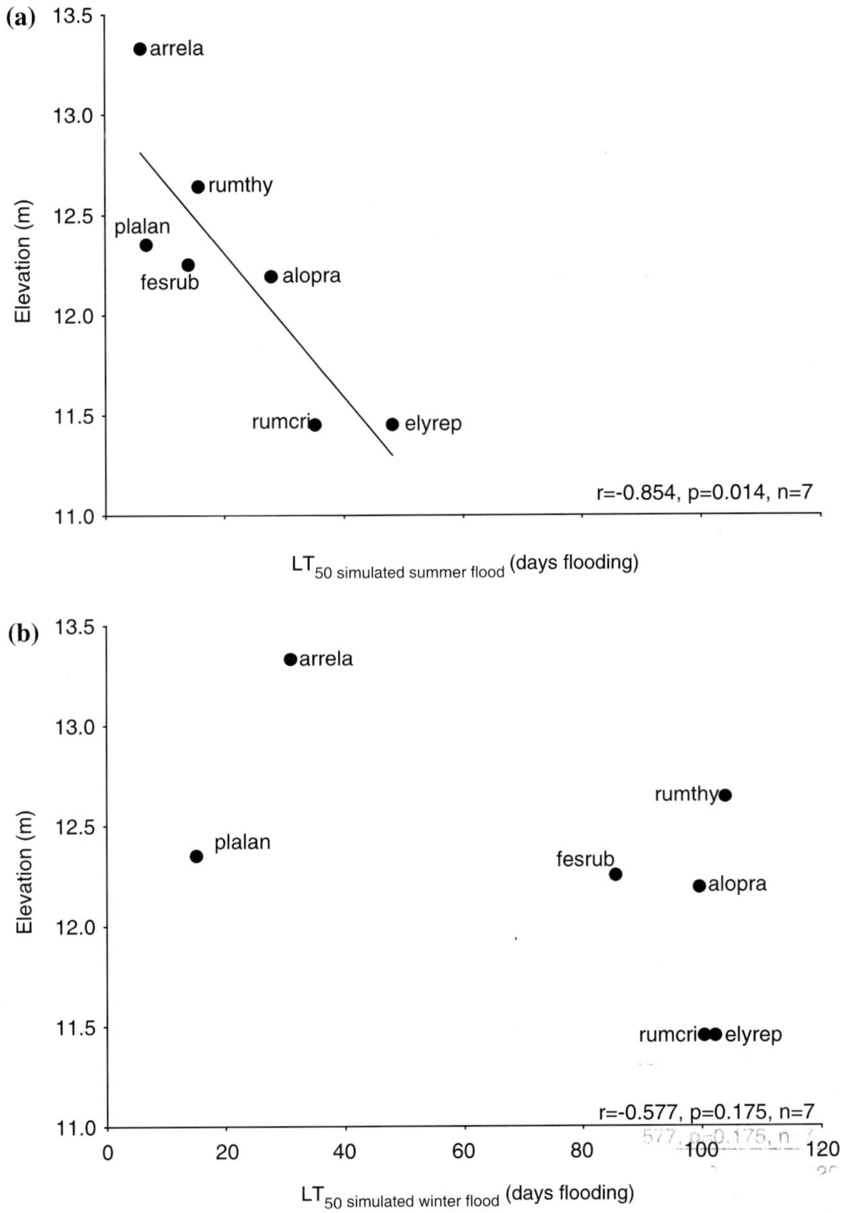

Figure 3. Relationship between species' lower distribution limits along the flooding gradient of the river Rhine in 2001 and the experimentally obtained values for flooding tolerance (LT_{50}) in (a) simulated summer floods and (b) simulated winter floods. For species abbreviations see Figure 2.

from low to high floodplain elevations whereas two very intolerant species (*Arrhenatherum elatius* and *Plantago lanceolata*) occurred only relatively high on the gradient.

We translated the species' lower distribution limits to the flooding durations in the winter of 2000/2001 and the summer of 1987 and correlated these flooding durations with their tolerance to simulated winter and summer floods, respectively. The flooding durations in the summer of 1987 and the winter of 2000/2001 were strongly correlated ($r^2 = 0.89$, $p < 0.001$) due to the strong relationship between flooding duration and elevation along the flooding gradient. Species' tolerance to simulated summer floods was highly correlated with the flooding durations at the lower distribution limits in the summer of 1987, the last severe summer flood in the study area (Fig. 4a). The flooding duration during the winter of 2000/2001 at species' lower distribution limits was not significantly correlated with their tolerance to simulated winter floods (Fig. 4b). Not all species occurred at elevations that were expected on the basis of their tolerance to simulated winter floods. Most striking was that some species that appeared to be tolerant to simulated winter floods (*Festuca rubra*, *Alopecurus pratensis* and *Rumex thyrsiflorus*) had their lower limits at positions that were only shortly flooded in the winter of 2000/2001 (Fig. 4b). On the other hand, *Plantago lanceolata*

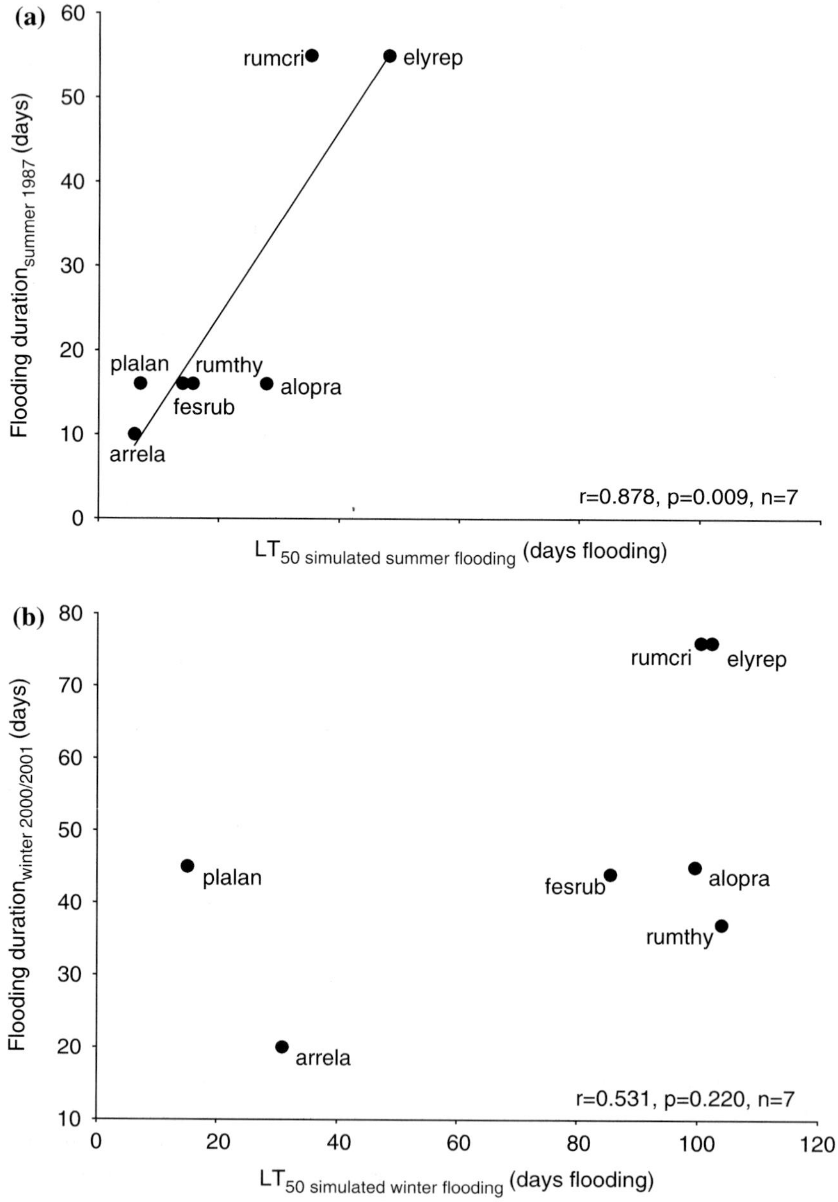

Figure 4. Relationship between flooding duration at species' lower distribution limits along the flooding gradient of the river Rhine in (a) the summer of 1987 and their flooding tolerance (LT_{50}) in simulated summer floods and (b) the flooding duration in the winter of 2000/2001 and their flooding tolerance in simulated winter floods. For species abbreviations see Figure 2.

occurred under longer flooding durations than was expected on the basis of its poor tolerance to winter floods. The very intolerant *Arrhenatherum elatius* only occurred under very short winter flooding durations, and its distribution seems to be limited by summer as well as winter flooding.

Discussion

The tolerance of plant species to complete submergence strongly depended on the simulated season of flooding. In accordance with our first hypothesis, all 10 species survived longer under the simulated winter flood than under the simulated summer flood. Our results thus corroborate those of Klimesová (1994) and Siebel (1998) who demonstrated that floods during the winter have a lower impact on survival of *Urtica dioica* plants than floods during the growing season. However, in contrast to our first hypothesis, the magnitude by which winter flooding enhanced tolerance was species-specific (Table 1). The specific response to flooding season may be related to interspecific differences in the use of stored carbohydrates during submergence, rather than to mitigating effects of higher oxygen concentrations in cold water (Van Eck et al., 2005a). Stored carbohydrates may prolong survival when submerged (Setter et al., 1997; Laan & Blom, 1990; Crawford, 2003; Nabben, 2001). Van Eck et al. (2005a) demonstrated that *Rumex crispus* had a conservative carbohydrate use at both low and high water temperatures. *Rumex acetosa*, although able to access its reserves in the roots, had a high rate of carbohydrate respiration in warm water but this rate slowed down at lower temperatures. This decreased rate of carbohydrate depletion probably explains why this species, like *Alopecurus pratensis*, *Festuca rubra* and *Rumex thyrsiflorus*, was able to extend its survival dramatically in simulated winter flood compared to summer floods in the present experiment. Accordingly, *Daucus carota* was not able to access its belowground carbohydrate reserves (Van Eck et al., 2005a) and its tolerance was also little enhanced in simulated winter floods. Hence, the findings of Van Eck et al. (2005a) together with the results presented here suggest that only species with an ability to mobilize and respire carbohydrates when submerged are able to prolong survival of flooding outside the growing season. Species that are intolerant to flooding regardless of season, such as *Arrhenatherum elatius*, *Daucus carota*, *Medicago falcata* and *Plantago lanceolata*, probably lack a physiological mechanism for accessing carbohydrate reserves when submerged.

Because seasonal effects of flooding are species-specific it may be expected that winter floods produce a different zonation in floodplain grasslands than summer floods. Accordingly, we found that species' lower limits were not equally well correlated with tolerance to summer and winter flooding. Surprisingly, however, it was the tolerance to summer floods that best explained lower distribution limits although the last summer flood in our study area occurred more than 14 years before our field data were collected. Moreover, a severe winter flood occurred immediately before the year in which field data were recorded but, in contrast to our second hypothesis, we found no significant correlation between species' distribution and tolerance to winter floods.

Our results further suggest that only the field distribution of *Arrhenatherum elatius*, the most intolerant species, was limited by winter flooding. Based on their tolerance to winter floods, species such as *Rumex thyrsiflorus*, *Alopecurus pratensis* and *Festuca rubra* could have occurred at much lower elevations than actually found. The poor correlation between flooding tolerance and lower limits suggests that other factors than flooding tolerance may be involved in determining lower distribution limits of these species in periods without severe summer floods (Van Eck et al., 2005b). Traits determining colonization ability may control the speed of migration down the flooding gradient as other studies indicate a strong relationship between colonization ability and abundance in disturbed habitats (Van der Sman et al., 1993; Collins et al., 1995; Henry et al., 1996; Burke & Grime, 1996). *Plantago lanceolata*, although very sensitive to winter flooding, did occur at relatively low floodplain elevations, suggesting that this species may rapidly colonise empty microsites created by winter floods. In addition to dispersal, winter floods may indirectly, through erosion and sedimentation, prevent successful establishment at lower positions (Van Eck et al., 2005b).

In conclusion, we have shown that species may either be tolerant to both winter and summer flooding, considerably more tolerant to winter flooding, or intolerant to flooding regardless of season. The species-specific tolerance to winter flooding suggests that these floods may also affect species distribution along elevation gradients in river floodplains. However, as shown here for grasslands along the river Rhine, the actual zonation may bear the signature of summer floods, because these floods, although less frequent, have a more dramatic impact on plant survival.

Acknowledgements

We thank I. Tellings, G.M. Bögemann and G. van der Weerden and his garden personnel for their practical support. M. van Kleunen for giving us statistical assistance with analysis of deviance. Data on water levels were kindly provided by the RIZA Waterdata-desk, the Netherlands. This is CWE-publication 419.

References

Breen, C. M., K. H. Rogers & P. J. Ashton, 1988. Vegetation processes in swamps and flooded plains. In Symoens, J. J. (ed.), Vegetation of Inland Waters Handbook of Vegetation Science 15/1. Kluwer Acadamic Publishers, Dordrecht: 223–247.

Burke, M. J. W. & J. P. Grime, 1996. An experimental study of plant community invisibility. Ecology 77: 776–790.

Carter, M. F. & J. B. Grace, 1990. Relationships between flooding tolerance, life history, and short-term competitive performance in three species of *Polygonum*. American Journal of Botany 77: 381–387.

Collins, S. L., S. M. Glenn & D. J. Gibson, 1995. Experimental analysis of intermediate disturbance and initial floristic composition: decoupling cause and effect. Ecology 76: 486–492.

Crawford, R. M. M., 2003. Seasonal difference in plant responses to flooding and anoxia. Canadian Journal of Botany 81: 1224–1246.

Crawford, R. M. M., C. E. Jeffree & W. G. Rees, 2003. Paludification and forest retreat in northern oceanic environments. Annals of Botany 91: 213–226.

Day, R. T., P. A. Keddy, J. McNeil & T. Carleton, 1988. Fertility and disturbance gradients: a summary model for riverine marsh vegetation. Ecology 69: 1044–1054.

He, J. B., G. M. Bögemann, H. M. van de Steeg, J. H. G. M. Rijnders, L. A. C. J. Voesenek & C. W. P. M. Blom, 1999. Survival tactics of *Ranunculus* species in river floodplains. Oecologia 118: 1–8.

Henry, C. P., C. Amoros & G. Bornette, 1996. Species traits and recolonization processes after flood disturbances in riverine macrophytes. Vegetatio 122: 13–27.

Johnson, C. M., P. R. Stout, T. C. Broyer & A. B. Carlton, 1957. Comparative chlorine requirements of different plants species. Plant and Soil 8: 337–353.

Klimesová, J., 1994. The effects of timing and duration of floods on growth of young plants of *Phalaris arundinacea* L. and *Urtica dioica* L: an experimental study. Aquatic Botany 48: 21–29.

Laan, P. & C. W. P. M. Blom, 1990. Growth and survival responses of *Rumex* species to flooding and submerged conditions: the importance of shoot elongation, underwater photosynthesis and reserve carbohydrates. Journal of Experimental Botany 41: 775–783.

Lenssen, J. P. M. & H. de Kroon, 2005. Abiotic constraints at the upper boundaries of two *Rumex* species on a freshwater flooding gradient. Journal of Ecology 93: 138–147.

McCullagh, P. & J. A. Nelder, 1991. Generalized Linear Models. Chapman & Hall, New York.

Mommer, L., H. de Kroon, R. Pierik, G. M. Bögemann & E. J. W. Visser, 2005. A functional comparison of acclimation to shade and submergence in two terrestrial plant species. New Phytologist 167: 197–206.

Nabben, R. H. M., 2001. Metabolic adaptations to flooding-induced oxygen deficiency and post-anoxia stress in *Rumex* species. PhD Thesis, University of Nijmegen.

Nilsson, C., M. Gardfjell & G. Grelsson, 1991. Importance of hydrochory in structuring plant communities along rivers. Canadian Journal of Botany 69: 2631–2633.

Pedersen, O., J. Borum, C. M. Duarte & M. D. Fortes, 1998. Oxygen dynamics in the rhizosphere of *Cymodocea rotundata*. Marine Ecology-Progress Series 169: 283–288.

Sand-Jensen, K. & H. Frost-Christensen, 1999. Plant growth and photosynthesis in the transition zone between land and stream. Aquatic Botany 63: 23–35.

SAS Institute, 2001. SAS Version 8.2. SAS Institute, Cary, North Carolina.

Setter, T. L., M. Ellis, E. V. Laureles, E. S. Ella, D. Senadhira, S. B. Mishra, S. Sarkarung & S. Datta, 1997. Physiology and genetics of submergence tolerance in rice. Annals of Botany 79: 67–77.

Siebel, H. N., 1998. Floodplain forest restoration. Tree seedling establishment and tall herb interference in relation to flooding and shading. PhD Thesis, University of Nijmegen.

Squires, L. & A. G. van der Valk, 1992. Water-depth tolerances of the dominant emergent macrophytes of the Delta Marsh, Manitoba. Canadian Journal of Botany 70: 1860–1867.

Studer-Ehrensberger, K., C. Studer & R. M. M. Crawford, 1993. Competition at community boundaries: mechanisms of vegetation structure in a dune–slack complex. Functional Ecology 7: 156–168.

Sykora, K. V., E. Scheper & F. van der Zee, 1988. Inundation and the distribution of plant communities on Dutch river dikes. Acta Botanica Neerlandica 37: 279–290.

Van der Sman, A. J. M., N. Joosten & C. W. P. M. Blom, 1993. Flooding regimes and life history characteristics of short-lived species in river forelands. Journal of Ecology 81: 121–130.

Van Eck, W. H. J. M., J. P. M. Lenssen, R. H. J. Rengelink, C. W. P. M. Blom & H. de Kroon, 2005a. Water temperature instead of acclimation stage and oxygen concentration determines responses to winter floods. Aquatic Botany 81: 253–264.

Van Eck, W. H. J. M., H. M. van de Steeg, C. W. P. M. Blom & H. de Kroon, 2004. Is tolerance to summer flooding correlated with distribution patterns in river floodplains? A comparative study of 20 terrestrial grassland species. Oikos 107: 393–405.

Van Eck, W. H. J. M., H. M. van de Steeg, C. W. P. M. Blom & H. de Kroon, 2005b. Recruitment limitation along disturbance gradients in river floodplains. Journal of Vegetation Science 16: 103–110.

Vervuren, P. J. A., S. M. J. H. Beurskens & C. W. P. M. Blom, 1999. Light acclimation, CO_2 response and long-term capacity of underwater photosynthesis in three terrestrial plant species. Plant, Cell and Environment 22: 959–968.

Vervuren, P. J. A., C. W. P. M. Blom & H. de Kroon, 2003. Extreme flooding events on the Rhine and the survival and distribution of riparian plant species. Journal of Ecology 91: 135–146.

Hydrobiologia (2006) 565:71–86
R.S.E.W. Leuven, A.M.J. Ragas, A.J.M. Smits & G. van der Velde (eds), Living Rivers: Trends and Challenges in Science and Management
DOI 10.1007/s10750-005-1906-6

Succession and rejuvenation in floodplains along the river Allier (France)

G.W. Geerling[1,5,*], A.M.J. Ragas[2,5], R.S.E.W. Leuven[2,5], J.H. van den Berg[3,5], M. Breedveld[1], D. Liefhebber[1] & A.J.M. Smits[1,4,5]

[1]*Institute for Science, Innovation and Society, Centre for Water and Society, Radboud University Nijmegen, Faculty of Science, P.O. Box 9010, 6500 GL Nijmegen, The Netherlands*
[2]*Department of Environmental Science, Radboud University Nijmegen, Faculty of Science, Institute for Wetland and Water Research, P.O. Box 9010, 6500 GL Nijmegen, The Netherlands*
[3]*Department of Physical Geography, Utrecht University, Faculty of Geographical Sciences, P.O. Box 80.115, 3508 TC Utrecht, The Netherlands*
[4]*Erasmus Centre for Sustainability and Management, Erasmus University Rotterdam, P.O. Box 1738, 3000 DR Rotterdam, The Netherlands*
[5]*Participant of the Netherlands Centre for River Studies (NCR), P.O. Box 177, 2600 MH Delft, The Netherlands*
*(*Author for correspondence: E-mail: Gertjan.Geerling@science.ru.nl)*

Key words: aerial photographs, Allier river, ecotope dynamics, floodplain vegetation, GIS, meandering, meta-climax, rejuvenation, steady state mosaic, succession, transition rate

Abstract

The spatio-temporal heterogeneity of a meandering part of the Allier river was studied by analysing ecotope composition and dynamics using a series of aerial images covering a period of 46 years (1954–2000). The ecotope dynamics was exemplified by two time series showing rejuvenating hydro-geomorphological processes, i.e., meander progression, meander cut-off and channel shift. The mean rejuvenation rate was 33.8 ha per 5 years for the 5.5 km long study area. The ecotope transition rates varied from 18% surface area change per 5 years to 58.7% surface area change per 5 years for pioneer vegetation. The combination of hydro-geomorphological processes and ecological succession resulted in a temporal diversity of the riparian area. In the year 2000 half of the total riparian landscape was 14 years or younger and 23% was not rejuvenated in 46 years. Eighty percent of the pioneer vegetation was found on young soils (<14 years) while more than 50% of the surface area of low dynamic ecotopes like bush and side channels was located on parts, which were stable for more than 46 years. Examining the relation between river stretch size and ecotope diversity showed that the ecotope diversity remained stable above a stretch size of 1.5 meander lengths for the years 1978, 1985 and 2000. The spatial and temporal analysis of the study area showed evidence supporting the steady state or meta-climax hypotheses, but influences of long-term processes on landscape composition were also found. Some implications for floodplain management are discussed.

Introduction

Since the late 80s, floodplains of highly regulated rivers are being reconstructed to increase flood protection and to follow society's call for strengthening riverine nature (Nienhuis & Leuven, 2001; Wolfert, 2001; Nienhuis et al., 2002; Lenders, 2003; Buijse et al., 2005; Van Stokkom et al., 2005). Plans involve geo-morphological interventions to increase the discharge capacity and to create semi-natural floodplains by stimulating natural processes like spontaneous succession, sedimentation, and to a lesser extent, erosion (Amoros, 2001; Prach & Pysek, 2001; Vulink, 2001; Wolfert, 2001).

The landscape unit pattern in natural river systems is shaped by a combination of two main driving forces: succession and rejuvenation.

Succession is the local transition of a landscape unit to another by changing species composition (Forman & Godron, 1986), while erosion in outer river bends and sedimentation in inner bends rejuvenates the vegetation types to a previous stage. In natural systems, the continuous disturbance of succession by rejuvenation processes results in a diverse landscape pattern with a high biodiversity (Amoros & Wade, 1996). However, semi-natural floodplains in regulated rivers generally lack natural rejuvenation mechanisms. This may result in a landscape pattern dominated by climax succession stages, which has a relatively low biodiversity and high hydraulic resistance (Bravard et al., 1986; Amoros & Wade, 1996; Baptist et al., 2004). This explains why river managers want to incorporate artificial rejuvenation measures in their management strategies (Smits et al., 2000). It is anticipated that clever application of artificial rejuvenation measures may increase biodiversity and safeguard flood protection goals (Buijse et al., 2005). However, to sensibly embed rejuvenation measures in river management, knowledge of the dynamics and the spatio-temporal heterogeneity of natural river systems is required (Ward et al., 2001). The present paper analyses succession and rejuvenation processes in a freely meandering river stretch in order to obtain information relevant for river management.

In a meandering system, the hydro-geomorphological processes associated with river channel migration rejuvenate the units that comprise the riparian landscape. Existing landscape units are rejuvenated while pioneer landscape units arise and go into succession. Landscape units are continuously present but shift in space, creating a spatio-temporally heterogeneous landscape pattern. If the system is in process equilibrium, the overall landscape unit dynamics must be stable at a certain scale level. This concept is called the steady-state mosaic (Forman & Godron, 1986) or meta-climax concept (Amoros & Wade, 1996). The dynamics and scale of the steady-state mosaic are largely controlled by flow and sediment regimes and the geological, climatic and biogeographical character of the river sector. For example, process equilibrium of a braided alpine river could be manifest within years in contrast with decades or more for a low gradient meandering channel (Van der Nat et al., 2003).

The aim of this paper is to determine the dynamics of landscape units in a freely meandering stretch of the river Allier (France) and the consequences for the spatio-temporal constitution of its riparian landscape. A time series of aerial photographs spanning 46 years was analysed to answer the following questions: 1. What are the transition rates of the different landscape units? 2. What is the spatio-temporal distribution of rejuvenation? 3. What is the surface area covered by the landscape units and how does it vary over time? 4. Can a river stretch size be determined, on which the landscape unit distribution is stable in all years?

Material and methods

Study site

The study site is a 6 km stretch of the river Allier, south of Moulins (France, Fig. 1). This is a meandering gravel river with lateral erosion in the outer bends and gravel point bars in the inner bends. Local sources state that before the transition to a nature area in the 1990s, the floodplains were subject to extensive grazing. It comprises about 500 ha of natural floodplain along a bit more than three meander lengths. The river is not used for navigation and the main channel in the research area is not regulated or excavated. These characteristics make it an interesting site to study meander processes in relation to riparian landscape composition and dynamics.

The Allier river's source is Lozère (1500 m altitude) located in the French 'Massif Centrale' (Wilbers, 1997). After 410 km, the river converges with the Loire river at Bec-d'Allier (186 m altitude). The Allier is a rain fed river with an unpredictable discharge course. The mean annual discharge is 160 $m^3 s^{-1}$ over the period 1850–1980 at Moulins (Gautier et al., 2000). Normally, peak discharges up to 1200 $m^3 s^{-1}$ (occurrence once every 10 years at Moulins) occur in winter and spring while the discharges are generally low in the summer with a minimum of 12 $m^3 s^{-1}$ (Gautier et al., 2000; Fig. 2).

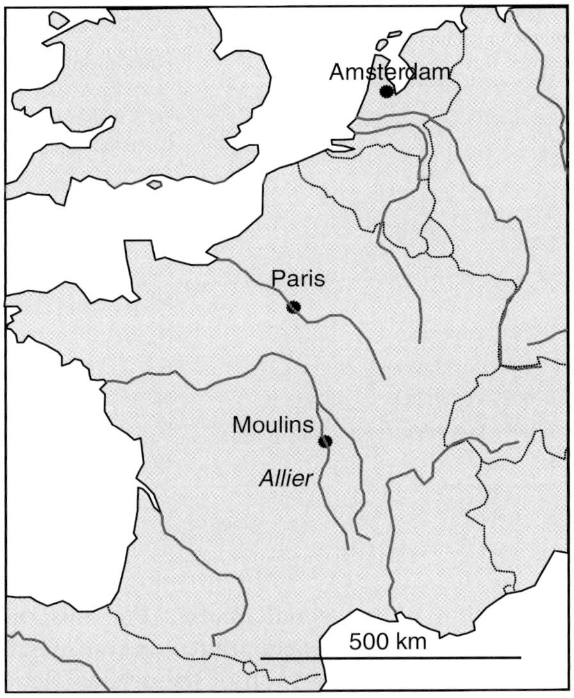

Figure 1. Location of the river Allier in Europe. The research area is located just south of Moulins. The north-west corner of the research area is (675330, 2170300) and the south-east corner is (678400, 2164550) in French national grid coordinates (Lambert zone II).

Preparing GIS maps

Based on a set of aerial photographs, maps were produced to analyse the landscape changes in the research area using GIS (Miller et al., 1995; Muller, 1997; Green & Hartley, 2000; Mendonca-Santos & Claramunt, 2001). The photographic material consisted of stereographic coverage of aerial images of the years 1954, 1960, 1967, 1978, 1985, 2000 and a non-stereographic set of 1992 (Photothèque-Nationale, 2003). The photographic scale varied between 1:25 000 and 1:14 500 and all images were taken in the summer (July/August). For the years 1954–1992 black and white photographs were available; the photographs of the year 2000 were true-colour. The 1992 photograph set was not mapped and only used to determine a sinuosity value.

Through a combination of field knowledge and expert knowledge on the interpretability of the available aerial image time series, a set of ecotope types was defined to classify landscape units (Table 1). A distinction is made between cultivated ecotopes (cultivated forest and agriculture) and natural ecotopes formed by river dynamics. An ecotope is a spatial unit of a certain extension (usually 0.25–1.5 ha), which is

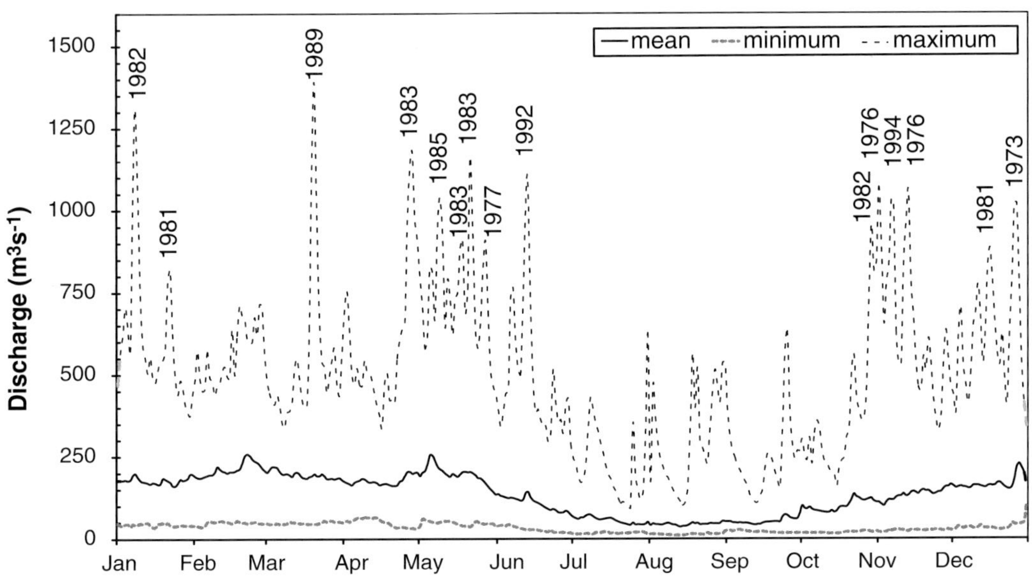

Figure 2. Minimum, mean and peak discharges of the river Allier at Moulins accumulated over the period 1968–2000. Peak discharges larger than 800 $m^3 s^{-1}$ are labelled with the year of occurrence (data: l'agence de l'eau Loire Bretagne, France).

Table 1. The mapped ecotopes (landscape units)

Ecotope (landscape unit)	Horizontal density	Human influence
Forest		C
Agriculture		C
Water, main channel		N
Bare soil (pointbar)		N
Pioneer vegetation		N
Grassland vegetation		N
Herbaceous vegetation		N
Bush (shrubs and trees <5 m)	Open canopy (20–60% coverage)	N
	Closed canopy (>60% coverage)	N
Forest (>5 m)	Open canopy (20–60% coverage)	N
	Closed canopy (>60% coverage)	N
Water, (closed) side channel		N

C, Cultivated landscape; N, Natural landscape.

homogenous as to vegetation structure and the main abiotic factors on site (Forman & Godron, 1986; Klijn & Udo de Haes, 1994; Lenders et al., 2001).

The aerial photographs were scanned and geo-referenced to a 1:25 000 topographical base map yielding rectified images of all years with a resolution between 2.1 and 2.5 m (IGN, 1990; Erdas, 1999; Mount et al., 2002). The maximum geo-reference error found relatively within the time series was about 10 m. In digitising ecotopes using aerial images two kinds of errors can be made: errors in outlining the ecotopes and errors in ecotope identification (Küchler & Zonneveld, 1988; ESRI, 2000).

First, the minimal mapping unit was defined as 40×40 m, i.e., 0.16 ha. The outline of the ecotopes was identified using colour, texture and vertical structure (explored using a stereoscope on the original images). ArcGIS 8.3 was used to manually digitise the outlines applying a fixed on-screen scale of 1:7500 (ESRI, 2000). To minimise overlay errors in the analysis phase, the 2000 map was produced first and used as a basis for the older maps. Only borders of polygons that shifted more than the relative geo-reference error of 10 m were considered ecotope outline changes and the polygons were redrawn.

For ecotope identification and evaluation of the digitised ecotope outlines, the stereoscope was used to exploit the original quality and vertical information of the aerial photos. For this, the arcGIS polyline maps were printed on transparencies and were placed on top of the original aerial images under a stereoscope (Topcon Model 3). This process resulted in ecotope maps for the years 1954–2000, which were subsequently used for the analysis.

GIS methods

All GIS analyses were performed using ArcGIS 8.3 and ArcGIS 9.0. For the raster calculations, the vector maps were rasterised to a 5×5 m grid.

To derive ecotope transition rates from the ecotope maps, transition matrices were produced of each map transition, e.g., 1954–1960, 1960–1967, and so on (Forman & Godron, 1986; Miller et al., 1995; Van der Nat et al., 2003; Narumalani et al., 2004). Transition matrices show to which new ecotopes an ecotope is transformed during the time span between two successive photographs. To be able to compare transition rates between all the maps, the percentage change of each ecotope was computed and standardised to a 5-year period to compensate for the variety in years between maps. In this analysis, the main channel and the adjacent pointbars (bare soil) were grouped because fluctuations in water level influenced their relative surface areas.

To visualise ecotope dynamics, a general ecotope succession scheme was developed, based on the transition matrices and field expertise (Fig. 3;

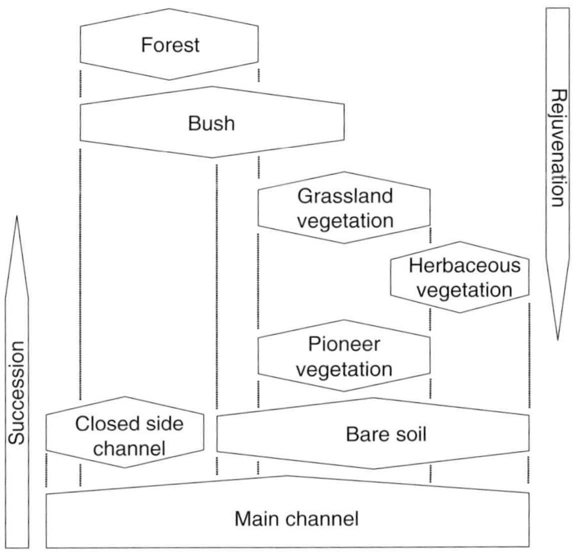

Figure 3. The succession scheme of the ecotopes along the river Allier.

Van den Berg & Balyuk, 2004). The ecotope transition matrices were simplified by classifying every possible ecotope transition into three categories: succession, rejuvenation or stability. The classification was based on the direction of change in the succession scheme (Fig. 3). Per ecotope the percentage area in succession, rejuvenation or remaining stable was computed for all transition periods. These percentages were visualised in triangular ternary plots. These plots are widely used in (soil) chemistry, to illustrate the composition of a three compound chemical mixture. In this paper, the axes show the area (as a percentage of the entire ecotope area) being stable, in succession, and in rejuvenation.

To investigate the age distribution of the ecotopes in the year 2000, a map was constructed showing the year of last rejuvenation since 1954 by combining the ecotope types main channel and bare soil (pointbars) of the years 1954–2000. This floodplain age map was overlaid with the ecotope map of the year 2000 to determine the age distribution of each ecotope type in 2000. Parts of the floodplain, which were not rejuvenated within the time span of the photographic survey, were assumed to be in succession for more than 46 years.

To investigate scale in relation to ecotope diversity, a method was developed analogous to determining the minimum area size of vegetation quadrats in field vegetation surveys. Here, the quadrat size is increased until the species composition becomes constant; this is the minimum quadrat size (Kent & Coker, 1994). To accomplish this with ecotope maps, the maps were cut into regular stretches perpendicular to the meandering direction of the river. The Shannon Index (SI) was used as landscape diversity measure, because it relates to the relative ecotope surface area distribution (McGarigal & Marks, 1995). The SI is high when all ecotope types occupy a similar area and decreases when this ecotope area distribution becomes more uneven. Starting upstream, the SI was calculated for the first 600 m stretch of the mapped area. Subsequently, the area was stepwise enlarged in downstream direction and the SI was repeatedly calculated yielding SI values for a growing area until the area covered the complete map surface. Fragstats 3.3 was used to calculate the SI (McGarigal & Marks, 1995).

Results

Ecotope maps

Figure 4 presents a time series demonstrating ecotope succession and rejuvenation caused by the hydro-geomorphological processes. The meander grew and moved northward in the years 1954, 1960, 1967. Between 1967 and 1978 a bridge was constructed on the downstream border of the research area which probably caused or facilitated the cut-off shown in the 1978 excerpt, and so creating a side channel. The cut-off resulted in a peak in the rejuvenation activity (Table 2) and a drop in sinuosity (Table 3), but as the meandering process continued, sinuosity reached its former values again in 1992–2000. The mean rejuvenation rate within the 5.5 km straight (3 meanders long) research area is 33.8 ha every 5 years (Table 2).

Figure 5 illustrates the influence of hydro-geomorphological processes on the spatial distribution of ecotopes, in this case the formation of a black poplar (*Populus nigra*) niche by a shift of the river channel in 1967 and 1978. The main channel shift left a depression in the landscape and simultaneously rejuvenated older succession stages across the stream. Subsequently, the depression (i.e., the former river channel) functioned as an

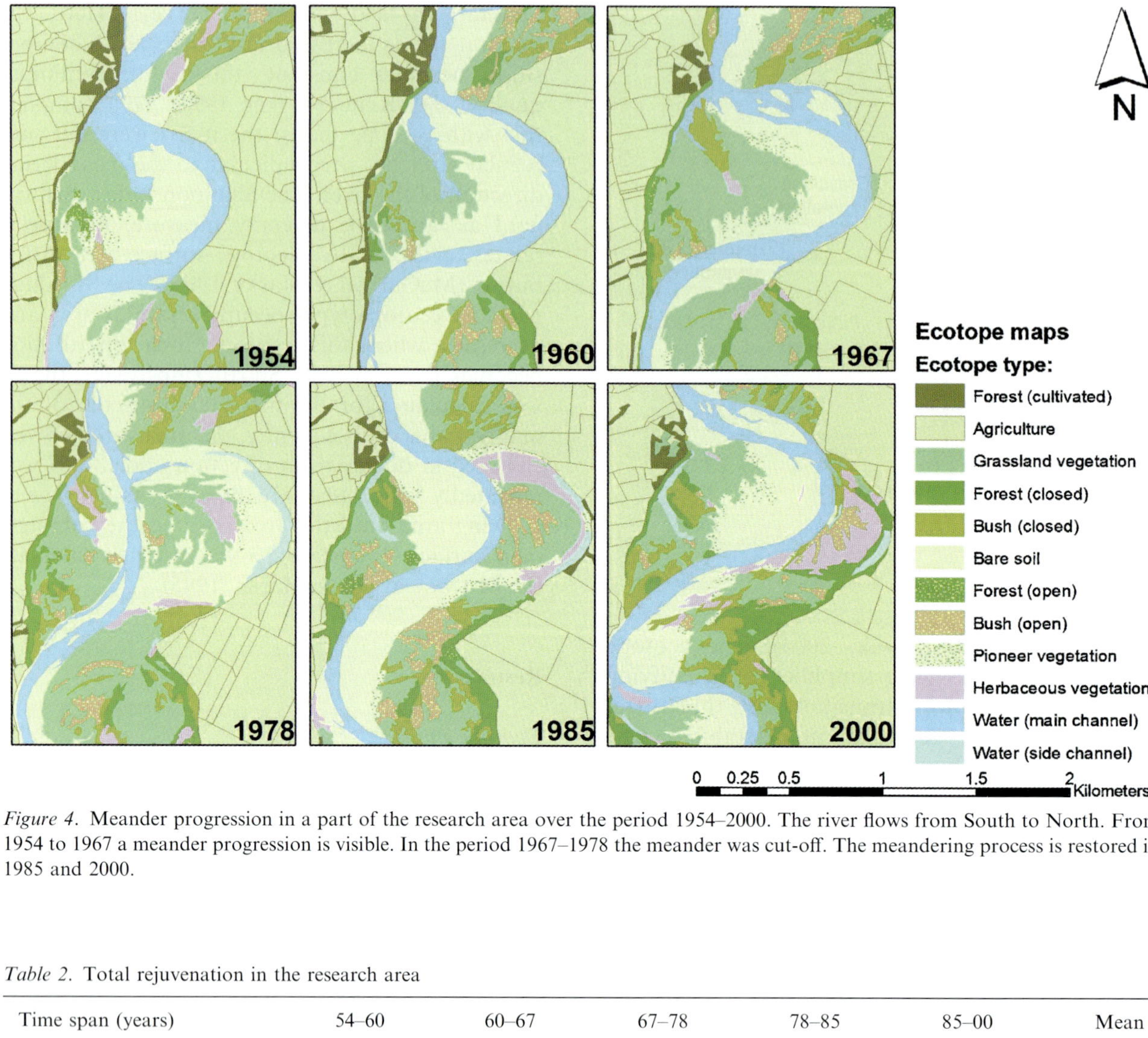

Figure 4. Meander progression in a part of the research area over the period 1954–2000. The river flows from South to North. From 1954 to 1967 a meander progression is visible. In the period 1967–1978 the meander was cut-off. The meandering process is restored in 1985 and 2000.

Table 2. Total rejuvenation in the research area

Time span (years)	54–60	60–67	67–78	78–85	85–00	Mean
Rejuvenation (ha)	31.5	57.8	68.9	72.9	80.1	
Rejuvenation (ha/5 year)	26.3	41.3	31.3	52.1	26.7	33.8

Table 3. Sinuosity of the studied river stretch

Year	Sinuosity
1954	1.35
1960	1.41
1967	1.45
1978	1.24
1985	1.27
1992	1.42
2000	1.47

environment for the settlement of black poplar. The small poplars grew from ecotope type bush to forest between the years 1985 and 2000.

Ecotope dynamics

An example of the ecotope transition matrices that were produced is shown in Table 4. The rows show to what extent (percentage area) the 1967 ecotopes (row headers) developed into different ecotopes in

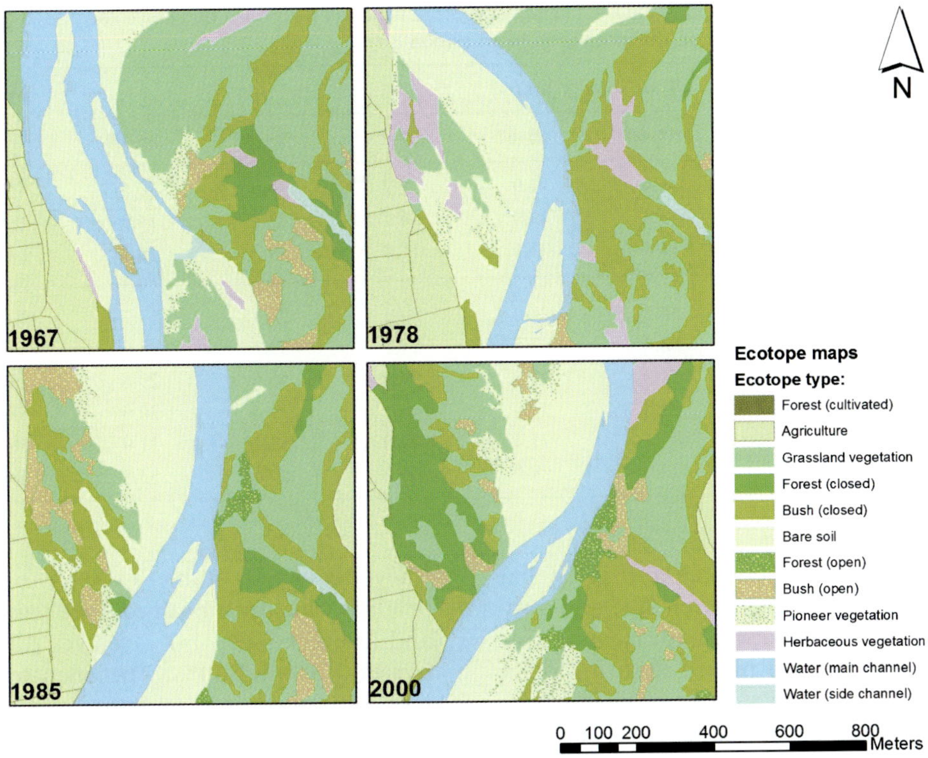

Figure 5. Meander shift rejuvenates ecotopes and creates niches for forest development over the period 1967–2000. The 1967–1978 shift rejuvenates ecotopes and creates niches for forest settlement in the former channels. In 1985 these channels are colonised by bush that grow to forest in the 1985–2000 period.

Table 4. Example of change matrix for one transition between the years 1967 and 1978, expressed as the percentage surface area change per ecotope type and total area for 1967

	Fcult	Ag	G	Fcl	Bcl	BS	Fo	Bo	P	H	MC	SC	Area (ha)
Fcult	36.98	47.11	0.53	1.43	6.90	0.00	0.00	0.00	0.05	6.99	0.00	0.00	51.59
Ag	1.18	97.28	0.86	0.18	0.27	0.02	0.02	0.00	0.02	0.12	0.04	0.01	954.96
G	0.00	1.39	62.20	0.88	2.17	19.99	0.44	2.92	3.50	0.85	5.30	0.36	148.72
Fcl	0.33	2.41	8.89	56.64	15.46	4.79	4.87	1.80	0.75	2.35	1.72	0.00	49.25
Bcl	0.00	0.60	18.30	10.63	45.79	1.41	0.10	13.20	0.56	2.16	7.20	0.05	59.56
BS	0.11	8.84	18.26	2.24	4.42	32.55	0.66	2.23	3.35	0.89	26.35	0.10	148.48
Fo	0.00	0.00	30.21	16.23	14.29	10.73	2.77	13.98	11.17	0.62	0.00	0.00	5.69
Bo	0.00	0.02	55.08	3.43	2.39	1.43	1.45	20.05	0.89	9.61	5.66	0.00	13.48
P	0.00	8.96	9.22	0.00	0.00	41.07	0.00	0.09	4.34	0.28	36.05	0.00	8.82
H	3.94	18.53	13.59	11.02	10.26	18.53	0.00	3.38	0.00	1.43	19.30	0.02	24.87
MC	0.08	8.43	23.64	2.04	6.86	21.65	0.49	1.91	3.53	0.05	31.10	0.21	57.45
SC	0.64	66.19	0.70	0.81	0.03	10.73	0.00	0.02	0.00	0.00	17.95	2.92	16.07

Fcult, Cultivated Forest; Ag, Agriculture; W & BS, Water and Bare soil; P, Pioneer vegetation; G, Grassland; H, Herbaceous vegetation; Bo, Open Bush; Fo, Open Forest; Fcl, Closed forest; Bcl, Closed bush; SC, Side channel.

1978 (column headers). Table 5 shows the ecotope transition rates for all time steps and standardised to a 5-year period. The four most dynamic ecotopes with more than 50% change per 5 years were open forest, open bush, pioneer vegetation, and herbaceous vegetation. Next to the surrounding

Table 5. Ecotope transition rates: percentage change to another ecotope for every map transition and standardised to a 5-year period. The data is numerically arranged based on the mean ecotope transition rate

Ecotope	Time span (years)					Mean	SD
	54–60	60–67	67–78	78–85	85–00		
Agriculture and cultivated forest	1.4	2.9	1.7	0.9	0.5	1.5	0.9
Main channel & bare soil	13.8	19.9	16.7	25.3	14.5	18.0	4.7
Forest (closed)	12.8	35.1	18.3	39.7	16.2	24.4	12.2
Grassland vegetation	47.3	13.5	20.7	40.9	19.3	28.3	14.8
Side channel	64.2	21.2	29.5	25.1	13.2	30.6	19.7
Bush (closed)	58.4	37.5	24.1	39.2	18.6	35.6	15.5
Bush (open)	58.3	46.5	39.7	56.6	29.6	46.1	12.0
Herbaceous vegetation	82.8	37.6	44.1	71.4	30.0	53.2	22.7
Forest (open)	81.1	71.4	44.0	52.4	31.9	56.1	20.0
Pioneer vegetation	77.5	68.0	44.2	70.6	33.3	58.7	18.9

SD, Standard deviation.

cultivated area, the main channel and point bar showed the lowest percentage of change and variability. Transition rates between the years 1954–1960 and 1978–1985 were higher than for other time spans.

The results of the visualisation of ecotope dynamics in ternary plots are presented in Fig. 6. Each data point represents the change of an ecotope in the period that lies between two successive maps. The most apparent example is the cultivated area, of which >95% of the surface area remained stable for each successive time span; all data of this ecotope type clearly show in the top corner of the ternary plot. The main channel and closed forest are opposites; their values lie, respectively on the succession axis and on the rejuvenation axis. Grassland and closed bush had a relatively low tendency for succession (<30%). They remained stable (>40%) or rejuvenated (>30%). The open bush ecotope varied in stability and succession, but rejuvenation remained constant around 40%. The open forest type, the pioneer vegetation and herbaceous vegetation showed low stability (<10%) and similar tendencies for succession and rejuvenation. The most diverse type in terms of succession, rejuvenation and stability was the side channel ecotope.

Floodplain and ecotope age

Figure 7 shows the year of last rejuvenation of the riparian area since 1954. Figure 8 shows the age distribution of the total floodplain area and of each ecotope in the year 2000. The age class >46 years consisted of the natural floodplain area that was not rejuvenated within the 46-year period

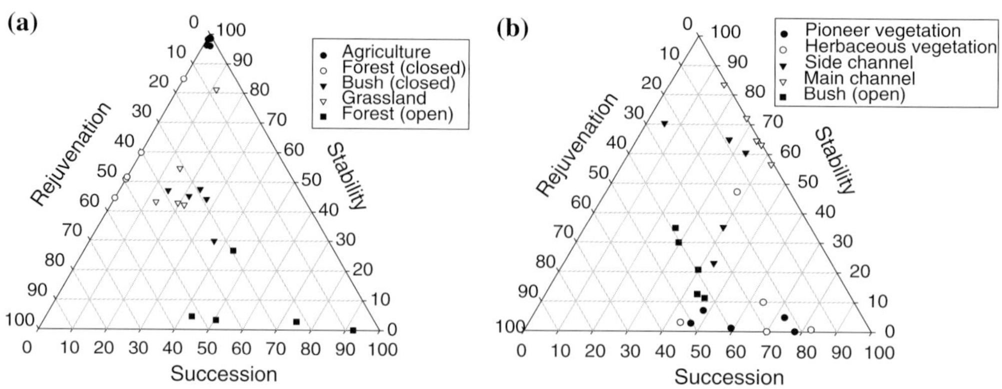

Figure 6. Ternary plots of ecotope stability, rejuvenation and succession.

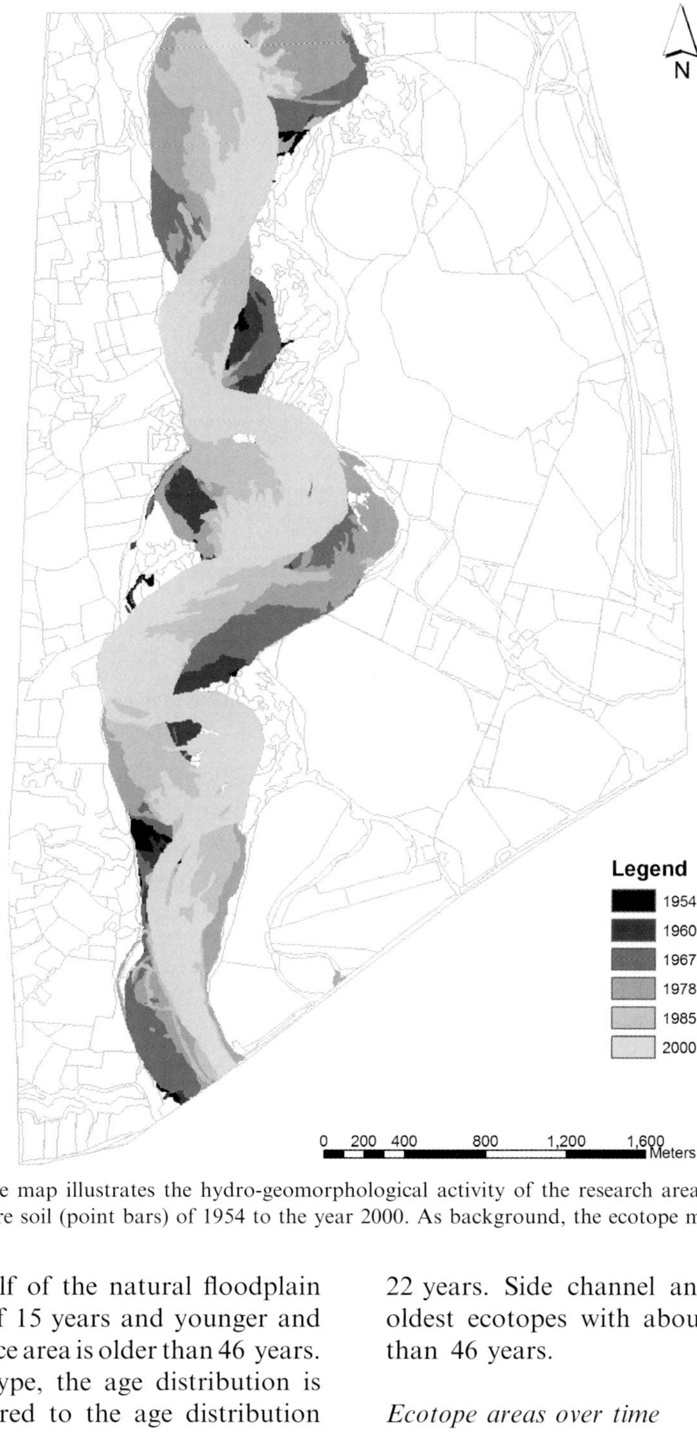

Figure 7. The floodplain age map illustrates the hydro-geomorphological activity of the research area by overlays of the ecotopes active main channel and bare soil (point bars) of 1954 to the year 2000. As background, the ecotope map of the year 2000 is used.

of the map series. Half of the natural floodplain consists of ecotopes of 15 years and younger and about 24% of the surface area is older than 46 years. Viewed per ecotope type, the age distribution is different when compared to the age distribution of the entire area. The youngest ecotope type is pioneer vegetation; more than 80% of its area is younger than 15 years. Grassland, herbaceous vegetation and open bush form an intermediate group with 50–60% of their area younger than 22 years. Side channel and closed bush are the oldest ecotopes with about half their area older than 46 years.

Ecotope areas over time

The temporal variation in the surface area coverage of different ecotope types is shown in Figure 9 and Table 6. The surface area of natural ecotopes (Table 1) vs. the surface area of cultivated

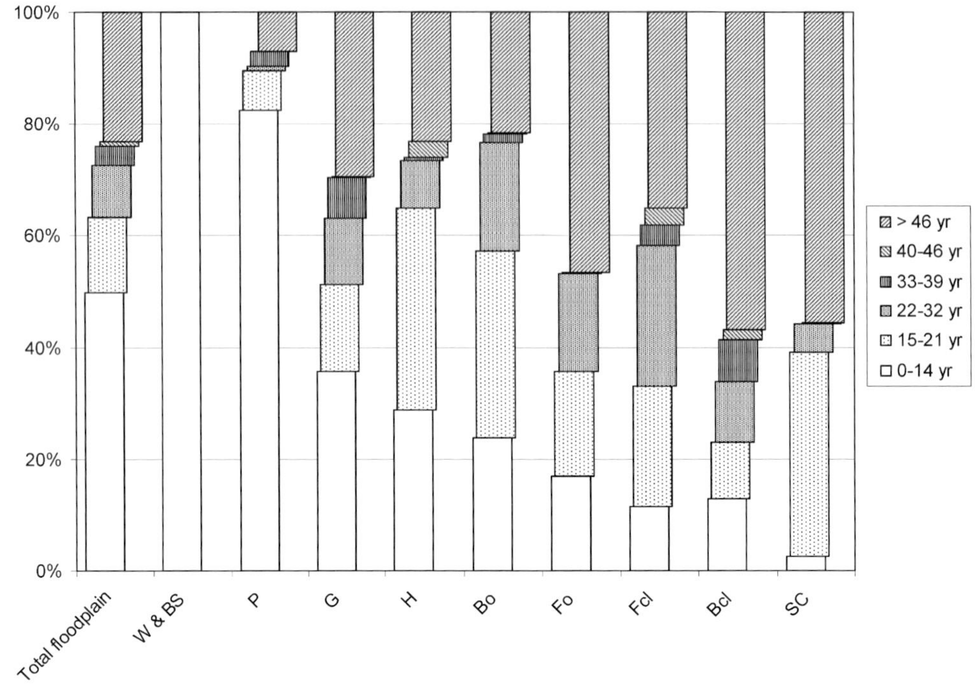

Figure 8. Floodplain and age distribution of natural ecotope types: W & BS, Water and Bare soil; P, Pioneer vegetation; G, Grassland; H, Herbaceous vegetation; Bo, Open bush; Fo, Open forest; Fcl, Closed forest; Bcl, Closed bush; SC, Side channel.

ecotopes changes on the local scale (Figs. 4 and 5) but fluctuates during the years at the river stretch scale only within a 10% range around a mean of 507 ha (see totals of Table 6). Grasslands and bare soil are the most variable, especially in the years 1954, 1960 and 1967, while for example the surface area of side channels is relatively stable. A decrease of open vegetation types like pioneer vegetation, grassland, herbaceous vegetation in favour of the closed types like

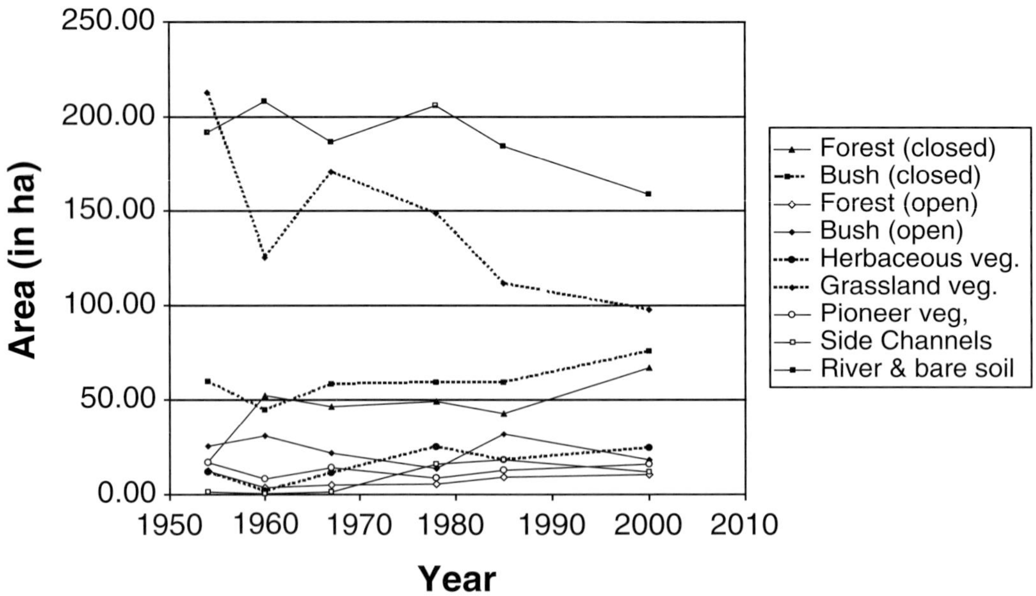

Figure 9. Total ecotope surface area (ha) over the period 1954–2000.

Table 6. The surface area of natural ecotopes and total natural floodplain (ha)

Ecotope	1954	1960	1967	1978	1985	2000
Forest (closed)	17.44	52.52	46.73	49.25	42.67	67.28
Bush (closed)	59.65	44.84	58.45	59.47	59.34	75.60
Forest (open)	12.20	3.63	4.91	5.67	9.12	10.62
Bush (open)	25.35	31.09	21.89	13.51	31.99	18.31
Herbaceous vegetation	11.91	1.94	11.50	24.89	18.08	24.56
Grassland vegetation	212.70	125.40	170.50	148.82	111.70	97.46
Pioneer vegetation	16.78	8.25	14.32	8.82	12.76	15.83
Side channel	1.58	0.61	1.39	16.05	18.45	11.86
Main channel and bare soil	191.80	208.25	186.50	205.91	184.27	158.78
Total	549.40	476.53	516.19	532.38	488.37	480.30

bush and forest is visible. In 1954, 79% of the research area was open, in 1978 76% and 64% in 2000. The drop in area of grassland vegetation between 1954 and 1960 was caused mainly by transition to agricultural area (34.5%, data not shown, but see Table 2 for years 1954 and 1960).

Ecotope diversity and scale

Figure 10 shows the landscape diversity of the study area, expressed as Shannon Index (SI), as a function of scale. The variation in SI values decreases when sliding from ecotope to river stretch scale. For the year 2000, the ecotope diversity remained stable if the floodplain surface area was about 250 ha, i.e., about 1.5 meander lengths. This seems to hold for the 1985 and 1978 results, but the 1954, 1960 and 1967 show an upward trend of SI values within the research area and no real stabilisation. An overall temporal trend of the SI values is also clearly visible, in time the overall landscape diversity is increasing.

Discussion

Mapping and GIS-analyses

The spatio-temporal heterogeneity of a meandering part of the Allier river was studied by analysing ecotope composition and dynamics using a series

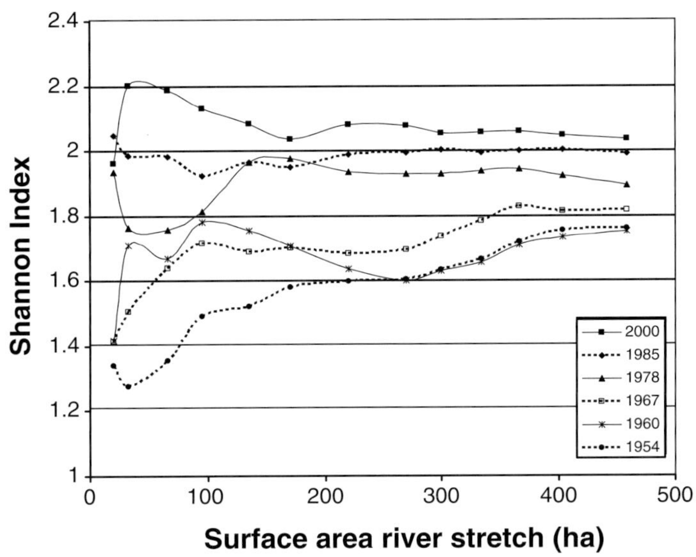

Figure 10. Landscape diversity in relation to the surface area of the river stretch that was used for calculation.

of aerial images covering a period of 46 years. Ecotopes were mapped starting with the aerial photograph of 2000 and retracing the changes in ecotope borders through time. This procedure worked well to overcome small geo-rectification differences of the different aerial photograph years. The overall quality of the aerial images was good but the quality and interpretability of early photos (1954, 1960) determined to some extent the resolution of the ecotope classification system.

The digitising process was optimised by a combination of digitising on screen and stereoscopic verification. In previous methods, the aerial images were viewed with a stereoscope and the ecotopes were traced on overlaid transparencies. Consequently, the minimal mapping unit depended on the trace-pen width. Subsequently the transparencies had to be scanned, geo-referenced and vectorised. Furthermore, before polygon vectorisation could start, the scan had to be checked and corrected manually for unclosed polygons using a drawing programme such as photoshop. This whole process was rather laborious and was shortened by digitising on screen. The verification and labelling of the on-screen digitising result was done by overlaying the digitised polygons (printed on transparencies) on top of the original aerial images under a stereoscope. In this way, the advantage of stereoscopic interpretation was kept.

Ecotope maps

The local dynamics are influenced by the succession speed of a particular ecotope and the local acting hydro-geomorphological processes. Figures 4 and 5 show the processes at work in the evolution of two small parts of the research area: rejuvenation of older succession stages by lateral erosion of outside bends, formation of new succession stages, formation of a side channel, and colonisation by vegetation of former channels. Figure 5 is a good example of the expansion and contraction events that steer riverine landscape heterogeneity (Tockner et al., 2000). The retracting water level followed the former channels in the point bar while seed dispersal took place and so steered the spatial distribution of vegetation settlement.

Ecotope dynamics

The mean ecotope transition rates (Table 5) follow the succession scheme illustrated in Figure 3 with dynamic ecotopes close to the main channel and less dynamic ecotopes to the climax stages, i.e., pioneer with the highest mean transition rate and closed forest with a relatively low mean transition rate. Two exceptions are grassland vegetation and open forest. Grassland is less dynamic than ecotope bush ecotope, probably because in the past the grasslands in the floodplains were used for grazing, so succession to open bush or open forest was inhibited. The open forest is relatively dynamic because in effect it is a mixed ecotope. Close to the river the ecotope type open forest consists of dynamic patches of young pioneer forest, so called softwood forest, and on well developed older stages it consists of low dynamic patches in succession to hardwood forest.

The ecotope transition rates in this study vary between 18 and 59% per 5 years. The mean rejuvenation rate is 33.8 ha per 5 years along the 5.5 km stretch of the study area. Studies presenting comparable values are scarce. As can be expected, the ecotope dynamics are lower when compared to dynamics in a braided alpine river where 80% of major landscape elements are rejuvenated within 3 years (Ward et al., 2001). A study on the river Ain (France) along a 40 km stretch of this river showed that rejuvenation rates decreased from about 100 ha per 10 years per 40 kms in the period 1945–1965 to 30 ha per 10 years per 40 kms in the period 1985–1991 (Marston et al., 1995). This river has a slightly lower mean annual discharge (130 $m^3 s^{-1}$) than the Allier. Between 1945 and 1991, the river dynamics decreased resulting in a single thread meandering river.

The transition rates of 1954–1960 and 1978–1985 transition are relatively high compared to the other years. In the period 1954–1960 the river channel was very active in the northern half of the research area. The limited availability of data on external pressures and influences that may explain this increase in activity, impede a satisfactory explanation. Possible explanations are listed below.

(1) A peak flow could be the cause, but discharge data on this period is not available for this

study, although in the Ubaye river in the Southern Alps about 400–500 km from the Allier catchment, a millennium flood is recorded in 1957 (Piégay & Salvador, 1997).

(2) An important factor is the sediment balance in the system; it can affect meander progression (Kondolf et al., 2002; Millar, 2005).

(3) The high activity could be a downstream geomorphological effect of the main channel running into a natural fixed bank and slowly passing this point in 1954–1967 (Fig. 4).

(4) The meander progression is increased when river banks consist of agricultural grounds (Micheli et al., 2004). The meander, shown in Figure 4, flows past agricultural area in the outer bend.

The increased dynamics in the 1978–1985 period can be attributed to the bridge effect (discussed later) and to the accumulation of major flood events in the early 80s (Fig. 2: January and December 1981; January and October 1982; April and May 1983; May 1985).

Floodplain and ecotope age

As a consequence of the spatial distribution of rejuvenation in the floodplain as shown in Figure 7, the ecotopes present are spatio-temporally distributed (Fig. 8). This spatio-temporal distribution is a characteristic of the steady-state mosaic or meta-climax. Figure 7 also shows the separate and combined effect of rejuvenation and succession. The floodplain age shows the age distribution caused by hydro-geomorphological processes and without ecotope succession. Due to ecotope succession the ecotope-age distribution of separate ecotopes is different as compared to total floodplain age composition. For example, half of the total riparian area is younger than 15 years; the ecotope closed-forest is almost for 90% situated on parts older than 15 years.

In Figure 8 the order of the succession scheme (Figs. 3 and 6) can be identified. Generally, the ecotopes having lower transition rates are relatively abundant on the older floodplain parts. Interesting is the ecotope-age distribution of open forest, which was classified as a dynamic ecotope with low stability based on the transition rates. However, seemingly contradicting the dynamic nature of this ecotope, more than 40% of the ecotope is found on older grounds. But, on older parts, the ecotope is a recent development because the older areas are being colonised by trees, i.e., in succession to (hardwood) forest stages via the open forest stage. Unfortunately, photo interpretation did not permit recognition of different types of open forest.

Ecotope areas over time

As shown on the local scale, ecotopes are dynamic (Table 4, Fig. 6), shifting in space through time (Figs. 4, 5 and 7). Within the river stretch or functional sector the overall ecotope distribution is less dynamic (Fig. 9), as assumed by the steady-state mosaic or meta-climax hypotheses (Forman & Godron, 1986; Amoros & Wade, 1996).

A true (theoretical) steady state (or meta-climax) within a stretch homogeneous in processes and environment would show as a stable ecotope distribution time series. However, our study shows a general trend in decrease of the proportion of open, low structure ecotopes towards an increase of structure rich ecotopes, such as forest and bush (Fig. 9). This trend in the ecotope distribution is caused by long-term changes of acting processes. Most probably a decrease of the grazing intensities. The area became a nature reserve in the 1994 and all grazing was phased out.

Another bias is the construction of the bridge near Chemilly, just south to the research area. Although the meander pattern recovered (Figure 4, Table 3), the exact influence of the bridge near Chemilly is not known. It can be hypothesised that what the shift accomplished is similar to a major flood event, though now induced by human intervention of narrowing the channel downstream by building a bridge and short cutting the first meander (Wilbers, personal communication) and simultaneously a flood occurrence in 1976 (1,020 $m^3 s^{-1}$). This channel shift created niches for various vegetation types, e.g., a poplar settlement. Together with lower grazing intensities, this can explain the increase in bush ecotope in 1985 and in 2000 the increase in forest ecotope (poplar becoming higher than 5 m) found in Figure 9.

In general, over medium time scales (10–100 years) most river systems can be viewed as

Millar, R. G., 2005. Theoretical regime equations for mobile gravel-bed rivers with stable banks. Geomorphology 64: 207–220.

Miller, J. R., T. T. Schulz, N. T. Hobbs, K. R. Wilson, D. L. Schrupp & W. L. Baker, 1995. Changes in the landscape structure of a southeastern Wyoming riparian zone following shifts in stream dynamics. Biological Conservation 72: 371–379.

Mount, N. J., P. M. Zukowskyj, R. M. Teeuw & T. A. Stott, 2002. Use of aerial photography and digital photogrammetry in the assessment of river channel destabilisation. In Leuven, R. S. E. W., I. Pourdevigne, & R. M. Teeuw (eds), Application of Geographic Information Systems and Remote Sensing in River Studies. Backhuys Publishers, Leiden: 41–62.

Muller, E., 1997. Mapping riparian vegetation along rivers: old concepts and new methods. Aquatic Botany 58: 411–437.

Narumalani, S., R. M. Deepak & R. G. Rothwell, 2004. Change detection and landscape metrics for inferring anthropogenic processes in the greater EFMO area. Remote Sensing of Environment 91: 478–489.

Nienhuis, P. H., A. D. Buijse, R. S. E. W. Leuven, A. J. M. Smits, R. W. J. de Nooij & E. M. Samborska, 2002. Ecological rehabilitation of the lowland basin of the river Rhine (NW Europe). Hydrobiologia 478: 53–72.

Nienhuis, P. H. & R. S. E. W. Leuven, 2001. River restoration and flood protection: controversy or synergism. Hydrobiologia 444: 85–99.

Petts, G. E. & C. Amoros, 1996. Fluvial Hydrosystems. Chapman & Hall, London.

Photothèque-Nationale, 2003. Aerial images of years 1954, 1960, 1967, 1978, 1985, 1992 and 2000, Photothèque Nationale, 2/4 avenue Pasteur. Saint-Mandé Cedex.

Piégay, H. & P.-G. Salvador, 1997. Contemporary floodplain forest evolution along the middle Ubaye river, Southern Alps, France. Global Ecology and Biogeography Letter 6: 397–406.

Prach, K. & P. Pysek, 2001. Using spontaneous succession for restoration of human-disturbed habitats: experience from Central Europe. Ecological Engineering 17: 55–62.

Smits, A. J. M., H. Havinga & E. C. L. Marteijn, 2000. New concepts in river and water management in the Rhine river basin: how to live with the unexpected?. In Smits, A. J. M., P. H. Nienhuis & R. S. E. W. Leuven, (eds), New Approaches to River Management. Backhuys Publishers, Leiden: 267–286.

Tockner, K., F. Malard & J. V. Ward, 2000. An extension of the flood pulse concept. Hydrological Processes 14: 2861–2883.

Van den Berg, J. H. & T. Balyuk, 2004. Interaction of Vegetation and Morphodynamics in Pointbars of the Lower Volga (Russia) and the Allier (France). Inter-university Centre for Geo-Ecological Research, Utrecht.

Van der Nat, D., K. Tockner, P. J. Edwards, J. V. Ward & A. M. Gurnell, 2003. Habitat change in braided floodplains (Tagliamento, NE-Italy). Freshwater Biology 48: 1799–1812.

Van Stokkom, T. C., A. J. M. Smits & R. S. E. W. Leuven, 2005. Flood defense in the Netherlands, a new era, a new approach. Water International 30: 76–87.

Vulink, J. T., 2001. Hungry Herds. Management of Temperate Lowland Wetlands by Grazing. PhD thesis, Rijkswaterstaat Directie IJsselmeergebied/Rijksuniversiteit Groningen.

Ward, J. V., K. Tockner, U. Uehlinger & F. Malard, 2001. Understanding natural patterns and processes in river corridors as the basis for effective river restoration. Regulated Rivers: Research & Management 17: 311–323.

Wilbers, A. W. E., 1997. The Allier, a river with two patterns. MSc Thesis Utrecht University, Utrecht (In Dutch).

Wolfert, H. P., 2001. Geomorphological Change and River Rehabilitation. PhD Thesis Wageningen University, Alterra Green World Research, Wageningen.

Hydrobiologia (2006) 565:87–103
R.S.E.W. Leuven, A.M.J. Ragas, A.J.M. Smits & G. van der Velde (eds), Living Rivers: Trends and Challenges in Science and Management
DOI 10.1007/s10750-005-1907-5

Airborne laser scanning as a tool for lowland floodplain vegetation monitoring

M.W. Straatsma* & H. Middelkoop
Department of Physical Geography, Utrecht University, P.O. Box 80115, 3508 TC Utrecht, The Netherlands
*(*Author for correspondence: Tel.: +31-302533888; Fax: +31-302531145; E-mail: m.straatsma@geo.uu.nl)*

Key words: airborne laser scanning, floodplain vegetation, vegetation structure, model input

Abstract

Monitoring of three-dimensional floodplain vegetation structure is essential for ecological studies, as well as for hydrodynamic modelling of rivers. Height and density of submerged vegetation and density of emergent vegetation are the key characteristics from which roughness parameters in hydraulic models are derived. Airborne laser scanning is a technique with broad applications in vegetation structure mapping, which therefore may be a promising tool in monitoring floodplain vegetation for river management applications. This paper first provides an introduction to the laser scanning technique, and reviews previous studies on the extraction of vegetation height and density of forests, low vegetation and meadows or unvegetated areas. Reliable predictions using laser scan data have been reported for forest height ($R^2 = 0.64$–0.98), parameters related to forest density, such as stem number, stem diameter, biomass, timber volume or basal area ($R^2 = 0.42$–0.93), and herbaceous vegetation height (summer condition; $R^2 = 0.75$–0.89). No empirical relations have been reported on density of herbaceous vegetation. Laser data of meadows and unvegetated areas show too much noise to predict vegetation structure correctly. In a case study for the lower Rhine river, the potential of laser scan mapping of vegetation structure was further explored for winter conditions. Three laser-derived metrics that are often reported in the literature have been applied to characterize local vertical distributions of laser reflections. The laser data clearly show the large structural differences both between and within vegetation units that currently are the basis of floodplain vegetation and roughness mapping. The results indicate that airborne laser scanning is a promising technique for extraction of 3D-structure of floodplain vegetation in winter, except for meadows and unvegetated areas.

Introduction

In the forthcoming decennia, the landscape of the lower Rhine and Meuse floodplains in the Netherlands may undergo major changes. The near-flooding events of 1993 and 1995 combined with the anticipated increase in peak flows due to climate change (Shabalova et al., 2003) have raised the awareness that the discharge capacity of the high-water bed of the rivers should be increased. While in previous decennia this has been done by raising the river dikes, a new flood management strategy has been adopted in recent years in the Netherlands, which involves accommodating more room for the river by lowering the floodplain surface, digging side channels, removing small embankments, and various other landscaping measures (Silva et al., 2001; Van Stokkom et al., 2005). At the same time, river management aims at ecological rehabilitation of the floodplains, which involves the restoration of various natural floodplain habitats, thereby enhancing natural biodiversity (Ward & Tockner, 2001; Nienhuis et al., 2002). To achieve these objectives – with safety as the primary boundary condition – a strategy of dynamic floodplain management is advocated (Nienhuis & Leuven, 2001). Accordingly, flood reduction measures will be carried out as an artificial form of floodplain rejuvenation, followed by a period of renewed sedimentation and vegetation development (Baptist et al., 2004). Over the years, this will lead to a higher floodplain level, as well as to an increase of hydrodynamic roughness exerted by the denser vegetation. By the time this would cause excessively high flood water levels, artificial rejuvenation of the floodplain will be undertaken again.

This management practice will lead to a high spatio-temporal variability of floodplain habitats, vegetation structure and related roughness, and demands adequate monitoring techniques. Spatial distribution of vegetation structure is also needed as input for ecological habitat modelling; for example, tree height and the full three dimensional structure of the forest are important parameters for bird habitat modelling (MacArthur & MacArthur, 1961). Detailed information of vegetation structure is needed to estimate the changes in hydrodynamic roughness of the floodplain. Furthermore, hydrodynamic roughness is a key variable required for modelling deposition of suspended sediment and associated contaminants within the floodplain.

Currently, vegetation mapping of the lower Rhine and Meuse floodplains is based on ecotopes. Ecotopes are 'spatial landscape units that are homogeneous as to vegetation structure, succession stage and the main abiotic factors that are relevant to plant growth' (Leuven et al., 2002). Mapping of ecotopes within the lower Rhine floodplain is based on visual interpretation and manual classification of vegetation units from aerial photographs, scale 1:10,000 (Jansen & Backx, 1998). Ecotopes are then linked to vegetation structural characteristics related to roughness using a lookup table (Van Velzen et al., 2003). This method, however, may become inadequate to monitor the spatio-temporal dynamics of vegetation roughness, since the procedures are time consuming and do not allow documentation of within-ecotope variation of vegetation roughness.

There is clearly a need for a more automated approach to assess hydrodynamic roughness of vegetated floodplain surfaces. Airborne remote sensing is regarded a useful technique for surveying and mapping floodplain vegetation. While successful attempts have been reported to map wetland vegetation using (hyper-) spectral remote sensing data (Thompson et al., 1998; Schmidt & Skidmore, 2003; Van der Sande et al., 2003), airborne laser scanning may provide a useful technique to map 3D vegetation structure as a measure of hydrodynamic roughness. Airborne laser scanning is an active remote sensing technique that involves high-resolution elevation measurement by means of laser pulses. Over the past years airborne laser scanning has become a widely used technique in the generation of digital elevation models (Oude Elberink et al., 2003) and in forest structure mapping, while experiments on mapping of shrub and herb vegetation have been reported as well. A few recent examples have reported on the application of laser scan data in lowland floodplains to map topography (Marks & Bates, 2000) and roughness, for use as input of hydrodynamic models (Cobby et al., 2001).

The objective of this paper is to review the potential of airborne laser scanning data as a tool for quantitative assessment of floodplain vegetation roughness, which is related to vegetation height and density. First we will introduce the parameters needed for hydrodynamic vegetation roughness. Secondly, we introduce the laser scanning method and review current applications related to vegetation structure, where we will make a distinction between (1) forest, (2) shrubs and herbaceous vegetation and (3) meadows and unvegetated areas. Finally, we present a case study of laser scanning of floodplain vegetation.

Hydrodynamic vegetation roughness

Hydrodynamic vegetation roughness refers to the resistance force exerted by vegetation on water flowing over or through it. Rough vegetation reduces water flow velocity and leads to higher water levels and thus increases flood risks. A meadow is hydrodynamically smooth, forests and dense shrubs are hydrodynamically rough (Chow, 1959). Two types of vegetation are distinguished with respect to roughness modelling; emergent and submerged. The parameter describing emergent vegetation, such as forest, is vegetation density (Petryk & Bosmajian, 1975). Hydrodynamic vegetation density is the sum of the projected plant areas (A) in the direction of the flow (F) per unit volume (cube) (Fig. 1). The unit is $m^2\ m^{-3}$, which reduces to m^{-1}. Under the assumption that vegetation consists of cylindrical elements, vegetation density is calculated as the product of number of stems per square metre and stem diameter, which also reduces to m^{-1}. The parameters describing submerged vegetation, such as grassland, reed, or herbs are vegetation height and density (Carollo et al., 2002). The spatial distribution of these vegetation characteristics is an essential input for hydrodynamic models. Therefore, the challenge is to estimate vegetation density for emergent vegetation and vegetation height and density for submerged vegetation from laser scanning data.

Airborne laser scanning technique

All airborne laser systems have three components in common (Wehr & Lohr, 1999; Fig. 2):

1. Differential Global Positioning System (dGPS) to locate the aircraft in space;
2. Inertial Navigation System (INS) to determine the orientation of the aircraft;
3. Laser Range Finder (LRF) to determine the distance between the aircraft and objects below.

Differential GPS consists of 2 components; a base station and a mobile GPS receiver. GPS is a satellite positioning system. The base station

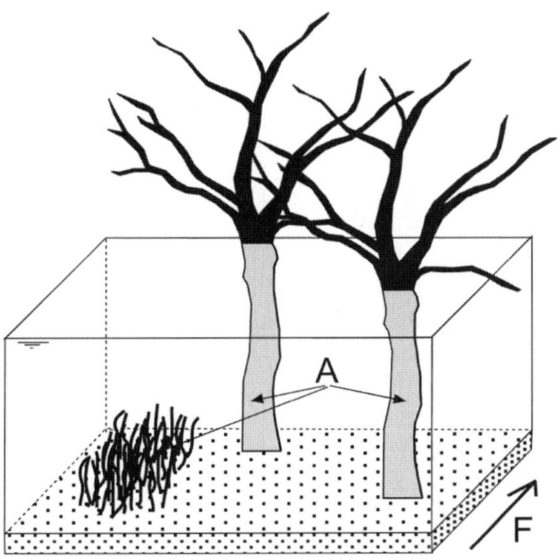

Figure 1. Hydrodynamic vegetation density in $m^2\ m^{-3}$: the sum of projected plant areas (*A*) in the direction of the water flow (*F*) per unit volume (cube).

remains immobile, hence the positioning error can be calculated, which enables correction of the positioning error of the aircraft. The INS measures acceleration of the aircraft along three orthogonal axes enabling the computation of the pitch, yawn and roll. Data of all three components are synchronized to enable the determination of the location and elevation of the measured object in a local coordinate system.

Two types of LRF exist: (1) Continuous wave (CW) ranging, in which the distance to the target is determined from the phase difference of the return pulse of a continuously emitted laser signal. Since this technique is not widely applied (Baltsavias, 1999b) it will not be discussed here. (2) Pulse LRF that fires discrete laser pulses of a few nanoseconds to the earth. The distance is calculated from the pulse travel time between emission and return of the pulse. The laser pulse typically has a wavelength in the near infrared (i.e. 1064 nm; Baltsavias, 1999a). Pulse frequencies range from 0.1 to 100 kHz.

The horizontal distribution of the laser hits on the ground depends on the scan pattern of the LRF. Early LRF systems used vertical profiling lasers, resulting in closely spaced points underneath the aircraft only. Modern systems deflect the laser pulses across track, resulting in a wider distribution of points below the aircraft. The scan pattern depends on the motion of the deflection device and aircraft speed (Wehr & Lohr, 1999). Figure 2 shows a saw-tooth scan pattern. The footprint size refers to the area on the surface that is illuminated by a single laser pulse: a larger footprint increases the chance of hitting more than one object with a single pulse.

The return signal can be recorded in different ways (Fig. 3): (1) waveform digitizing (Blair et al., 1999; Means et al., 1999; Wagner et al., 2004) and (2) discrete return (Baltsavias, 1999b). A waveform-digitizing scanner records the full reflection profile of the returned laser pulse in a series of equally spaced time bins. Conversely, a discrete return system records only discrete levels in the vertical profile. Typically, discrete return laser scanners record first pulse (FP) and last pulse (LP), but some systems are able to record up to 6 intermediate pulses (IMP). Last pulse is more likely to represent the ground surface, first and intermediate pulse will be reflected from the canopy. However, a minimum vertical object separa-

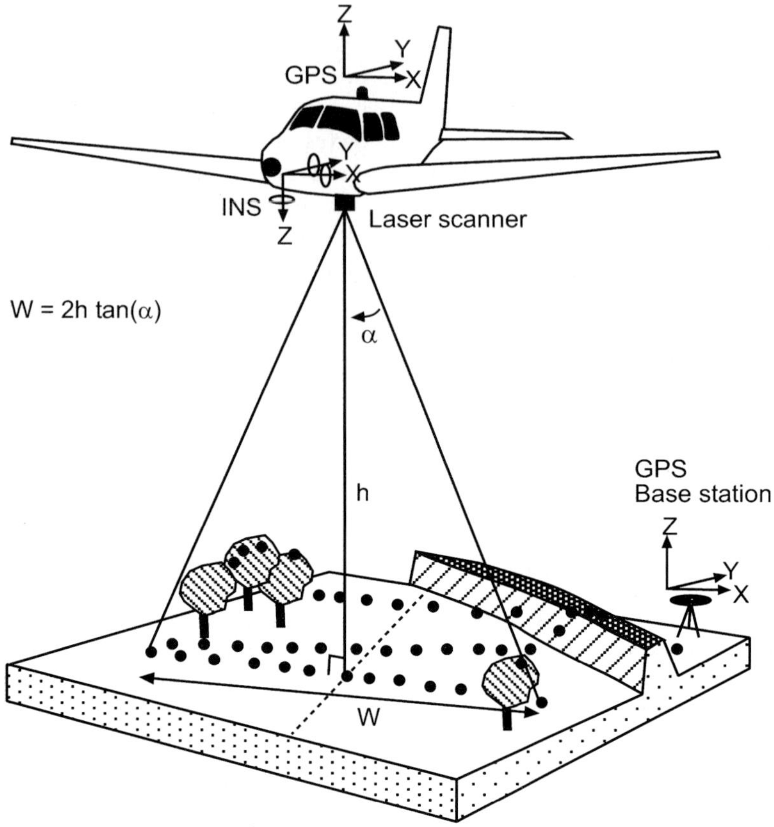

Figure 2. Components of an airborne laser scanning system: (1) GPS for positioning in space plus a base station for error correction, (2) INS, Inertial Navigation System, for recording the orientation of the aircraft in space, (3) Laser scanner for measuring the distance between the aircraft and the objects below it. When combined, a 3D point cloud of laser hits is generated.

tion of about 1.5 m is needed to register first and last pulse separately, otherwise these two pulses will merge into one (Baltsavias, 1999a). Echo detection refers to the reflection intensity at which a pulse is registered. This determines which height of the returned waveform is recorded (Katzenbeisser, 2003; Fig. 3). The reflection intensity for herbs in winter is slowly increasing towards the ground surface. The exact height recorded by the laser scanner depends on the detection threshold that is used. Varying this threshold may cause large differences in detected vegetation height.

Waveform digitizing scanners often use large footprints, typically with 10–25 m diameters (Blair et al., 1999; Lefsky et al., 1999a). In recent years, little waveform-digitizing data was available. NASA deployed such scanners in preparation for the space borne Vegetation Canopy Lidar (VCL), but they were not commercially available (Baltsavias, 1999b). Recently, a commercial waveform-digitizing scanner has become operational (Hug et al., 2004), which opens up new possibilities in vegetation mapping. Blair & Hofton (1999) successfully modelled large-footprint waveforms using small-footprint first pulse data that indicates that no major differences exist between the two types of scanners.

The minimum detectable object size is of specific importance when measuring senescent herbaceous vegetation. Ritchie et al. (1993) estimated the minimum detectable vegetation element, leaves in their case, to be 2–3 cm, which is still larger than the stalk diameter of herbs in winter. Baltsavias (1999a) listed a number of parameters influencing the minimum object size that can be detected: these include reflectivity, laser power, detector sensitivity and laser wavelength. Flying height also determines the minimum detectable object size.

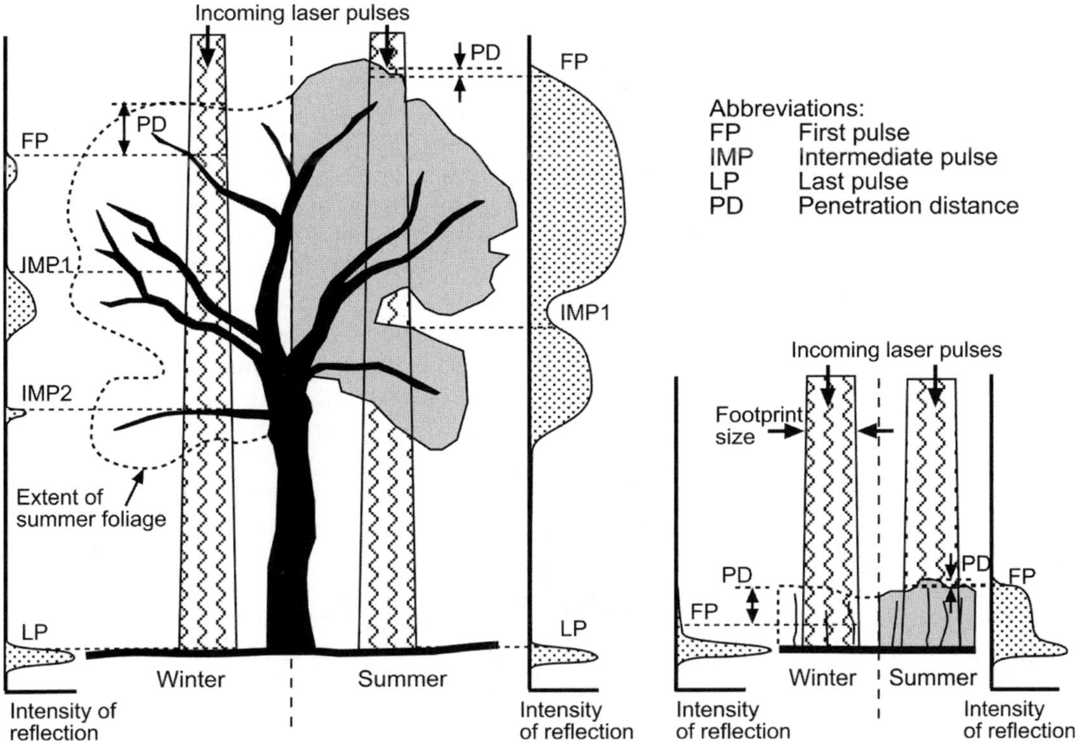

Figure 3. Seasonal effects on laser reflection from trees (left panel) and herbaceous vegetation (right panel). Reflection intensity is higher and penetration depth into the canopy is smaller in summer for both forests and herbaceous vegetation.

Due to beam divergence, the footprint will increase with flying altitude, while the energy per unit area will decrease. Nevertheless, Næsset (2004) found no significant difference in forest canopy height estimation when flying altitude varied by 60%, while using the same laser scanner. Conversely, Yu et al. (2004) did report that tree height was progressively under-estimated when the flying altitude increased by a factor 2 to 4.75. Water reflects little energy in the near-infrared (Lillesand & Kiefer, 1994). This usually leads to large errors in the height measurement of vegetation emerging above a water body because the water level is not detected properly and the water depth is unknown (Hopkinson et al., 2004a). This is particularly important for aquatic vegetation such as reed. However, also using the traditional methods the correct water depth will be difficult to predict.

Laser scanning monitoring of floodplain vegetation should be carried out in the period with the highest chance of flooding, which is winter in the Netherlands, but the floodplain should not be inundated during data acquisition, nor should the floodplains be snow-covered as this will cover part of the vegetation. The presence of leaves may strongly affect the laser return signal (Fig. 3). Leafs increase the echo intensity and limit the penetration distance into the vegetation before a significant return is generated. Although the effects of seasonal variations in foliage on the representation of the three-dimensional vegetation structure by laser scanning are widely known, still little efforts have been undertaken to quantify the consequences for the shape of the returned waveform or the vertical point distribution.

Review: airborne laser scanning of vegetation

The literature review is focused on the prediction of vegetation height and density of three main structural vegetation types in floodplains: (1) forest, (2) herbaceous vegetation and shrubs and (3) meadows and unvegetated areas. However, the first step is ground surface reconstruction to enable the calculation of the point distribution relative to the ground surface.

Calculation of relative vertical point distribution

The calculation of the vertical point distribution above the ground surface requires a Digital Terrain Model (DTM) describing the ground surface elevation. To determine the surface elevation using laser scanning, the laser data reflected from the ground surface must be separated from the laser data reflected from vegetation or other objects above the ground surface (Kilian et al., 1996; Kraus & Pfeifer, 1998; Vosselman, 2000). Sithole & Vosselman (2004) compared eight different algorithms used to segment the laser point cloud in ground surface and non-ground surface points. Most filter algorithms calculate a measure of discontinuity based on a local neighbourhood. They conclude that all methods perform well in smooth landscapes, but produce errors in rough terrain with dense vegetation. Still, vegetation in smooth terrain can deteriorate the quality of the resulting DTM (Pfeifer et al., 2004). Hodgson & Bresnahan (2004) report absolute errors in the DTM under various vegetation types of less than 20 cm. Since floodplain topography is relatively smooth, except ditches, embankments and eroding shorelines, it is expected that laser-based DTMs of the floodplain surface will be sufficiently accurate as a basis for vegetation height determination.

Once a DTM is derived from the laser data, the height above the ground surface can be calculated for each individual point. The result is a three-dimensional point cloud representing heights above ground level. This can be converted to a vertical point height distribution characterized by various statistics such as percentiles, measures of central tendency and shape. These statistics are subsequently related to vegetation structure characteristics, which are determined by field reference data (Lefsky et al., 1999a; Næsset, 2002).

Forest vegetation structure

Over the last 20 years many papers – mostly originating from forestry research – have reported on extracting forest properties, such as stand height, biomass, timber volume, stem number and stem diameter, from airborne laser scanning data. Lefsky et al. (2002) and Lim et al. (2003) provide reviews on this subject. Forest stand characteristics have been successfully extracted from laser data over a wide range of forest types and climate zones, from the tropics (Drake et al., 2002) and temperate zone (Ritchie et al., 1993; Means et al., 1999; Lefsky et al., 1999a, b) to the boreal zone (Nilsson, 1996; Næsset, 1997a, b; Magnussen & Boudewyn, 1998; Næsset, 2002; Morsdorf et al., 2004). Initial interest was in forest stand characteristics, but recently single tree mapping and tree species identification are topics of research (Brandtberg et al., 2003; Holmgren & Persson 2004). Most studies considered forest under leaf-on conditions.

Since the hydrodynamic roughness of forest is high, extracting floodplain forest stand characteristics from laser scanning data is highly relevant for river management. We will focus on those characteristics of forests that relate to hydrodynamic vegetation density.

Forest height

Estimates of forest height are mostly derived from percentiles of the laser data. It is then assumed that the vertical leaf area distribution in forests is the same as the vertical distribution of laser points (Magnussen & Boudewyn, 1998). Consequently, the *n*-percentile of the laser vertical point distribution would correspond to the height above the ground below which *n* percent of the leaf area occurs. The good correspondence between leaf area percentiles and laser data percentiles has been confirmed by various field sampling studies where height distributions were measured for a range of canopy types (Ritchie et al., 1993; Næsset, 1997a; Means et al., 1999; Næsset & Økland, 2002). Laser estimates were less than 6% different from field and laser estimates of canopy height. Table 1 shows explained variances based on regression analyses from research on forest height prediction.

Forest height can be expressed by different properties, including (1) mean tree height, (2) dominant height, i.e. the arithmetic mean of the 100 largest trees, (3) maximum tree height in a limited sized plot, or (4) Lorey's mean tree height, which is the weighted mean of the canopy height based on the basal area of individual trees, and is used to account for the influence of larger trees on the mean canopy height. The latter three parameters focus on harvestable timber and ignore small trees. The quality of the prediction of forest stand height depends on the way tree height is expressed, as well as on the percentiles used for prediction.

Table 1. Overview of explained variances obtained in the prediction of forest canopy height using various laser-derived statistics as independent variable and field canopy heights as dependent variable

Reference	Footprint size (m)	Height (R^2)	Number of reference data	Remarks
Ritchie et al. (1993)	small	0.98	8	laser profiler
Næsset (1997a)	0.13–0.17	0.94	36	laser scanner
Magnussen & Boudewyn (1998)	small	0.64	36	laser scanner
Means et al. (1999)	10.4	0.95	24	laser scanner
Lefsky et al. (1999a)	10.4	0.78	48	laser scanner
Næsset & Bjerknes (2001)	0.21	0.83	39	laser scanner
Næsset (2002)	0.21	0.74–0.95	144	laser scanner
Næsset & Økland (2002)	0.18	0.91	37	laser scanner
Hopkinson et al. (2004b)	small	0.86	54	laser scanner
Persson et al. (2002)	0.3–3.7	0.99	135	individual trees
Brandtberg et al. (2003)	0.1	0.68	48	individual trees
Morsdorf et al. (2004)	0.30	0.92	918	individual trees

Regression models obtained in a number of studies explain on average more than 80% of the variation in tree height measured in the field (Table 1). The studies do not agree on a single laser-derived statistic to predict forest height. Moreover, the correlation between mean tree height and laser derived statistics becomes weaker when the higher trees intercept most of the laser pulses, so that the smaller trees are not detected by laser scanning (Popescu et al., 2002). Næsset (2002), Næsset & Økland (2002) and Holmgren & Jonsson (2004) demonstrated that the overall quality of laser scanning prediction of forest height is better than conventional methods, such as manual interpretation of aerial stereo photos.

Individual tree height estimation has recently become topic of research. Persson et al. (2002) delineated individual coniferous trees using local maxima in a smoothed canopy model. Brandtberg et al. (2003) studied deciduous trees in leaf-off condition and used a scale-space technique which smoothes the canopy model at various scales. By identifying height clusters at various scales, the image was segmented to identify individual trees. Morsdorf et al. (2004) identified coniferous trees using cluster analyses on the raw 3D laser scanning data.

Forest vegetation density

Several characteristics of forest stands relate to vegetation density, such as stem number, stem diameter, basal area, biomass, or timber volume. Table 2 gives an overview of the prediction results for these characteristics using laser-derived parameters based on the vertical distribution of laser points as independent variables. Dependent variable was always field reference data. Number of stems can also be determined using the methods discussed in the section on individual tree delineation. This is likely to give a minimum value for the stem number, because overtopped trees and undergrowth vegetation are difficult to detect (Maltamo et al., 2004). Vegetation density should be computed over the lower part of the vegetation that will be inundated by the water. Hence, information is needed on specific height increments as provided by Lefsky et al. (1999b) and Næsset (2002).

Lefsky et al. (1999b) reconstructed the canopy height profile (CHP) from large footprint, waveform digitizing, laser-scanning data. The CHP is the 'surface area of all canopy material, woody and foliage, as a function of height'. CHP is based on method of MacArthur & Horn (1969) who assumed that the intensity of light, which travels through a forest canopy, shows an extinction curve depending on the occlusions from the vegetation. This was validated by Aber (1979). Their method computes the leaf area index (LAI) over a specified height interval. For laser scanning data the following formula can be used to compute the LAI between heights $h1$ and $h2$:

Table 2. Overview of hydrodynamic vegetation density and related forest stand properties based on linear regression analyses

Reference	FP size[a] (m)	D_v^a (R^2)	N^a (R^2)	D^a (R^2)	BA[a] (R^2)	B/TV[a] (R^2)	Np[a]
Nilsson (1996)	0.75–3.0		–	–	–	0.78	27
Næsset (1997b)	0.13–0.16		–	–	–	0.46–0.89	36
Means et al. (1999)	10.4		–	–	0.88	0.96	24
Lefsky et al. (1999a)	10.4		0.85	0.61	0.87	0.91	22
Lefsky et al. (1999b)	10.4		–	–	0.69	0.81	48
Næsset & Bjerknes (2001)	0.21		0.42	–	–	–	39
Næsset (2002)	0.21		0.5–0.7	0.39–0.78	0.69–0.89	0.8–0.93	144
Drake et al. (2002)	25		–	0.93	0.72	0.93	25
Asselman (2002)	0.17	0.6					30
Holmgren et al. (2003)	Small					0.82–0.9	?

Independent variable was a laser-derived parameter based on the vertical distribution of laser points.
[a]FP size = laser footprint size, D_v = hydrodynamic vegetation density, N = stem number, D = stem diameter, BA = basal area, B/TV = biomass/timber volume, Np = number of plots.

$$\mathrm{LAI}_{h1-h2} = \ln\left(\frac{N_{h2}}{N_{h1}}\right) \tag{1}$$

in which N_{h1} and N_{h2} are the number of points below heights $h1$ and $h2$. Lefsky et al. (1999b) found excellent agreement between ground and laser measurements of the CHP. Estimates of CHP and LAI are however *vertically oriented* descriptors, while hydrodynamic vegetation density is a measure of *horizontal* obstruction. Næsset (2002), using small footprint data, used canopy density statistics based on the vertical point distribution. In this case canopy densities are 'the proportions of the laser hits above the 0, 10, ... ,90 percentiles of the height distributions to the total number of pulses'. These statistics proved useful in predicting mean diameter by basal area, stem number, basal area and volume. A combination of two canopy densities, one at ground level and the second at the height of maximum water level during flood, might be strongly correlated to the hydrodynamic vegetation density. However, this method does not compensate for occlusions from the top of the canopy.

The methods and results reported from these forestry studies, however, do not explicitly relate to hydrodynamic vegetation density, as this was not the focus of these studies. Extraction of vegetation density associated to hydraulic roughness of vegetation from laser-scanning data has been reported only by Asselman (2002). She found the 90-percentile to be negatively correlated with floodplain vegetation density that included forest, reeds and shrubs. Apparently, tall trees more efficiently shade the ground layer and limit understory growth. However, tree height alone cannot explain all variation in vegetation density as understory vegetation does occur when the canopy is not closed. In conclusion we could say that laser scanning is a promising technique to map vegetation density because many related parameters have been predicted accurately. However many prediction methods remain to be tested for this application: LAI based on MacArthur & Horn (1969), canopy density based on Næsset (2002) and individual tree delineation.

Structure of herbaceous vegetation and shrubs

Compared to forestry research, few studies have been reported on the extraction of vegetation structure of low vegetation such as reed, natural grassland, herbaceous vegetation or low shrubs. Again we will discuss vegetation height and density as the hydraulically important variables to extract from the laser data.

Height of herbaceous vegetation

Vegetation height is the primary structural characteristic that determines the hydraulic roughness of submerged vegetation. Table 3 gives an overview of research on low vegetation types, in which empirical relations were established to predict vegetation height from laser-scanning data. Ritchie et al. (1993) studied low (0.1–0.3 m) desert shrubs using a laser profiler. In a similar approach

Table 3. Overview of papers on vegetation height of low vegetation with regression equations

Reference	Footprint size (m)	Height (R^2)	Number of plots	Vegetation type	Height range in the field (m)
Ritchie et al. (1993)	small	0.75	7	Desert shrubs	0.13–0.27
Davenport et al. (2000)	0.15–0.23	0.89	18	crops $h < 1$ m	0–0.9
Cobby et al. (2001)	0.2	0.8	55	grassland and crops	0–1.2
Hopkinson et al. (2004b)	small	0.77	14	aquatic, grass, herbs, low shrubs	0–1.25

as used for forested areas, they established empirical relations between percentiles of laser data and vegetation height measured in the field. The 95 percentile of the vertical distribution of the laser points explained 75% of the variation of the vegetation height. Weltz et al. (1994) and Ritchie et al. (1996) compared the vertical point distribution of laser data with the vertical distribution of vegetation height measured in field plots, and found good agreement between the histograms of the laser data and vegetation height distributions.

Various studies used the standard deviation of the laser data in a local window as a predictor for vegetation height. Davenport et al. (2000), and Cobby et al. (2001) used this statistic for grasslands and agricultural crops, Hopkinson et al. (2004b) for shrubs, aquatic marshland vegetation, grassland and herbs. The regression equations established in these studies varied greatly. Cobby et al. (2001) used a log-linear regression, which did not give satisfactory results on the data of Hopkinson et al. (2004b). Moreover, the slope of the regression equation of Hopkinson et al. (2004a) was three times higher than the one from Davenport et al. (2000). The high regression slope reported by Davenport et al. (2000) might be due to higher density of the crops when compared to the natural vegetation studied by Hopkinson (2004b). Dense crops create a continuous canopy cover and most laser pulses will reflect of the top of the canopy thereby limiting the variation in the vertical distribution. In a comparative study, Hopkinson et al. (2004a) concluded that vegetation height estimation of aquatic vegetation is associated with the largest errors. This is due to the low reflectivity of the water, preventing accurate determination of the ground surface elevation; hence this vegetation height should be regarded as a minimum value. These results indicate that, for individual laser data sets, the explained variance is high. However, the empirical relations differ greatly and therefore must be determined for each survey using ground reference data.

Asselman (2002) estimated vegetation height of meadows and herbaceous vegetation in a floodplain area under winter conditions from laser data with a point density of 10 points m^{-2}. She calculated laser-derived vegetation height as the difference between the actual point height and the minimum value in a local 1 m × 1 m window. She recommended using the median value of the laser data height distribution as predictor of vegetation height for meadows and herbaceous vegetation. However, large scatter was present and no regression analyses were carried out.

When compared to forestry studies (Table 1), results on low vegetation in summer (Table 3) are of slightly lower predictive quality. This, however, is a major achievement given that the range of vegetation heights that are predicted is much smaller. Estimating herb vegetation height under winter conditions remains a challenge.

Vegetation density and coverage of herbaceous vegetation

To date, little progress has been reported on the estimation of density of low vegetation using airborne laser scanning data. Asselman (2002) found only very poor relations between height distribution of laser scan data and the vegetation density of low floodplain vegetation. Therefore she established a lookup table to assign vegetation density values to areas with a specific vegetation height range to enable roughness computation. Mason et al. (2003) got around this problem by using vegetation height only as input for their hydrodynamic model. Vegetation coverage can be seen as a parameter that is related sideways to vegetation density, as correct estimation of vegetation coverage is a prerequisite for vegetation density prediction. Weltz et al. (1994) concluded for arid and semi-arid vegetation that laser altimetry data consistently overestimated coverage of vegetation lower than 0.3 m and underestimated vegetation cover of vegetation higher than 0.5 m, due to noise in the laser signal and an inaccurate estimation of higher vegetation. Ritchie et al. (1993) concluded that laser estimates of vegetation coverage for vegetation higher than 1 m were consistent with field reports. However, while vegetation cover can be predicted accurately using laser data, predictive models for vegetation density still have not yet been established. Vegetation density determination from laser scanning data thus remains a challenge.

Meadows and unvegetated areas

Meadows and unvegetated areas are both considered as the same land cover type in most papers reporting on laser scanning. The vertical precision of the laser height measurements, expressed as the standard deviation of unvegetated flat terrain, varies between 0.04 m (Davenport et al., 2000) and 0.07 m (Hopkinson et al., 2004a). The accuracy of laser-derived DTMs of unvegetated areas is 10–20 cm according to theoretical considerations (Huising & Gomes Perreira, 1998), later confirmed by empirical analysis (Hodgson & Bresnahan, 2004). The resulting height range in laser scan data of flat, unvegetated areas will be thus in the order of 20 cm. With this precision, accurate estimation of vegetation height and density for meadows is unlikely using discrete return laser systems. No studies have reported on the use of waveform digitizing laser scanners to estimate very short vegetation or unvegetated areas. Large footprint waveform digitizing lasers are also not suitable for this task, as it will be impossible to distinguish between small irregularities in the ground surface elevation and the vegetation. However, small footprint waveform digitizing laser scanners might improve the vegetation height estimates of low vegetation because ground elevation differences can be ignored over such small areas in case of relatively flat floodplains. Moreover, the errors from GPS and INS do not affect the digitized waveform.

Discussion and conclusion of the review

Airborne laser scanning is an emerging technique with broad applications in vegetation structure mapping. Vegetation height of forests and herbaceous vegetation in summer can be measured reliably. Parameters related to vegetation density have been studied by many studies in forestry, and can be determined almost equally accurately well as forest height. Still, different laser-derived statistics are used to predict the same forest characteristic, and when the same statistic is used in different areas, regression coefficients with vegetation height and density vary a lot. The same holds for the vegetation height of low vegetation. Extraction of vegetation height and density of herbaceous vegetation under winter conditions still remains to be investigated. This indicates that in the near future field data will remain necessary to establish regression models to predict vegetation structure using airborne laser data. The reasons for this are all the parameters which influence the minimum detectable object as listed in section 3. A solution could be a physically based model in which ray tracing of laser pulses is combined with a 3D vegetation model (Kay & Kajiya, 1986). However, such a physically based model would not be trivial as many parameters should be taken into account like: flying height, scan pattern and angle, beam divergence, laser power, detection threshold of the receiver, 3D structure and reflectivity of woody vegetation, reflectivity and orientation of leaves, soil reflectivity and multipath effects of the laser pulse in the vegetation.

Case study: vegetation mapping of Dutch floodplains

As a first assessment of applicability of laser scanning data for prediction of floodplain vegetation structure in winter conditions, we evaluated different statistics of laser data of a floodplain along the Lower Rhine in the Netherlands.

The 'Duursche Waarden' floodplain study area

The 'Duursche Waarden' floodplain (Fig. 4) along the IJssel river, the smallest distributary of the river Rhine in the Netherlands, was used as test area for vegetation mapping using airborne laser altimetry. The floodplain consists of a large meander along the concave bank of the main channel, and includes two man-made side channels, and a small river dune. The area is partly used as meadow and arable land, while large areas have become nature area. The vegetation comprises (Koppejan, 1998): (1) softwood forest willow, (*Salix alba, Salix viminalis*), poplar (*Populus nigra, Populus* × *canadensis*), (2) hardwood forest with oak (*Quercus robur*), ash (*Fraxinus excelsior*) and a small pine stand (*Pinus sylvestris*) on a river dune, together with (3) reed marshes (*Phragmites australis*), and (4) herbaceous vegetation with sedge (*Carex hirta*), sorrel (*Rumex obtusifolius*), nettle (*Urtica dioica*), thistle (*Cirsium arvense*) and clover (*Trifolium repens*). Within the floodplain, some gravel and clay pits occur. At the upstream side the floodplain is protected from low-magnitude floods by a minor dike. Inundation of the floodplain usually occurs in winter, so surveys of vegetation structure related to hydrodynamic roughness were also undertaken during the winter season.

Data collection

The laser data was collected in winter (March 11, 2001). First pulse, small footprint data was collected from low flying heights (~80 m). Point densities varied between 5 and 12 points m^{-2}. The right panel in Figure 4 shows height differences in the laser data within moving window with a 1 m circular radius. Simultaneously with the laser data, vegetation height and density were measured in the field on 39 plots of size 200 m^2 or larger in the same floodplain (Fig. 4c). Vegetation density was determined from the product of number of stalks or stems per unit area and the average stalk or stem diameter. The average stem diameter for each plot was based on 30 individual measurements using a sliding gauge. Vegetation height was based on 30 individual measurements using a measuring rod. Vegetation height of forests was not measured as the trees reach above the maximum flood water levels.

Vegetation structure in laser data

Figure 4c shows the height differences in moving local circular window with a 1 m radius, which is based on the unsegmented laser scanning point cloud. Obvious similarities are present in the distribution of forest ecotopes and high values in the

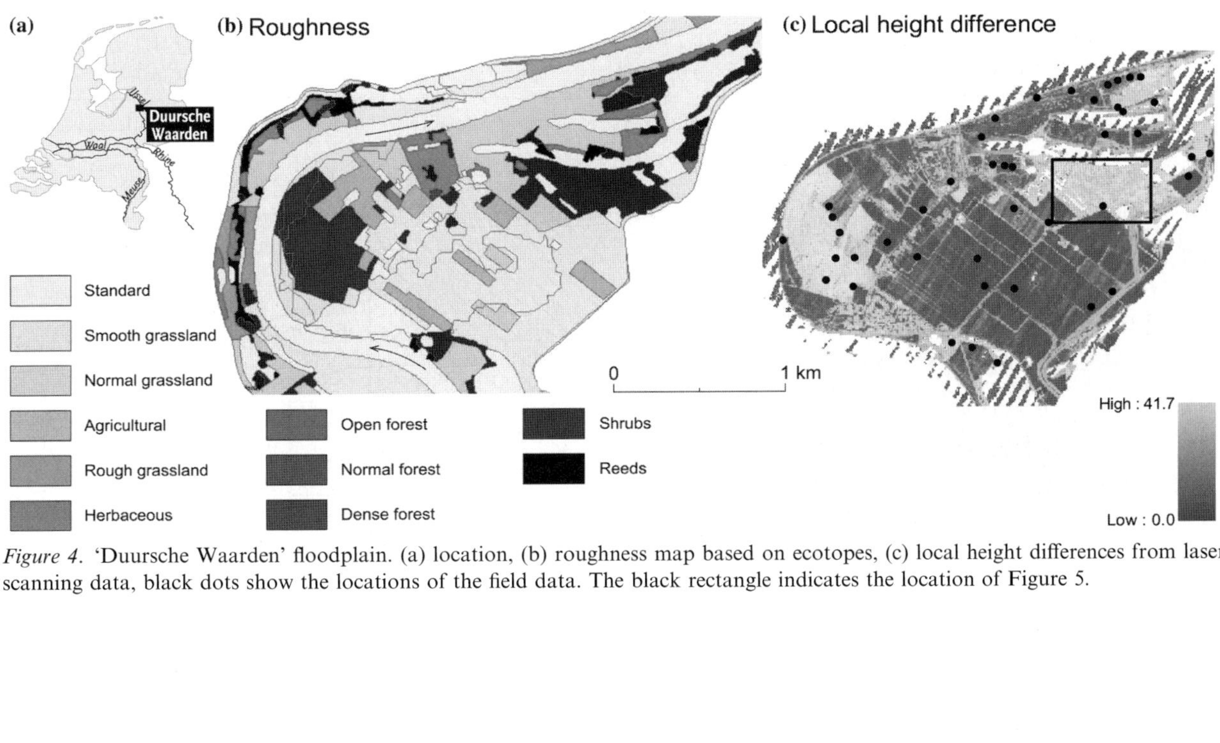

Figure 4. 'Duursche Waarden' floodplain. (a) location, (b) roughness map based on ecotopes, (c) local height differences from laser scanning data, black dots show the locations of the field data. The black rectangle indicates the location of Figure 5.

laser image. Shrubs and young forests show intermediate values and grasslands are characterized by small local height differences. However, when zooming in, both field and laser data show large variation within specific ecotopes. Table 4 compares the vegetation structural characteristics based on the lookup table of Van Velzen et al. (2003), which are used in hydrodynamic modelling, with the range of field values of vegetation height and density and the range of three selected laser-derived parameters. Both laser and field data show variation of more than one order of magnitude within a specific ecotope. An example of this difference is given in Figure 5, where the ecotope 'forest wetland' is classified as having a vegetation density of 0.023 $m^2\ m^{-3}$, but includes areas of high and low vegetation with a vegetation density in the field ranging between 0.019 and 0.19 $m^2\ m^{-3}$.

Figure 6 shows laser point clouds and local histograms of field plots with variable size. Four typical types of vegetation within the floodplain are depicted. Maximum inundation depth during flooding is 5 m at these locations. The difference between open (Fig. 6a) and dense (Fig. 6b) forests is clearly visible in the laser point cloud (middle column) and the resulting histogram (right). Also the LAI values, computed over the 0.5–2.5 m height range, increase with increasing vegetation density. LAI is 0.015 for open forest and 0.028 for dense forest. The ground surface is also clearly recognizable. Laser scanning also detected herbaceous vegetation in winter (Fig. 6c). Here, point density progressively decreases with height above the ground, which indicates that stalks are recognizable. Most laser returns (up to 90%) represent the ground surface. Therefore, the median value of a laser point cloud is not a good parameter to predict vegetation height. The gradual decrease in point density with height indicates that it is difficult to separate laser points reflected from the ground surface from points reflected from vegetation. The point cloud and resulting histogram of a meadow is indicated in Figure 6d. The histogram shows an almost perfect Gaussian error distribution. Standard deviation is 4 cm and the range is 30 cm.

Table 4. Comparison of model input to observed values of vegetation height and density and laser-derived parameters

Hydrodynamic model input			Range of field values		Range of selected laser-derived parameters		
Ecotype type[a]	Height (m)[b]	Density (m^{-2}/m^{-3})[b]	Height (m)	Density (m^{-2}/m^{-3})	Std Dev. (m)	D_{95}[c]	LAI[d] 10^{-2} (m^{-1})
Hardwood forest	–	0.023	–	0.005–0.056	1.8–5.5	10.3–22.4	0.3–3.8
Softwood forest	–	0.023	–	0.015–0.19	4.4–8.5	5.4–27.9	1.4–19
Softwood shrubs	6	0.13	5–7	0.08–0.11	0.7–2.3	2.7–7.3	9–20
Natural grassland	0.1	12	0.1–0.7	0.0003–0.065	0.03–0.11	0.16–0.38	1.3–16
Meadow	0.06	45	0.02–0.07	–	0.015–0.035	0.13–0.20	0.7–3.2

[a]Based on Jansen & Backx (1998), [b]Based on Van Velzen et al. (2003), [c]95 percentile of the relative height distribution, [d]leaf area index.

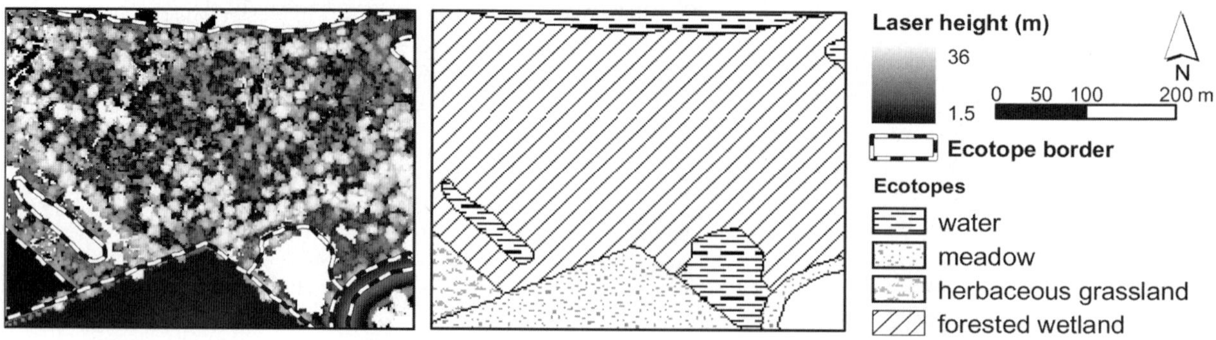

Figure 5. Detail of the Duursche Waarden floodplain: laser image (left) and ecotope map (right). The laser image shows much more detail than the ecotope map.

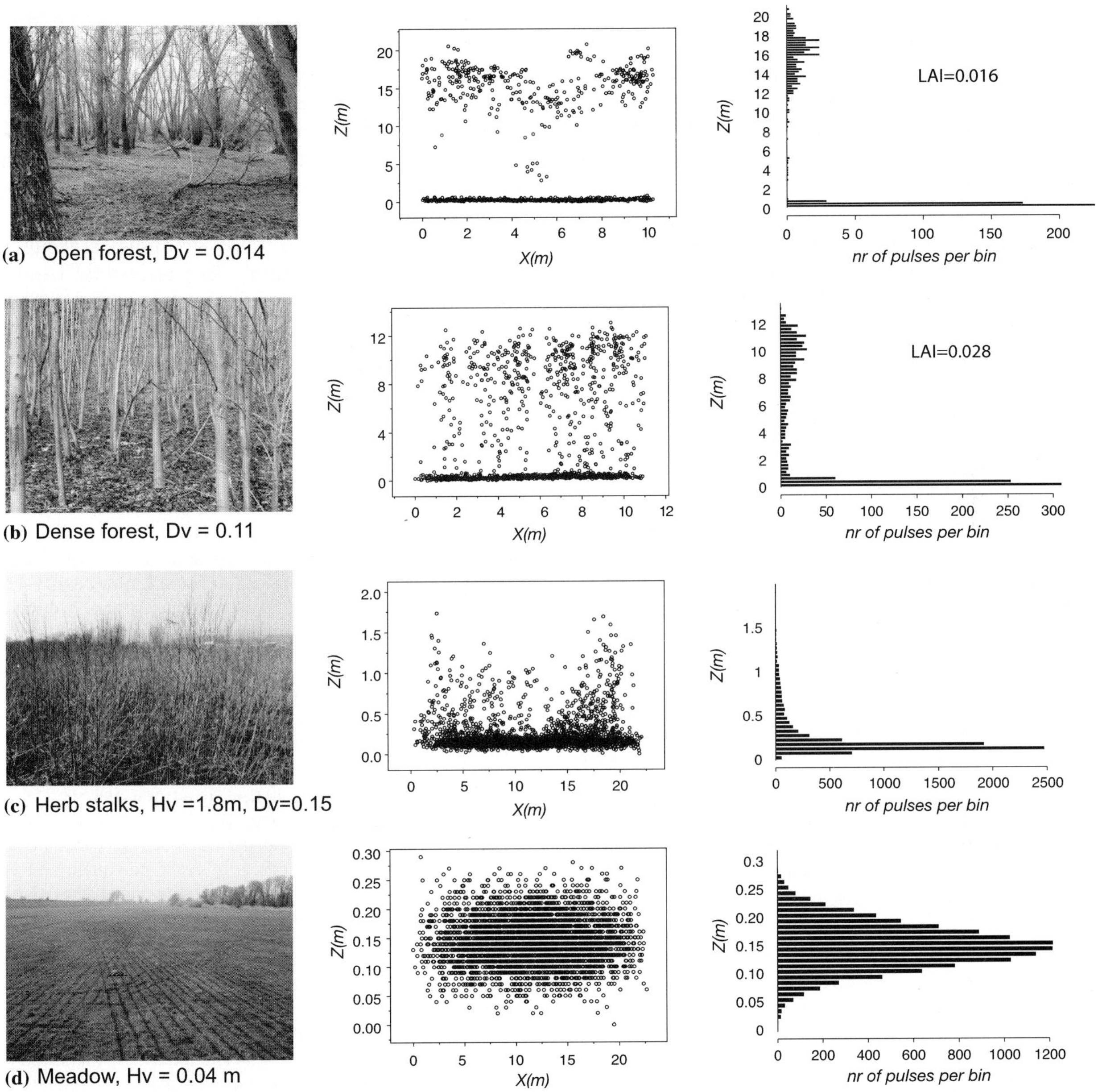

Figure 6. Examples of the influence of vegetation characteristics on the vertical point distribution. Open forest (a) shows less laser hits just above the ground surface than dense forest (b). Herb stalks are detectable (c). Meadows show a Gaussian error distribution (d). Dv = vegetation density, Hv = vegetation height, Z(m) = height above local zero plane (m), X(m) = local easting (m), LAI = Leaf Area Index (m^{-1}).

Prospects: the future of airborne laser scanning in floodplain management

Floodplain vegetation structure is spatially heterogeneous, and will vary in time due to vegetation succession (Baptist et al., 2004; Jesse, 2004). Airborne laser scanning is a promising tool to extract vegetation structural characteristics. Typical differences in forest structure within floodplains are easily recognizable from local differences in the laser point cloud, but also the ability to determine vegetation density of forests seems likely. However, a first step in data processing should be the classification of vegetation types, because the empirical relations are vegetation-type specific. Forest, shrubs and herbaceous vegetation are discernable from vegetation heights derived from laser data, but meadows and agricultural lands are not. Therefore, boundaries between vegetation types should be generated using an object-based classification algorithm based on spectral and laser data, for example as in Hay et al. (2003).

Vegetation density of floodplain forests could be predicted using the LAI method of MacArthur & Horn (1969), a canopy density based on Næsset (2002) or individual tree delineation. Herbaceous vegetation in winter can be detected by laser scanning. Since previous studies focused on crops and grasslands in summer condition, research of floodplain vegetation roughness should focus on detailed investigation of the relation between the laser scanning data and field measurements of vegetation height and density of herbaceous vegetation under winter conditions. Special attention should be paid to the reconstruction of DTMs under these vegetation types, since a large proportion of the laser points are reflected from the ground surface. Due to the noise in the currently available laser scanning data, it seems impossible to determine vegetation height of floodplain meadows. Full waveform digitizing data may improve these estimates. Thus, from the literature review and the first field assessment for the lower Rhine floodplain airborne laser scanning data seems a promising tool for hydrodynamic roughness determination for vegetation types other than meadows. Compared to spectral data, laser scanning offers the ability to map the 3D structure of vegetation, avoiding the use of lookup tables with their inherent data loss. Combining laser scanner data with spectral data may provide an adequate solution to classify unvegetated areas and meadows. In this case, a lookup table still needs to be used to convert land cover to roughness values.

Management of floodplains involves comparing costs of different techniques. Laser scanning mapping of vegetation structure is more expensive than traditional airborne photography. Hartmann et al. (2004) conclude that a combination of multispectral remote sensing and laser scanning is five times more expensive for roughness determination than the traditional method. Increased scan frequency of the laser system would allow acquisition of the laser data from higher altitudes, covering larger areas in a single flight strip while preserving the required point density. In the future, such technical developments may reduce acquisition cost. Laser altimetry data provides useful information for various other applications like floodplain morphology, 3D shape of groynes, location of shorelines and break lines (Bollweg et al., 2004). For example, for groyne monitoring and floodplain DTM generation, laser scanning is cheaper than conventional methods like terrestrial height measurements or analytic photogrammetry. This indicates that laser scanning is particularly suitable for floodplain management if multiple end products are derived from the laser data.

References

Aber, J. Ds., 1979. A method for estimating foliage-height profiles in broad leaved forests. Journal of Ecology 67: 35–40.

Asselman, N. E. M., 2002. Laser altimetry and hydraulic roughness of vegetation, further studies using ground truth. Q3139, WL|Delft Hydraulics, Delft.

Baltsavias, E. P., 1999a. Airborne laser scanning: basic relations and formulas. ISPRS Journal of Photogrammetry and Remote Sensing 54: 199–214.

Baltsavias, E. P., 1999b. Airborne laser scanning: existing systems and firms and other resources. ISPRS Journal of Photogrammetry & Remote Sensing 64: 164–198.

Baptist, M. J., W. E. Penning, H. Duel, A. J. M. Smits, G. W. Geerling, G. E. M. van der Lee & J. S. L. Van Alphen, 2004. Assessment of the effects of cyclic floodplain rejuvenation on flood levels and biodiversity along the Rhine River. River Research and Applications 20: 285–297.

Blair, J. B. & M. A. Hofton, 1999. Modelling laser altimeter return waveforms over complex vegetation using high-resolution elevation data. Geophysical Research Letters 26: 2509–2512.

Blair, J. B., D. L. Rabine & M. A. Hofton, 1999. The laser vegetation imaging sensor: a medium-altitude, digitisation-only, airborne laser altimeter for mapping vegetation and topography. ISPRS Journal of Photogrammetry & Remote Sensing 54: 115–122.

Bollweg, A., R. Brügelmann, H. van Dijk & E. Vaessen, 2004. Laseraltimetriedata met hoge puntdichtheid voor rivierbeheer. AGI-GAR-2004011, Adviesdienst Geoinformatie en ICT, Delft (in Dutch).

Brandtberg, T., T. A. Warner, R. E. Landenberger & J. B. McGraw, 2003. Detection and analysis of individual leaf-off tree crowns in small footprint, high sampling density lidar data from the eastern deciduous forest in North America. Remote Sensing of Environment 85: 290–303.

Carollo, F. G., V. Ferro & D. Termini, 2002. Flow velocity measurements in vegetated channels. Journal of Hydraulic Engineering 128: 664–673.

Chow, V. T., 1959. Open channel hydraulics. McGraw Hill, New York.

Cobby, D. M., D. C. Mason & I. J. Davenport, 2001. Image processing of airborne scanning laser altimetry data for improved river flood modelling. ISPRS Journal of Photogrammetry & Remote Sensing 56: 121–138.

Davenport, I. J., R. B. Bradbury, G. R. F. Anderson, J. R. Hayman, J. R. Krebs, D. C. Mason, J. D. Wilson & N. J. Veck, 2000. Improving bird population models using airborne remote sensing. International Journal of Remote Sensing 21: 2705–2717.

Drake, J. B., R. O. Dubayah, D. B. Clark, R. G. Knox, J. B. Blair, M. A. Hofton, R. L. Chazdon, J. F. Weishampel & S. Prince, 2002. Estimation of tropical forest structural characteristics using large-footprint lidar. Remote Sensing of Environment 79: 305–319.

Hartmann, R., R. Brügelmann & A. Bollweg, 2004. Vegetatieclassificatie uit laser en optische RS data. AGI/0109/GAR002, Adviesdienst Geoinformatie en ICT, Delft (in Dutch).

Hay, G. J., T. Blaschke, D. J. Marceau & A. Bouchard, 2003. A comparison of three image-object methods for the multiscale analyses of landscape structure. ISPRS Journal of Photogrammetry & Remote Sensing 57: 327–345.

Hodgson, M. E. & P. Bresnahan, 2004. Accuracy of airborne lidar-derived elevation: empirical assessment and error budget. Photogrammetric Engineering & Remote Sensing 70: 331–339.

Holmgren, J., M. Nilsson & H. Olsson, 2003. Estimation of tree height and stem volume on plots using airborne laser scanning. Forest Science 49: 419–428.

Holmgren, J. & A. Persson, 2004. Identifying species of individual trees using airborne laser scanner. Remote Sensing of Environment 90: 415–423.

Holmgren, J. & T. Jonsson, 2004. Large scale airborne laser scanning of forest resources in Sweden. In Thies, M., B. Koch, H. Spiecker & H. Weinacker (eds), Laser-scanners for Forest and Landscape Assessment. International Archives of Photogrammetry and Remote Sensing and Spatial Information Sciences, Volume 36, part 8/W2: 157–160.

Hopkinson, C., L. E. Chasmer, G. Zsigovics, I. F. Creed, M. Sitar, P. Treitz & R. V. Maher, 2004a. Errors in lidar ground elevation and wetland height estimates. In Thies, M., B. Koch, H. Spiecker & H. Weinacker (eds), Laser-scanners for Forest and Landscape Assessment. International Archives of Photogrammetry and Remote Sensing and Spatial Information Sciences, Volume 36, part 8/W2: 108–113.

Hopkinson, C., K. Lim, L. E. Chasmer, P. Treitz, I. F. Creed & C. Gynan, 2004b. Wetland grass to plantation forest – estimating vegetation height from the standard deviation of lidar frequency distributions. In Thies, M., B. Koch, H. Spiecker & H. Weinacker (eds), Laser-scanners for Forest and Landscape Assessment. International Archives of Photogrammetry and Remote Sensing and Spatial Information Sciences, Volume 36, part 8/W2: 288–294.

Hug, C., A. Ullrich & A. Grimm, 2004. Litemapper-5600: A waveform-digitizing lidar terrain and vegetation mapping system. In Thies, M., B. Koch, H. Spiecker & H. Weinacker (eds), Laser-scanners for Forest and Landscape Assessment. International Archives of Photogrammetry and Remote Sensing and Spatial Information Sciences, Volume 36, part 8/W2: 24–29.

Huising, E. J. & L. M. Gomes Pereira, 1998. Errors and accuracy estimates of laser data acquired by various laser scanning systems for topographic applications. ISPRS Journal of Photogrammetry & Remote Sensing 53: 245–261.

Jansen, B. J. M. & J. J. G. M. Backx, 1998. Ecotope mapping Rhine Branches-east 1997. 98.054, Rijksinstituut voor Integraal Zoetwaterbeheer en Afvalwaterbehandeling, Lelystad (in Dutch).

Jesse, P., 2004. Hydraulic resistance in (nature) development. RIZA werkdocument, 2003.124X, Rijksinstituut voor Integraal Zoetwaterbeheer en Afvalwaterbehandeling, Arnhem (in Dutch).

Katzenbeisser, R., 2003. Technical Note On: Echo Detection. Toposys GMBH, Biberach.

Kay, T. L. & J. T. Kajiya, 1986. Ray tracing complex scenes. Computer Graphics 20: 269–278.

Kilian, J., N. Haala & M. Englich, 1996. Capture and evaluation of airborne scanner data. International Archives of Photogrammetry and Remote Sensing 1031 B3: 383–388.

Koppejan, H., 1998. Toelichting bij de vegetatie- en structuurkaart Duursche Waarden 1996 op basis van false-color luchtfoto's. MDGAT 9716, Rijkswaterstaat Meetkundige Dienst, Delft (in Dutch).

Kraus, K. & N. Pfeifer, 1998. Determination of terrain models in wooded areas with airborne laser scanner data. ISPRS Journal of Photogrammetry and Remote Sensing 53: 193–203.

Lefsky, M. A., W. B. Cohen, S. A. Acker, G. G. Parker, T. A. Spies & D. Harding, 1999a. Lidar remote sensing of the canopy structure and biophysical properties of Douglas-Fir Western Hemlock forests. Remote Sensing of Environment 70: 339–361.

Lefsky, M. A., D. Harding, W. B. Cohen, G. Parker & H. H. Shugart, 1999b. Surface lidar remote sensing of basal area and biomass in deciduous forests of Eastern Maryland, USA. Remote Sensing of Environment 67: 83–98.

Lefsky, M. A., W. B. Cohen, G. Parker & D. Harding, 2002. Lidar remote sensing for ecosystem studies. BioScience 52: 19–30.

Leuven, R. S. E. W., Y. Gerig, I. Poudevigne, G. W. Geerling, L. Kooistra & B. G. W. Aarts, 2002. Cumulative impact assessment of ecological rehabilitation and infrastructure facilities in floodplains along the middle reach of the River Waal. In Leuven, R. S. E. W., I. Poudevigne & R. M. Teeuw (eds), Application of Geographic Information Systems and Remote Sensing in River Studies. Backhuys Publishers, Leiden: 201–216.

Lillesand, T. M. & R. W. Kiefer, 1994. Remote Sensing and Image Interpretation. John Wiley & sons, New York.

Lim, K., P. Treitz, M. Wulder, B. St-Onge & M. Flood, 2003. Lidar remote sensing of forest structure. Progress in Physical Geography 27: 88–106.

MacArthur, R. H. & H. S. Horn, 1969. Foliage profile by vertical measurements. Ecology 50: 802–804.

MacArthur, R. H. & J. W. MacArthur, 1961. On bird species diversity. Ecology 42: 594–598.

Magnussen, S. & P. Boudewyn, 1998. Derivations of stand height from airborne laser scanner data with canopy-based quantile estimators. Canadian Journal of Forest Research 28: 1016–1031.

Maltamo, M., K. Eerikainen, J. Pitkanen, J. Hyyppa & M. Vehmas, 2004. Estimation of timber volume and stem density based on scanning laser altimetry and expected tree size distribution functions. Remote Sensing of Environment 90: 319–330.

Marks, K. & P. Bates, 2000. Integration of high-resolution topographic data with floodplain flow models. Hydrological Processes 14: 2109–2122.

Mason, D. C., D. M. Cobby, M. S. Horrit & P. Bates, 2003. Floodplain friction parameterization in two-dimensional river flood models using vegetation heights derived from airborne scanning laser altimetry. Hydrological Processes 17: 1711–1732.

Means, J. E., S. A. Acker, D. J. Harding, J. B. Blair, M. A. Lefsky, W. B. Cohen, M. E. Harmon & W. A. McKee, 1999. Use of large-footprint scanning airborne lidar to estimate forest stand characteristics in the Western Cascades of Oregon. Remote Sensing of Environment 67: 298–308.

Morsdorf, F., E. Meier, B. Kotz, K. I. Itten, M. Dobbertin & B. Allgower, 2004. LIDAR-based geometric reconstruction of boreal type forest stands at single tree level for forest and wildland fire management. Remote Sensing of Environment 92: 353–362.

Næsset, E, 1997a. Determination of mean tree height of forest stands using airborne laser scanner data. ISPRS Journal of Photogrammetry & Remote Sensing 52: 49–56.

Næsset, E., 1997b. Estimating timber volume of forest stands using airborne laser scanner data. Remote Sensing of Environment 61: 246–253.

Næsset, E., 2002. Predicting forest stand characteristics with airborne scanning laser using a practical two-stage procedure and field data. Remote Sensing of Environment 80: 88–99.

Næsset, E., 2004. Effects of different flying altitudes on biophysical stand properties estimated from canopy height and density measured with a small-footprint airborne scanning laser. Remote Sensing of Environment 91: 243–255.

Næsset, E. & K. -O. Bjerknes, 2001. Estimating tree heights and number of stems in young forest stands using airborne laser scanner data. Remote Sensing of Environment 78: 328–340.

Næsset, E. & T. Okland, 2002. Estimating tree height and tree crown properties using airborne scanning laser in a boreal nature reserve. Remote Sensing of Environment 79: 105–115.

Nienhuis, P. H. & R. S. E. W. Leuven, 2001. River restoration and floodprotection: controversy or synergism. Hydrobiologia 444: 85–99.

Nienhuis, P. H., A. D. Buijse, R. S. E. W. Leuven, A. J. M. Smits, R. J. W. de Nooij & E. M. Samborska, 2002. Ecological rehabilitation of the lowland basin of the river Rhine (NW Europe). Hydrobiologia 478: 53–72.

Nilsson, M., 1996. Estimating tree heights and stand volume using an airborne lidar system. Remote Sensing of Environment 56: 1–7.

Oude Elberink, S., G. Brand & R. Brügelmann, 2003. Quality improvement of laser altimetry DEM's. In Maas, H. -G., G. Vosselman & A. Streilein (eds), 3-D Reconstruction from Airborne Laserscanner and InSAR Data. International Archives of Photogrammetry and Remote Sensing and Spatial Information Sciences 34 3/W13: 51–58.

Persson, Å., J. Holmgren & U. Söderman, 2002. Detecting and measuring individual trees using an airborne laser scanner. Photogrammetric Engineering & Remote Sensing 68: 925–932.

Petryk, S. & G. Bosmajian, 1975. Analysis of flow through vegetation. Journal of Hydraulics Divisions 101: 871–884.

Pfeifer, N., B. Gorte & S. Oude Elberink, 2004. Influences of vegetation on laser altimetry – analyses and correction approaches. In Thies, M., B. Koch, H. Spiecker & H. Weinacker (eds), Laser-scanners for Forest and Landscape Assessment. International Archives of Photogrammetry and Remote Sensing and Spatial Information Sciences 36 8/W2: 283–287.

Popescu, S. C., R. H. Wynne & R. F. Nelson, 2002. Estimating plot-level tree heights with lidar: local filtering with a canopy-height based variable window size. Computers and Electronics in Agriculture 37: 71–95.

Ritchie, J. C., D. L. Evans, D. Jacobs, J. H. Everitt & M. A. Weltz, 1993. Measuring canopy structure with an airborne laser altimeter. Transactions of ASAE 36: 1235–1238.

Ritchie, J. C., M. Menenti & M. A. Weltz, 1996. Measurements of land surface features using an airborne laser altimeter: the HAPEX-sahel experiment. International Journal of Remote Sensing 17: 3705–3724.

Schmidt, K. S. & A. K. Skidmore, 2003. Spectral discrimination of vegetation types in a coastal wetland. Remote Sensing of Environment 85: 92–108.

Shabalova, M. V., W. P. A. van Deursen & T. A. Buishand, 2003. Assessing future discharge of the river Rhine using regional climate model integrations and a hydrological model. Climate Research 23: 233–246.

Silva, W., F. Klijn & J. Dijkman, 2001. Room for the Rhine branches in The Netherlands. RIZA-report 2001.031, Delft Hydraulics report R3294, Rijksinstituut voor Integraal Zoetwaterbeheer en Afvalwaterbehandeling, WL | Delft Hydraulics, Arnhem, Delft.

Sithole, G. & G. Vosselman, 2004. Experimental comparison of filter algorithms for bare-Earth extraction from airborne laser scanning point clouds. ISPRS Journal of Photogrammetry & Remote Sensing 59: 85–101.

Thompson, A. G., R. M. Fuller & J. A. Eastwood, 1998. Supervised versus unsupervised methods for classification of coasts and river corridors from airborne remote sensing. International Journal of Remote Sensing 19: 3423–3431.

Van der Sande, C. J., S. M. de Jong & A. P. J. de Roo, 2003. A segmentation and classification approach of IKONOS-2 imagery for land cover mapping to assist flood risk and flood damage assessment. International Journal of Applied Earth Observation and Geoinformation 4: 217–229.

Van Stokkom, H. T. C., A. J. M. Smits & R. S. E. W. Leuven, 2005. Flood defense in The Netherlands A new era, a new approach. Water International 30: 76–87.

Van Velzen, E. H., P. Jesse, P. Cornelissen & H. Coops, 2003. Flow friction of vegetation in floodplains part 1. 2003.028, Rijksinstituut voor Integraal Zoetwaterbeheer en Afvalwaterbehandeling, Arnhem (in Dutch).

Vosselman, G., 2000. Slope based filtering of laser altimetry data. International Archives of Photogrammetry and Remote Sensing, Amsterdam 33 B3/2: 935–942.

Wagner, W., A. Ulrich, T. Melzer, C. Briese & K. Kraus, 2004. From single pulse to full waveform digitizing laser scanners: potential and practical challenges. International Archives of Photogrammetry and Remote Sensing, Amsterdam 35 B3: www.isprs.org/istanbul/comm3/papers/267.pdf.

Ward, J. V. & K. Tockner, 2001. Biodiversity: towards a unifying theme for river ecology. Freshwater Biology 46: 807–819.

Wehr, A. & U. Lohr, 1999. Airborne laser scanning – an introduction and overview. ISPRS Journal of Photogrammetry & Remote Sensing 54: 68–82.

Weltz, M. A., J. C. Ritchie & H. Dale Fox, 1994. Comparison of laser and field measurements of vegetation height and canopy cover. Water Resources Research 30: 1311–1319.

Yu, X., J. Hyyppä & H. Hyyppä, 2004. Effects of flight altitude on tree height estimation using airborne laser scanning. In Thies, M., B. Koch, H. Spiecker & H. Weinacker (eds), Laser-scanners for Forest and Landscape Assessment. International Archives of Photogrammetry and Remote Sensing and Spatial Information Sciences 36 8/W2: 96–101.

Hydrobiologia (2006) 565:105–120
R.S.E.W. Leuven, A.M.J. Ragas, A.J.M. Smits & G. van der Velde (eds), Living Rivers: Trends and Challenges in Science and Management
DOI 10.1007/s10750-005-1908-4

Reproduction, growth, and migration of fishes in a regulated lowland tributary: potential recruitment to the river Meuse

B.J.A. Pollux[1,2,*], A. Korosi[3], W.C.E.P. Verberk[4,5], P.M.J. Pollux[2] & G. van der Velde[5]

[1]*Department of Plant–Animal Interaction, Centre for Limnology, Netherlands Institute of Ecology, P.O. Box 1299, 3600 BG Maarssen, The Netherlands*

[2]*Department of Aquatic Ecology & Environmental Biology, Institute for Wetland and Water Research, Radboud University Nijmegen, Toernooiveld 1, 6525 ED Nijmegen, The Netherlands*

[3]*Department of Cellular Animal Physiology, Radboud University Nijmegen, Toernooiveld 1, 6525 ED Nijmegen, The Netherlands*

[4]*Bargerveen Foundation, P.O. Box 9010, 6500 GL Nijmegen, The Netherlands*

[5]*Department of Animal Ecology & Ecophysiology, Institute for Wetland and Water Research, Radboud University Nijmegen, Toernooiveld 1, 6525 ED Nijmegen, The Netherlands*

*(*Author for correspondence: Tel.: +31-24-3652134; Fax: +31-24-3652521; E-mail: b.pollux@science.ru.nl)*

Key words: larvae, juveniles, migration, nursery, residents, transients

Abstract

Many European rivers are characterized by a canalized main channel, steep stony embankments, the absence of aquatic vegetation, regulated flow dynamics, reduced hydrological connectivity to the floodplains and a lack of spawning and nursery areas for many fish species. In such regulated rivers, tributaries may be particularly important for recruitment of fish populations in the main channel. This paper describes the reproduction, growth and migration of fishes in the Everlose Beek, a regulated lowland tributary stream of the river Meuse (The Netherlands), using bi-weekly sampling from January to December 2002. A total of 8615 fishes were caught, belonging to 13 different species. The fish species were classified into three groups, viz., residents, migrants and transients, based on the presence of various life-stages in the tributary. Size-frequency data suggest that each group uses the Everlose Beek differently: (i) Stone loach (*Barbatula barbatula*), Gudgeon (*Gobio gobio*) and Three-spined stickleback (*Gasterosteus aculeatus*) were resident species using the tributary as a spawning, nursery and adult habitat; (ii) Bream (*Abramis brama*), Roach (*Rutilus rutilus*), Rudd (*R. erythrophthalmus*), Tench (*Tinca tinca*), and Pike (*Esox lucius*) were migratory species, using the tributary as a spawning area, as well as a nursery habitat during their first year of growth, but migrating towards the river Meuse typically at a length of 5–15 cm; and (iii) Bleak (*Alburnus alburnus*), Sunbleak (*Leucaspius delineatus*), Carp (*Cyprinus carpio*), Crucian carp (*Carassius carassius*), and Perch (*Perca fluviatilis*) were transient species, characterized by an absence of reproduction, and the occurrence in very low densities of >age-1 juveniles and adults only. Lowland tributaries, such as the Everlose Beek, can contribute to the recruitment of particularly migrant species, hence contributing to fish populations of the regulated river Meuse.

Introduction

Natural river systems consist of a main (braided) channel, providing longitudinal linkage within the corridor (River Continuum Concept; Vannote et al., 1980) and transversal linkages to floodplain water bodies (Flood Pulse Concept; Junk et al., 1989). Along the longitudinal axis, however, the canalization of many north-west European lowland rivers has resulted in a severe loss of habitat heterogeneity. Since spatial and temporal variation in structural habitat complexity is one of the most important conditions for the existence of well balanced aquatic riverine communities, this has generally resulted in impoverished communities in the main channel (Gorman & Karr, 1978; Schlosser, 1991; Jungwirth et al., 1993; Townsend & Hildrew, 1994; Jurajda, 1995; Nilsson et al., 1997; Ward et al., 2002; Aarts & Nienhuis, 2003). Moreover, European large rivers have an important transport function. The intense shipping traffic has, both directly and indirectly, detrimental impacts on the fish fauna. The continuous wave action has a direct negative impact on the survival of fish eggs and larvae (Arlinghaus et al., 2002; Wolter & Arlinghaus, 2003). Indirectly, the wave action and heavy eutrophication together with the resulting turbidity of the water column, prevent the establishment of aquatic vegetation in the main channel (Admiraal et al., 1993). Since, macrophyte beds serve as obligate spawning substrates for phytophilic species, as shelter for juveniles and as rich foraging habitats for adults, shipping traffic also (indirectly) exerts an effect on the fish fauna by rendering the main channel unsuitable for many fish species (Arlinghaus et al., 2002; Wolter & Arlinghaus, 2003). Along the transversal axis of the main river channel, the physical and ecological interactions (Schiemer, 1985; Ward, 1989; Calow & Petts, 1994; Van den Brink, 1994; Allan, 1995; Welcomme, 1995; Petts & Amoros, 1996; Ward et al., 2002) have become disrupted by the placement of dikes, dams and weirs (Dynesius & Nilsson, 1994). This regulation of flow has led to a decline in natural water level variations. As a consequence, floodplain lakes and other off-channel water bodies have effectively become isolated from the main channel and the ecological functioning has become disrupted (Bain et al., 1988; Ward & Stanford, 1995; Van den Brink et al., 1996; Aarts et al., 2004). The reduced habitat heterogeneity, the intense shipping traffic and the disrupted ecological function of floodplains have had severe adverse consequences for the availability of spawning and nursery habitats for many fish species in large rivers (Jungwirth et al., 1993; Vriese et al., 1994; Jurajda, 1995; Copp, 1997; Jurajda, 1999).

When studying fish ecology in regulated rivers, the importance of lowland tributaries is commonly overlooked, as attention is focused on the main channel and its immediate off-channel habitats, such as flood-plains or man-made secondary side channels (Simons et al., 2001; Buijse et al., 2002; Grift et al., 2003). However, in regulated rivers with reduced availability of spawning and nursery habitats, potential recruitment from tributaries may be particularly important. Unfortunately, natural free-flowing lowland tributaries are hardly found anymore in the Netherlands (Verdonschot & Nijboer, 2002) and other Western European countries (Wolter, 2001, Verdonschot & Nijboer, 2002; Nienhuis et al., 200b). Nevertheless, although most lowland tributaries are regulated (Verdonschot & Nijboer, 2002), they are often characterized by abundant aquatic vegetation and extensive or even no shipping traffic. As a result, regulated lowland tributaries often harbour a diverse fish fauna (Delmastro, 1982; Steinberg, 1992; Vandelannoote et al., 1998; Crombaghs et al., 2000), from which fishes can be recruited for populations in the main river, either through drift of larvae (Robinson et al., 1998; De Graaf et al., 1999; Reichard et al., 2002) or through migration of juveniles (Borcherding et al., 2002). Surprisingly, however, although most lowland streams have been influenced by man, little is known about the ecological function of such regulated stream ecosystems. Very few studies have focused on the ecology of fishes in regulated lowland streams, and quantitative data describing ecological links between regulated tributaries and the main channel are largely lacking.

To gain a better insight into the ecology of fishes in regulated tributaries and their links to the main river channel, size-frequency data were collected from the fish fauna of the Everlose Beek, a regulated lowland tributary of the river Meuse (the Netherlands), during January–December 2002, using a single standardized sampling technique. The first objective of the present study was to determine whether, and if, which species use the lowland stream as (i) a reproduction habitat, (ii) a nursery area for larval and juvenile 0+ stages, and (iii) as an adult habitat. Since, recruitment of fishes is largely determined by survival and growth over the first year (Kirjasniemi & Valtonen, 1997), our second objective was to study the growth of 0+ fishes in the Everlose Beek. We furthermore hypothesized, that due to the shallow nature of the Everlose Beek, it would not necessarily provide a suitable habitat for larger (adult) individuals. Therefore, as a third objective, we assessed whether certain species display ontogenetic migration, from the tributary stream towards the river Meuse.

Materials and methods

Study area

The western bank of the river Meuse, in the province of Limburg (The Netherlands), features many regulated lowland tributary streams, with a total estimated length of >500 km, which discharge into the river (Fig. 1). The tributaries are characterized by an unnatural hydrology because they are

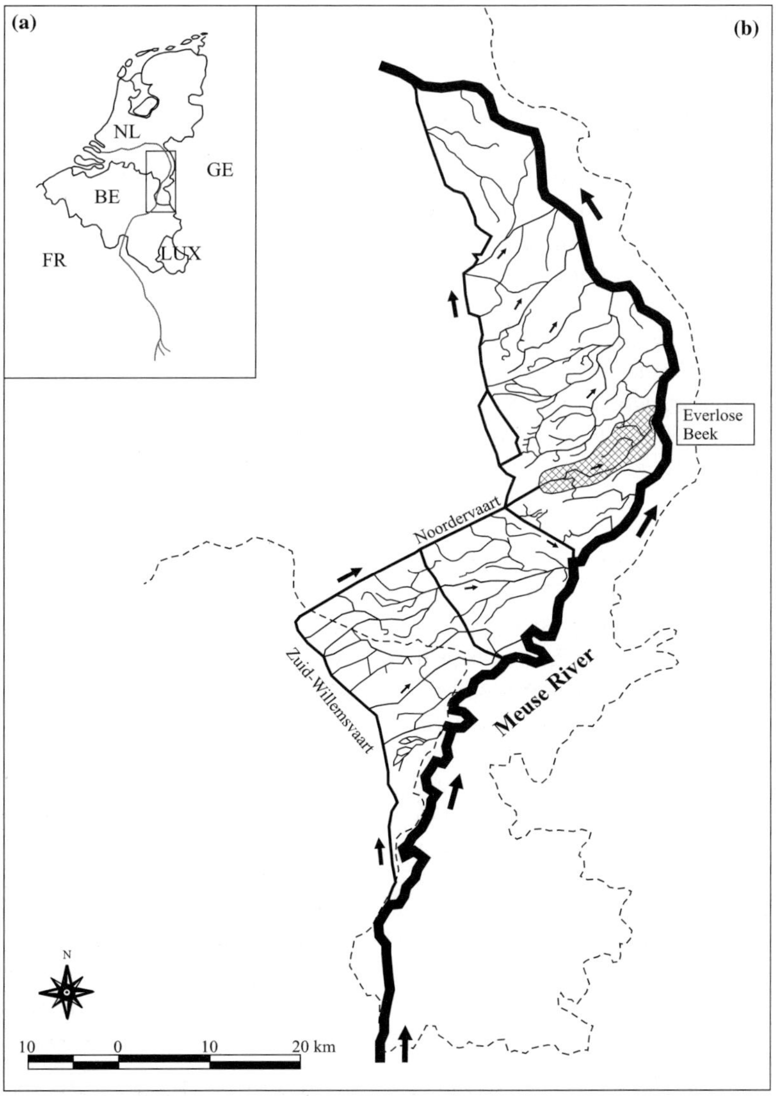

Figure 1. (a) The river Meuse, flowing through France, Belgium and the Netherlands, where it discharges into the North Sea. (b) Map of the river Meuse in the Province of Limburg (The Netherlands), showing the lowland tributary streams on its western bank. These lowland streams are supplied by water originally deriving from the river Meuse. The black arrows indicate the flow direction and the hatched out area indicates the study area, the Everlose Beek.

indirectly supplied with water from the river Meuse itself, via canals. Water enters the Zuid-Willemsvaart canal near Maastricht, which in turn supplies water to the tributaries, ultimately discharging back into the Meuse further downstream (Fig. 1). Fishes can only access these streams from upstream reaches near Maastricht, whereas movement in the tributaries is restricted to unidirectional migration in a downstream direction, due to the presence of a large number of weirs.

The Everlose Beek was chosen because it is a typical regulated lowland tributary of the Meuse, closely resembling other lowland streams in the Province of Limburg (Fig. 1) in terms of morphology, hydrology, vegetation, and fish fauna (Crombaghs et al., 2000). It is characterized by a shallow water depth (0.1–0.6 m), low stream velocities (0.1–0.5 m s^{-1}), a width of 1–10 m, a variety of substrates (silt, mud, sand, and stony bottoms), and locally dense vegetation (featuring, e.g. *Potamogeton perfoliatus, P. natans, Myriophyllum spicatum, Ceratophyllum demersum, Elodea canadensis, Ranunculus circinatus, Sparganium emersum, Sagittaria sagttifolia, Sium latifolium, Glyceria maxima, Mentha aquatica*, and *Myosotis scorpioides*).

Sampling procedure

Due to the shallowness of the tributary, fishes are most effectively sampled with hand nets (Crombaghs et al., 2000). Sampling gear consisted of a small round dip net (diameter 17.5 cm; mesh size 1.0×1.0 mm^2) for collecting fish embryos and larvae, and two types of D-shaped landing nets (60×40 cm^2; mesh sizes 1.0×1.0 mm^2 and 70×50 cm^2; 3.0×3.0 mm^2) for collecting larvae, juveniles, and adults. During 2002, fishes were sampled weekly to biweekly (a total of 36 sampling days). On each sampling day, several river stretches (mean ± SE of 4.5 ± 0.3 locations per sampling day) were sampled according to the following standardized procedure. Two fishermen were positioned in the stream, facing upstream, holding the straight side of the D-shaped nets on the bottom. A third person waded towards them from an upstream position, while holding the D-shaped net on the bottom. This allowed mobile fish species that normally swim away at the first sign of danger (typically in a downstream direction), to be captured. Next, the three fishermen walked the same stretch again in the upstream direction, specifically sampling the more structurally complex habitats, such as dense vegetation, overhanging tree roots and stony bottoms (using their feet to turn over the stones, while keeping the mouth of the net behind the stones facing upstream). This allowed them to capture the remaining fish species, which utilize complex habitats for shelter rather than escaping by swimming away quickly. Embryos, larvae and small juveniles were taken to the laboratory for identification, using the key by Pinder (2001). Larger juveniles (>4 cm) and adults were identified and measured in the field, and subsequently released. Fork length was measured to the nearest mm for fishes smaller than 20 cm and to the nearest cm for larger fish. Fish densities were expressed as numbers of fish collected per 100 m^2 of sampled area.

Growth model

The growth of 0+ fish was followed by regularly sampling the fish fauna from the Everlose Beek. Observed lengths were plotted against sampling date, showing the length increase over time. The average growth of 0+ fishes in a population was described by the Gompertz equation (Molls, 1997; Gamito,1998):

$$L_{(t)} = K * e^{-e^{-r(t-M)}}$$

In this equation, $L(t)$ represents the length of the fish at time t ($t=0$ being the time of first appearance of larvae), K the upper asymptotic growth, M the time of maximum growth and r a growth related parameter. For each sampling date, the average length was calculated and regressed against predicted values of the Gompertz equation, by means of nonlinear regression using STATISTICA version 6.0 (StatSoft Inc., Tulsa, Oklahoma, U.S.A.). The program uses a least squares estimation procedure (Levenberg–Marquardt) to minimize the sum of squared deviations of the observed values from those predicted by the model (StatSoft Inc., Tulsa, Oklahoma, U.S.A.).

Classification

Fish species were classified into different groups, based on the absence or presence of different life-stages. The analysis was based on a distinction between five different life-stages (Table 1): (i) larval stage, defined as length at birth up to the length at ontogenetic transition to the juvenile stage, (ii) 0+ juvenile stage, defined as the length at ontogenetic transition to the juvenile stage up to the average length attained after one year, (iii) >1+ juvenile stage, defined as the average length attained after one year up to the length at sexual maturity, (iv) small adults and (v) large adults. Length ranges of these life-stages for each species were mainly based on ranges found in the Netherlands (Table 1). Species were grouped by means of cluster analysis, using Bio Diversity Professional Beta 1 (McAleece, 1997). The classification, based on presence–absence data of the different life-stages (Table 1), was performed by calculating Jaccard coefficients. The complete linkage algorithm was used, as this algorithm is most suitable for the derivation of discrete groups (Jongman et al., 1995).

Results

Seasonal dynamics

A total of 8615 fishes belonging to 13 different species were caught in the Everlose Beek. Monthly length-frequency distribution data for the eight dominant species are presented in Fig. 2, showing seasonal variation in fish densities and length classes. These eight species, i.e. Three-spined stickleback, Stone loach, Gudgeon, Roach, Bream, Tench, Rudd, and Pike, showed fish densities that were high during the summer and low during the winter, primarily caused by a sudden increase in the number of larvae during spring and summer. Juveniles and adults of Three-spined stickleback, Stone loach, and Gudgeon were present throughout the year. Roach, Bream, Tench, Rudd, and Pike were mainly represented by larvae and juveniles. Small adults were occasionally captured, typically in spring and in low densities, whereas larger adults were notably absent throughout the year (Fig. 2). The remaining five species that were caught in the Everlose Beek, i.e. Bleak, Sunbleak, Carp, Crucian carp, and Perch, were rarely captured (not shown in Fig. 2), and the absence of larvae and juveniles suggests that these species did not reproduce in the tributary. For these species, no seasonal trend in either densities or size classes could be inferred.

Reproduction and growth

Larvae and 0+ juveniles were observed for eight of the 13 species. The presence of recently hatched

Table 1. Size-classes (in cm) representing the five life-stages for each species, used for the Jaccard clustering

Species	Life Stages				
	I	II	III	IV	V
Stone loach (*Barbatula barbatula*)	0.35–1.8	1.9–4.5	4.6–5.5	5.6–8.0	8.1–10.0
Gudgeon (*Gobio gobio*)	0.5–1.3	1.4–5.0	5.1–8.0	8.1–10.0	10.1–14.0
Three-spined stickleback (*Gasterosteus aculeatus*)	0.4–1.5	1.6–4.0	*	4.1–5.0	5.1–6.0
Bream (*Abramis brama*)	0.45–1.9	2.0–4.5	4.6–15.0	15.1–40.0	40.1–85.0
Pike (*Esox lucius*)	0.9–2.6	2.7–20.0	20.1–35.0	35.1–60.0	60.1–120.0
Tench (*Tinca tinca*)	0.35–1.9	2.0–4.5	4.6–17.0[5]	17.1–45.0	45.0–70.0
Roach (*Rutilus rutilus*)	0.6–1.7	1.8–6.0	6.1–17.0	17.1–30.0	30.1–40.0
Rudd (*Rutilus erythrophthalmus*)	0.45–1.7	1.7–6.0	6.1–15.0	15.1–30.0.	30.1–45.0
Bleak (*Alburnus alburnus*)	0.5–1.4	1.5–5.5	5.6–8.5	8.6–15.0	15.1–25.0
Sunbleak (*Leucaspius delineatus*)	0.45–1.5	1.6–2.0[6]	2.1–5.0	5.1–7.5	7.5–10.0
Carp (*Cyprinus carpio*)	0.5–1.7	1.8–10.0	10.1–40.0[7]	40.1–70.0	80.1–100.0
Crucian carp (*Carassius carassius*)	0.5–1.7	1.8–4.0	4.1–15.0	15.1–35.0	35.1–50.0
Perch (*Perca fluviatilis*)	0.5–2.1	2.2–7.0	7.1–15.0	15.1–30.0	30.1–45.0

*Three-spined stickleback attained sexual maturity in its first year. Life-stages: I. Larval stage[1]; II. 0+juveniles[1,2]; III. > 1 +juveniles[2,3,4]; IV. Small adults[3,4]; V. Large adults[3,4] (length ranges are based on: [1]Pinder (2001), [2]Pollux et al. (2004), [3]De Nie (1996), [4]Crombaghs et al. (2000), [5]Brylińska et al. (1999), [6]Gozlan et al. (2003) and [7]Baruš et al. (2002)).

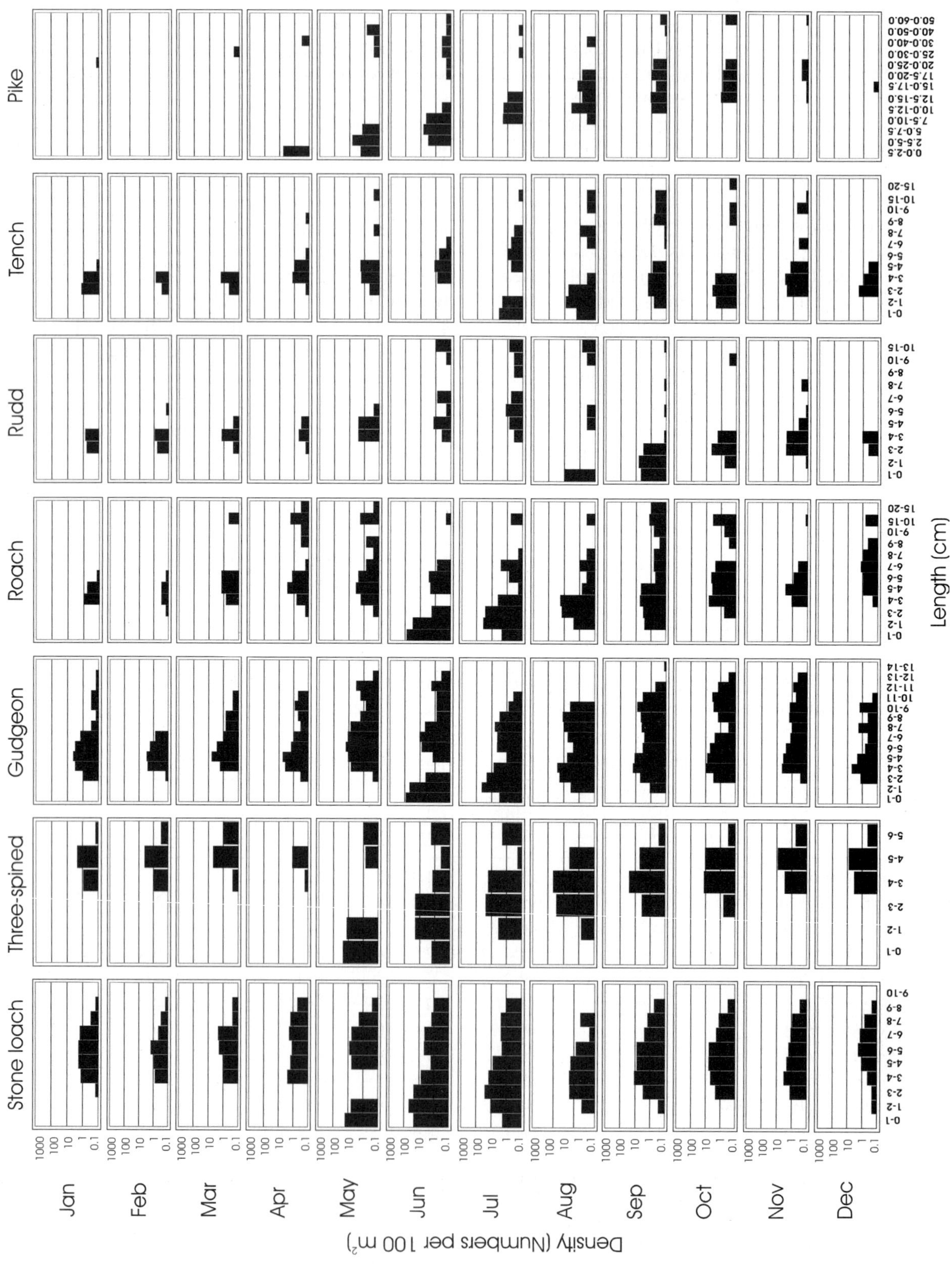
Stone loach
Three-spined
Gudgeon
Roach
Rudd
Tench
Pike
Jan
Feb
Mar
Apr
May
Jun
Jul
Aug
Sep
Oct
Nov
Dec
Density (Numbers per 100 m²)
Length (cm)

Figure 2. Monthly size-frequency distributions of Stone loach (*Barbatula barbatula*), Three-spined stickleback (*Gasterosteus aculeatus*), Gudgeon (*Gobio gobio*), Roach (*Rutilus rutilus*), Rudd (*Rutilus erythrophthalmus*), Tench (*Tinca tinca*), Pike (*Esox lucius*) and Bream (*Abramis brama*) from the Everlose Beek in 2002. Fork length on the *x*-axis (only observed size classes are presented) and numbers of fish per 100 m^2 on the *y*-axis (note the log-scale).

larvae (Fig. 3a), suggests that reproduction occurred in the Everlose Beek (cf. Nunn et al., 2002). Plots of observed lengths against sampling dates (Fig. 3a) show differences between species with respect to the beginning and duration of larval occurrences. Larvae of Pike were observed earliest (from April), those of Rudd last (until September). Stone loach had the longest period in which larvae were found (during May–July). Average growth curves and Gompertz growth equations are presented in Figure 3b. The consistent growth patterns, derived from regularly sampling larvae and juveniles, indicate that the eight species use the Everlose Beek as a nursery area during their first year.

Classification and migration

The cluster analysis, based on the presence of life-stages in the Everlose Beek (Table 1) revealed three different groups (Fig. 4). Group 1 consists of the three smallest species (maximum lengths ranging from 5 to 15 cm), for which all life-stages (from larvae to adults) were found in the tributary (Fig. 5). This group represents the numerically dominant species, accounting for 69% of all collected fishes in this study (Fig. 6c). The species in Group 2 are characterized by the predominant presence of larvae and juveniles (Fig. 5), suggesting that when they reach a certain size these species leave the tributary and migrate to the main river channel. The cut-off lengths at the *x*-axis (indicated by a black arrow) give an approximate indication of the size at migration (Cocheret de la Morinière et al., 2002): 5–15 cm for Bream, Tench, Rudd and Roach (coinciding with the size at sexual maturity) and 20–60 cm for Pike (Fig. 5). This group accounts for 30% of all collected fishes in this study (Fig. 6c). Group 3 consists of species of which predominantly larger juveniles and adults were captured in the tributary (Fig. 5). Here, the absence of larvae and small juveniles suggests that these species originate from upstream areas. This group represents the numerically rare species, accounting for less than 1% of all collected fishes in this study (Fig. 6c).

Discussion

Fish species can use tributaries during different life-stages (as larvae, juveniles, adults) for different reasons (e.g. reproduction, growth, food, shelter) and at different periods (e.g. different seasons). The results of the present study show that the fish species of the Everlose Beek can be roughly classified into three groups, viz. residents, migrants and transients, each group using the tributary in a different way (Fig. 6).

Residents (Group I)

Residents are species that complete their whole life cycle in the tributary, using it as a spawning, nursery and adult habitat (Fig. 6a). They are among the most abundant species (comprising 69% of the total catch in the Everlose Beek (Fig. 6c)). The results show that Stone loach, Gudgeon and Three-spined stickleback in the Everlose Beek are resident species. Although it cannot be excluded that some individuals in the Everlose Beek move downstream, either by drift or migration, the monthly size-frequency distributions show that fish densities of residents remain stable throughout the year, though gradually

Figure 3. Growth of Stone loach (*Barbatula barbatula*), Three-spined stickleback (*Gasterosteus aculeatus*), Gudgeon (*Gobio gobio*), Roach (*Rutilus rutilus*), Bream (*Abramis brama*), Rudd (*Rutilus erythrophthalmus*), Tench (*Tinca tinca*), and Pike (*Esox lucius*): (a) 0-age fishes captured in the Everlose Beek during January–December 2002. Each black dot represents one or more 0-age fishes, the gray dots on the *x*-axis indicate the sampling dates, the dashed lines indicate length at hatching (Pinder, 2001), the continuous lines indicate length at transition from larval to juvenile stage (Pinder, 2001) and N signifies the total number of 0-age fishes that were captured. Note the different scales on the *y*-axis for each species. (b) Average 0-age growth curves calculated by regressing average fish length at each collection date (black dots) against predicted values of the Gompertz growth function.

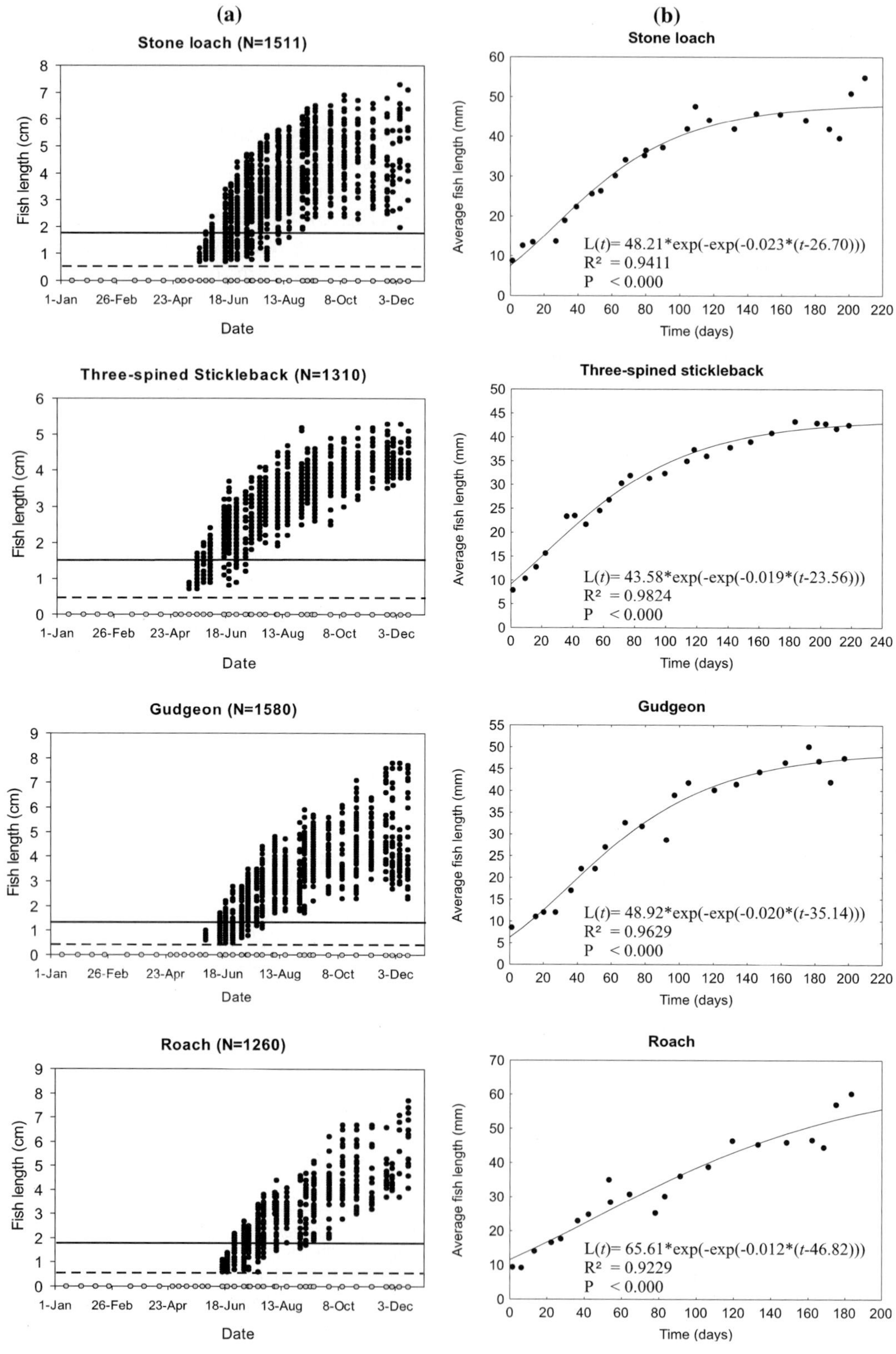
(a)
(b)
Stone loach (N=1511)
Fish length (cm)
Date
1-Jan
26-Feb
23-Apr
18-Jun
13-Aug
8-Oct
3-Dec
Stone loach
Average fish length (mm)
Time (days)
L(t)= 48.21*exp(-exp(-0.023*(t-26.70)))
R² = 0.9411
P < 0.000
Three-spined Stickleback (N=1310)
Three-spined stickleback
L(t)= 43.58*exp(-exp(-0.019*(t-23.56)))
R² = 0.9824
P < 0.000
Gudgeon (N=1580)
Gudgeon
L(t)= 48.92*exp(-exp(-0.020*(t-35.14)))
R² = 0.9629
P < 0.000
Roach (N=1260)
Roach
L(t)= 65.61*exp(-exp(-0.012*(t-46.82)))
R² = 0.9229
P < 0.000

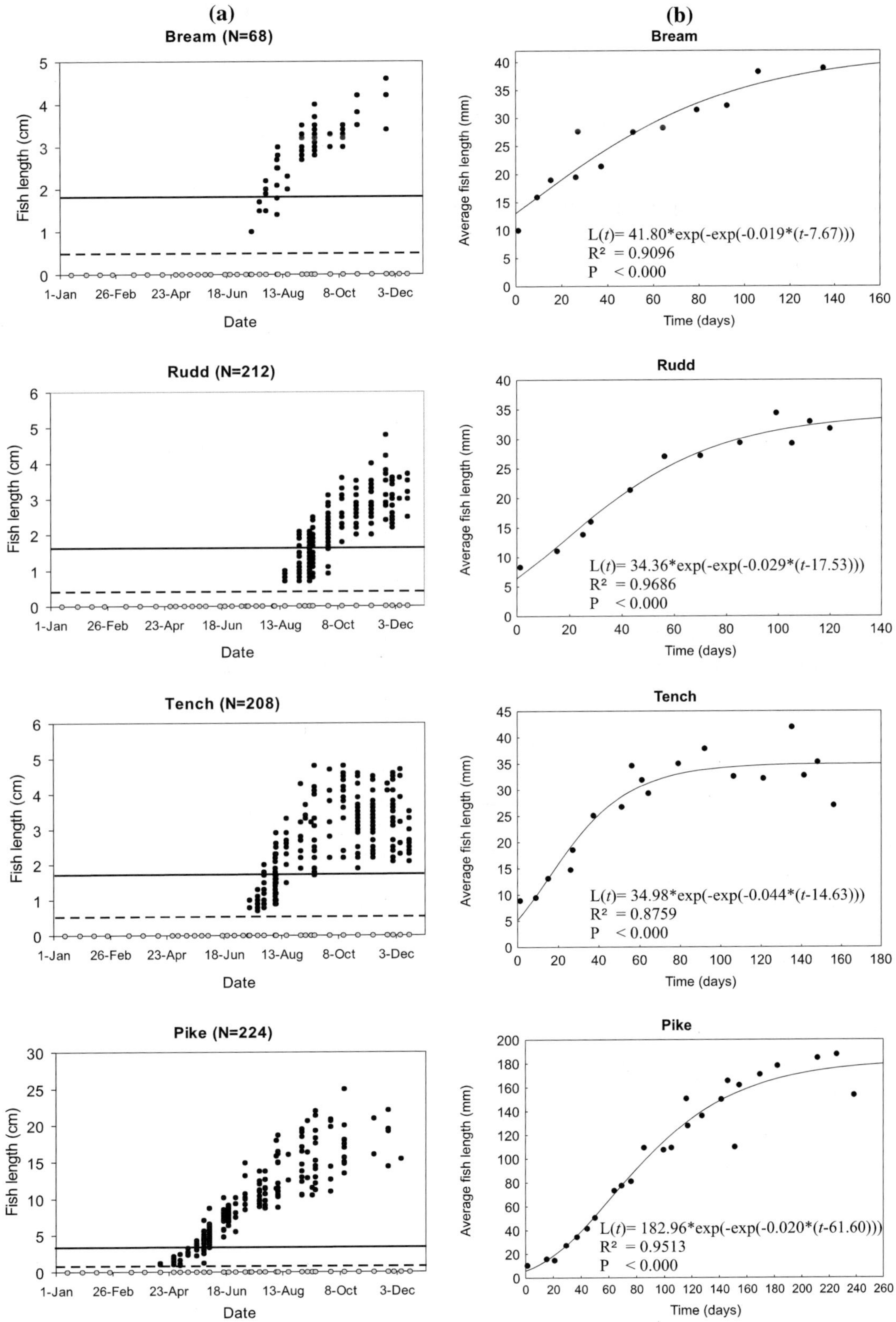

Figure 3. (Continued)

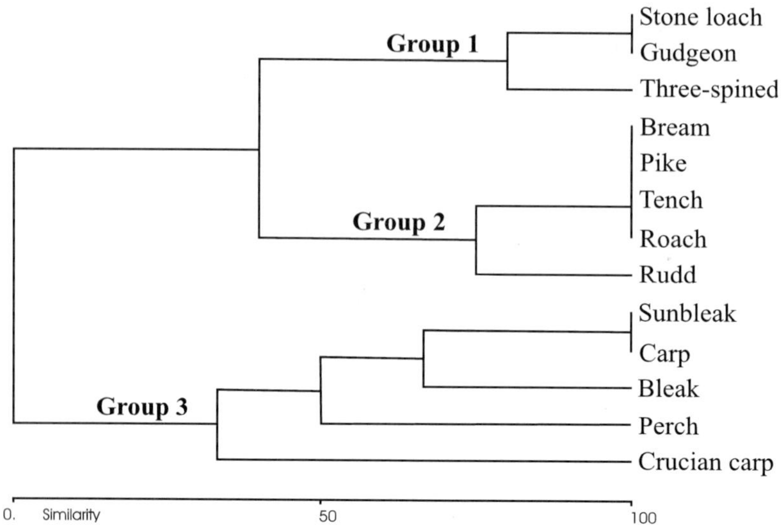

Figure 4. Complete linkage of presence–absence data for the different life-stages (Table 1), using Jaccard coefficients.

decreasing somewhat during fall and winter, indicating winter mortality rather than migration. Other studies have reported that resident species, such as Gudgeon, occupy a 'home range' (Gerking, 1953) and exhibit only restricted movements over small geographical distances, although a relatively small part of the population may sometimes display more migratory behaviour (Stott, 1961, 1967; Bănărescu,1999). In the Everlose Beek, residents are species which find suitable spawning and nursery habitats in the slow-flowing and densely vegetated lowland streams (i.e. phytophilic, phytolithophylic or psammophilic species according to Balon, 1975, 1981). Moreover, they will most likely be small species (with a maximum size range of approximately 5–15 cm), enabling them to find sufficient food and wintering habitats in the shallow streams (see below). Apart from the Stone loach, Gudgeon and Three-spined stickleback found in the Everlose Beek (in the present study) this group may also include species like Nine-spined stickleback (*Pungitius pungitius*), Spined loach (*Cobitis taenia*), and Eastern mudminnow (*Umbra pygmaea*), which have not been observed in the Everlose Beek but have been found in some of the other tributaries shown in Fig. 1 (Crombaghs et al., 2000).

Migrants (Group II)

Migrant species are characterized by the absence of larger adult size-classes (Fig. 5). Although it is known that the sampling technique, deployed in this study, tends to underestimate the presence of larger size-classes, compared to e.g. electro-fishing (Dorenbosch et al., 2000), it is unlikely that the absence of adult migrant individuals >15 cm is solely due to sampling selectivity. The capture of large adult individuals of highly mobile transient species, such as Carp (~30–40 cm), Crucian carp (~35 cm) and Perch (~20–40 cm) argues against this. Moreover, the capture of a specific size-class of Bream (~30–40 cm; Fig. 5), during a restricted period of the year (i.e., spring Fig. 2), strongly suggests that Bream >15 cm are simply absent during most of the year. Therefore, it is suggested that the absence of individuals >15 cm is not due to sampling selectivity, but the result of ontogenetic migration. Larvae and juveniles of migrant species use the tributary as nursery habitat, however, at some moment during their life cycle, typically as juveniles or small adults, they leave the tributary and migrate to the River Meuse (Fig. 6a), resulting in a low abundance or even absence of adults in the tributary during most of the year (Fig. 6d). The results suggest, that this ontogenetic migration occurs at a length of approximately 5–15 cm for Bream, Roach, Rudd and Tench, and around 25 cm for Pike.

Ontogenetic migration is generally associated with (i) ontogenetic changes in resource use (i.e. diet shifts), (ii) reduced protection from piscivores due to increased body size (i.e. the aquatic

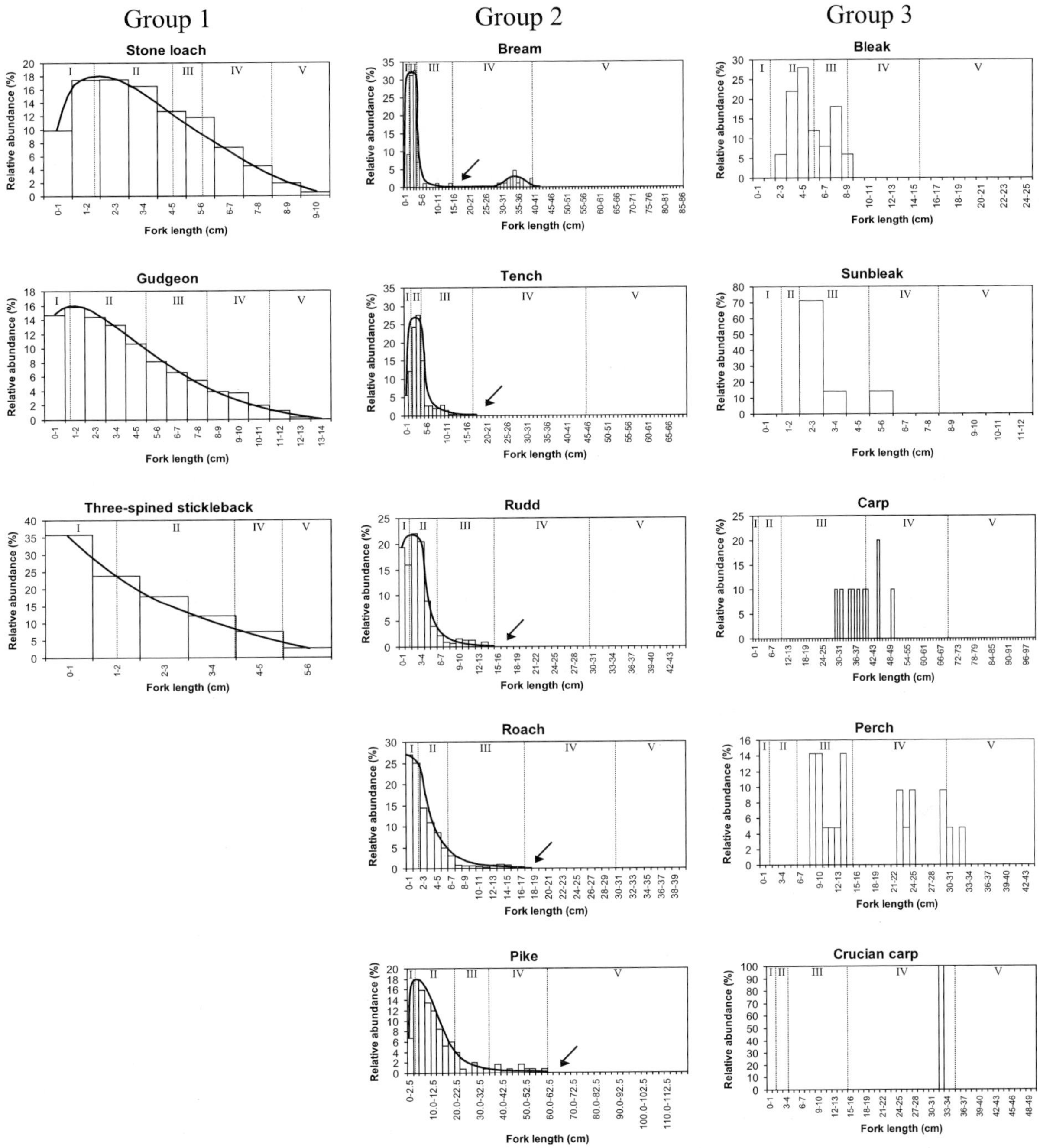

Figure 5. Relative abundance (%) of size-classes found in the Everlose Beek during 2002. The *x*-axes give the length ranges found in the Netherlands (based on De Nie, 1996; Crombaghs et al., 2000). The vertical dashed lines represent the borders of the different life-stages (see Table 1). The black arrows in Group 2 indicate the approximate length at ontogenetic migration from the tributary to the River Meuse.

vegetation no longer provides sufficient protection) or (iii) avoidance of intra-specific competition (Persson & Crowder, 1998; Cocheret de la Morinière et al., 2003). The ontogenetic migration from the tributaries to the main river is most likely initiated by seasonal environmental changes in the tributaries. During summer, the lowland tributaries on the western bank of the river Meuse are (like

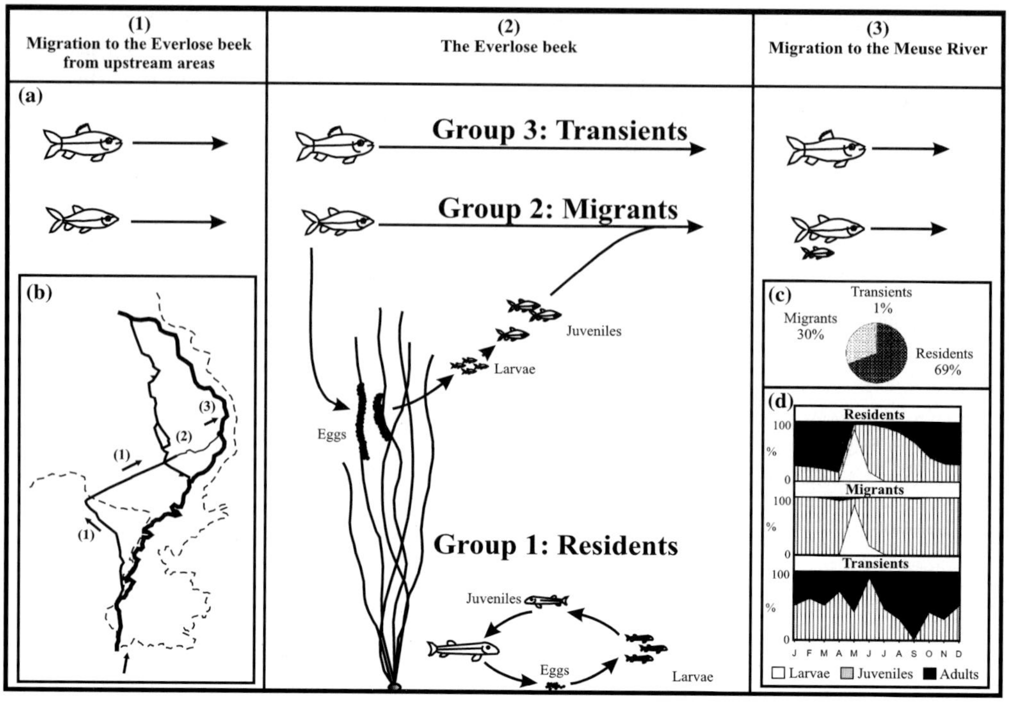

Figure 6. Schematic representation of the way in which resident, migrant, and transient fish species utilize the lowland tributary streams on the western bank of the river Meuse. (a) Residents complete their entire life-cycle in the tributary streams, migrants are born in the tributary streams but display ontogenetic migration towards the river Meuse and transients merely pass through the systems of canals and tributary streams without reproducing within the streams. (b) Adult migrant and transient fishes originate from upstream areas (i.e., the Zuid-Willemsvaart or Noordervaart canals), migrate towards the Everlose Beek and ultimately migrate farther downstream towards the river Meuse, displaying only unidirectional migration due to the presence of weirs. (c) Total catch (%) of resident, migrant and transient species in the Everlose Beek. (d) Composition (%) of life-stages (i.e. larvae, juveniles and adults) throughout the year (note the absence of adults for migrant species and the absence of larvae for transient species).

many lowland streams in north-west Europe) characterized by abundant aquatic vegetation, which provides structurally complex habitats offering refuge and rich foraging areas for small fishes (Rozas & Odum, 1988; Diehl & Koinijów, 1998; Grenouillet & Pont, 2001; Grenouillet et al., 2001). However, during the late fall and winter, the aboveground macrophyte biomass rapidly disappears, reducing refuge potential for fishes. In addition, flow velocity increases and water temperatures display larger and more rapid fluctuations, causing severe metabolic stress in fishes (Schlosser, 1991). Consequently, fishes (particularly larger >1+ individuals) migrate from the shallow tributaries in the fall and move to deeper water in the main channel which has a lower flow velocity and shows smaller temperature fluctuations (Nikolsky, 1963; Northcote 1978; Schlosser, 1991). Smaller individuals, up to a size of 15 cm, survive the winter (just like residents) in small coves in the riverbank, depressions in the riverbed, among overhanging tree roots or, like Tench and Rudd, under leaves or in mud.

Transients (Group III)

Transients are characterized by very low densities (Fig. 6c) and the absence of larvae in the tributary (Fig. 6d). Transient species most likely originate from upstream areas, i.e. the Zuid-Willemsvaart or Noordervaart canals (Fig. 6b), and do not use the Everlose Beek for their reproduction (Fig. 6a). They may dwell in the tributary for a while but ultimately migrate further downstream towards the Meuse, most likely during late fall and winter, when conditions in the tributaries change rapidly and densities of these species are lowest.

The results suggest that Bleak, Sunbleak, Perch, Carp, and Crucian carp can be considered to be transient species. These transients are species for which the shallow, slow-flowing, lowland tributaries do not provide suitable spawning and nursery habitats or optimal adult habitats. Carp and Crucian carp prefer stagnant water bodies, such as isolated floodplain lakes, both for their reproduction and as adult habitats (Baruš et al., 2002), while adult Perch and Bleak prefer the deeper and more open waters of the main channel of the river Meuse (Lelek & Buhse, 1992), where they find ample spawning and nursery habitats (Vriese et al., 1994). The Sunbleak constitutes a notable exception, finding its optimal habitat in slow-flowing, densely vegetated ditches, canals and narrow streams (Lelek & Buhse, 1992). The Everlose Beek therefore provides a very suitable habitat for this species. In the present study, very few Sunbleak were caught and reproduction was not observed. An earlier study by Akkermans (1996), however, found a large number of Sunbleak during fish sampling (using the same standardized methodology), including larvae and small juveniles (Akkermans, pers. comm.). Populations of Sunbleak are known to show extreme fluctuations in size across years, with densities varying from sporadic in one year to explosively abundant in another (Lelek & Buhse, 1992; Crombaghs et al., 2000). Therefore, despite the absence of larvae in 2002, the Sunbleak is more likely to be a resident species, since it is able to reproduce in the Everlose Beek, has a small adult size (<10 cm) enabling it to survive the winter in the shallow tributaries, and is also known elsewhere, as a non-migratory species (http://www.fishbase.org).

Potential recruitment from lowland tributaries

During the last two centuries, modifications to the geomorphology of the large lowland rivers Rhine and Meuse have resulted in steep and fortified stony embankments and a severe lack of aquatic vegetation in the main channel (Nienhuis et al., 1998; Rant, 2001; Nienhuis et al., 2002a, b), leading to a greatly reduced or even absent recruitment potential for limnophilic and rheophilic species (Vriese et al., 1994). In addition, former floodplain areas directly adjacent to regulated rivers have been turned into agricultural land, greatly reducing the total floodplain area along large rivers (Nienhuis et al., 2002a, b; Buijse et al., 2002) agree with the change. Simultaneously, flow regulation by means of dams and weirs has led to a decline in natural water level variations effectively leading to the isolation of the remaining floodplain water bodies (Bain et al., 1988; Dynesius & Nilsson, 1994; Ward & Stanford, 1995; Aarts et al., 2004). The few remaining floodplain lakes often have a depaupered fish fauna, dominated by a few eurytopic species, such as Bream, White bream and Roach (Van den Brink et al., 1996; Grift et al., 2001; Buijse et al., 2002). Newly created man-made habitats, such as gavel-pit lakes and excavated secondary side-channels connected to the main river, are mainly used by eurytopic species, to a lesser degree by rheophilic species, yet rarely by limnophilic species (Neumann et al., 1994; Staas & Neumann, 1994, 1996; Simons et al., 2001; Grift et al., 2003). Not surprisingly, the fish fauna of the large lowland rivers Rhine and Meuse in the Netherlands is currently dominated by eurytopic species (Van der Velde et al., 1990; Admiraal et al., 1993; Van den Brink et al., 1996; Raat, 2001).

What is surprising, however, is that despite the apparent lack of spawning and nursery areas for limnophilic species in the main channel, floodplain lakes, gravel-pit lakes and secondary side-channels, a few limnophilic species, particularly Pike, Tench and Rudd are still found in the river Meuse, and although they are generally found in small numbers, their occurrences have been quantified as common to locally common by Admiraal et al. (1993) and Crombaghs et al. (2000), respectively. This has given rise to the hypothesis that the recruitment sources for these predominantly phytophylic spawners are situated elsewhere in the river basin. The results of the present study suggest that regulated lowland tributaries may act as such recruitment sources, preventing the total extinction of these species in the River Meuse.

In the Netherlands, the Meuse is connected to over a 100 lowland streams (Maris et al., 2003). The lowland streams on the western bank of the Meuse in the Province of Limburg (Fig. 1) alone already have an estimated total length of over 500 km, with an average width of approximately 5–10 m (the eastern bank of the river Meuse also features a number of tributary streams, not shown in Fig. 1). Although most of these lowland streams are to some degree regulated (natural free flowing lowland rivers and streams can hardly be found in the Netherlands anymore; Verdonschot & Nijboer, 2002), they comprise an area of considerable size consisting of slow-flowing, shallow-water habitats with locally abundant vegetation, hence providing suitable spawning and nursery habitats for many species. The present study shows that phytophylic (e.g. Rudd, Tench, Pike, and Three-spined stickleback), psammophilic (Stone loach and Gudgeon), and polyphilic (Roach and Bream) spawners can reproduce in these regulated streams. A few species spend their entire lives in the shallow streams (the residents), but other species display ontogenetic migration to the river Meuse (the migrants). We suggest that regulated lowland tributaries, such as the Everlose Beek, may function as important recruitment sources for many limnophilic species in large rivers. We further suggest that the persistence of the severely reduced populations of limnophilic species in heavily modified rivers in north-western Europe, such as the river Meuse, may be attributed to recruitment from regulated lowland tributary streams.

Acknowledgement

This is publication 3586 of the Netherlands Institute of Ecology (NIOO-KNAW), and publication 397 of the Centre for Wetland Ecology (CWE).

References

Aarts, B. G. W. & P. H. Nienhuis, 2003. Fish zonations and guilds as the basis for assessment of ecological integrity of large rivers. Hydrobiologia 500: 157–178.

Aarts, B. G. W., F. W. B. van den Brink & P. H. Nienhuis, 2004. Habitat loss as the main cause of the slow recovery of fish faunas of regulated large rivers in Europe: the transversal floodplain gradient. River Research and Applications 20: 3–23.

Admiraal, W., G. van der Velde, H. Smit & W. G. Cazemier, 1993. The rivers Rhine and Meuse in The Netherlands: present state and signs of ecological recovery. Hydrobiologia 265: 97–128.

Akkermans, R. W., 1996. De verspreiding van het Vetje in Limburg. Natuurhistorisch Maandblad 85: 38–41 (in Dutch).

Allan, J. D., 1995. Stream Ecology: Structure and Function of Running Waters. Chapman & Hall, London.

Arlinghaus, R., C. Engelhardt, A. Sukhodolov & C. Wolter, 2002. Fish recruitment in a canal with intensive navigation: implications for ecosystem management. Journal of Fish Biology 61: 1386–1402.

Bain, M. B., J. T. Finn & H. E. Booke, 1988. Streamflow regulation and fish community structure. Ecology 69: 382–392.

Balon, E. K., 1975. Reproductive guilds of fishes: proposal and definition. Journal of the Fisheries Research Board of Canada 32: 821–864.

Balon, E. K., 1981. Additions and amendments to the classification of reproductive styles in fishes. Environmental Biology of Fishes 6: 377–389.

Bănărescu, P. M., 1999. *Gobio gobio* (Linnaeus, 1758). In Bănărescu, P. M. (ed.), The Freshwater Fishes of Europe Vol. 5/I Cyprinidae 2 Part I. Aula-Verlag GmbH, Wiebelsheim: 24–46.

Baruš, V., M. Peňáz & K. Kohlmann, 2002. *Cyprinus carpio* (Linnaeus, 1758). In Bănărescu, P. M. & H. J. Paepke (eds), The Freshwater Fishes of Europe Vol. 5/III Cyprinidae 2 Part III. Aula-Verlag GmbH, Wiebelsheim: 85–179.

Borcherding, J., M. Bauerfeld, D. Hintzen & D. Neumann, 2002. Lateral migrations of fishes between floodplain lakes and their drainage channels at the Lower Rhine: diel and seasonal aspects. Journal of Fish Biology 61: 1154–1170.

Buijse, A. D., H. Coops, M. Staras, L. H. Jans, G. J. van Geest, R. E. Grift, B. W. Ibelings, W. Oosterberg & F. C. J. M. Roozen, 2002. Restoration strategies for river floodplains along large lowland rivers in Europe. Freshwater Biology 47: 889–907.

Brylińska, M., E. Bryliński & M. Bnińska, 1999. *Tinca tinca* (Linnaeus, 1758). In Bănărescu, P. M. (ed.), The Freshwater Fishes of Europe Vol. 5/I Cyprinidae 2 Part I. Aula-Verlag GmbH, Wiebelsheim: 248–281.

Calow, P. & G. E. Petts, 1994. The Rivers Handbook: Hydrological and Ecological Principles Vols. 1–2. Blackwell Science Ltd, Oxford.

Cocheret de la Morinière, E., B. J. A. Pollux, I. Nagelkerken & G. van der Velde, 2002. Post-settlement life-cycle migration patterns and habitat preference of coral reef fish that use seagrass and mangrove habitats as nurseries. Estuarine, Coastal and Shelf Science 55: 309–321.

Cocheret de la Morinière, E., B. J.A. Pollux, I. Nagelkerken, M. A. Hemminga, A. H. L. Huiskes & G. van der Velde, 2003. Ontogenetic dietary changes of coral reef fishes in the mangrove-seagrass-reef continuum: stable isotopes and gut-content analysis. Marine Ecology Progress Series 246: 279–289.

Copp, G. H., 1997. Microhabitat use of fish larvae and 0+ juveniles in a highly regulated section of the river Great Ouse. Regulated Rivers: Research & Management 13: 267–276.

Crombaghs, B. H. J. M., R. W. Akkermans, R. E. M. B. Gubbels & G. Hoogerwerf, 2000. Vissen in Limburgse beken: De Verspreiding en Ecologie van Vissen in Stromende Wateren in Limburg. Stichting Natuurpublicaties Limburg, Maastricht (in Dutch).

De Graaf, G. J., A. F. Born, A. M. K. Uddin & S. Huda, 1999. Larval fish movement in the River Lohajang, Tangail, Bangladesh. Fisheries Management and Ecology 6: 109–120.

Delmastro, G., 1982. I pesci del bacino del Po. Museo Civico di storia Naturale di Carmagnola, CLESAC, Milano (in Italian).

De Nie, H. W., 1996. Atlas van de Nederlandse Zoetwatervissen. Media Publishing Int, BV, Doetinchem (in Dutch).

Diehl, S. & R. Kornijów, 1998. Influence of submerged macrophytes on trophic interactions among fish and macroinvertebrates. In Jeppensen, E. M. Søndergaard, M. Søndergaard, & K. Christoffersen (eds), The Structuring Role of Submerged Macrophytes in Lakes. Springer-Verlag, New York: 24–46.

Dorenbosch, M., G. van der Velde & B. H. J. M. Crombaghs, 2000. Schepnet versus elektrisch net: een vergelijking tussen twee vismethoden. Natuurhistorisch Maandblad 89: 62–66 (in Dutch).

Dynesius, M. & C. Nilsson, 1994. Fragmentation and flow regulation of river systems in the northern third of the world. Science 266: 753–762.

Gamito, S., 1998. Growth models and their use in ecological modelling: an application to a fish population. Ecological Modelling 113: 83–94.

Gerking, S. D., 1953. Evidence for the concepts of home range and territory in stream fishes. Ecology 34: 347–365.

Gorman, O. T. & J. R. Karr, 1978. Habitat structure and stream fish communities. Ecology 59: 507–515.

Gozlan, R. E., A. C. Pinder, S. Durand & J. Bass, 2003. Could the small size of sunbleak, *Leucaspius delineatus* (Pisces, Cyprinidae) be an ecological advantage in invading British waterbodies? Folio Zoologica 52: 99–108.

Grenouillet, G. & D. Pont, 2001. Juvenile fishes in macrophyte beds: influence of food resources, habitat structure and body size. Journal of Fish Biology 59: 939–959.

Grenouillet, G., D. Pont & J. M. Olivier, 2001. Linking zooplankton and juvenile fish assemblages in a large lowland river: influence of submerged macrophytes. Archiv für Hydrobiologie 151: 383–404.

Grift, R. E., A. D. Buijse, W. L. T. van Densen, M. A. M. Machiels, J. Kranenbarg, J. G. P. Klein Breteler & J. J. G. M. Backx, 2003. Suitable habitats for 0-group fish in rehabilitated floodplains along the lower river Rhine. River Research and Applications 19: 353–374.

Grift, R. E., A. D. Buijse, J. G. P. Klein Breteler, W. L. T. van Densen, M. A. M. Machiels & J. J. G. M. Backx, 2001. Migration of bream between the main channel and floodplain lakes along the lower River Rhine during the connection phase. Journal of Fish Biology 59: 1033–1055.

Jongman, R. H. G., C. J. F. ter Braak & O. F. R. van Tongeren, 1995. Data Analysis in Community and Landscape Ecology. Cambridge University Press, Cambridge.

Jungwirth, M., O. Moog & S. Muhar, 1993. Effects of river bed restructuring on fish and benthos of a fifth order stream, Melk, Austria. Regulated Rivers: Research & Management 8: 195–204.

Junk, W. J., P. B. Bayley & R. E. Sparks, 1989. The flood pulse concept in river–floodplain systems. Canadian Special Publication of Fisheries and Aquatic Sciences 106: 110–127.

Jurajda, P., 1995. Effect of channelization and regulation on fish recruitment in a flood plain river. Regulated Rivers: Research & Management 10: 207–215.

Jurajda, P., 1999. Comparative nursery habitat use by 0+ fish in a modified lowland river. Regulated Rivers: Research & Management 15: 113–124.

Kirjasniemi, M. & T. Valtonen, 1997. Winter mortality of young-of-the-year pikeperch (*Stizostedion lucioperca*). Ecology of Freshwater Fish 6: 155–160.

Lelek, A. & G. Buhse, 1992. Fische des Rheins-früher und heute. Springer-Verlag, Berlin (in German).

Maris, M., J. J. Bakhuizen, H. Bakker & B. Peters, 2003. De betekenis van het grens- en zandmaasproject voor de visfauna in de Maas en zijbeken. Natuurhistorisch Maandblad 92: 274–279 (in Dutch).

McAleece, N., 1997. Biodiversity Professional Beta 1. The Natural History Museum London and The Scottisch Association for Marine Science, UK.

Molls, F., 1997. Populationsbiologie der Fischarten einer niederrheinischen Auenlandschaft Reproductionserfolge, Lebenszyklen, Kurzdistanzwanderungen. PhD thesis, University of Cologne, Cologne (in German).

Neumann, D., C. Seidenberg-Busse, A. Petermeier, S. Staas, F. Molls & J. Rutschke, 1994. Gravel-pit lakes connected with the river Rhine as a reserve for high productivity of plankton and young fish. Water Science & Technology 29: 267–271.

Nienhuis, P. H., J. P. Bakker, A. P. Grootjans, R. D. Gulati & V. N. de Jonge, 2002a. The state of the art of aquatic and semi-aquatic ecological restoration projects in the Netherlands. Hydrobiologia 478: 219–233.

Nienhuis, P. H., A. D. Buijse, R. S. E. W. Leuven, A. J. M. Smits, R. J. W. de Nooij & E. M. Samborska, 2002b. Ecological rehabilitation of the lowland basin of the river Rhine (NW Europe). Hydrobiologia 478: 53–72.

Nienhuis, P. H., R. S. E. W. Leuven & A. M. J. Ragas (eds), 1998. New Concepts for Sustainable Management of River Basins. Backhuys Publishers, Leiden.

Nikolsky, G. V., 1963. The Ecology of Fishes. Academic Press, London.

Nilsson, C., R. Jansson & U. Zinko, 1997. Long-term responses of river-margin vegetation to water-level regulation. Science 276: 798–800.

Northcote, T. G., 1978. Migratory strategies and production in freshwater fishes. In Gerking, S. D. (ed.), Ecology of Freshwater Fish Production. Blackwell Scientific Publications, Oxford: 326–259.

Nunn, A. D., I. G. Cowx & J. P. Harvey, 2002. Recruitment patterns of six species of cyprinid fishes in the lower River Trent, England. Ecology of Freshwater Fish ll: 74–84.

Persson, L. & L. B. Crowder, 1998. Fish–habitat interactions mediated via ontogenetic niche shifts. In Jeppensen, E. M. Søndergaard, M. Søndergaard & K. Christoffersen (eds), The Structuring Role of Submerged Macrophytes in Lakes. Springer-Verlag, New York: 3–23.

Petts, G. E. & C. Amoros, 1996. Fluvial Hydrosystems. Chapman & Hall, London.

Pinder, A. C., 2001. Keys to larval and juvenile stages of coarse fishes from fresh waters in the British Isles Freshwater Biological Association. Scientific Publication 60: 1–136.

Pollux, B. J. A., A. Korosi, W. C. E. P. Verberk & P. M. J. Pollux, 2004. Voortplanting, groei en migratie van vissen in de Everlose beek. Natuurhistorisch Maandblad 93: 1–8 (in Dutch).

Raat, A. J. P., 2001. Ecological rehabilitation of the Dutch part of the River Rhine with special attention to the fish. Regulated Rivers: Research & Management 17: 131–144.

Reichard, M., P. Jurajda & M. Ondrackova, 2002. Interannual variability in seasonal dynamics and species composition of drifting young-of-the-year fishes in two European lowland rivers. Journal of Fish Biology 60: 87–101.

Robinson, A. T., R. W. Clakson & R. E. Forrest, 1998. Dispersal of larval fishes in a regulated river tributary. Transactions of the American Fisheries Society 127: 772–786.

Rozas, L. P. & W. E. Odum, 1988. Occupation of submerged aquatic vegetation by fishes: testing the roles of food and refuge. Oecologia 77: 101–106.

Schiemer, F., 1985. Die Bedeutung Von Augewässern als Schutzzonen für die Fishfauna. Oesterreichische Wasserwirtschaft 37: 239–245.

Schlosser, I. J., 1991. Stream fish ecology: a landscape perspective. BioScience 41: 704–712.

Simons, J. H. E. J., C. Bakker, M. H. I. Schrop, L. H. Jans, F. R. Kok & R. E. Grift, 2001. Man-made secondary channels along the river Rhine (The Netherlands); results of post-project monitoring. Regulated Rivers: Research & Management 17: 473–491.

Staas, S. & D. Neumann, 1994. Reproduction of fish in the lower river Rhine and connected gravel-pit lakes. Water Science & Technology 29: 311–313.

Staas, S. & D. Neumann, 1996. The occurrence of larval and juvenile 0+ fish in the Lower River Rhine. Archiv für Hydrobiologie Supplement 113, Large Rivers 10: 325–332.

Steinberg, L., 1992. Fische unserer Bäche und Flüsse. Verbreitung, Gefährdung und Schutz in Nordrhein-Westfalen, Ministerium für Umwelt, Raumordnung und Landwirtschaft des Landes Nordrhein-Westfalen, Düsseldorf (in German).

Stott, B., 1961. Movement of coarse fish in rivers. Nature 190: 737–738.

Stott, B., 1967. The movements and population densities of roach (*Rutilus rutilus* (L.)) and gudgeon (*Gobio gobio* (L.)) in the River Mole. Journal of Animal Ecology 36: 407–423.

Vandelannoote, A., R. Yseboodt, B. Bruylants, R. Verheyen, J. Coeck, C. Belpaire, G. Van Thuyne, B. Denayer, J. Beyens, D. De Charleroy, J. Maes & P. Vandenabeele, 1998. Atlas van de Vlaamse beeken riviervissen. Water-Energik-vLario (WEL), Wijnegem (in Dutch).

Van den Brink, F.W.B., 1994. Impact of hydrology on floodplain lake ecosystems along the lower Rhine and Meuse. PhD thesis, University of Nijmegen, Nijmegen.

Van den Brink, F. W. B., G. van der Velde, A. D. Buijse & A. G. Klink, 1996. Biodiversity in the lower Rhine and Meuse River-Floodplains: its significance for ecological river management. Netherlands Journal of Aquatic Ecology 30: 129–149.

Van der Velde, G., F. W. B. van den Brink, M. van der Gaag & P. J. M. Bergers, 1990. Changes in numbers of mobile macroinvertebrates and fish in the river Waal in 1987, studied by sampling the cooling-water intakes of a power plant: first results of a Rhine biomonitoring project. In Kinzelbach, R. & G. Friedrich (eds), Biologie des Rheins. Gustav Fischer Verlag, Stuttgart: 325–342.

Vannote, R. L., G. W. Minshall, K. W. Cummins, J. R. Sedell & C. E. Cushing, 1980. The river continuum concept. Canadian Journal of Fisheries and Aquatic Sciences 37: 130–137.

Verdonschot, P. F. M. & R. C. Nijboer, 2002. Towards a decision support system for stream restoration in the Netherlands: an overview of restoration projects and future needs. Hydrobiologia 478: 131–148.

Vriese, F. T., S. Semmekrot & A. J. P. Raat, 1994. Assessment of spawning and nursery areas in the River Meuse. Water Science & Technology 29: 297–299.

Ward, J. V., 1989. The four-dimensional nature of lotic ecosystems. Journal of the North American Benthological Society 8: 2–8.

Ward, J. V. & J. A. Stanford, 1995. Ecological connectivity in alluvial river ecosystems and its disruption by flow regulation. Regulated Rivers: Research & Management 11: 105–119.

Ward, J. V., K. Tockner, D. B. Arscott & C. Claret, 2002. Riverine landscape diversity. Freshwater Biology 47: 517–539.

Welcomme, R. L., 1995. Relationships between fisheries and the integrity of river systems. Regulated Rivers: Research & Management 11: 121–136.

Wolter, C. & R. Arlinghaus, 2003. Navigation impacts on freshwater fish assemblages: the ecological relevance of swimming performance. Reviews in Fish Biology and Fisheries 13: 63–89.

Wolter, C., 2001. Conservation of fish species diversity in navigable waterways. Landscape and Urban Planning 53: 135–144.

Hydrobiologia (2006) 565:121–133 © Springer 2006
R.S.E.W. Leuven, A.M.J. Ragas, A.J.M. Smits & G. van der Velde (eds), Living Rivers: Trends and Challenges in Science and Management
DOI 10.1007/s10750-005-1909-3

Amphibian diversity and nestedness in a dynamic floodplain river (Tagliamento, NE-Italy)

K. Tockner[1,2,*], I. Klaus[1], C. Baumgartner[3] & J.V. Ward[1]
[1]*Department of Limnology, EAWAG/ETH, 8600 Dübendorf, Switzerland*
[2]*Institute of Ecosystem Studies (IES), Box AB, Millbrook, 12545 NY, USA*
[3]*Alluvial Zone National Park, 2304, Orth/Donau, Austria*
*(*Author for correspondence: E-mail: Klement.Tockner@eawag.ch)*

Key words: braided river, *Bufo bufo*, conservation, disturbance, indicator species, island, large wood, population, parafluvial pond, *Rana latastei*, restoration

Abstract

Amphibian distribution and assemblage structure were investigated along the last morphologically intact river corridor in Central Europe (Tagliamento). Thirteen taxa were identified with *Rana latastei* and *Bufo bufo* being the predominant species. In the main study reach, a 2 km^2 dynamic island-braided floodplain in the middle section of the river, 130 water bodies were delineated that were situated either in the active floodplain (82 sites) or in the adjacent riparian forest (48 sites). Results demonstrated that the active floodplain increased appreciably the available habitat for amphibians, despite frequent disturbances by floods or droughts. Amphibian richness within a given habitat was significantly correlated with distance from vegetated islands, fish density, and water temperature. In the active floodplain, species distribution was highly predictable, exhibiting nearly perfect nestedness, suggesting that selective colonisation and extinction processes predominated. The degree of nestedness was much higher than in the adjacent riparian forest or in regulated floodplains in Central Europe. Results clearly emphasise that amphibians can exploit the entire hydrodynamic gradient, except the main channel. In the active floodplain, vegetated islands and large woody debris are important, directly and indirectly, in maintaining both habitat and amphibian diversity and density in this gravel-bed river.

Introduction

Globally, amphibian populations have declined over the past several decades and continue to do so (Houlahan et al., 2000; Stuart et al., 2004). Possible underlying causes of the decline are changes in climate, increased exposure to UV-B radiation, increased prevalence of diseases, acidification, water pollution, habitat fragmentation, and habitat loss (Leuven et al., 1986; Alford & Richards, 1999; Kiesecker et al., 2001). Therefore, many amphibian species are listed as threatened or endangered, regionally and globally (Beebee, 1996; Nöllert & Nöllert, 1992).

Natural floodplains are highly dynamic environments with floods as the primary agent of disturbance. Their high species diversity and landscape-corridor function place floodplains high on the conservation agenda (Ward et al., 1999a; Hughes & Rood, 2001; Tockner & Stanford, 2002). However, today they are among the most endangered ecosystems worldwide. In Europe, for example, more than 90% of the former floodplains either disappeared or they are functionally extinct (Tockner et al., 2006). Amphibians are generally considered as ‘indicators’ of stable floodplain ponds with a low degree of hydrological connectivity, or as indicators of temporary waters lacking fish predators (e.g., Waringer-Löschenkohl & Waringer, 1990; Joly & Morand, 1994; Morand & Joly, 1995; Wellborn et al., 1996; Skelly, 1997; Tockner et al., 1999; Kuhn et al., 2001). Limited information is available, however, about amphibian populations in dynamic gravel-bed rivers, since such rivers were never the main focus of amphibian research (e.g., Beebee, 1996). One reason was that nearly all formerly dynamic floodplain rivers in developed countries were regulated during the last two centuries (Petts et al., 1989; Nilsson et al., 2005; Tockner & Stanford, 2002). Further, flood prone gravel-bed rivers were not expected to provide hospitable habitats for amphibians. Kuhn (1993), however, demonstrated that species such as *Bufo bufo* exhibited a pronounced reproduction plasticity that allowed this species to exploit dynamic and ephemeral habitats in gravel-bed rivers.

Understanding patterns and processes in natural river corridors is a prerequisite for a sustainable conservation and management of their biodiversity (e.g., Ward & Tockner, 2001). The Tagliamento River in NE-Italy, the last morphologically intact river corridor in the Alps, offered the rare opportunity to investigate amphibian populations under natural environmental conditions. In the main investigation area, an island-braided floodplain, we compared amphibian population density, diversity and nestedness in the active floodplain (area that extends laterally to the lower limit of persistent vegetation and is frequently modified by floods) and the adjacent riparian forest (periodically inundated by the river). In the present paper, we link habitat heterogeneity with the composition and distribution of amphibians and identify indicator species for different floodplain habitats. In addition, we present an empirical example of nestedness in the dynamic floodplain system and compare it with more regulated riverine floodplains. Measure of nestedness provides a quantitative indicator of the degree of community 'order' in fragmented systems (e.g., Atmar & Patterson, 1993; Patterson & Atmar, 2000).

Materials and methods

The Fiume Tagliamento

The Fiume Tagliamento in NE-Italy (46° N, 12° 30′ E; Fig. 1) is the last large gravel-bed river in the Alps that has escaped intensive river management (Müller, 1995; Ward et al., 1999b; Tockner et al., 2003). More than 70% of the catchment area (2580 km^2) is located within the southern fringe of the Alps, with Mt. Coglians as the highest peak (2781 m a.s.l.). The Tagliamento is a 7th order river, characterised by a flashy hydrological regime, with highest discharges during spring and autumn. The main-stem corridor covers about 150 km^2. The corridor is characterised by a high number of vegetated islands within the active zone (652 islands $>$0.01 ha), numerous gravel bars (952), a considerable habitat diversity and a continuous riparian woodland along the margins of the active channel (Tockner & Ward, 1999). For a complete description of the catchment and longitudinal geomorphic features see Ward et al. (1999b), Arscott et al. (2000, 2002), Gurnell et al. (2001), and Tockner et al. (2003).

We investigated the amphibian fauna in six geomorphic reaches along the entire corridor, with detailed studies in an island-braided reach in the middle section of the river. The six geomorphic reaches are: constrained headwater streams (Reach I, 1005–1200 m a.s.l.), headwater island-braided floodplain (Reach II, 705 m a.s.l.), bar-braided floodplain (Reach III, 200 m a.s.l.), island-braided lowland reach (Reach IV, 180 m a.s.l.), braided-to-meandering transitional floodplain (Reach V, 19 m a.s.l.) and meandering floodplain (Reach VI, 5 m a.s.l.).

The main investigation focused on a 2 km^2 island-braided floodplain in Reach IV (river-km 80; Figs 1 & 2). There, the floodplain was separated into the active area frequently inundated and reworked by floods and the adjacent riparian forest only inundated during annual floods. Along the left bank, hillslope forests of Monte Ragogna bordered the active floodplain. Along the right bank, the riparian forest extends laterally to a distance of about 0.5–1 km. Bare gravel, aquatic habitats and vegetated islands were the main landscape elements of the active floodplain. The Tagliamento is characterised by a dynamic flood regime, with turnover rates of aquatic habitats in Reach IV as high as 50 % during a single flood season (Arscott et al., 2002; Van der Nat et al., 2003).

Sampling methods and data analyses

Along the mainstem corridor, all water bodies in the six reaches were sampled in March, May, and July 2000, periods without severe floods. In the detailed investigation of the floodplain in Reach IV, the exact location and area of aquatic habitats

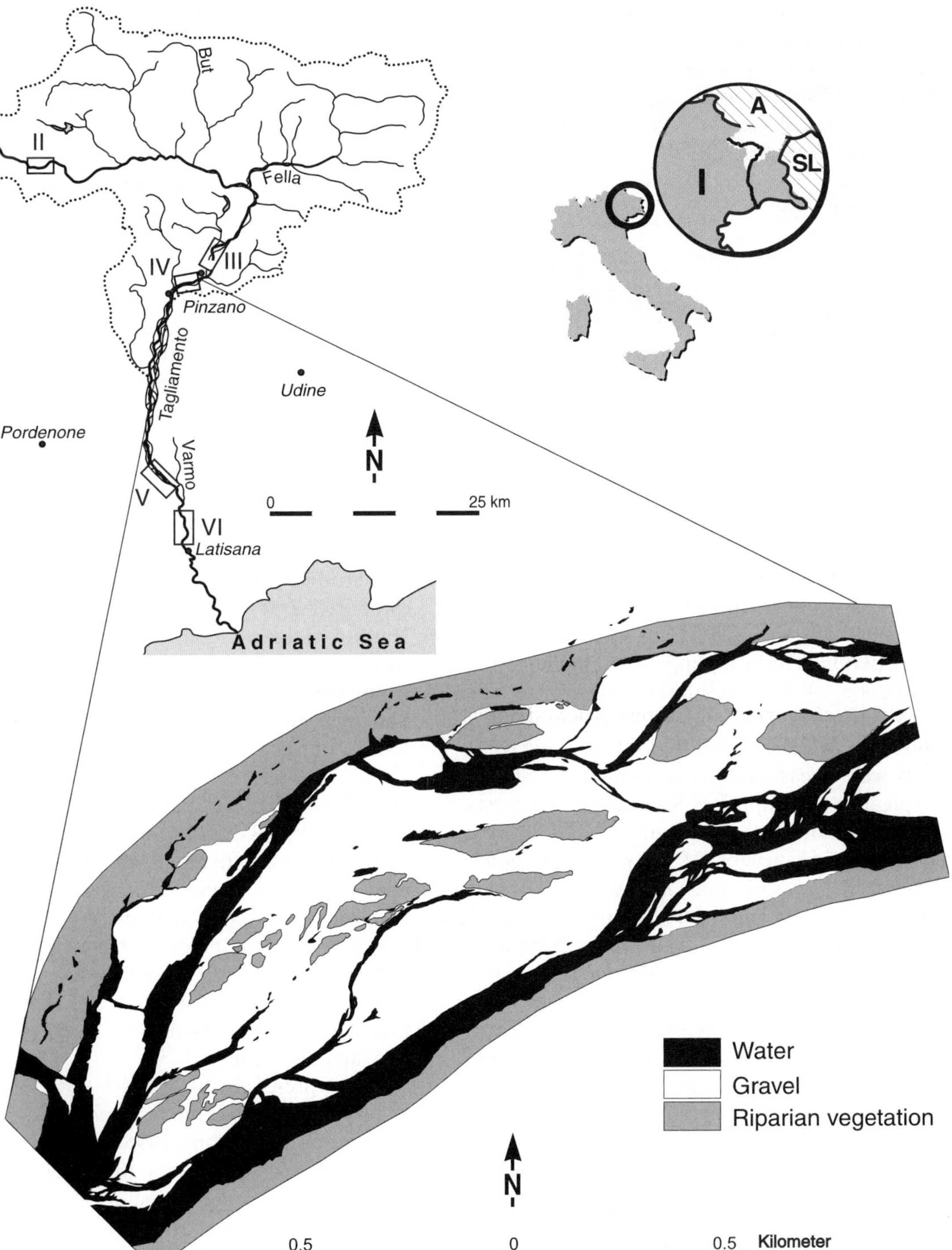

Figure 1. Tagliamento catchment, locations of study reaches I–VI, and map of the main study area (Reach IV; mapping date: May 2000). Most ponds investigated are too small to be shown at this scale. The river flows from right to left.

Figure 2. The main investigation area in the mid-section of Reach IV. Active floodplain width is up to 900 m (Photo: D. Arscott, May 2000).

within the active floodplain and the adjacent riparian forest were measured using a differential-GPS (Global Positioning System). Individual water bodies were marked (130 in all) and numbered in the field. Between 7 March and 15 July 2000 all water bodies were repeatedly (totally eight times) visited throughout the amphibian breeding period. This sampling period included the larval period of all amphibians known to use Central European floodplains as spawning sites (Nöllert & Nöllert, 1992). Further, adjacent uplands were occasionally investigated for species not typical for floodplains. Species presence was based on eggs or larvae encountered in a specific water body. Egg masses of brown frogs (*Rana dalmatina, R. latastei*, and *R. temporaria*) and common toad (*Bufo bufo*) were quantified for each water body. Since it was difficult at the beginning to separate egg clutches of *R. dalmatina* and *R. latastei*, density data of brown frogs were lumped in the analyses. The number of egg strings of *B. bufo* has to be considered as an approximate value since egg strings from several females are sometimes inextricably entangled on the same spawning support or have been broken. Densities were standardised for both aquatic area and terrestrial habitat area (riparian forest and vegetated islands; egg masses per ha). The riparian forest extended laterally over a distance of ca. 500 m from the active floodplain to a main road and a railway that follow the course of the river. For brown frogs and the common toad a lateral terrestrial breeding migration distance of 500 m was considered as appropriate, although migration distance can be up to 1500 m (C. Baumgartner, unpubl. data).

Because it was not possible to distinguish the forms of green frogs expected for the Tagliamento corridor (*Rana lessonae, R. klepton esculenta*; Günther & Plöttner, 1994, Lapini et al., 1999) from eggs or larvae, they were grouped as a single taxon in all analyses. Maximum water depth (m), specific conductance (μS; portable meter), water temperature (°C) and the presence of fish (four abundance classes based on visual examinations: absent, rare, abundant, dominant) were recorded during each field campaign. Oxygen (% saturation), sediment composition (relative proportion of silt, sand, fine and coarse gravel, coarse particulate organic matter), density of riparian vegetation (four cover classes: <10%, 10–25%, 25–75% and >75%), macrophyte cover (four classes as for riparian vegetation), accumulations of large woody debris (LWD) (four cover classes as for riparian vegetation), and actual surface area (m^2) were estimated every second visit (four times totally). Rank-correlation analyses (Spearman-rank test) were used to test the relationship between environmental factors (average values), taxa richness and egg mass density as a surrogate for adult female population size.

Two-Way INdicator SPecies ANalysis (TWINSPAN), a complex clustering method (Hill, 1979), was used to identify indicator species characterising individual sampling sites (based on presence/absence). Samples were ordinated using reciprocal averaging and clusters were arranged subsequently in a hierarchical procedure. The habitat association between species was also calculated. For each pair of species the contingency table of presence/absence was used to calculate a Chi-squared value.

The presence of nested distributions was estimated using the 'nestedness calculator' (Atmar & Patterson, 1995). Its metric, 'system temperature' (*T*), reflects the extent of order in a presence-absence species matrix. Perfectly nested communities have a $T=0$, and maximally disordered ones a $T=100$. Analyses were carried out on presence/absence matrices that have been packed into a state of minimum unexpectedness. The characteristic '*T*' of randomised matrices was calculated through Monte Carlo simulations (500 iterations per test) and compared with

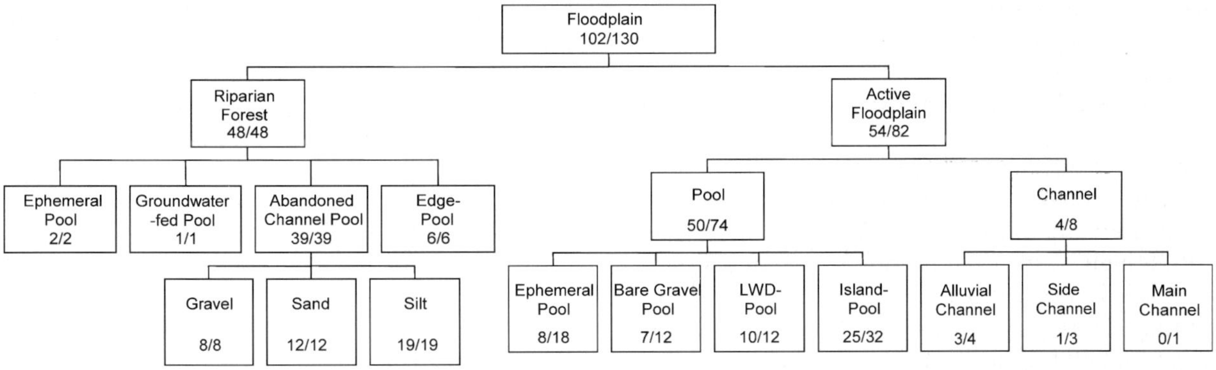

Figure 3. Floodplain water bodies separated into different types based on their position and morphology. Numbers indicate water bodies with brown frogs (*Rana latastei, R. temporaria, R. dalmatina*) and/or *Bufo bufo* (first value) and the total number of water bodies in each category (second value). LWD: Large woody debris.

observed 'temperatures'. The nestedness calculator also produces a measure of unexpected presence or absence of species. Species with many such deviations from perfect nestedness cause idiosyncratic patterns due to their 'species temperatures' that are much higher than the overall matrix temperature. Such patterns can be explained with respect to extinction and recolonisation, and may reflect exclusions or the presence of unique geomorphic features on some 'islands' (Atmar & Patterson, 1993). Nestedness values were compared with values calculated from more regulated floodplains along the Danube and the Rhône rivers. There, data were collected in an intensity comparable to the present study (Joly & Morand, 1994; C. Baumgartner, unpubl. data).

Results

Habitat identification and spawning sites

In the main investigation area (Reach IV), 130 water bodies were identified, 1/3 of which were located in the riparian forest and 2/3 in the active corridor (Fig. 3). The average surface area of lentic water bodies (floodplain ponds) was 340 m^2, without a significant difference between those located in the two parts of the floodplain. Based on

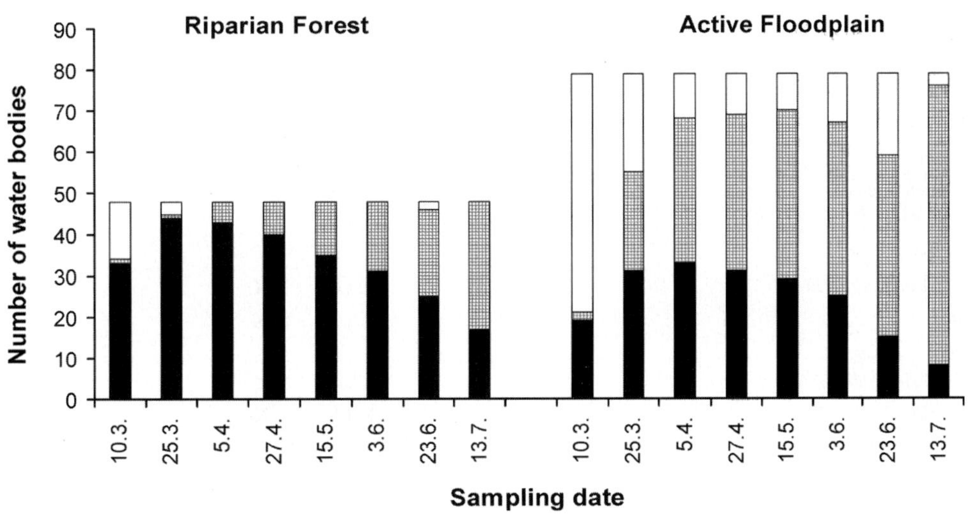

Figure 4. The total number of water bodies in the active floodplain and the adjacent riparian forest of Reach IV and their proportion with egg masses/larvae (black), without egg masses/larvae (hatched), and those that were dry (white) during each sampling date. Egg masses/larvae were from brown frogs *(Rana latastei, R. temporaria, R. dalmatina)* and/or the common toad (*Bufo bufo*).

their location and their formation, ponds were divided into several classes and subclasses. Within the active plain, more than 60% of all water bodies were associated with LWD or vegetated islands. Most ponds in the riparian forest were permanent; however, many water bodies in the active floodplain were temporary. In the active plain, many ponds were dry at the end of winter but were filled during high spring flow in mid-March and slowly contracted during subsequent months (Fig. 4). All riparian forest water bodies and two third of the water bodies within the active plain were used as spawning habitats by amphibians (eggs/larvae present). Within the active plain, ponds associated with large wood (LWD) and vegetated islands were preferred as spawning sites (cf. Fig. 3). Main and side channel habitats were almost devoid of amphibians. During the investigation period the number of water bodies utilised by amphibians decreased (Fig. 4).

Species diversity and number of egg mass clutches

Ten taxa were identified along the corridor of the Tagliamento river. Three additional species (*Salamandra salamandra, Pelobates fuscus* and *Triturus alpestris*) were observed in adjacent uplands. Along the entire corridor, the number per

Table 1. Distribution of amphibians along the Tagliamento (reaches I–VI)

Reach	I	II	III	IV	V	VI	Adjacent upland
Altitude (m a.s.l.)	1050	705	165	140	20	5	
Average slope (%)	5.5–19.5	2.5	1	1	0.5	< 0.5	
Geomorphic type	Constrained	Island-braided	Bar-braided	Island-braided	Braided-anastomosed	Meandering	
Width of active floodplain (m)	30	260	830	1000	830	250	
Triturus vulgaris (Smooth Newt)				X			
Triturus carnifex (Alpine Warty Newt)				X			
Triturus alpestris (Alpine Newt)							X
Salamandra salamandra (Fire Salamander)							X
Bombina variegata (Yellow-bellied Toad)		X					
Bufo bufo (Common Toad)		X	X	X (A)	X		
Bufo viridis (Green Toad)			X	X (A)		X	
Hyla intermedia (Italian Tree Frog)				X (A)			
Rana latastei (Italian Agile Frog)			X	X (A)	X		
Rana temporaria (Common Frog)	X	X	X	X (A)	X		
Rana dalmatina (Agile Frog)			X	X (A)	X		
Rana lessonae and *R. kl. esculenta (Green Frogs)*				X (A)	X	X	
Pelobates fuscus (Common Spadefoot)							X

X: species recorded. Reach IV: A: Species found in the active floodplain. For comparison, we included for reach IV data from three sampling dates (March, May, July) as for all other reaches.

reach ranged from 1 species (Reach I) to 9 taxa (Reach IV) (Table 1).

In the main study area (Reach IV), eight taxa were identified in the riparian forest and seven taxa in the active plain. *Bufo bufo* and *Rana latastei* were the most common species in active and riparian floodplain habitats. *Bufo viridis* was restricted to the active corridor; *Triturus vulgaris* and *T. carnifex* were only observed in the riparian forest. Average species richness was higher in riparian forest water bodies (maximum of six species per water body, mostly 2–3 species per pond) compared to the active floodplain (mostly 1–2 species per pond, with a maximum of five taxa; Fig. 5). The only parameters tested that were significantly correlated with species richness were surface water temperature (active floodplain and riparian forest, positive correlation), distances to vegetated islands (active floodplain, negative correlation), and fish density (active floodplain, positive correlation; Table 2).

In the riparian forest, total number of egg mass clutches (as a surrogate for adult female population density) of brown frogs combined (*R. latastei*, *R. dalmatina*, *R. temporaria*) and common toad (*B. bufo*) were as high as 1850 and 700, respectively (Fig. 6). In the active plain, the total number of egg clutches was 350 for brown frogs and 65 for the common toad, respectively. The maximum number of egg masses per individual water body in the floodplain (active plain and riparian forest) was 150 for brown frogs and 200 for *B. bufo*. Total and average (per hectare surface water) egg mass density of *B. bufo* and brown frogs was between six and ten times lower in the active floodplain compared to the riparian forest. Egg mass density of brown frogs was correlated with pond size (positive correlation) in the riparian forest, and to temperature (positive), fish density (positive) and distance to islands (negative) in the active plain. Density of *B. bufo* was correlated with pond size (riparian forest), vegetation cover and fish density (active plain; Table 2). Based on the area of surrounding vegetated terrestrial habitats (vegetated islands and riparian forest), within a certain distance to ponds (see methods), egg mass density measured 23 egg masses per hectare and was very similar in the active plain and the adjacent riparian forest.

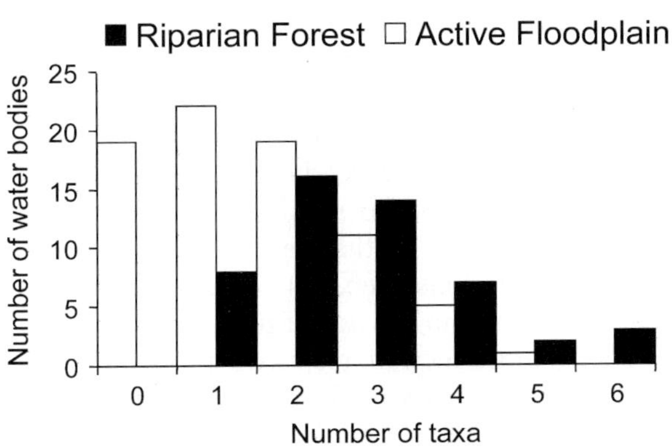

Figure 5. Frequency distribution of floodplain waters (active and riparian forest habitats) with different taxa richness.

Indicator species and species association

Based on TWINSPAN analyses of amphibian species, active floodplain and riparian forest habitats were clearly distinct (Fig. 7). *Bufo viridis* was an indicator species of bare gravel ponds, *R. temporaria* and *B. bufo* characterised island- and LWD-associated waters in the active floodplain; however, *T. vulgaris* and *R. dalmatina* primarily occurred in isolated ponds in the riparian forest dominated by fine sediments and a dense vegetation cover. Ponds along the margin of the active corridor were colonised by species characteristic of both the active plain and the riparian forest (mainly *B. bufo* and *R. latastei*; Fig. 7). No positive spatial species associations were observed;

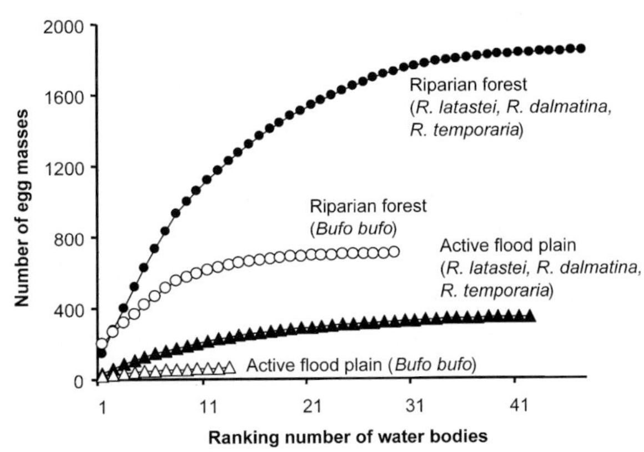

Figure 6. Cumulative number of egg-masses of brown frogs (*Rana latastei, R. temporaria, R. dalmatina*) and common toad (*Bufo bufo*) versus number of water bodies sampled from the active floodplain and the adjacent riparian forest. Ponds are arranged according to densities of egg masses, from the highest to the lowest.

Table 2. Rank-correlations of species diversity and egg mass density (egg masses of brown frogs, *Rana temporaria*, *R. latastei* & *R. dalmatina*, and the common toad, *Bufo bufo*) with various extrinsic factors

	Active floodplain ($n=82$)	Riparian forest ($n=48$)
Taxa richness ($S=10$)		
Temperature (°C)	$R=0.41$ ($p<0.001$)	$R=0.29$ ($p=0.02$)
Fish density (1–4)	$R=0.40$ ($p<0.001$)	$R=0.05$ ($p=0.71$)
Distance from vegetated islands (m)	$R=-0.34$ ($p=0.02$)	N. A.
Egg mass density: Brown frogs (*Rana latastei, R. temporaria, R. dalmatina*)		
Area (ha)	$R=0.12$ ($p=0.30$)	$R=0.59$ ($p<0.001$)
Temperature (°C)	$R=0.36$ ($p<0.001$)	$R=0.27$ ($p=0.05$)
Fish density (1–4)	$R=0.40$ ($p<0.001$)	$R=0.19$ ($p=0.18$)
Distance from vegetated islands (m)	$R=-0.38$ ($p<0.01$)	N. A.
Egg mass density: *Bufo bufo*		
Area (ha)	$R=0.05$ ($p=0.65$)	$R=0.36$ ($p=0.01$)
Fish density (1–4)	$R=0.30$ ($p=0.007$)	$R=0.12$ ($p=0.39$)
Vegetation cover (1–4)	$R=0.42$ ($p=0.002$)	$R=-0.16$ ($p=0.26$)

N.A. not applicable. Only significant correlations are shown.

however, significant negative spatial associations occurred between *R. dalmatina, R. latastei, R. temporaria* and *B. bufo* (data not shown).

Nestedness

Nestedness-analyses indicated that amphibians were distributed as non-random assemblages (Table 3). Observed 'system temperatures' ranged from 5.0 (active floodplain) to 11.7 (riparian forest) and were significantly lower than estimated values produced by Monte Carlo simulations ($p< <0.001$). A few species showed unexpected presence and absence patterns (idiosyncratic species). In the riparian forest, *R. latastei* and *T. vulgaris* had 'temperatures' that were higher compared to the average. In the active floodplain, *B. viridis* had slightly higher temperatures compared to the average of all other taxa, demonstrating that these species tended to be absent from species-rich sites. 'System temperature' in the active floodplain of the Tagliamento was much lower compared to regulated floodplains along the Danube and Rhône rivers (Table 3).

Discussion

Floodplains along the Tagliamento are characterised by a very high density and diversity of aquatic habitats, including ephemeral, lentic and lotic water bodies (Fig. 2; Arscott et al., 2000; Karaus et al., 2005). In gravel-bed rivers, lentic water bodies represent only a small proportion of total aquatic area; however, based on their number and variety, ponds are key features for maintaining aquatic invertebrate diversity in many floodplain systems (Homes et al., 1999; Ward et al., 1999a; Arscott et al., 2005; Karaus et al., 2005). Arscott et al. (2000), who compared aquatic habitat diversity in the six geomorphic reaches along the Tagliamento, found the highest habitat diversity in the island-braided floodplain (Reach IV). This is also the reach with the highest amphibian diversity (Table 1). Total species richness of amphibians along the Tagliamento was similar to values reported for the largest Central European rivers such as the Rhône and Danube (Ward et al., 1999a; Table 3). This suggests that dynamic gravel-bed rivers, such as the Tagliamento, are focal areas for amphibians. Moreover, endangered species such as *R. latastei* (endemic to northern Italy) developed large populations in the Tagliamento floodplain (based on egg mass density). *Bufo bufo*, the most frequent species in the active plain, exhibits a pronounced reproduction plasticity that allows this species to explore dynamic and ephemeral habitats (Kuhn, 1993). This includes spawning synchronisation with the hydrology, a fast metamorphosis, and the formation of small spawning aggregations.

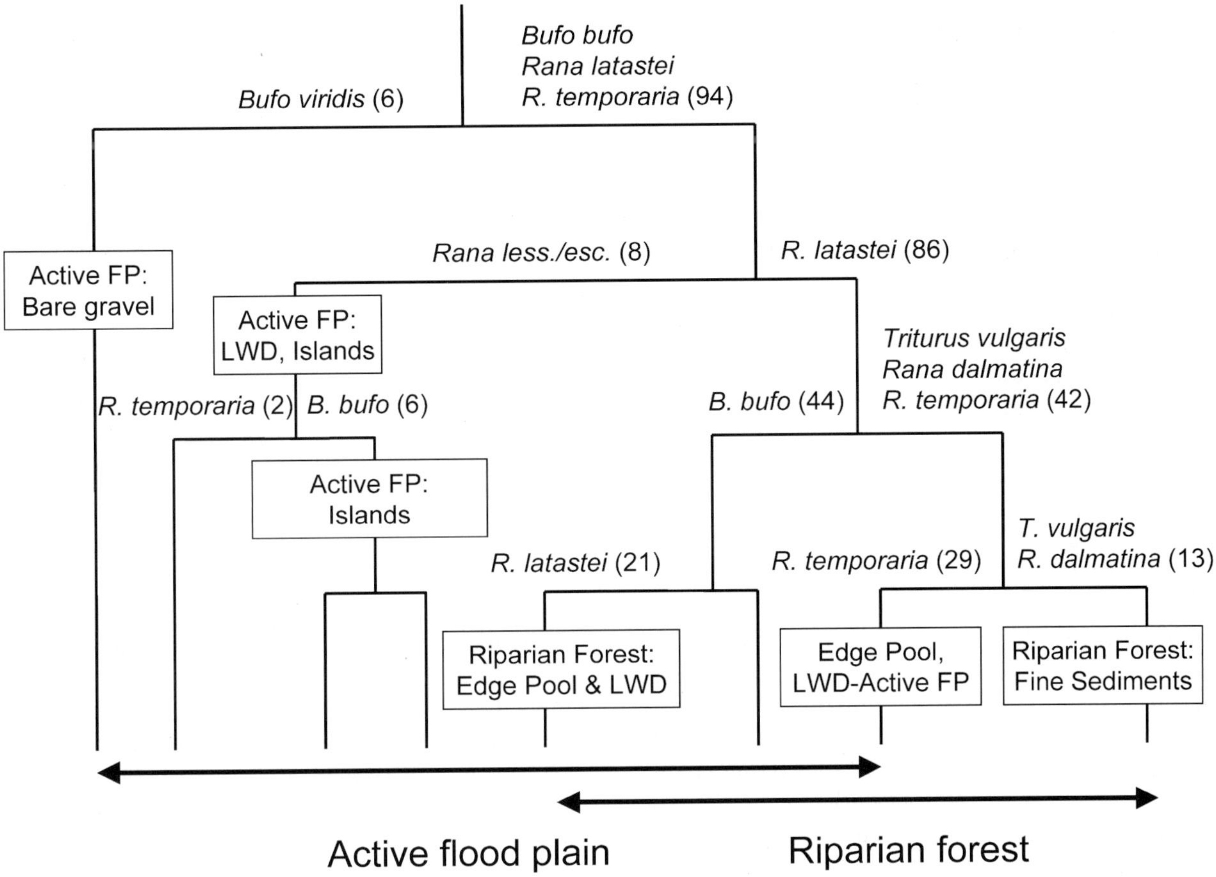

Figure 7. TWINSPAN separation of floodplain waters and the indicator species responsible for the separation. LWD: Large woody debris.

The present results demonstrate that most species were able to utilise active floodplain habitats as spawning sites, although frequency of species occurrence, egg mass density, and the average number of species per water body were lower than in the adjacent riparian forest. None of the measured parameters, except water temperature, influenced amphibian diversity in both parts of the floodplain. For example, a significant positive relationship between egg mass density and pond size was found for the riparian forest but not for the active plain. This finding corresponds to data from many other wetlands, such as depression wetlands in the southern United States (Snodgrass et al., 2000) and ponds in Switzerland (Oertli et al., 2002) where no relationship between amphibian species richness and habitat size were observed. Pond size matters, not for species richness but for egg mass density, at least in the riparian forest. This implies that species distribution patterns are determined by other characteristics of water bodies, such as hydrological connectivity (Morand & Joly, 1995), pond age (Merovich & Howard, 2000), presence of predators (Werner & McPeek, 1994), vegetation structure and cover (Healy et al., 1997), or the presence of vegetated islands and large woody debris (LWD; this study).

During their life cycle most amphibian species depend on both aquatic and terrestrial habitats. In the active floodplain, most water bodies colonised by amphibians were associated with vegetated islands and LWD. Hence, vegetated islands and large accumulations of woody debris were expected to control species numbers and egg mass density as demonstrated by the significant correlation between species richness and the distance to vegetated islands (so called 'wood benefit' *sensu*

Table 3. Nestedness analysis for the Tagliamento (active floodplain, riparian forest; Reach IV) compared with data for floodplains along the Danube (Austria) and the Rhône (France). Regelsbrunn (Danube) is the most dynamic of these additional floodplains, Obere Lobau (Danube) the most isolated one

	N sites	*N* taxa	Fill (%)	System *T*	Generated *T* (mean ± SD)	Significance
Tagliamento (Reach IV)						
Riparian Forest	50	9	30.4	11.7	53.2 ± 6.1	< 0.001
Active Floodplain	58	7	29.0	5.0	49.9 ± 6.9	< 0.001
Rhône (Jons, Upper Rhône)	17	10	28.7	12.9	45.1 ± 8.7	< 0.001
Danube (Regelsbrunn)	56	10	28.2	16.1	52.9 ± 5.0	< 0.001
Danube (Orth)	28	7	37.7	17.4	50.5 ± 7.2	< 0.001
Danube (Untere Lobau)	73	9	29.3	10.5	55.5 ± 5.3	< 0.001
Danube (Obere Lobau)	68	9	30.3	18.3	56.6 ± 5.6	< 0.001
Danube (Obere Lobau, only isolated sites)	29	9	36.0	31.5	50.9 ± 6.6	< 0.001

'Fill' represents the presences (%) in the site-taxon matrix, 'System *T*' is the system temperatures observed, and 'Generated *T*' is the system temperatures generated by Monte Carlo randomisations (500 iterations each). Data from the Rhône: Morand & Joly (1995). Danube: C. Baumgartner (unpubl. data).

Gurnell & Petts, 2002). Further, egg mass density of brown frogs per ha vegetated island was similar to densities in the riparian forest and was much higher than in regulated floodplains. The average density of egg masses of brown frogs in regulated floodplains along the Danube was about 3.5 per ha floodplain forest (Pintar et al., 1997; Baumgartner et al., 1997); this was one-order-of magnitude lower than in the present study. We may expect that LWD and vegetated islands play a pivotal role for (i) providing potential terrestrial habitats for adults (e.g., hibernation), (ii) creating aquatic habitats, and (iii) facilitating the subsequent colonisation by amphibians (Gurnell et al., 2005). LWD increases the permanency of water bodies (deep scour ponds at the apex of wood accumulations; e.g., Abbe & Montgomery, 1996), provides shelter during flood events, reduces predation by fish and birds, increases food availability by providing surfaces for biofilm development, and serves as a stable structure to attach egg masses (e.g., by brown frogs, *B. bufo*). The unexpected positive relationship between fish density and amphibian diversity in the active floodplain demonstrated that LWD may facilitate the coexistence of otherwise mutually exclusive groups. We did not investigate fish in more detail, but we observed that ponds were primarily colonised by fish larvae and young fish which most probably do not feed extensively on amphibian larvae. In addition, large wood and extensive shallow areas provide shelter for amphibians against predation.

Amphibians showed highly ordered distributional patterns (low 'system temperature'), especially in the active floodplain. This high degree of nestedness implies that a few common species tended to be ubiquitous and rare species tended to occur only at species-rich sites. As in many other systems (e.g., Patterson & Brown, 1991; Hecnar & Mc'Closkey, 1997), both extinction and selective colonisation may have contributed to structuring the nested assemblages of floodplain amphibians. The high turnover rate of aquatic water bodies ('selective' extinction of amphibians caused by flood fill/scour or drying) and the species preference of ponds associated with LWD and islands (selective colonisation) can explain the very high degree of nestedness in the active floodplain. Indeed, nestedness in the active plain was considerably higher than that reported in most investigations of aquatic and terrestrial areas (e.g., 294 examples listed in Atmar & Patterson, 1995; Table 3). Similar high degrees of nestedness have only been described for fish communities in Australian desert springs (Kodric-Brown & Brown, 1993). Those authors concluded that a common biogeographic history, similar contemporary environments and hierarchical ecological relationships among species were necessary to create such a deterministic assembly structure. In Swedish streams, however, Malmqvist & Hoffsten (2000) calculated 'system temperatures' that ranged from 12.6 for Simuliidae to 29.2 for Plecoptera. This lower degree of nestedness in streams (higher 'system temperature') was expected to result from a higher probability of exchange rates between sampling sites (high degree of connectivity) and/or by high dispersal properties of species (e.g. drifting invertebrates, wind-dispersed plants; e.g. Kadmon, 1995). The much higher 'system temperature' calculated for regulated floodplains (Table 3) may result from a stable degree of connectivity, either between individual water bodies or by the presence of a closed floodplain forest matrix that may facilitate the exchange of larval and adult amphibians.

Nestedness has also major implications for the development of conservation strategies since a high degree of nestedness (low 'system temperature') means that the protection of the most diverse habitats is required in order to also conserve rare species. Hecnar & Mc'Closkey (1997), for example, concluded from their investigation of the amphibian fauna in 118 ponds in SW Ontario (Canada) that single large reserves were preferable to several small reserves for the conservation of temperate pond-dwelling amphibian assemblages. Alford & Richards (1999) concluded from their extensive review that most amphibians exhibited a metapopulation structure (Marsh & Trenham, 2001). Since in the active channel along the Tagliamento the exact location of sink and source habitats considerably changes with floods (half-life expectancy of ponds is less than 7 months, Van der Nat et al., 2003), the preservation of an entire floodplain complex is required for maintaining intact amphibian populations (De Nooij et al., 2006). Our results also demonstrate that amphibians in the active floodplain are differently organised than in the riparian forest (based on degree of nestedness), and that most likely different variables control the distribution and density in the active and passive floodplain sections (Table 2).

Natural floodplains are distinctive landscapes with respect to their natural richness and their bioproduction. River canalisation has restricted the active part of floodplains to a narrow ribbon of riparian vegetation along the main river channel. Elimination of islands and the reduced diversity of floodplain ponds have detrimental effects on both aquatic and terrestrial organisms (e.g. Stanford et al., 1996; Karaus et al., 2005). In regulated rivers, amphibians were eliminated from the active corridor and restricted in their occurrence to the isolated parts of the floodplain. Re-establishing hydrological connectivity between the river and its floodplain has been generally regarded as having negative effects on amphibian species. The present investigation, however, demonstrates that even one of the most dynamic river corridors in Europe (Tagliamento) provides extensive habitats for diverse amphibian communities. Consequently, there is no general conflict between maintaining/creating high fluvial dynamics and high amphibian diversity and density. This is supported by recent studies along the Danube (Austria), where the reconnection of formerly isolated floodplains did not affect amphibian diversity and density in a negative way (C. Baumgartner, unpubl. data).

Acknowledgements

This research was supported by a grant (0-20572-98) from the 'Forschungskommission' of the Swiss Federal Institute of Technology (ETH-Zurich). We wish to thank D. van der Nat for providing GIS-maps; and J. Kuhn, L. Indermaur and D. Arscott for thoughtful comments on earlier versions of the manuscript. Very helpful comments by R.S.E.W. Leuven and 3 anonymous reviewers are gratefully acknowledged. Parts of this work were presented at the INTECOL conference in Utrecht, 2004. This work has been completed while K. Tockner was a sabbatical visitor at the Institute of Ecosystem Studies (IES), Millbrook, NY.

References

Abbe, T. B. & D. R. Montgomery, 1996. Large woody debris jams, channel hydraulics and habitat formation in large rivers. Regulated Rivers: Research & Management 12: 201–221.

Alford, R. A. & S. J. Richards, 1999. Global amphibian declines: a problem in applied ecology. Annual Revue of Ecology and Systematics 30: 133–165.

Arscott, D. B., K. Tockner & J. V. Ward, 2000. Aquatic habitat diversity along the corridor of an Alpine floodplain river (Fiume Tagliamento, Italy). Archiv für Hydrobiologie 149: 679–704.

Arscott, D. B., K. Tockner & J. V. Ward, 2002. Geomorphic dynamics along a braided-river corridor in the Alps (Fiume Tagliamento, NE Italy). Ecosystems 5: 802–814.

Arscott, D. B., K. Tockner & J. V. Ward, 2005. Lateral organization of aquatic invertebrates along the corridor of a braided floodplain river. Journal of the North American Benthological Society 24: 934–954.

Atmar, W. & B. D. Patterson, 1993. The measure of order and disorder in the distribution of species in fragmented habitats. Oecologia 96: 373–382.

Atmar, W. & B. D. Patterson, 1995. The Nestedness Temperature Calculator, a Visual BASIC Program Including 294 Presence–Absence Matrices. AICS Research Inc. University Park, New Mexico, and the Filed Museum, Chicago.

Baumgartner, C., A. Waringer-Löschenkohl & M. Pintar, 1997. Bedeutung der Konnektivität für die Springfroschpopulationen der Donauauen. Rana, Sonderheft 2: 159–162.

Beebee, T. J. C., 1996. Ecology and Conservation of Amphibians. Chapman & Hall, London.

De Nooij, R. J. W., W. C. E. P. Verberk, H. J. R. Lenders, R. S. E. W. Leuven & P. H. Nienhuis, 2006. The importance of hydrodynamics for protected and endangered biodiversity of lowland rivers. Hydrobiologia 565: 153–162.

Günther, R. & J. Plöttner, 1994. Morphometric, enzymological and bioacoustic studies in Italian water frogs (Amphibia, Ranidae). Zoologica Poloniae 39: 387–415.

Gurnell, A. M. & G. E. Petts, 2002. Island-dominated landscapes of large floodplain rivers, a European perspective. Freshwater Biology 47: 581–600.

Gurnell, A. M., G. E. Petts, D. M. Hannah, B. P. G. Smith, P. J. Edwards, J. Kollmann, J. V. Ward & K. Tockner, 2001. Island formation along the gravel-bed Fiume Tagliamento, Italy. Earth Surface Processes and Landforms 26: 31–62.

Gurnell, A. M., K. Tockner, G. E. Petts & P. J. Edwards, 2005. Large wood delivers biocomplexity along river corridors. Frontiers in Ecology and Environment 3: 377–382.

Healey, M., D. Thompson & A. Robertson, 1997. Amphibian communities associated with billabong habitats on the Murrumbidgee floodplain, Australia. Australian Journal of Ecology 22: 270–278.

Hecnar, S. J. & R. T. Mc'Closkey, 1997. Patterns of nestedness and species association in a pond-dwelling amphibian fauna. Oikos 80: 371–381.

Hill, M. O., 1979. TWINSPAN – FORTRAN program for arranging multivariate data in an ordered two-way table by classification of individuals and attributes, Cornell University, Ithaca.

Homes, V., D. Hering & M. Reich, 1999. The distribution and macrofauna of ponds in stretches of an Alpine floodplain differently impacted by hydrological engineering. Regulated Rivers: Research & Management 15: 405–417.

Houlahan, J. E., C. S. Findlay, B. R. Schmidt, A. H. Meyer & L. Kuzmin, 2000. Quantitative evidence for global amphibian population declines. Nature 404: 752–755.

Hughes, F. M. R. & S. B. Rood, 2001. Floodplains. In Warren, A. & J. R. French (eds), Habitat conservation: managing the physical environment editors. John Wiley & Sons, Chichester: 105–121.

Joly, P. & A. Morand, 1994. Theoretical habitat templets, species traits, and species richness: amphibians in the Upper Rhone River and its floodplain. Freshwater Biology 31: 455–468.

Kadmon, R., 1995. Nested species subsets and geographic isolation: a case study. Ecology 76: 458–465.

Karaus, U., L. Alder & K. Tockner, 2005. "Concave islands": diversity and dynamics of parafluvial ponds in a gravel-bed river. Wetlands 25: 26–37.

Kiesecker, J. M., A. R. Blaustein & L. K. Belden, 2001. Complex causes of amphibian population declines. Nature 410: 681–684.

Kodric-Brown, A. & J. H. Brown, 1993. Highly structured fish communities in Australian desert springs. Ecology 74: 1847–1855.

Kuhn, J., 1993. Fortpflanzungsbiologie der Erdkröte *Bufo b. bufo* (L.) in einer Wildflussaue. Zeitschrift für Ökologie und Naturschutz 2: 1–10.

Kuhn, J., H. Laufer & M. Pintar (eds), 2001. Amphibien in Auen. Zeitschrift für Feldherpetologie Special Issue 8: 1–264.

Lapini A., N. Dall' Asta, S. Bressi, P. Dolce & P. Pellarini, 1999. Atlante corologico degli Anfibi e dei Rettili del Friuli-Venezia Giulia. Pubblicazione n. 43. Edizioni del Museo Friulano di Storia Naturale, Udine.

Leuven, R. S. E. W., C. den Hartog, M. M. C. Christiaans & W. H. C. Heijligers, 1986. Effects of water acidification on the distribution pattern and the reproductive success of amphibians. Experientia 42: 495–503.

Malmqvist, B. & P.-O. Hoffsten, 2000. Macroinvertebrate taxonomic richness, community structure and nestedness in Swedish streams. Archiv für Hydrobiologie 150: 29–54.

Marsh, D. M. & P. C. Trenham, 2001. Metapopulation dynamics and amphibian conservation. Conservation Biology 15: 40–49.

Merovich, C. E. & J. H. Howard, 2000. Amphibian use of constructed ponds on Maryland's eastern shore. The Journal of the Iowa Academy of Science 107: 151–159.

Morand, A. & P. Joly, 1995. Habitat variability and space utilization by the amphibian communities of the Upper-Rhone floodplain. Hydrobiologia 300/301: 249–257.

Müller, N., 1995. River dynamics and floodplain vegetation and their alterations due to human impact. Archiv für Hydrobiologie Supplement 101: 477–512.

Nilsson, C., C. A. Reidy, M. Dynesius & C. Revenga, 2005. Fragmentation and flow regulation of the world's large river systems. Science 308: 405–408.

Nöllert, A. & C. Nöllert, 1992. Die Amphibien Europas. Franck-Kosmos, Stuttgart.

Oertli, B., D. Auderset Joyse, E. Castella, R. Juge, D. Cambin, D. & J.-B. Lachavanne, 2002. Does size matter? The relationship between pond area and biodiversity. Biological Conservation 104: 59–70.

Patterson, B. D. & J. H. Brown, 1991. Regionally nested patterns of species composition in granivorous rodent assemblages. Journal of Biogeography 18: 395–402.

Patterson, B. D. & W. Atmar, 2000. Analyzing species composition in fragments. Zoological Monographs 46: 9–24.

Petts, G. E., H. Moller & A. L. Roux, 1989. Historical Changes of Large Alluvial Rivers, Western Europe. John Wiley & Sons, Chichester, UK.

Pintar, M., C. Baumgartner & A. Waringer-Löschenkohl, 1997. Verbreitung des Springfrosches in Augebieten der niederösterreichischen Donau. Rana 2: 153–158.

Skelly, D. K., 1997. Tadpole communities. Pond permanence and predation are powerful forces shaping the structure of tadpole communities. American Scientist 85: 36–45.

Snodgrass, J. W., M. J. Komorosky, A. L. Bryan Jr. & J. Burger, 2000. Relationships among isolated wetland size, hydroperiod, and amphibian species richness: implications for wetland regulations. Conservation Biology 14: 414–419.

Stanford, J. A., J. V. Ward, W. J. Liss, C. A. Frissell, R. N. Williams, J. A. Lichatowitch & C. C. Coutant, 1996. A general protocol for restoration of regulated rivers. Regulated Rivers: Research & Management 12: 391–413.

Stuart, S. N., J. S. Chanson, N. A. Cox, B. E. Young, A. L. Rodrigues, D. L. Fischman & R. W. Waller, 2004. Status and trends of amphibian declines and extinctions worldwide. Science 306: 1783–1786.

Tockner, K. & J. V. Ward, 1999. Biodiversity along riparian corridors. Archiv für Hydrobiologie, Supplement 115: 293–310.

Tockner, K., F. Schiemer, C. Baumgartner, G. Kum, E. Weigand, I. Zweimüller & J. V. Ward, 1999. The Danube restoration project: species diversity patterns across connectivity gradients in the floodplain system. Regulated Rivers: Research & Management 15: 245–258.

Tockner, K. & J. A. Stanford, 2002. Riverine floodplains: present state and future trends. Environmental Conservation 29: 308–330.

Tockner, K., J. V. Ward, D. B. Arscott, P. J. Edwards, J. Kollmann, A. M. Gurnell, G. E. Petts & B. Maiolini, 2003. The Tagliamento River: a model ecosystem of European importance. Aquatic Sciences 65: 239–253.

Tockner, K., S. E. Bunn, G. Quinn, R. J. Naimann, J. A. Stanford & C. Gordon, 2006. Floodplains: critically threatened ecosystems In Polunin, N. C. (ed.), The State of the World's Waters. Cambridge University Press, Cambridge. In press.

Van der Nat, D., K. Tockner, P. J. Edwards, J. V. Ward & A. M. Gurnell, 2003. Habitat change in braided rivers (Tagliamento, NE-Italy). Freshwater Biology 48: 1799–1812.

Ward, J. V., K. Tockner & F. Schiemer, 1999a. Biodiversity of floodplain river systems: ecotones and connectivity. Regulated Rivers: Research & Management 15: 125–139.

Ward, J. V., K. Tockner, P. J. Edwards, J. Kollmann, G. Bretschko, A. M. Gurnell, G. E. Petts & B. Rossaro, 1999b. A reference river system for the Alps: The "Fiume Tagliamento". Regulated Rivers: Research & Management 15: 63–75.

Ward, J. V. & K. Tockner, 2001. Biodiversity: towards a unifying theme for river ecology. Freshwater Biology 46: 807–819.

Waringer-Löschenkohl, A. & J. Waringer, 1990. Zur Typisierung von Augewässern anhand der Litoralfauna (Evertebraten, Amphibien). Archiv für Hydrobiologie, Supplement 84: 73–94.

Wellborn, G. A., D. K. Skelly & E. E. Werner, 1996. Mechanisms creating community structure across a freshwater habitat gradient. Annual Revue of Ecology and Systematics 27: 337–363.

Werner, E. E. & M. A. McPeek, 1994. Direct and indirect effects of predators on two anuran species along an environmental gradient. Ecology 75: 1368–1382.

Hydrobiologia (2006) 565:135–152
R.S.E.W. Leuven, A.M.J. Ragas, A.J.M. Smits & G. van der Velde (eds), Living Rivers: Trends and Challenges in Science and Management
DOI 10.1007/s10750-005-1910-x

Modelling recolonisation of heterogeneous river floodplains by small mammals

S. Wijnhoven[1,]*, G. van der Velde[2], R.S.E.W. Leuven[3] & A.J.M. Smits[1,4]
[1]*Centre for Water and Society, Faculty of Science, Radboud University Nijmegen, P.O. Box 9010, 6500 GL Nijmegen, The Netherlands*
[2]*Department of Animal Ecology and Ecophysiology, Institute for Wetland and Water Research, Radboud University Nijmegen, P.O. Box 9010, 6500 GL Nijmegen, The Netherlands*
[3]*Department of Environmental Science, Institute for Wetland and Water Research, Radboud University Nijmegen, P.O. Box 9010, 6500 GL Nijmegen, The Netherlands*
[4]*Water Management and Sustainability, Faculty of Social Sciences, Erasmus University Rotterdam, P.O. Box 1738, 3000 DR Rotterdam, The Netherlands*
*(*Author for correspondence: Tel.: +31 24 3652427; Fax: +31 24 3653030; E-mail: S.Wijnhoven@science.ru.nl)*

Key words: dispersal behaviour, landscape patterns, multiple regression models, recolonisation, riverine ecosystems, small mammals

Abstract

Riverine landscapes are characterised by recurrent flooding events and successional landscape mosaics with high habitat heterogeneity, providing species-specific patterns of suitable and unsuitable biotopes. Landscape characteristics, like distance, barriers and their specifications (e.g. cumulative barrier width, barrier number) and the spatial arrangement of suitable habitat areas, are expected to affect the dispersal of animals in landscapes. The distribution of voles, shrews and mice in a floodplain was monitored for 2 years using live traps. Recolonisation was found to be a slow process, resulting in a heterogeneous distribution of small mammals in the floodplains. *Microtus arvalis* was found just after the floods in low densities and on or near non-flooded areas only. From summer to autumn densities gradually increased, and specimens could be observed on larger distances from the non-flooded areas. The density development pattern of *Crocidura russula* was similar to that of *M. arvalis*, but densities increased faster. In contrast, *Clethrionomys glareolus* and *Sorex araneus* could immediately be observed in the former flooded areas just after the floods, but throughout the year these species were not trapped on distances further than 120 m from the non-flooded areas. *Micromys minutus* and *Apodemus sylvaticus* were trapped only occasionally in spring and summer after which those species could be found throughout the floodplain in larger densities in autumn. To analyse the influence of landscape characteristics on recolonisation, the floodplain was classified into suitable, marginal and unsuitable habitat landscape units for each of the small mammal species, based on the trapping results. Landscape characteristics relating to monitoring sites were measured from an aerial photograph using a geographic information system. After that, presence and recolonisation time of small mammals at monitoring sites could be described by multiple regression models based on these measured landscape characteristics. The predictive power of these models was tested in another floodplain by determining the species distribution after 35 weeks. Multiple regression models appeared to be useful in analysing recolonisation patterns and determining the importance of landscape characteristics for recolonisation by small mammals after flooding events. Available distribution data suggest three different types of recolonisers: (1) Gradual, density induced colonisers; (2) Active dispersers and (3) Long-distance dispersers after a lag. Results of regression models confirmed that *M. arvalis* could be characterised as a type 1 and *A. sylvaticus* as a type 3 species. The classification of the other species was not possible due to the relatively short time available for recolonisation.

Introduction

Small mammals (e.g. voles, shrews and mice) in floodplains have to cope with periodic floods. Flooding has a strong impact on their populations, resulting in high mortality, and restricting their presence to refugia on elevated terrains after inundation (Pachinger & Haferkorn, 1998; Andersen et al., 2000). After the water has retreated, the floodplain landscape has to be recolonised from these sources (Robinson et al., 2002). Small mammals are important animals in riverine ecosystems, because of their feeding and burrowing activities and their role in food webs (Wijnhoven et al., 2005). They are important prey items for several endangered or protected birds of prey and carnivorous mammals (Erlinge et al., 1983; Jongbloed et al., 1996; Hanski et al., 2001; Leuven et al., 2005), for which many floodplains are assumed to function as important conservation areas (Andersen et al., 2000; Wike et al., 2000; Robinson et al., 2002; Van den Brink et al., 2003; De Nooij et al., 2004). As predators predominantly forage in areas with abundant prey, it is important to understand the temporal and spatial distribution patterns of small mammals.

Several small mammal species are known to form potentially fast-growing populations (Erlinge et al., 1983; Delattre et al., 1999; Hanski et al., 2001). However, connectivity of landscapes is very important for a rapid dispersal through, or recolonisation of, such landscapes (Zhang & Usher, 1991; Wolff, 1999). Riverine landscapes are dynamic, and biologically and spatially complex, characterised by a successional landscape mosaic with high habitat heterogeneity (Robinson et al., 2002). Characteristics of the landscape elements between small mammal source populations and unoccupied suitable habitats, like distance, the presence and characteristics of barriers and the spatial arrangement of suitable habitat areas, are expected to affect the dispersal (in terms of speed and direction) of animals (Van Apeldoorn et al., 1992; Diffendorfer et al., 1999; Peles et al., 1999; Matthiopoulos, 2003). The role of various landscape characteristics in guiding dispersal and recolonisation may be species-specific, while landscape suitability may also differ for various species (Kozakiewicz, 1993; Bowers & Barrett, 1999; Robinson et al., 2002). Unsuitable areas can function as temporary or permanent barriers to dispersal (Bondrup-Nielsen, 1985; Peles et al., 1999), whereas relatively small suitable parts can influence the connectivity (e.g. by forming corridors or stepping-stones; Kozakiewicz, 1993; Wolff, 1999) and therefore the recolonisation speed. Models based on landscape characteristics have been used to study the impacts of habitat fragmentation on species distribution patterns (Vos & Chardon, 1998; Verboom et al., 2001; Verbeylen et al., 2003), and such models might be useful for floodplain conservation and management (Wolff, 1999).

The factors that determine the recolonisation process by small mammal species in various diversified floodplains are expected to be similar, although the relative importance of landscape characteristics may depend on the species. In this study, we investigated: (1) The relative importance of various landscape characteristics for the recolonisation of floodplains by small mammals; (2) the existence of different patterns of dispersal behaviour (species-specific recolonisation patterns); (3) the usefulness of multiple regression models of floodplain recolonisation by small mammals as tools to analyse the importance of various landscape characteristics; and (4) the usefulness of multiple regression models in predicting the time to recolonisation and the presence of species at certain locations after flooding events.

Materials and methods

Research areas

The research project took place at the Afferdensche en Deestsche Waarden (ADW) and Millingerwaard (MW). Both floodplains are located on the left bank of the river Waal, the main branch of the river Rhine in the Netherlands (Fig. 1). The ADW, with an area of 280 ha, is situated 20 km west of the town of Nijmegen and consists of conservation areas and farmland. The ADW includes elevated areas left over from former brick factories, as well as clay excavations, small water bodies and side channels of the river. Similar structures can also be found in the MW, situated 20 km east of Nijmegen. The MW includes a larger section (including the entire 41.2 ha research

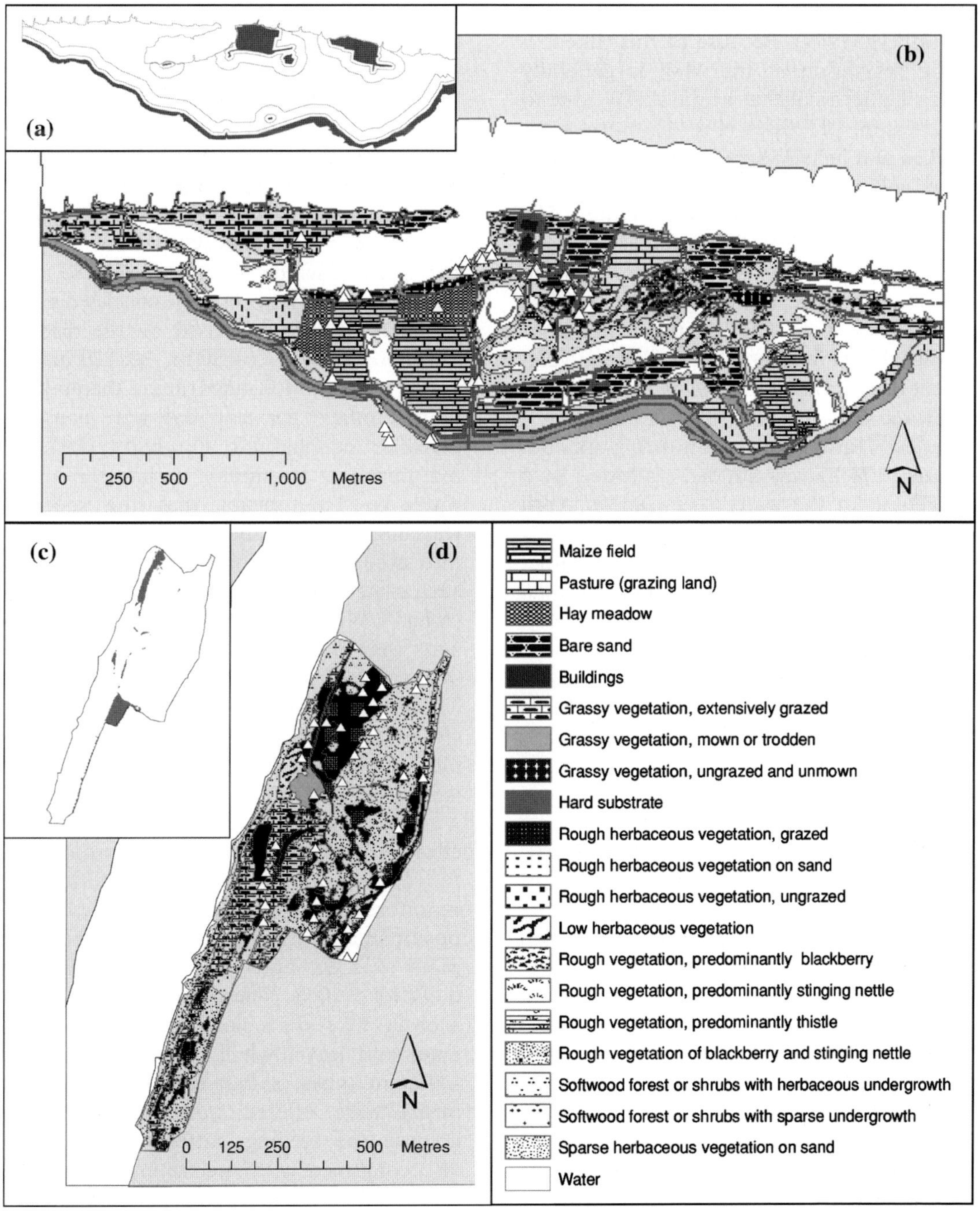

Figure 1. Characteristics of the research areas. (a) Locations of the non-flooded areas (shaded) in the Afferdensche en Deestsche Waarden floodplain (ADW), with the borders of the distance zones (0–30 m; 30–120 m; > 120 m) indicated as lines. (b) Classification of the ADW into small mammal ecotopes. (c) Locations of the non-flooded areas (shaded) in the research area at the Millingerwaard floodplain (MW). (d) Classification of MW into small mammal ecotopes. The small mammal monitoring plots are shown as triangles.

area) designated as conservation area than the ADW. Both floodplains are included in ecological rehabilitation programmes involving natural and semi-natural wetlands, grasslands, shrub-dominated vegetation and softwood forests. Ecological rehabilitation measures started in 1990 at the MW

(Helmer & Smeets, 1990), and after 1995 at the ADW (Zandberg, 1999). Because of this time difference, the vegetation at the MW is generally denser and rougher than that at the ADW, due to natural succession processes (Fig. 1). Both areas are embanked and include areas inside and outside the summer dikes (the dikes closest to the river that protect land against summer flooding). The summer dikes in the MW nowadays mostly resemble natural elevations like river dunes. Both research areas are regularly inundated. The flooding frequency at the ADW is more than once a year, while the MW is not flooded every year, due to its elevation. Small mammals in the ADW were monitored between April 2001 and January 2003 (Fig. 2). During this monitoring period, approximately 194 ha was flooded (followed by a gradually retreat of the water) between 16 April 2001 and 5 May 2001, between 24 February 2002 and 30 March 2002 and between 14 November 2002 and 5 December 2002. Approximately 126 ha was also flooded from 29 January 2002 to 24 February 2002. In the MW, small mammals were monitored between a flood event just before the spring of 2002 (after which the water had retreated by 18 March) and one in the first week of 2003.

Measuring small mammal distribution

In the ADW, the small mammal distribution patterns were monitored using trap lines with 10 live traps each, distributed over 34 selected monitoring plots, as indicated in Fig. 1, and as described in Wijnhoven et al. (2005). The live traps were checked every 4 h for 72 h after a 2-day prebaiting period, and were monitored at 7, 13, 24 and 34 weeks after the water had retreated in 2001, and at 2, 6, 16, 27 and 36 weeks after the water had retreated in 2002. During weeks 33 and 34 after the water had retreated in 2002, some of the monitoring sites were flooded again. Small mammals could only survive in refugia on the elevated non-flooded areas during inundation of the remaining parts of the area. In the 2 years of live trapping, nine small mammal species (voles, shrews, mice and mustelids) were trapped in the ADW, viz. *Microtus arvalis* (Pallas 1771) (Common vole), *Clethrionomys glareolus* (Schreber 1780) (Bank vole), *Sorex araneus* (Linnaeus 1758) (Common shrew), *Crocidura russula* (Hermann 1780) (White-toothed shrew), *Micromys minutus* (Pallas 1771) (Harvest mouse), *Apodemus sylvaticus* (Linnaeus 1758) (Wood mouse), *Microtus agrestis* (Linnaeus 1761) (Short-tailed field vole), *Sorex minutus* (Linnaeus 1766) (Pygmy shrew) and *Mustela nivalis* (Linnaeus 1766) (Weasel). Between 46 and 247 individuals were trapped for each of the first six of these species. The other three species were trapped only occasionally (less than 10 individuals per species), which was insufficient to analyse the recolonisation patterns. The species densities were calculated for zones based on the distance from non-flooded areas (0–30 m, 30–120 m, >120 m), and for the non-flooded areas themselves. Trapping numbers per trap line were converted into densities per habitat suitability class (suitable, marginal, or unsuitable habitats, as shown in Table 1). This meant that the species-specific trapping range of the trap lines had to be taken into account. The total area per habitat class in each zone was used to calculate the total density (n ha^{-1}; Wijnhoven et al., 2005). The trapping results were used to calculate the time to recolonisation for 68 situations (2 years of monitoring at 34 sites) assuming that the first trapping of a species at a monitoring site corresponds to the recolonisation time (in weeks after the retreat of the water) for that site.

In an attempt to cover a larger area with less effort, the populations of *M. arvalis*, *C. glareolus*, *S. araneus* and *C. russula* at the MW were monitored using 'food traps'. These traps were positioned in lines of five across the monitoring plots (Fig. 1), and were checked and refreshed daily for 3 days. These traps were constructed in such a way that small mammals could freely enter and leave, while larger animals could not. Different types of bait (apple, carrot and tinned meat) were offered together in each trap to analyse the feeding patterns. The method was also combined with live trapping several times at the ADW, and observations were made in terrariums, allowing species to be identified from the feeding patterns. The specific feeding patterns were as follows. *M. arvalis* had a preference for apple (often gone), ate most of the meat and also ate carrot. *C. glareolus* preferred carrot, ate apple and ate some meat. *S. araneus* ate all the meat and took tiny pieces of apple. *C. russula* preferred carrot and ate small pieces

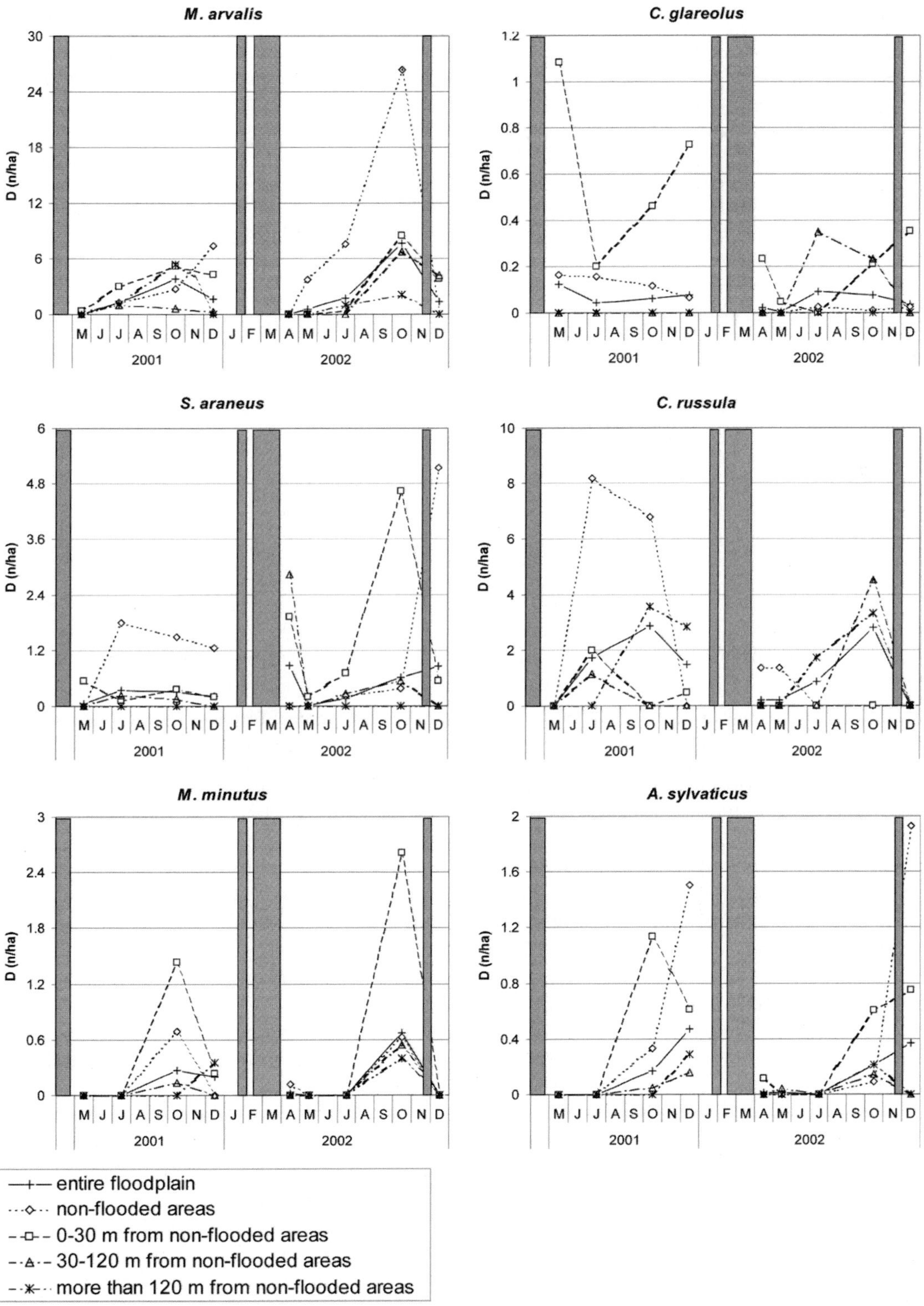

Figure 2. Species densities (n/ha^{-1}) at various distances from non-flooded areas in the Afferdensche en Deestsche Waarden floodplain through the years 2001 and 2002. Periods of inundation of the flooding parts are indicated in grey.

Table 1. Suitability of riverine ecotopes for six small mammal species, determined from the 2001 trapping results in the Afferdensche en Deestsche Waarden floodplain (derived from Wijnhoven et al., 2005)

Small mammal ecotopes	*M. arvalis*	*C. glareolus*	*S. araneus*	*C. russula*	*M. minutus*	*A. sylvaticus*
Maize field	U	U	U	U	U	U
Softwood forest or shrubs with herbaceous undergrowth	U	M	M	U	U	M
Softwood forest or shrubs with sparse undergrowth	U	U	U	U	U	S
Rough vegetation of blackberry and stinging nettle	U	S	M	M	M	S
Rough vegetation, predominantly blackberry	U	M	M	U	M	S
Sparse herbaceous vegetation on sand	U	U	U	U	U	U
Hay meadow	M	U	U	U	U	U
Grassy vegetation, ungrazed and unmown	S	U	U	M	M	U
Water	U	U	U	U	U	U
Grassy vegetation, mown or trodden	M	U	U	M	U	U
Rough herbaceous vegetation, ungrazed	S	S	S	S	S	M
Pasture (grazing land)	M	U	U	U	U	U
Grassy vegetation, extensively grazed	M	U	U	S	U	U
Bare sand	U	U	U	U	U	U
Rough herbaceous vegetation on sand	U	U	M	U	S	U
Rough herbaceous vegetation, grazed	S	U	M	S	M	M
Low herbaceous vegetation	M	U	U	U	U	U
Buildings	U	U	U	U	U	U
Rough vegetation, predominantly stinging nettle	U	U	U	U	U	M
Rough vegetation, predominantly thistle	U	U	U	U	U	M
Hard substrate	U	U	U	U	U	U

Suitability of ecotopes: U, unsuitable ecotope; M, marginal ecotope; S, suitable ecotope. Trap lines were assumed to be situated in a suitable ecotope when a species was present during each trapping period after the first observation. If the species was only occasionally present, the trap line was assumed to be in a marginal ecotope. If a species was only present in a trap line located on a non-flooded part in December, while there were neighbouring ecotopes which were assumed to be suitable, the trap line was assumed to be in a marginal habitat. In all other cases, the trap lines were assumed to be situated in unsuitable ecotopes.

of meat and apple. *A. sylvaticus* ate meat and apple and took tiny pieces of carrot. *M. agrestis* ate carrot and apple in similar amounts, and also ate from the meat. No data were available on the feeding patterns of *M. minutus*. When one of the vole species was present, all the bait in a trap was sometimes gone. As we made 15 observations per monitoring site, we were also able to ascertain if more than one species had been present, as there were always cases in which only one of them visited a trap. The characteristic pattern of *M. agrestis* was not observed in the MW, and we did not find the pattern characterising *A. sylvaticus*, although the species was assumed to be present. Monitoring sites were located at distances (shortest linear distance) of 0, 50, 100, 150, 250 and 400 m from non-flooded areas, covering the entire research area in the MW. In total, 42 monitoring sites were analysed, yielding the distribution of four frequently observed species in the MW, 35 weeks after inundation.

Landscape characteristics

The landscape structure of the two research areas was classified into 21 structure classes to analyse the landscape characteristics (Fig. 1). The structure classes were based on vegetation and soil characteristics and were called small mammal ecotopes (Wijnhoven et al., 2005). In the ADW, this classification was based on intensive monitoring of vegetation development during fieldwork. The MW was classified into small mammal ecotopes using the vegetation map by De Ronde (2003). On the basis of the small mammal ecotope maps, species-specific habitat suitability maps according to the classification shown in Table 1 were drawn for the six small mammal species investigated.

All research plots were positioned using a GARMIN GPS 12 Personal Navigator. To measure the values of the landscape characteristics of each of the monitoring sites, we defined the recolonisation routes. We assumed that species had taken the shortest recolonisation route through suitable and marginal habitats from potential sources of recolonisation to the monitoring sites. When no direct route through suitable or marginal habitats was available, unsuitable habitats (potential barriers) were assumed to be crossed, and we used the shortest cumulative distance (when more than one potential barrier had to be crossed). As potential sources of recolonisation we defined those non-flooded areas on which species were indeed observed, or on which species were assumed to be present, as they were caught in suitable habitats that were located within 10 m from non-flooded areas and were directly connected by suitable habitats. The source of recolonisation was defined as the potential source with the shortest recolonisation route. All measurements of distances and areas between recolonisation sources and monitoring sites were done in ArcMap 8.0.

Landscape characteristics such as distances, surface areas and suitability of landscape structures have often been used in various forms in regression models predicting or evaluating species distributions (King et al., 1998; Mörtberg, 2001; Carignan & Villard, 2002; Chase et al., 2003; Chardon et al., 2003). We analysed a number of landscape characteristics for the various small mammal species. Distance (D) is the length in metres of the recolonisation route. Cumulative barrier width (W_{bar}) (referred to below as barrier width) is the summed length in metres of the recolonisation route through unsuitable habitats. Number of barriers (N_{bar}) is the number of barriers counted along the recolonisation route. Area of suitable and marginal habitat influencing recolonisation (A_{hab}) was calculated by first taking the shortest linear distance between the source and the monitoring site as the radius of a circle with the source as its centre, and then calculating the area of suitable and marginal habitats within this circle in square metres. Habitat suitability (S_{hab}) of the monitoring site was scored as 1, 2 or 3 for suitable, marginal and unsuitable habitats, respectively, as shown in Table 1.

Calculations and statistics

To compare the importance of landscape characteristics for the small mammal species in the two research areas, five models were calculated for each species, based on the monitoring results in the ADW. Two logistic regression models of the type $\ln(Y) = aX_1 + bX_2 + cX_3 + \cdots$, calculated using SPSS 11.5, predicted the presence (P) of species, with presence being validated as 1, and absence as 0 (Table 2). Comparable models have been used by Mörtberg (2001), Carignan & Villard (2002) and Chardon et al. (2003) for various species. Model I included the data of all monitoring sites, while Model II included only the data of monitoring sites in suitable and marginal habitats. The other three models were linear regression models of the type $Y = aX_1 + bX_2 + cX_3 + \cdots$, calculated using Microsoft Excel 2000, predicting the recolonisation time (RT) in weeks. These linear models were based on the data of suitable and marginal habitats only. Linear models have been used in several studies on species distributions, for instance those by King et al. (1998) and Chase et al. (2003), and yielded stronger regressions (in terms of R^2) than logistic models.

Monitoring sites where species were never observed were excluded from Model III. In the other two models, the absence of species was scored with a recolonisation time of 45 (Model IV) and 90 (Model V) weeks, respectively. Although these values are arbitrary, being merely used to analyse their effect on the model, 45 weeks was approximately the time between two successive floods and therefore the period available for recolonisation. When regression equations are based on data of suitable and marginal habitats only, this will automatically result in a weaker regression for habitat suitability. All measured landscape characteristics were included in the model, including those which did not show significant regression, as we wanted to analyse changes in regression factors between models. We report the significance of the calculated regression coefficients based on the calculated t-values, as well as the positive or negative influence of the characteristic on recolonisation. To evaluate the importance of the landscape

Table 2. Regression coefficients of multiple regression models (least squares method) for the recolonisation of monitoring sites in the Afferdensche en Deestsche Waarden floodplain. Two models explain the presence ($P=1$) or absence ($P=0$) of six small mammal species based on landscape characteristics, calculated by logistic regression from the total data set (Model I) or from data for suitable and marginal habitats only (Model II). Three models calculate recolonisation time using linear regression for suitable and marginal habitats only. In model III, the absence of species is omitted ($A=/$), in model IV recolonisation time (RT) for the absence of species is set on 45 weeks ($A=45$), in model V on 90 weeks ($A=90$). If the regression coefficient is found significant this is indicated in bold underlined ($p<0.05$) or underlined ($p<0.25$)

		D	W_{bar}	N_{bar}	A_{hab}	S_{hab}	c	n	R^2
M. arvalis									
I	ln(P)	-4.63×10^{-3}	-1.11×10^{-5}	-5.76×10^{-1}	-2.24×10^{-6}	-6.16×10^{-1}	**1.91**	68	0.25
II	ln(P)	-4.62×10^{-3}	**0.20×10^{-1}**	**-3.31**	-2.19×10^{-7}	-1.17	3.21	47	0.15
III	RT	4.37×10^{-2}	-1.53×10^{-1}	3.09	-1.34×10^{-4}	6.38	8.47	27	0.14
IV	RT	3.35×10^{-2}	**-1.13**	**20.8**	-4.24×10^{-5}	9.97	8.82	47	0.29
V	RT	7.76×10^{-2}	**-2.94**	**49.9**	-4.04×10^{-5}	18.8	4.73	47	0.30
C. glareolus									
I	ln(P)	-1.06×10^{-2}	-1.76×10^{-2}	1.29×10^{-1}	1.26×10^{-4}	**-1.61**	**3.30**	68	0.36
II	ln(P)	2.11×10^{-2}	-1.12×10^{-2}	-1.72	4.95×10^{-6}	6.92×10^{-1}	5.56×10^{-1}	35	0.21
III	RT	4.80×10^{-1}	3.22×10^{-2}	-59.2	-2.61×10^{-3}	-9.87	24.5	16	0.14
IV	RT	1.14×10^{-2}	9.62×10^{-2}	4.66	-1.74×10^{-4}	-4.60×10^{-1}	**22.1**	35	0.40
V	RT	3.32×10^{-2}	1.93×10^{-1}	12.3	-3.38×10^{-4}	-2.65×10^{-1}	31.0	35	0.46
S. araneus									
I	ln(P)	-7.71×10^{-3}	-2.76×10^{-2}	1.41×10^{-1}	1.64×10^{-5}	-4.71×10^{-1}	1.78	68	0.28
II	ln(P)	-6.79×10^{-3}	-1.99×10^{-2}	5.99×10^{-2}	1.71×10^{-5}	-4.39×10^{-1}	1.60	48	0.10
III	RT	-1.36×10^{-1}	1.54×10^{-1}	6.89×10^{-1}	1.18×10^{-3}	6.54	3.80	24	0.18
IV	RT	1.90×10^{-2}	1.67×10^{-1}	-3.13×10^{-1}	4.17×10^{-5}	8.85	10.2	48	0.21
V	RT	5.52×10^{-2}	3.71×10^{-1}	-3.21×10^{-1}	-5.40×10^{-5}	15.5	14.7	48	0.19
C. russula									
I	ln(P)	-6.04×10^{-3}	-1.52×10^{-2}	4.39×10^{-1}	3.42×10^{-5}	-4.57×10^{-1}	-3.97×10^{-2}	68	0.07
II	ln(P)	-6.84×10^{-3}	-1.66×10^{-2}	8.36×10^{-1}	3.65×10^{-5}	5.48×10^{-1}	-1.81	50	0.05
III	RT	8.27×10^{-3}	-8.80×10^{-2}	-9.17×10^{-2}	-7.34×10^{-6}	-14.4	**39.4**	15	0.46
IV	RT	2.82×10^{-2}	4.43×10^{-2}	-3.24	-1.49×10^{-4}	-6.38	**46.7**	50	0.06
V	RT	8.00×10^{-2}	1.88×10^{-1}	-10.8	-4.25×10^{-4}	-11.0	**86.8**	50	0.08
M. minutus									
I	ln(P)	7.34×10^{-3}	-1.28×10^{-2}	-4.64×10^{-1}	-9.59×10^{-5}	-1.08	1.75	68	0.18
II	ln(P)	2.09×10^{-3}	1.36×10^{-2}	-5.52×10^{-1}	-6.83×10^{-5}	-2.62×10^{-1}	4.42×10^{-1}	49	0.09
III	RT	5.52×10^{-2}	-6.60×10^{-2}	-3.35	3.37×10^{-4}	-4.50	**30.4**	16	0.48
IV	RT	3.89×10^{-2}	-1.20×10^{-1}	2.08	1.36×10^{-4}	-3.07×10^{-1}	**35.5**	49	0.16
V	RT	6.35×10^{-2}	-3.07×10^{-1}	3.40	3.45×10^{-4}	2.72	**54.6**	49	0.14
A. sylvaticus									
I	ln(P)	-6.27×10^{-4}	-3.37×10^{-3}	**-9.75×10^{-1}**	3.08×10^{-5}	**-1.54**	**4.43**	68	0.31
II	ln(P)	-9.39×10^{-3}	3.25×10^{-2}	**-1.01**	9.30×10^{-5}	**-2.55**	**5.75**	47	0.19
III	RT	3.56×10^{-2}	-1.24×10^{-1}	-4.88	4.44×10^{-5}	8.32	**19.9**	24	0.20
IV	RT	3.36×10^{-2}	-1.17×10^{-1}	1.58	-2.50×10^{-4}	**10.5**	**17.1**	47	0.28
V	RT	1.00×10^{-1}	-3.43×10^{-1}	9.39	-9.38×10^{-4}	**29.0**	-14.2	47	0.34

Landscape characteristics measured between the sources of recolonisation and the monitoring sites: D, distance (m); W_{bar}, cumulative barrier width (m); N_{bar}, number of barriers; A_{hab}, area of intervening suitable and marginal habitat (m^2); S_{hab}, habitat suitability of monitoring site (suitable habitat, 1; marginal habitat, 2; unsuitable habitat, 3); c, constant; n indicates sample size; R^2 indicates the explained variance by the models.

characteristics for the recolonisation process we consider the significant related parameters of the models which explain more than 25% of variance in the data as potentially important. The relation of these landscape characteristics should be consistent (negative or positive relation) in those models which explain more than 25% of the variance.

The predictive power of the regression models was evaluated by applying the regression equations to the spatial data for the MW and comparing the predicted recolonisation data with the measured data for each of the monitoring sites. A predicted value ≥0.5 (for models I and II), and a value <35 (for models III, IV and V) was interpreted as presence (Table 3). The predictive power was the percentage of predictions that corresponded to the observations. Average values ± standard deviations for various landscape characteristics were calculated for the different study sites, data sets and species (Table 4), as they can explain observed differences in recolonisation and/or differences in model outcomes.

Results

Figure 2 shows the development of the densities of small mammal species in the ADW. Differences between 2001 and 2002 were observed, but general trends in population development (e.g. total densities and timing of density peaks) of the species were similar for the 2 years. The densities of *M. arvalis* increased gradually in spring, and faster towards autumn. The densities

Table 3. Comparison between measured and predicted recolonisation of the Millingerwaard floodplain

			M. arvalis		*C. glareolus*		*S. araneus*		*C. russula*	
			Measured	Model	Measured	Model	Measured	Model	Measured	Model
Model I	Total data, $P=1$, $A=0$	Average	0.595	2.87	0.238	0.684	0.262	1.83	0.195	1.84×10^2
		SD	0.497	4.67	0.431	4.19	0.445	1.62	0.401	7.48×10^2
		n		42		42		42		41
		Predictive power		57.1		76.2		40.5		46.3
Model II	Habitat data, $P=1$, $A=0$	Average	0.636	1.70×10^8	0.278	1.98	0.333	2.18	0.235	1.94×10^3
		SD	0.492	7.97×10^8	0.461	7.03	0.485	1.38	0.437	7.20×10^3
		n		22		18		31		33
		Predictive power		77.3		50.0		32.3		42.4
Model III	Habitat data, $P=$w, $A=/$	Average		19.5		−27.2		34.8		16.8
		SD		2.58		45.1		26.3		7.96
		n		22		18		31		33
		Predictive power		63.6		38.9		64.5		21.2
Model IV	Habitat data, $P=$w, $A=45$	Average		26.8		26.7		32.3		31.9
		SD		13.6		4.43		3.37		7.79
		n		22		18		31		33
		Predictive power		77.3		27.8		48.4		51.5
Model V	Habitat data, $P=$w, $A=90$	Average		40.3		46.8		53.1		55.5
		SD		36.0		10.7		5.3		21.4
		n		22		18		31		33
		Predictive power		63.6		55.6		71.0		66.7

P, presence of a species, *A*, absence of a species; the first two models are based on presence–absence data; in the other three models, the first observation of a species at a monitoring site is given as recolonisation time (RT) in weeks (w), with the absence of species either ignored ($A=/$), or set at 45 or 90 weeks. Predictions are based on the multiple regression models derived from the Afferdensche en Deestsche Waarden floodplain data for various small mammal species. Table shows average ± standard deviation for the measurements and the models, and the number of measurements included (*n*). The predictive power is shown as the percentage of predictions that are similar to the trapping results; when the predicted value $x\geq0.5$, the prediction is equal to *P*, and when $x<0.5$, the prediction is equal to *A* in models I and II; when $x\geq35$, the prediction is equal to *P*, and when $x<35$, x is similar to *A* in models III, IV and V.

Table 4. Values (average ± standard deviation) of the landscape characteristics data for the various small mammal species in the Afferdensche en Deestsche Waarden floodplain (ADW) and the Millingerwaard floodplain (MW) used to predict recolonisation

		ADW_t		MW_t		ADW_h		MW_h	
		Average	±SD	Average	±SD	Average	±SD	Average	±SD
M. arvalis	D	219.2	185.8	95.5	91.2	138.5	131.2	83.7	92.9
	W_{bar}	21.4	30.7	12.2	18.8	4.4	8.7	5.6	14.0
	N_{bar}	1.1	1.3	0.9	0.9	0.5	0.7	0.4	0.6
	A_{hab}	39217.1	41164.2	15083.4	24556.6	32022.4	43853.4	12736.9	23746.5
	S_{hab}	1.8	0.9	2.1	0.9	1.3	0.5	1.4	0.5
C. glareolus	D	248.5	242.1	171.9	141.6	169.2	207.6	160.9	129.3
	W_{bar}	109.6	124.8	32.1	34.0	59.0	69.1	17.1	15.2
	N_{bar}	1.4	1.1	1.4	0.9	0.8	0.9	1.1	0.8
	A_{hab}	15412.5	18252.7	18322.7	20914.0	9974.3	14419.5	20226.0	22587.6
	S_{hab}	2.1	0.9	2.2	1.0	1.3	0.5	1.1	0.3
S. araneus	D	124.6	126.0	126.1	111.1	93.2	104.4	134.5	107.9
	W_{bar}	53.2	82.4	7.6	11.5	24.8	33.1	6.3	12.0
	N_{bar}	1.4	1.6	0.7	0.6	1.4	1.8	0.5	0.5
	A_{hab}	7054.8	9970.4	27975.9	34325.1	6732.6	11024.6	29633.2	34657.3
	S_{hab}	2.1	0.7	2.2	0.5	1.7	0.5	2.0	0.2
C. russula	D	235.5	169.3	135.7	92.3	222.0	182.3	139.2	84.2
	W_{bar}	47.3	42.4	7.2	22.7	34.8	37.9	6.6	24.8
	N_{bar}	1.9	1.2	0.6	0.5	1.7	1.2	0.5	0.5
	A_{hab}	22999	22303.4	51592.5	53129.9	23501.6	24051.8	49128.4	50031.1
	S_{hab}	1.8	0.8	1.8	0.7	1.3	0.5	1.6	0.5
M. minutus	D	152.3	125.7	n.a.	n.a.	120.0	104.8	n.a.	n.a.
	W_{bar}	41.6	54.8	n.a.	n.a.	21.4	25.4	n.a.	n.a.
	N_{bar}	1.7	1.6	n.a.	n.a.	1.6	1.7	n.a.	n.a.
	A_{hab}	12491.0	11501.3	n.a.	n.a.	10179.9	10644.5	n.a.	n.a.
	S_{hab}	2.1	0.7	n.a.	n.a.	1.7	0.5	n.a.	n.a.
A. sylvaticus	D	227.7	231.5	n.a.	n.a.	161.2	178.2	n.a.	n.a.
	W_{bar}	50.2	47.7	n.a.	n.a.	34.6	36.5	n.a.	n.a.
	N_{bar}	2.1	1.5	n.a.	n.a.	1.7	1.3	n.a.	n.a.
	A_{hab}	17109.2	18733.5	n.a.	n.a.	12635.9	15720.7	n.a.	n.a.
	S_{hab}	2.1	0.7	n.a.	n.a.	1.8	0.4	n.a.	n.a.

D, distance (m); W_{bar}, cumulative barrier width (m); N_{bar}, number of barriers; A_{hab}, area of intervening suitable and marginal habitat (m^2); S_{hab}, habitat suitability (1, suitable habitat; 2, marginal habitat; 3, unsuitable habitat); n.a., not available. Table shows values of monitoring data for all sites (ADW_t and MW_t) or for the sites in the suitable and marginal habitats only (ADW_h and MW_h).

of *M. arvalis* were generally highest on non-flooded areas throughout the year (especially in 2002), while densities in the other zones increased later in the year, an increase which was related to the distance to non-flooded areas. The densities of *C. glareolus* were highest immediately after the retreat of the water in the zone at a distance of 0–30 m from the non-flooded areas. Their total densities in the ADW were relatively stable throughout the year, and *C. glareolus* was not trapped at distances of more than 120 m from the non-flooded areas. *S. araneus* was also present immediately after the water's retreat in the lower flooded parts, after which densities fluctuated. This species was not observed at more than 120 m from the non-flooded areas either. In spring, *C. russula* was only trapped in non-flooded areas, after which the species gradually appeared at greater distances from these areas, while its total densities increased. Both *M. minutus* and *A. sylvaticus* were not trapped or only occasionally until autumn, when they appeared with greatest densities in the 0–30 m zone, but also at larger distances. In October 2002, both species were observed at more than 120 m from the non-flooded parts, and they were also observed in this zone in December 2001. *M. arvalis*, *C. russula* and *M. minutus* showed a decrease in total densities towards December, while the densities of *A. sylvaticus* were still increasing, and the densities of *C. glareolus* and *S. araneus* were at least stable.

Combining Table 1 and Figure 1 shows that the research areas were of different quality (in terms of suitability and connectivity) for the different species. This is also shown in Table 4, which evaluates the landscape characteristics in relation to monitoring sites. Some of the landscape characteristics, like cumulative barrier width, barrier number and recolonisation distance, generally had higher values in the ADW than in the MW. The area of suitable habitats between the source of recolonisation and a randomly chosen monitoring site was, however, similar for the two floodplains, or larger at the MW than at the ADW. The barrier width was greater for *C. glareolus* than for the other species, and recolonisation distances were also large for this species, as well as for *C. russula* and *A. sylvaticus*. The recolonisation distance for *M. arvalis* was large in the ADW, but small in the MW, unlike what we found for *S. araneus*. The area of suitable habitats in the vicinity was large for *C. russula* in both research areas, and also large for *M. arvalis* in the ADW. This was also reflected in the monitoring sites, which were often found in suitable habitats (i.e., with low S_{hab}).

When data for unsuitable habitats were excluded, the presence or recolonisation time of *M. arvalis* was most related to barrier number and barrier width (Table 2). Recolonisation time was also related to habitat suitability for the models IV and V which explain more than 25% of the variance in the data. Significant relations were found between habitat suitability and the presence of *C. glareolus* (model I), and the number of barriers and recolonisation time (model V). These models explained 36 and 46% of the variance in data, respectively. However relations between these landscape characteristics and the presence or recolonisation time were only found in a singular occasion. Only one model (I) appeared to explain more than 25% of the variance in the presence data of *S. araneus*, at which a significant relation was found with barrier width. For *C. russula* this was only the case for model III where a significant relation between recolonisation time and habitat suitability was found, and for *M. minutus* for the same model where relations with distance and the number of barriers were found. The models I, IV and V explained more than 25% of the variance of presence and recolonisation time data for *A. sylvaticus*. For this species, habitat suitability was of significance in all of the three models, and the number of barriers, the barrier width and the distance were of significance in two of the three models.

It was found that the variance explained by the models (comparison between model I and II) improved for all species when the data of the unsuitable habitats were also taken into account. For *M. arvalis*, *C. glareolus* and *A. sylvaticus* it is found that the variance in recolonisation time explained by the model is much larger when absence data are included in the model. For *C. russula* and *M. minutus* this is just the other way around.

The presence–absence models for the ADW had greater predictive power for the MW observations for *M. arvalis* and *C. glareolus* than for the other species (Table 3). The predictive power for *M. arvalis* was greatest when unsuitable habitats were excluded, while that for *C. glareolus* was greatest when these were included. However, there was a huge variance in the models for *M. arvalis* and *C. russula* in particular. All of the species were expected to occur on more MW monitoring sites than where they were actually observed. Except for *M. arvalis*, the predictive power of the models decreased when unsuitable habitats were excluded. The predictive power of each of the models for *M. arvalis* based on recolonisation time was more than 60%, the greatest predictive power being shown by Model IV. The predictive power of the models for *S. araneus* and *C. russula* was greatest when absence meant a recolonisation time of 90 weeks.

Discussion

Small mammal distributions

Small mammals, like other animal species, are not homogeneously distributed over a floodplain. Although this fact is well known to ecologists (Leuven & Poudevigne, 2002; Poudevigne et al., 2002; Robinson et al., 2002; De Nooij et al., 2004), it is rarely taken into account in impact assessments of physical reconstruction and floodplain management plans. At best, these tend to take account of habitat suitability (Kooistra et al., 2005) for different species, assuming that species will be present when habitat structures and properties (e.g. vegetation, substrate, moisture and size) are suitable. However, frequent inundations are a serious problem for several species, which have to return after a flood. The recolonisation of habitats which are at first sight suitable habitats may then be a slow process, especially when landscapes are patchy, and lack large connected areas of suitable habitats. Figure 2 shows that the recolonisation of the ADW floodplain by small mammals was a slow process in both years of monitoring. The densities of all species were generally highest on the non-flooded areas or in the neighbouring zones (at distances of 0–30 m), and after 7–9 months, several species were still not present at distances of more than 120 m from the non-flooded areas, or were present there in lower densities than in other zones.

Landscape characteristics and recolonisation

We selected five landscape characteristics which we expected to potentially affect recolonisation. The distance between a source population and a monitoring site is assumed to be important, as each species has its action radius, and the potential to disperse over a certain distance in time (Wolff, 1999; Wijnhoven et al., 2005). Distance is also assumed to be positively related to recolonisation time (Diffendorfer et al., 1999), and negatively related to the 'logarithm of presence' $\ln(P)$, in the absence of interference by other factors. For *A. sylvaticus*, increasing distance was indeed significantly related to absence or an increase in recolonisation time (in 2 of the 3 models which explain more than 25% of the variance in the data), and the same was true for *M. minutus* (in model III; Table 2). For all other species, other landscape characteristics appeared to be more important in determining recolonisation, and in fact, *A. sylvaticus* also showed stronger regressions with other landscape characteristics. This indicates that distance in itself is not the major factor explaining the relatively slow recolonisation.

Potential barriers (unsuitable habitats) which have to be crossed can slow down a recolonisation process. How long it will take before barriers are crossed is related to barrier characteristics, and increasing densities within the source area can function as a trigger to cross barriers (Montgomery et al., 1991). It is often assumed that individuals (active dispersers more so than non-active dispersers) regularly sally outside suitable habitats, which is often encouraged by crowding (Gaines & McClenaghan, 1980; Bondrup-Nielsen, 1985; Bondrup-Nielsen & Karlsson, 1985; Van Apeldoorn et al., 1992; Diffendorfer et al., 1999). These movements through unsuitable habitats are more likely to result in the crossing of a partial barrier when its width is smaller. However, barriers can also accelerate recolonisation (distance per time; Peles et al., 1999), as individuals may travel across them faster or in a straighter line, and are less tempted to settle in these unsuitable habitats. The number of barriers can therefore have either a positive or a negative effect on recolonisation speed. Barrier width was found to be a significant factor for *M. arvalis* and *A. sylvaticus* (2 of the 3 models which explain more than 25% of the variance in the data) as wider barriers delayed recolonisation. By contrast, the recolonisation speed of *S. araneus* (in the only model which explains more than 25% of the variance in the data) was positively related to barrier width. The number of barriers, which also had implications for barrier width as we measured it, was positively related to recolonisation time (in the sense of delaying recolonisation) for *M. arvalis* (2 of the 3 models) and *C. glareolus* (1 of the 3 models) and for *A. sylvaticus* (2 of the 3 models) and *M. minutus* (the only model which explains more than 25% of the variance in the data). In the last case, model III showed a significant effect, which indicates that the number of barriers was not important for the areas that were recolonised, but was problematic for those monitoring sites that were not recolonised.

Another important characteristic is the surface area of available suitable habitats in the vicinity, as it can influence the direction and distance of dispersal. It is assumed that individuals are more likely to disperse in the direction of a connected suitable habitat than through an unsuitable habitat. If the source of recolonisation is a large connected area of suitable habitat, it will take longer before individuals are forced to disperse through unsuitable habitats due to crowding (Boyce & Boyce, 1988; Briner et al., 2005), resulting in a positive relation between the available area of suitable habitat and the recolonisation time. A larger area of suitable habitat can also mean an increase in connectivity over longer distances, by offering corridors or stepping-stones. This should result in a negative relation between the area of suitable habitat and the recolonisation time. A larger area of suitable habitats only had a significant accelerating effect on *A. sylvaticus* (1 of 3 models) and therefore did not seem to be a good descriptive factor for recolonisation the way we measured it.

Finally, habitat suitability will be of importance, as suitable habitats will probably be colonised sooner than marginal habitats with the same connectivity. We found that suitable habitats were colonised significantly sooner than marginal and unsuitable habitats for *A. sylvaticus* (all of the 3 models), *M. arvalis* (2 of the 3 models) and *C. glareolus* (1 of the 3 models), except for *C. russula*, for which we found the opposite regression for the only model which explain more than 25% of the variance in the data. This probably means that habitat suitability has not been very accurately described for *C. russula*, or that marginal habitats are not colonised later than suitable habitats at similar connectivity, but that suitability is only reflected by the densities it can carry.

The fact that the constant factor in several regression equations was also significant means that not all important landscape characteristics for recolonisation were included, or that the variation in the models of recolonisation time was relatively small, which would be expected to improve if recolonisation was monitored for longer than approximately 36 weeks. Especially for *C. glareolus* only the constant factor appeared to be significantly related in the three relevant models, which indicates that the recolonisation of the habitats occurred independent of the landscape characteristics measured by us. This could be the result of the fact that suitable habitats near the sources were colonised soon, while others were not reached throughout the monitoring period.

Types of recolonisation

Based on the density development patterns, we expect that three types of recolonisers can be distinguished, which should be reflected in the importance of the landscape characteristics determining the recolonisation of floodplains by small mammal species. The three different types distinguished are schematised in Figure 3. For species of the first type, the 'gradual, density induced colonisers', population densities first have to increase in suitable areas (sources) before barriers are crossed. When densities increase, individuals are forced to move into unsuitable habitat searching for suitable habitats, at which suitable habitats in the surroundings will be colonised first. Each extra barrier will delay recolonisation, while wider barriers to a certain extent can accelerate recolonisation, as individuals will not settle in unsuitable habitats. Just after a flood these species are only observed on and near non-flooded areas after which the floodplain will be colonised gradually. Species of type 2, the 'Active dispersers', frequently sally outside the suitable habitats. They can be found at several sites within a short time, and colonisation is often driven by habitat quality at a particular moment. Their dispersal is not impeded by areas of unsuitable habitats or potential barriers to a certain extent. A typical recolonisation pattern for these species involves rapid recolonisation of several adjacent sites, followed by a very slow or gradual colonisation of sites with poor connectivity. For species of type 3, the 'Long-distance dispersers after a lag', small stretches of unsuitable habitats initially prevent colonisation of suitable habitats. After longer periods, suitable habitats at large distances are also colonised. A certain increase in the population density in source areas is necessary before individuals start to cross barriers. Once dispersal through unsuitable habitats has been initiated, these species can travel over large distances. Therefore, recolonisation time is typically negatively related to barrier width.

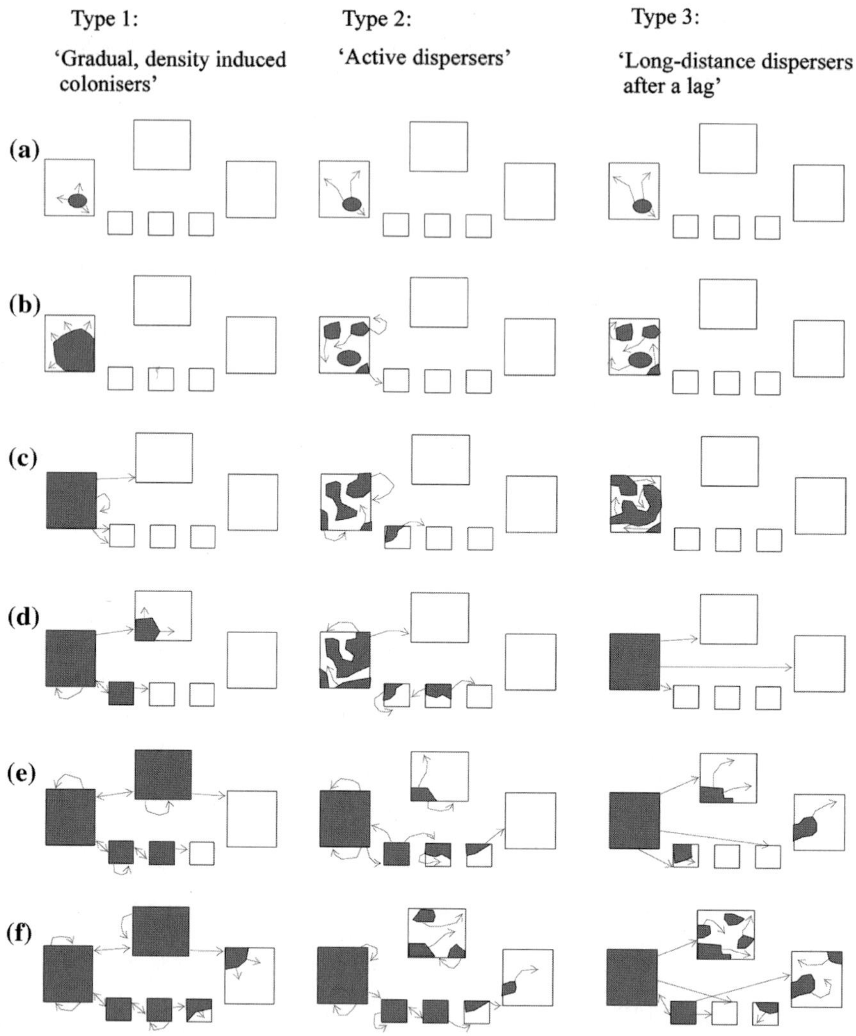

Figure 3. Three types of recolonisation by small mammal species shown in six comparable time intervals for each of the types, as indicated by letters. The arrows indicate the dispersal movements between suitable habitats (rectangles), in which the populated parts are shown in grey. Unsuitable habitats in between suitable patches can form temporary barriers to recolonisation.

The presence of *M. arvalis* was related to barrier number and barrier width and is a typical species of type 1. *M. arvalis* and also *C. russula* populations grew faster than those of the other species, gradually recolonising areas at larger distances. Both species recolonised areas at distances of more than 120 m. Therefore the density development pattern of *C. russula* also looked like type 1, but this could not be confirmed by the regression analyses. The fact that a negative relation between habitat suitability and recolonisation time was observed for *C. russula* indicates that the habitat is not described well or that the connectivity of the suitable habitats was poor.

Based on the density development patterns *C. glareolus* and *S. araneus* are expected to belong to type 2. They show a rapid recolonisation of several adjacent sites, followed by a very slow or gradual colonisation of sites with poor connectivity. Fast recolonisation of adjacent monitoring sites could be the result of favourable conditions in the lower parts just after the flood, probably relating to food availability and quality. Fresh plant shoots and relatively immobile earthworms are plentiful in the formerly inundated parts, and invertebrates are abundant in the debris at the flood mark. In addition, competition is minimal and there are plenty of shelter opportunities just after inundation. The total densities of *C. glareolus* and *S. araneus* were relatively stable throughout the year, and both species were not observed in about 50% of the total area (viz. the zone at a distance of more than 120 m from the non-flooded areas). It seems that for these species the research area had a poor connectivity and there was a lack of time for complete recolonisation. Although a positive relation between barrier width and recolonisation time was found for *S. araneus*, more data on distributions in floodplains are actually necessary to discriminate if *S. araneus* is indeed a type 2 species. Also the classification of *C. glareolus* in one of the recolonisation types is unclear from the regression models. For *C. glareolus* the number of barriers and recolonisation time are related, which is in line with publications referring to this species as a species with low dispersal activity, not very frequently crossing barriers (Bondrup-Nielsen & Karlsson, 1985; Kozakiewicz, 1993). However this could also be the result of several suitable habitats laying out of reach for this species within the research period.

For species of type 3, small stretches of unsuitable habitats initially prevent colonisation of suitable habitats. After longer periods, suitable habitats at large distances are also colonised. A certain increase in the population density in source areas is necessary before individuals start to cross barriers. Once dispersal through unsuitable habitats has been initiated, these species can travel over large distances. Therefore, recolonisation time is typically negatively related to barrier width. *A. sylvaticus* is a typical representative of this type. Also *M. minutus* shows a similar density development pattern, but cannot be classified due to the poor regressions. Poor trappability during a large part of the year may also influence the pattern for this species (Lange et al., 1994). As a result, *M. minutus* suddenly appeared at several monitoring sites in autumn, making it look like a type 3 species, although it was most probably already present earlier in the year at some monitoring sites.

Predictive capacity of the regression models

Although regression models based on ecological data, and especially trapping results of small mammals, show a great deal of unexplained variance (low R^2 values, as shown in Table 2), they show valuable information, especially in relation to heterogeneously structured landscapes or species with different recolonisation strategies. Specifically, these models can give valuable information on the relative importance of landscape characteristics from different levels of significance, and show trends in positive or negative influence on species recolonisation. We decided to include all landscape characteristics in the regression models (even those which were not found to be significant), as the regression models were based on the data of only one floodplain with its own characteristics. Retaining all parameters in the model can give valuable information about the importance of landscape characteristics, when predictions for another floodplain (MW) are combined with its landscape characteristics. The R^2 values show that the absence of *C. russula* and *M. minutus* from several sites was not adequately explained by the landscape characteristics, as the regressions were much stronger when absence data were omitted. It is also shown that unsuitable habitats often explained the absence of species, as the regressions of model I were generally much better than those of model II. The presence and recolonisation time appeared to be described best by the chosen landscape characteristics for *M. arvalis* and *A. sylvaticus*, as for those species the constant value was not the only or best descriptive parameter, and the different models showed similar results.

The presence or absence of species at a monitoring site is easier to measure than the time until recolonisation, and this may be sufficient for several applications. The use of several monitoring sessions reduces the risk of missing a species which is actually present. Predicting the presence of species will be less accurate when floods occur in various seasons, as the development of small mammal populations will vary over the year (Southern, 1965; Montgomery, 1989), and the impact on species survival is different (Van der Velde et al., 2004). As the flooding intervals in the ADW and the MW were similar in the years in which we monitored them, we feel justified in using the presence–absence models of the ADW to make predictions for the MW. Except for individuals dispersing or occasionally sallying out of the suitable habitats, species are not expected to occur in unsuitable habitats. Therefore, we expect the predictive power of a model to increase when the trap lines in unsuitable habitats are ignored. Having monitoring sites in unsuitable habitats in the dataset will increase the deviation of the regression coefficients. This is actually not the case for all species which indicates that the unsuitable habitats also often had a poor connectivity in those cases.

Predictions of the presence of species in variously structured floodplains with different flooding regimes actually necessitate the use of regression models based on recolonisation time. A problem of these regression models is that of handling data on species absence. If these data are ignored, less information about decelerated recolonisation is incorporated. These regression models are expected to underestimate the actual recolonisation time in a landscape. It is possible to allocate a value to the absence data which is larger than the maximum recolonisation time recorded. Some uncertainty will remain, however, which can lead to underestimation or overestimation. The relative importance of model parameters for predictions about different species will not be changed by varying the value of the recolonisation time at monitoring sites where species are absent. In ideal situations, the regression models are based on a much longer monitoring time than is available in the area of prediction. This makes it unnecessary to add the value of the recolonisation time to monitoring sites that are not recolonised, as the predictive power of model III will be improved. The predictive power of similar models such as IV and V will also be improved, as a larger value for recolonisation time (>90) can be given to the sites where species are absent.

Species-specific landscape characteristics should be compared between the two research areas to interpret the predictive capacity of models, as values of the landscape characteristics in the area of prediction should lie within the range of values for the monitoring area. The predictions for the presence of *M. arvalis* in the MW were reasonable (predictive power of 57.1 and 77.3%; Table 3), while the landscape characteristics for *M. arvalis* in the ADW and the MW were also similar. This was especially the case for the most important factors determining the presence of *M. arvalis*, viz., barrier number and width (Table 4). Only the distances and the area of suitable habitats in the MW seem to be smaller, which does not greatly influence the predictions. The presence of all species, but especially that of *S. araneus* and *C. russula*, in the MW was, however, overestimated by the models (leading to a poor predictive power of between 32.3 and 46.3%). In general, barriers in the MW are narrower and greater areas of suitable habitats are present than in the ADW, for all species except for *M. arvalis*. Barriers may prevent the recolonisation of the MW by *C. glareolus* and *C. russula* to a greater extent than expected. For *S. araneus*, the barrier number and width are smaller in the MW, while the area of suitable habitat is larger. Barriers may prevent recolonisation by this species as well, or the much larger suitable area may slow down the recolonisation process. While the predictive power of the recolonisation model of *M. arvalis* was good (77.3%) when the recolonisation time for the suitable habitats where the species was assumed to be absent was set at 45 weeks (Table 3), the models for the other species had better predictive power at a mean recolonisation time of 90 weeks. This suggests that most of the suitable habitats for *M. arvalis* are inhabited when there is no flooding for a whole year, while for the other species, the complete recolonisation of the MW is probably a matter of years.

Conclusions

The results of this study confirm that the recolonisation of floodplains by small mammal species after flooding events is a relatively slow process. Floodplains in which linear distances between non-flooded areas and potential habitats never exceed 1 km, are found to be not entirely recolonised within 1 year. As the time between two successive floods is generally not long enough for a complete recolonisation of the floodplain, species distribution within floodplains is generally far from homogeneous, even when only suitable habitats are considered. The recolonisation patterns are species-specific. Three general types were distinguished based on recolonisation patterns. Species behaviour towards landscape characteristics as shown by regression equations confirmed the positioning of *M. arvalis* and *A. sylvaticus* in these types.

This study also showed that multivariate regression models are useful in determining the importance of landscape characteristics for recolonisation. The predictive power of the models looks promising, but can be improved using data from a few other research areas, especially those where longer recolonisation times are available. This would improve especially the value of models based on recolonisation time.

Acknowledgements

We would like to thank S. Finote, M.B. Oostendorp, M.J.J.A. Verbruggen, C. Doukoure, A. Lequien, J.T.H.P. Tonk, J.F. Smets and M.J.M. Stassen for their valuable help with the collection of the field data. Thanks are due to E. Schellekens (Arcadis) and the research institute Alterra for lending us the live traps, to J. Klerkx for his linguistic comments, to the State Forestry Services (SBB) and Stichting Ark for their permission to do research in the 'Afferdensche en Deestsche Waarden' and 'Millingerwaard' floodplains, and to M. Reich, S.J. Ormerod, E. Krywko, A.M.J. Ragas and P.H. Nienhuis for their useful comments on the manuscript. This study was financially supported by NWO-SSEO and NWO-LOICZ, and is CWE publication 405.

References

Andersen, D. C., K. R. Wilson, M. S. Miller & M. Falck, 2000. Movement patterns of riparian small mammals during predictable floodplain inundation. Journal of Mammalogy 81: 1087–1099.

Bondrup-Nielsen, S., 1985. An evaluation of the effects of space use and habitat patterns on dispersal in small mammals. Annales Zoologici Fennici 22: 373–383.

Bondrup-Nielsen, S. & F. Karlsson, 1985. Movements and spatial patterns in populations of *Clethrionomys* species: A review. Annales Zoologici Fennici 22: 385–392.

Bowers, M. A. & G. W. Barrett, 1999. Synthesis: A review of the science and prescriptions for the future. In Barrett, G. W. & J. D. Peles (eds), Landscape Ecology of Small Mammals. Springer Verlag, New York, 313–337.

Boyce, C. C. K. & J. L. Boyce III, 1988. Population biology of *Microtus arvalis*. III. Regulation of numbers and breeding dispersion of females. Journal of Animal Ecology 57: 737–754.

Briner, T., W. Nentwig & J. -P. Airoldi, 2005. Habitat quality of wildflower strips for common voles (*Microtus arvalis*) and its relevance for agriculture. Agriculture, Ecosystems and Environment 105: 173–179.

Carignan, V. & M. -A. Villard, 2002. Effects of variations in micro-mammal abundance on artificial nest predation in conifer plantations and adjoining deciduous forests. Forest Ecology and Management 157: 255–265.

Chardon, J. P., F. Adriaensen & E. Matthysen, 2003. Incorporating landscape elements into a connectivity measure: a case study for the Speckled wood butterfly (*Pararge aegeria* L.). Landscape Ecology 18: 561–573.

Chase, J. F., J. J. Walsh, A. Cruz, J. W. Prather & H. M. Swanson, 2003. Spatial and temporal activity patterns of the brood parasitic brown-headed cowbird at an urban/wildland interface. Landscape and Urban Planning 64: 179–190.

Delattre, P., B. de Sousa, E. Fichet-Calvet, J. P. Quéré & P. Giraudoux, 1999. Vole outbreaks in a landscape context: evidence from a six year study of *Microtus arvalis*. Landscape Ecology 14: 401–412.

De Nooij, R. J. W., H. J. R. Lenders, R. S. E. W. Leuven, G. de Blust, N. Geilen, B. Goldschmidt, S. Muller, I. Poudevigne & P. H. Nienhuis, 2004. BIO-SAFE: assessing the impact of physical reconstruction on protected and endangered species. River Research and Applications 20: 299–313.

De Ronde, I., 2003. Vegetatie ontwikkeling in de Millingerwaard gebaseerd op herhaalde vegetatie karteringen 1994–2002. MSc thesis, Wageningen University, Wageningen (in Dutch).

Diffendorfer, J. E., M. S. Gaines & R. D. Holt, 1999. Patterns and impacts of movements at different scales in small mammals. In Barrett, G. W. & J. D. Peles (eds), Landscape Ecology of Small Mammals. Springer Verlag, New York, 63–88.

Erlinge, S., G. Göransson, L. Hansson, G. Högstedt, O. Liberg, I. N. Nilsson, T. Nilsson, T. von Schantz & M. Sylvén, 1983. Predation as a regulating factor on small rodent populations in southern Sweden. Oikos 40: 36–52.

Gaines, M. S. & L. R. McClenaghan Jr., 1980. Dispersal in small mammals. Annual Review of Ecology and Systematics 11: 163–196.

Hanski, I., H. Henttonen, E. Korpimäki, L. Oksanen & P. Turchin, 2001. Small-rodent dynamics and predation. Ecology 82: 1505–1520.

Helmer, W. & P. J. A. M. Smeets, 1990. Natuur- en landschapsherstel in de Gelderse Poort: uitgewerkt voor de Ooijpolder en de Millingerwaard. (2nd edn), Direction Forestry and Landscape Development, Utrecht (in Dutch).

Jongbloed, R. H., T. P. Traas & R. Luttik, 1996. A probabilistic model for deriving soil quality criteria based on secondary poisoning of top predators II. Calculations for dichlorodiphenyltrichloroethane (DDT) and cadmium. Ecotoxicology and Environmental Safety 34: 279–306.

King, D. I., C. R. Griffin & R. M. DeGraaf, 1998. Nest predator distribution among clearcut forest, forest edge and forest interior in an extensively forested landscape. Forest Ecology and Management 104: 151–156.

Kooistra, L., M. A. J. Huijbregts, A. M. J. Ragas, R. Wehrens & R. S. E. W. Leuven, 2005. Spatial variability and uncertainty in ecological risk assessment: a case study on the effect of cadmium for the little owl in a Dutch river floodplain. Environmental Science and Technology 39: 2177–2187.

Kozakiewicz, M., 1993. Habitat isolation and ecological barriers – the effect on small mammal populations and communities. Acta Theriologica 38: 1–30.

Lange, R., P. Twisk, A. van Winden & A. van Diepenbeek, 1994. Zoogdieren van West Europa. KNNV-uitgeverij, Utrecht (in Dutch).

Leuven, R. S. E. W. & I. Poudevigne, 2002. Riverine landscape dynamics and ecological risk assessment. Freshwater Biology 47: 845–865.

Leuven, R. S. E. W., S. Wijnhoven, L. Kooistra, R. J. W. de Nooij & M. A. J. Huijbregts, 2005. Toxicological constraints for rehabilitation of riverine habitats: a case study for metal contamination of floodplain soils along the Rhine. Archiv für Hydrobiologie, Supplement 155: 657–676.

Matthiopoulos, J., 2003. The use of space by animals as a function of accessibility and preference. Ecological Modelling 159: 239–268.

Montgomery, W. I., 1989. Population regulation in the wood mouse *Apodemus sylvaticus*. II. Density dependence in spatial distribution and reproduction. Journal of Animal Ecology 58: 477–494.

Montgomery, W. I., W. L. Wilson, R. Hamilton & P. McCartney, 1991. Dispersion in the wood mouse, *Apodemus sylvaticus*: variable resources in time and space. Journal of Animal Ecology 60: 179–192.

Mörtberg, U. M., 2001. Resident bird species in urban forest remnants; landscape and habitat perspectives. Landscape Ecology 16: 193–203.

Pachinger, K. & J. Haferkorn, 1998. Comparison of the small mammal communities in floodplain forests at the Danube and Elbe rivers. Ekológia (Bratislava) 17: 11–19.

Peles, J. D., D. R. Bowne & G. W. Barrett, 1999. Influence of landscape structure on movement patterns of small mammals. In Barrett, G. W. & J. D. Peles (eds), Landscape Ecology of Small Mammals. Springer Verlag, New York, 41–62.

Poudevigne, I., D. Alard, R. S. E. W. Leuven & P. H. Nienhuis, 2002. A system approach to river restoration: a case study in the lower Seine valley, France. River Research and Applications 18: 239–247.

Robinson, C. T., K. Tockner & J. V. Ward, 2002. The fauna of dynamic riverine landscapes. Freshwater Biology 47: 661–677.

Southern, H. N., 1965. Handbook of British Mammals. Mammal Society of the British Isles. Blackwell Scientific Publications, Oxford.

Van Apeldoorn, R. C., W. T. Oostenbrink, A. van Winden & F. F. van der Zee, 1992. Effects of habitat fragmentation on the bank vole, *Clethrionomys glareolus*, in an agricultural landscape. Oikos 65: 265–274.

Van den Brink, N. W., N. M. Groen, J. de Jonge & A. T. C. Bosveld, 2003. Ecotoxicological suitability of floodplain habitats in The Netherlands for the little owl (*Athene noctua vidalli*). Environmental Pollution 122: 127–134.

Van der Velde, G., R. S. E. W. Leuven & I. Nagelkerken, 2004. Types of river ecosystems. In: Fresh surface water. Water Sciences, Engineering and Technology Resources. Encyclopedia of life support systems (EOLSS). UNESCO, EOLSS Publishers, Oxford: 1–29 (http://www.eolss.net/).

Verbeylen, G., L. de Bruyn, F. Adriaensen & E. Matthysen, 2003. Does matrix resistance influence red squirrel (*Sciurus vulgaris*). Landscape Ecology 18: 791–805.

Verboom, J., R. Foppen, P. Chardon, P. Opdam & P. Luttikhuizen, 2001. Introducing the key patch approach for habitat networks with persistent populations: an example for marshland birds. Biological Conservation 100: 89–101.

Vos, C. C. & J. P. Chardon, 1998. Effects of habitat fragmentation and road density on the distribution pattern of the moor frog *Rana arvalis*. Journal of Applied Ecology 35: 44–56.

Wijnhoven, S., G. van der Velde, R. S. E. W. Leuven & A. J. M. Smits, 2005. Flooding ecology of voles, mice and shrews: the importance of geomorphological and vegetational heterogeneity in river floodplains. Acta Theriologica 50: 453–473.

Wike, L. D., F. D. Martin, H. G. Hanlin & L. S. Paddock, 2000. Small mammal populations in a restored stream corridor. Ecological Engineering 15: S121–S129.

Wolff, J. O., 1999. Behavioral model systems. In Barrett, G. W. & J. D. Peles (eds), Landscape Ecology of Small Mammals. Springer Verlag, New York, 11–40.

Zandberg, B., 1999. Afferdensche en Deestsche Waarden; Inrichtingsplan. Report 99.001. –Directorate-General of Public Works and Water Management, Arnhem (in Dutch).

Zhang, Z. & M. B. Usher, 1991. Dispersal of wood mice and bank voles in an agricultural landscape. Acta Theriologica 36: 239–245.

Hydrobiologia (2006) 565:153–162
R.S.E.W. Leuven, A.M.J. Ragas, A.J.M. Smits & G. van der Velde (eds), Living Rivers: Trends and Challenges in Science and Management
DOI 10.1007/s10750-005-1911-9

The importance of hydrodynamics for protected and endangered biodiversity of lowland rivers

R.J.W. de Nooij[1,3,*], W.C.E.P. Verberk[2,3,4], H.J.R. Lenders[1,3], R.S.E.W. Leuven[1,3] & P.H. Nienhuis[1,4]

[1]*Department of Environmental Science, Institute for Wetland and Water Research, Faculty of Science, Radboud University Nijmegen, P.O. Box 9010, 6500 GL Nijmegen, The Netherlands*
[2]*Department of Animal Ecology and Ecophysiology, Institute for Wetland and Water Research, Faculty of Science, Radboud University Nijmegen, P.O. Box 9010, 6500 GL Nijmegen, The Netherlands*
[3]*Netherlands Centre for Nature Studies (NCN), P.O. Box 9010, 6500 GL Nijmegen, The Netherlands*
[4]*Bargerveen Foundation, P.O. Box 9010, 6500 GL Nijmegen, The Netherlands*
*(*Author for correspondence: E-mail: R.deNooij@science.ru.nl)*

Key words: animal and plant species, ecological rehabilitation, flood defence, hydrodynamics, legally protected and endangered species, river management

Abstract

This paper examines the relationship between protected and endangered riverine species (target species) and hydrodynamics in river-floodplain ecosystems, combining ecological and policy-legal aspects of biodiversity conservation in river management. The importance of different hydrodynamic conditions along a lateral gradient was quantified for various taxonomic groups. Our results show that (i) target species require ecotopes along the entire hydrodynamic gradient; (ii) different parts of the hydrodynamic gradient are important to different species, belonging to different taxonomic groups; (iii) in particular low-dynamic parts are important for many species and (iv) species differ in their specificity for hydrodynamic conditions. Many species of higher plants, fish and butterflies have a narrow range for hydrodynamics and many species of birds and mammals use ecotopes along the entire gradient. Even when focussing only on target species, the entire natural hydrodynamic gradient is important. This means that the riverine species assemblage as a whole can benefit from measures focussing on target species only. River reconstruction and management should aim at re-establishing the entire hydrodynamic gradient, increasing the spatial heterogeneity of hydrodynamic conditions.

Introduction

Natural river-floodplain ecosystems exhibit a hydrodynamic gradient from the main channel to inundation-free areas. A wide variety of riverine habitats exists along this gradient, in space and time, created by the dynamic interaction of water, sediment and biota, leading to high biodiversity (Bayley, 1995; Ward et al., 2002). Species characteristic of the river-floodplain ecosystem (hereafter termed riverine species) have adapted their life histories to match riverine conditions. However, many riverine species have become rare and endangered in the Rhine and Meuse catchments, as a consequence of the dramatic changes in river-floodplain ecosystems (e.g. the construction of dikes, dams, groynes and weirs, conversion of floodplains to agricultural land, water pollution and invasive species). These modifications have greatly reduced the spatial heterogeneity, as well as the variation in hydrodynamic conditions along the lateral gradient in the floodplain, but also along the longitudinal and vertical dimensions (Ward & Tockner, 1999; Aarts et al., 2004).

Ecological rehabilitation aims at restoring riverine biodiversity by rehabilitating hydrodynamic and morphodynamic processes in river-floodplain ecosystems, and introducing semi-natural grazing regimes (Nienhuis et al., 2002). Ecological rehabilitation also includes improvement of water chemistry and the remediation of toxic river sediments (Leuven et al., 2005).

Large-scale reconstruction measures are being prepared and implemented in river basins of north-western Europe for the purpose of flood defence, ecological rehabilitation and infrastructural improvements (Van Stokkom et al., 2005). These measures will have far-reaching consequences for the physical structure and dynamics, and hence for the ecological functioning, of river-floodplain ecosystems (Nienhuis et al., 1998). Political and legal goals state the importance of ecological rehabilitation and provide regulations and time horizons. According to the European Water Framework Directive (Council Directive 2000/60/EC), for natural rivers a good ecological status and for heavily modified waters a good ecological potential have to be achieved by 2015.

The legislative framework for nature protection in Europe consists of the Habitats Directive and the Birds Directive. Significant negative impacts of human activities on species and habitats protected by these directives are not allowed, unless (i) there are no alternative solutions and (ii) there are imperative reasons of overriding public interest that demand these activities (Council Directive 79/409/EEC; Council Directive 92/43/EEC). Even if these two conditions have been met, the negative impacts on protected habitats have to be compensated for. River managers are therefore obliged to take protected species into account in their effect assessments for spatial planning, physical reconstruction and management (e.g. Environmental Impact Assessments and Strategic Environmental Assessments). Another important and widely used instrument in species conservation is that of Red Lists. In this paper, we use the term target species to refer to both legally protected and red-listed species.

Attuning the aims of flood defence, ecological rehabilitation and nature protection requires tools that integrate policy and legislation goals with ecological knowledge about target species in river-floodplain ecosystems. Expressing both actual and potential biodiversity values offers opportunities to assess the impacts of physical reconstruction on biodiversity. The BIO-SAFE model (Spreadsheet Application For Evaluation of BIOdiversity) is such a tool (Lenders et al., 2001; De Nooij et al., 2004, 2005).

The theoretical relation between hydrodynamics and biodiversity is well-known and is exemplified in the flood pulse concept (e.g. Junk et al., 1989). Empirical evidence supports this concept and shows that different taxonomic groups utilize the gradient differently (Van den Brink, 1994; Van den Brink et al., 1996; Ward & Tockner, 2001; Chovanec et al., 2005). However, to what extent the relation between hydrodynamics and biodiversity is also valid for target species is largely unknown. Maximizing ecological benefits of floodplain reconstruction, and minimizing conflicts between river management and nature protection, require knowledge on the response of protected and red-listed riverine species to river dynamics. The BIO-SAFE model integrates available knowledge about habitat demands of these target species for the rivers Rhine and Meuse with their political and legal status. This model was used to answer the following questions:

- How important are different parts of the hydrodynamic gradient for target species of different taxonomic groups in river-floodplain ecosystems?
- How specifically do target species utilize the various parts of the hydrodynamic gradient?
- What are the implications of the response of riverine target species to hydrodynamic conditions for river management?

Materials and methods

Model description

BIO-SAFE is a valuation model which links ecotopes to riverine target species listed in the European Habitats Directive, the European Birds Directive, the Conventions of Bern and Bonn and Red Lists. Ecotopes are defined as spatial units of a certain extent, which are relatively homogeneous in terms of vegetation structure, succession stage and the main abiotic site factors that are relevant to plant growth (Klijn & Udo de Haes, 1994). BIOSAFE describes the habitat of riverine target species in terms of riverine ecotopes, derived from the Water Ecotope Classification published by Van der Molen et al. (2003), which includes the River Ecotope System (RES) by Rademakers & Wolfert (1994). The classification by Van der Molen et al. (2003) for rivers is based on vegetation structure and composition, inundation frequency (hydrodynamics), morphodynamics and land use. In BIOSAFE, ecotopes are distinguished at four levels of scale (1:100,000; 1:50,000; 1:25,000; 1:10,000). At the finest level of scale (1:10,000), 60 different ecotopes are distinguished. River engineers, landscape ecologists and landscape designers use ecotopes in hydraulic models, landscape ecology and landscape design, making the concept of ecotopes a suitable tool for communication between the various disciplines active in river management. The model incorporates both natural ecotopes and man-made ecotopes.

Species were selected based on their occurrence in river-floodplain ecosystems. This included species characteristic of the current situation, but also of natural river-floodplain ecosystems. Selected species were subsequently attributed to the different ecotopes. The information on species and habitats was derived from a thorough literature survey, supplemented by expert knowledge. Taxonomic groups included in the model are higher plants, birds, herpetofauna (amphibians & reptiles), mammals, fish, butterflies and odonates (dragonflies & damselflies). In linking species to ecotopes the habitat demands of all life cycle stages were considered. Species in other taxonomic groups were either not listed as target species or were not characteristic of the Rhine and Meuse river-floodplain ecosystems. Table 1 lists the numbers of species included in each taxonomic group.

To each species, values were assigned on the basis of its policy and legislation status. Through the linkage of species to ecotopes, values were assigned to ecotopes as well (Fig. 1). An explanation of the species selection process, the value assignment, the development of the ecotope typology and the linkage of species to ecotopes, including a full description of the functionalities of the BIOSAFE model, can be found in Lenders et al. (2001) and De Nooij et al. (2001, 2004).

Data analysis and calculations

In this study, species selection, linkage to ecotopes and value assignment concern lowland river-floodplain ecosystems in the Netherlands and the Dutch political-legislative context. All analyses were carried out at the finest level of scale, as preliminary analyses had indicated that ecotopes defined at coarser levels were less accurate in describing the habitats of many species. The ecotopes were classified into seven different hydrodynamic classes along the hydrodynamic gradient (hydroclasses, Table 1). Note that an ecotope can occur in more than one hydroclass, such as the ecotope called natural levee pasture, which can occur under flooding conditions with a total inundation time of less than 100 days per year (i.e. hydroclasses 5 and 6; Table 1).

Based on the values assigned to the species, the linkage of the species to the ecotopes and the relation between hydrodynamic conditions and the occurrence of ecotopes (Table 1), we quantified the biodiversity potential of a hydrodynamic class (Hydroclass Importance) (Fig. 1). This procedure was applied to all taxonomic groups combined (Fig. 2) and to each taxonomic group separately (Fig. 3). For all taxonomic groups combined, the total score in all hydroclasses was set at 100% for each taxonomic group (in order to weight the different taxonomic groups equally, i.e. irrespective of the total number of species in a taxonomic group). In order to calculate the importance for each ecotope, the species scores were first summed per taxonomic group, resulting in the *Potential Taxonomic group Biodiversity constant* (PTB):

$$\mathrm{PTB} = \sum \text{species scores (for all species per taxonomic group)} \quad (1)$$

Subsequently, the score of a species was assigned to the ecotopes it was linked to (species score). This species score was divided by the PTB, resulting in the *Species specific Ecotope Importance* ($\mathrm{SEI}_{\mathrm{ecotope\,i}}$) which is the species' relative contribution to the maximum potential value of that ecotope.

$$\mathrm{SEI}_{\mathrm{ecotope\,i}} = \text{species score}/\mathrm{PTB} \quad (2)$$

For each taxonomic group, the SEI values of all species were summed per hydroclass (Table 1) in order to

Table 1. Numbers of target species and ecotopes (between brackets) included in BIO-SAFE for each taxonomic group and per hydrodynamic class. Criteria for the classification of hydroclasses were modified from Van der Molen et al. (2003). A full list of the species included can be found in De Nooij et al. (2001)

Hydro-class	Criterion	Higher plants ($n=136$)	Birds ($n=60$)	Herpetofauna ($n=9$)	Mammals ($n=9$)	Fish ($n=20$)	Butterflies ($n=17$)	Odonates ($n=6$)	Total ($n=257$)
1	Deep water (>1.5 m)	0 (0)	22 (2)	3 (1)	3 (2)	14 (2)	0 (0)	0 (0)	42 (2)
2	Permanently flooded (<1.5 m)	4 (6)	31 (9)	3 (7)	5 (9)	18 (9)	3 (1)	4 (5)	68 (10)
3	River bank	18 (7)	36 (10)	3 (5)	8 (9)	14 (10)	3 (1)	4 (6)	86 (11)
4	Flooded >100 d $year^{-1}$	25 (12)	56 (13)	9 (12)	8 (13)	8 (4)	6 (6)	6 (10)	118 (13)
5	Flooded 20–100 d $year^{-1}$	105 (22)	53 (22)	9 (20)	8 (20)	6 (3)	17 (13)	5 (14)	203 (23)
6	Flooded <20 d $year^{-1}$	112 (23)	53 (23)	9 (22)	8 (22)	6 (6)	15 (12)	5 (14)	208 (25)
7	Never flooded*	54 (13)	46 (12)	9 (11)	7 (12)	6 (5)	10 (5)	5 (6)	137 (14)

*Note that in the absence of flooding, aquatic ecotopes can still be present.

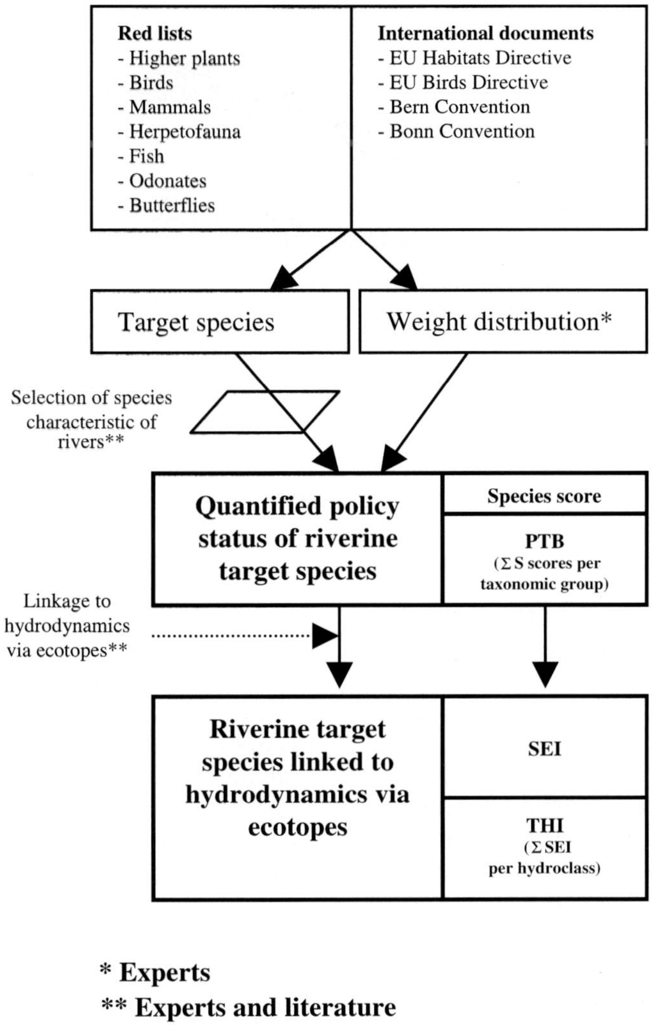

Figure 1. Schematic overview of the construction of BIO-SAFE and quantification of the potential of hydrodynamic conditions for target species. S score: Species-specific score, quantifying policy relevance; PTB: Potential Taxonomic group Biodiversity constant; SEI: Species-specific Ecotope Importance; relative contribution of species to the maximum potential value for that ecotope; THI: Taxonomic group Hydroclass Importance; potential value of the hydrodynamic class.

calculate the *Taxonomic group Hydroclass Importance* (THI). This is defined as the potential value of that hydrodynamic class as a habitat for riverine target species belonging to a certain taxonomic group.

$$\mathrm{THI}_{\mathrm{hydroclass\,k}} = \sum \mathrm{SEI} \quad (3)$$

(for all species per taxonomic group occurring in ecotopes with hydrodynamic class k)

Per hydrodynamic class, ecotopes were only used in the calculation if (i) they occurred under those hydrodynamic conditions and (ii) they were used by at least one species in that taxonomic group (see Table 1), in order to exclude ecotopes which are unsuitable for that taxonomic group (e.g. fish do not occur in terrestrial ecotopes).

Results

When all taxonomic groups of target species are combined, the biodiversity potential shows an increase with decreasing hydrodynamics, until class 6, where an optimum is reached (Fig. 2). This

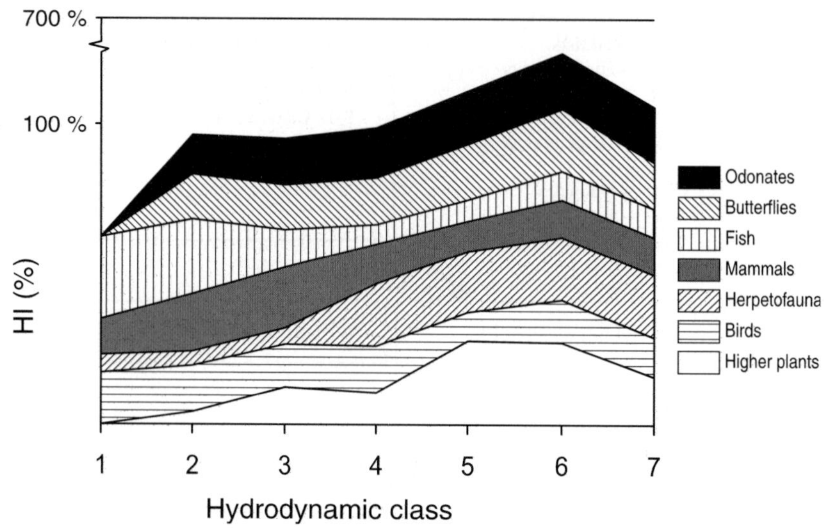

Figure 2. Hydroclass Importance (HI), for all species groups combined, along the hydrodynamic gradient. The contribution of each taxonomic group was set at 100%. 1: highest degree of hydrodynamics, 7: lowest degree of hydrodynamics.

class also harbours the largest number of target species (Table 1). The contribution of the different taxonomic groups to the potential differs markedly along the hydrodynamic gradient. The results per taxonomic group are given in Figure 3.

For higher plants, the low-dynamic parts (classes 5 and 6) are most important, although a number of plants are confined to class 3, viz., strapwort (*Corrigiola litoralis*), tall pepperwort (*Lepidium graminifolium*) and bur medick (*Medicago minima*). Many plant species have a narrow range (high specificity) along the hydrodynamic gradient.

Fish show a pattern opposite to that of plants, in which the high-dynamic parts (classes 1 and 2) are most important. Sturgeon (*Acipenser sturio*), twaite shad (*Alosa fallax*) and lampern (*Lampetra fluviatilis*) are restricted to the high-dynamic parts. Species with a broad range are predominantly found in the low-dynamic parts (classes 5–7); these include weatherfish (*Misgurnus fossilis*), eel (*Anguilla anguilla*) and crucian carp (*Carassius carassius*).

For birds, the entire hydrodynamic gradient is important. Species with a narrow range (high specificity), such as black stork (*Ciconia nigra*), curlew (*Numenius arquata*) and redshank (*Tringa totanus*) are mainly found in the low-dynamic parts (classes 4–7). Many bird species have a broad range for the hydrodynamic gradient; these include waterfowl like several duck species, black tern (*Chlidonias niger*) and common tern (*Sterna hirundo*).

The extremely dynamic parts (class 1) are of no importance to butterflies, while the high-dynamic parts (classes 2 and 3) are important to brown argus (*Aricia agestis*), queen of spain frittillary (*Issoria lathonia*) and glanville frittillary (*Melitaea cinxia*). Species restricted to the low-dynamic parts (4–7) include scarce large blue (*Maculinea teleius*), chequered skipper (*Carterocephalus palaemon*) and silver-washed frittillary (*Argynnis paphia*).

For herpetofauna, it is the low-dynamic parts (classes 4–7) that are the most important by far. The high-dynamic parts (1–3) are only used by grass snake (*Natrix natrix*), lake frog (*Rana ridibunda*) and edible frog (*Rana kl. esculenta*). These species occur across the entire hydrodynamic gradient.

Odonates, like butterflies, are absent from the extremely dynamic parts. Gomphidae use almost the entire gradient, while green hawker (*Aeshna viridis*) and hairy dragonfly (*Brachytron pratense*) are restricted to the low-dynamic parts (classes 4–7).

For mammals, the entire hydrodynamic gradient is important, although compared to birds, the high-dynamic parts (classes 2 and 3) are more important to them. Species with a narrow range for the hydrodynamic gradient include red deer (*Cervus elaphus*), root vole (*Microtus oeconomus*) and water shrew (*Neomys fodiens*). Species with a broad range for the hydrodynamic gradient include beaver (*Castor fiber*), otter (*Lutra lutra*) and pond bat (*Myotis dasycneme*).

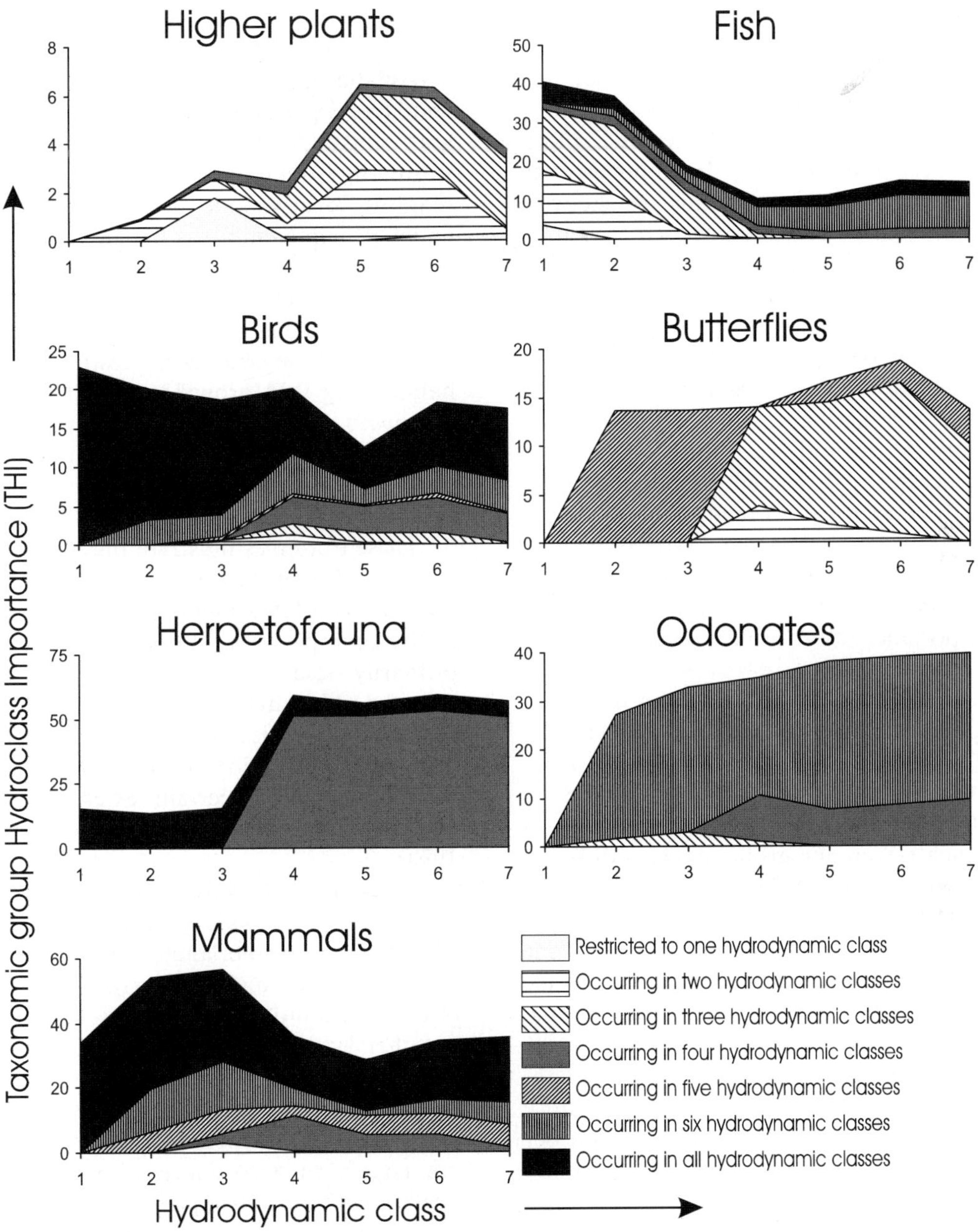

Figure 3. Taxonomic group Hydroclass Importance (THI) for the different species groups along the hydrodynamic gradient. Species were grouped according to their specificity for hydrodynamic conditions, i.e., the number of different hydrodynamic classes in which they can be found. 1: highest degree of hydrodynamics, 7: lowest degree of hydrodynamics.

Discussion

Traditionally, river management has focussed mainly on the chemical and physical aspects of river-floodplain ecosystems, such as water quality, water quantity and flow power. Safety and economic development (in terms of flood defence, food and drinking water quality, infrastructure, agriculture, mineral extraction, etc.) were always the main goals. River managers today are faced with highly modified river-floodplain ecosystems, with large numbers of species that are nowadays protected and/or endangered and therefore require special attention. The challenge in the near future is to reconstruct and manage these river-floodplain ecosystems in a way that reconciles flood risk management, infrastructural works and economic development with ecological rehabilitation (Nienhuis & Leuven, 2001), within legal boundaries imposed by nature conservation legislation. River-floodplain ecosystems are structured by hydrodynamics, morphodynamics and vegetation succession. River managers should be able to influence these processes in such a way as to achieve the various goals of river management. This requires knowledge about the causal relationships between physical processes and desired endpoints. In the present study, target species (endpoints of nature conservation) were linked to hydrodynamics. Our results therefore integrate the ecological significance of hydrodynamic conditions with the relative importance of riverine species in policy and legislation.

Although target species occur along the entire gradient, the potential for these species increases gradually with decreasing dynamics and reaches an optimum when the inundation frequency is between 0 and 20 days per year (class 6). In addition, each taxonomic group shows a different distribution along the gradient and responds to hydrodynamics in a typical and different fashion. For some groups (e.g., rheophilous fish), it is the high-dynamic parts that are more important, while for other groups (e.g. higher plants, herpetofauna, butterflies), the most important parts are those with low dynamics. Many species of birds, mammals and odonates predominantly use the entire gradient, while most plant species and – to a lesser extent – butterflies and fish are specifically bound to one or two hydrodynamic classes.

Species found along the entire gradient may be indifferent to hydrodynamics, but may also specifically utilize different parts of the gradient for different activities (e.g. foraging, breeding and resting) and therefore depend on the entire gradient in order to successfully complete their lifecycle (Verberk & Esselink 2003). Usually, the latter situation is the case. For example, black terns (*Chlidonias niger*) are very specific as regards their breeding site, using floating rafts of terrestrializing vegetation, such as those formed by water soldier (*Stratiotes aloides*), but forage in a broad range of ecotopes (from the main channel and lakes to marshland vegetation), feeding mainly on small fish and large aquatic insects, such as dragonflies. The grass snake (*Natrix natrix*) is also found along the entire hydrodynamic gradient, but selectively requires low-dynamic ecotopes with conditions making them suitable as hatcheries for their eggs, while many other ecotopes are suitable foraging or resting sites. The water bat (*Myotis daubentonii*) selectively catches its prey above open water, and is specific with regard to its resting and breeding habitat (e.g. old growth trees). The river darter (*Gomphus flavipes*) specifically uses submerged sandy river banks during its larval stages, whereas upon emerging, the adult uses a variety of terrestrial ecotopes for maturation, roosting and foraging.

These examples illustrate the limitations of the concept of ecotopes in describing habitats when applied to mobile animals. This was to be expected, as the ecotope classification concept we used was primarily based on factors structuring plant communities. Although many important causal factors for animal species are not incorporated in the ecotope classification applied in this study, this concept can to a certain extent be applied to quantify biodiversity potential (Lenders et al., 1998).

Our study dealt only with riverine target species of various taxonomic groups of the Rhine and Meuse in the Netherlands. These river-floodplain ecosystems are highly managed lowland rivers, potentially limiting the extrapolation of the results to other systems. However, species and ecotopes incorporated in the model were derived from both natural and modified river-floodplain ecosystems (see Materials and methods). Moreover, the results are largely in accordance with those of studies dealing with various aquatic taxonomic groups of river-floodplain ecosystems along the lower Rhine and Meuse (Van den Brink, 1994; Van den Brink et al., 1996), with total species richness in the Danube (Ward & Tockner, 2001), with a weighted biodiversity score in the Danube (Chovanec et al., 2005) and with fish biodiversity in general (Aarts et al., 2004). This is a strong indication that, for river-floodplain ecosystems, (i) our results may have generic meaning and (ii) the riverine species assemblage as a whole can benefit from measures focussing on creating suitable conditions for target species only. These results are important in the light of the implementation of many measures as required by the EU's Water Framework Directive (WFD).

Our results show that for target species the entire hydrodynamic gradient found in a natural riverine landscape is important. Our results also show that both aquatic and terrestrial ecotopes in the low-dynamic parts of the hydrodynamic gradient are particularly valuable. The importance of low-dynamic aquatic parts was also highlighted by a study by Van den Brink et al. (1996).

Measures aimed at flood defence, which include lowering of floodplains and river dike diversion (winter bed enlargement), may provide opportunities as well as threats for (protected) biodiversity. The habitat demands of riverine species in relation to hydrodynamics should set the boundary conditions for physical reconstruction and management aimed at combining safety goals with ecological rehabilitation and nature protection (Nienhuis & Leuven, 2001; Van Stokkom et al., 2005). Increasing the opportunities for target species requires enlargement of the winter bed (i.e., more space for the gradient to develop), a prerequisite running counter to current reconstruction plans. Because space is scarce, river managers are looking for room for water discharge in the vertical dimension, by riverbed deepening and floodplain lowering (Nienhuis & Leuven, 1998). This causes higher levels of hydrodynamics between the dikes, and thus does not result in restoration of the hydrodynamic gradient, but in a loss of low-dynamic parts (and the creation of a 'bathtub' situation). Our results indicate that these parts are of vital importance, so that situations with limited space require tailor-made designs. These designs need to combine our results on the importance of different hydrodynamic conditions for riverine species with more specific knowledge about their demands in terms of size and configuration of habitat elements (e.g., ecotopes; Wiens, 2002). For example, extra space for low-dynamic ecotopes can be created by over-dimensioning of flood defence measures such as lowering of floodplain and widening or digging of secondary channels. Tailor-made designs also require a sound inventory and assessment of the actual situation and potentials specific for that location.

In conclusion, even when focussing only on target species, no part of the natural hydrodynamic gradient can be neglected in reconstruction and restoration designs. River reconstruction and management should aim at enlarging the winter bed in order to re-establish the entire hydrodynamic gradient. When this is not possible, spatial and temporal heterogeneity of hydrodynamic conditions should be maximised within the spatial limits.

Acknowledgements

The authors would like to thank Mr. P. Voskamp for his explorative research on this subject and Prof. Dr. G. van der Velde, Dr. F.W.B. van den Brink, J.H.H.M. Klerkx MSc and an anonymous referee for useful comments on an earlier draft of this paper. This study was financed by the Faculty of Science of the Radboud University Nijmegen within the framework of a programme for stimulation of innovative research. This is NCN publication Number 2006-01.

References

Aarts, B. G. W., F. W. B. van den Brink & P. H. Nienhuis, 2004. Habitat loss as the main cause of the slow recovery of fish faunas of regulated large rivers in Europe: the transversal floodplain gradient. River Research and Applications 20: 3–23.

Bayley, P. B., 1995. Understanding large river-floodplain ecosystems. BioScience 145: 153–158.

Chovanec, A., J. Waringer, M. Straif, W. Graf, W. Reckendorfer, A. Waringer-Löschenkohl, H. Waidbacher & H. Schultz, 2005. The Floodplain Index – a new approach for assessing the ecological status of river/floodplainsystems according to the EU Water Framework Directive. Archiv für Hydrobiologie Supplement 155, Large Rivers 15: 169–185.

De Nooij, R. J. W., D. Alard, G. de Blust, N. Geilen, B. Goldschmidt, V. Huesing, H. J. R. Lenders, R. S. E. W. Leuven, K. M. Lotterman, S. Muller, P. H. Nienhuis & I. Poudevigne, 2001. Development and application of BIO-SAFE, a policy and legislation based model for assessment of impacts of flood prevention measures on biodiversity in river basins. Final report IRMA-SPONGE project 11. NCR publication 11-2001. Netherlands Centre for River Studies, Delft (http://www.ncr-web.org/downloads/pub11.pdf).

De Nooij, R. J. W., R. S. E. W. Leuven, H. J. R. Lenders, G. de Blust, N. Geilen, B. Goldschmidt, S. Muller, P. H. Nienhuis & I. Poudevigne, 2004. BIO-SAFE: assessing the impact of physical reconstruction on protected and endangered species. River Research and Applications 20: 299–313.

De Nooij, R. J. W., H. W. H. Hendriks, R. S. E. W. Leuven, H. J. R. Lenders & P. H. Nienhuis, 2005. Evaluation of floodplain rehabilitation: a comparison of ecological and policy based biodiversity assessment. Archiv für Hydrobiologie Supplement 155, Large Rivers 15: 413–424.

Junk, W. J., P. B. Bayley & R. E. Sparks, 1989. The flood-pulse concept in river-floodplain systems. Canadian Special Publication of Fisheries and Aquatic Sciences 106: 110–127.

Klijn, F. & H. A. de Udo Haes, 1994. A hierarchical approach to ecosystems and its implications for ecological land classification. Landscape Ecology 9: 89–104.

Lenders, H. J. R., R. S. E. W. Leuven, P. H. Nienhuis, K. D. Oostinga & P. J. M. van den Heuvel, 1998. Ecological rehabilitation of floodplains along the middle reach of the river Waal: a prosperous future for fauna target species? In Nienhuis, P. H., R. S. E. W. Leuven, & A. M. J. Ragas (eds), New Concepts for Sustainable Management of River Basins. Backhuys Publishers, Leiden: 35–52.

Lenders, H. J. R., R. S. E. W. Leuven, P. H. Nienhuis, R. J. W. de Nooij & S. A. M. van Rooij, 2001. BIO-SAFE: A method for evaluation of biodiversity values on the basis of political and legal criteria. Landscape and Urban Planning 55: 119–135.

Leuven, R. S. E. W., S. Wijnhoven, L. Kooistra, R. J. W. de Nooij & M. A. J. Huijbregts, 2005. Toxicological constraints for rehabilitation of riverine habitats: a case study for metal contamination of floodplain soils along the Rhine. Archiv für Hydrobiologie Supplement 155, Large Rivers 15: 657–676.

Nienhuis, P. H., R. S. E. W. Leuven & A. M. J. Ragas, 1998. New concepts for sustainable management of river basins. Backhuys publishers, Leiden.

Nienhuis, P. H. & R. S. E. W. Leuven, 1998. Ecological concepts for the sustainable management of lowland river basins: a review. In Nienhuis, P. H. R. S. E. W. Leuven & A. M. J. Ragas (eds), New Concepts for Sustainable Management of River Basins. Backhuys Publishers, Leiden: 7–33.

Nienhuis, P. H. & R. S. E. W. Leuven, 2001. River restoration and flood protection: controversy or synergism? Hydrobiologia 444: 85–99.

Nienhuis, P. H., A. D. Buijse, R. S. E. W. Leuven, A. J. M. Smits, R. J. W. de Nooij & E. M. Samborska, 2002. Ecological rehabilitation of the lowland basin of the river Rhine (NW Europe). Hydrobiologia 478: 53–72.

Rademakers, J. G. M. & H. P. Wolfert, 1994. Het Rivier-Ecotopen-Stelsel; een indeling van ecologisch relevante ruimtelijke eenheden ten behoeve van ontwerp- en beleidsstudies in het buitendijkse rivierengebied. Publicaties en rapporten van het project 'Ecologisch Herstel Rijn en Maas'. RIZA, Lelystad (in Dutch, with English summary).

Van den Brink, F. W. B., 1994. Impact of hydrology on floodplain lake ecosystems along the lower Rhine and Meuse. Ph.D. Thesis. University of Nijmegen, Nijmegen.

Van den Brink, F. W. B., G. van der Velde, A. D. Buijse & A. G. Klink, 1996. Biodiversity in the lower Rhine and Meuse river-floodplains: its significance for ecological river management. Netherlands Journal of Aquatic Ecology 30: 129–149.

Van der Molen, D. T., N. Geilen, J. J. G. M. Backx, B. J. M. Jansen & H. P. Wolfert, 2003. Water Ecotope Classification for integrated water management in the Netherlands. European Water Management Online (http://www.ewaonline.de/journal/2003_03.pdf).

Van Stokkom, H. T. C., A. J. M. Smits & R. S. E. W. Leuven, 2005. Flood defense in the Netherlands: a new era, a new approach. Water International 30: 76–87.

Verberk, W. C. E. P. & H. Esselink, 2003. Restoring fauna diversity requires an integration of animal ecology and landscape ecology. Landschap 20: 3–7 (in Dutch, with English summary).

Ward, J.V., K. Tockner & F. Schiemer, 1999. Biodiversity of floodplain river ecosystems: ecotones and connectivity. Regulated Rivers: Research and Management 15: 125 139.

Ward, J. V. & K. Tockner, 2001. Biodiversity: towards a unifying theme for river ecology. Freshwater Biology 46: 807–819.

Ward, J. V., K. Tockner, D. B. Arscott & C. Claret, 2002. Riverine landscape diversity. Freshwater Biology 47: 517–539.

Wiens, J. A., 2002. Riverine landscapes: taking landscape ecology into the water. Freshwater Biology 47: 501 515.

Part Two
Rehabilitation of Riverine Ecosystems

Hydrobiologia (2006) 565:165–186
R.S.E.W. Leuven, A.M.J. Ragas, A.J.M. Smits & G. van der Velde (eds), Living Rivers: Trends and Challenges in Science and Management
DOI 10.1007/s10750-005-1912-8

Biogeochemical constraints on the ecological rehabilitation of wetland vegetation in river floodplains

L.P.M. Lamers[1,2,*], R. Loeb[1], A.M. Antheunisse[3], M. Miletto[4], E.C.H.E.T. Lucassen[1,2], A.W. Boxman[1], A.J.P. Smolders[1,2] & J.G.M. Roelofs[1,2]
[1]*Department of Aquatic Ecology and Environmental Biology, Institute for Wetland and Water Research, Radboud University Nijmegen, Toernooiveld 1, 6525 ED Nijmegen, The Netherlands*
[2]*B-Ware Research Centre, Toernooiveld 1, 6525 ED Nijmegen, The Netherlands*
[3]*Landscape Ecology, Institute of Environmental Biology, Utrecht University, P.O. Box 80084, 3508 TB Utrecht, The Netherlands*
[4]*Department of Microbial Wetland Ecology, Netherlands Institute of Ecology (NIOO-KNAW), Centre for Limnology, Rijksstraatweg 6, 3631 AC Nieuwersluis, The Netherlands*
*(*Author for correspondence: E-mail: L.Lamers@science.ru.nl)*

Key words: biodiversity, biogeochemistry, eutrophication, flooding, nature conservation, review, water management

Abstract

The European policy for river management during peak discharge periods is currently changing from exclusion strategies (reinforcement of dykes) to allowing a more natural situation by creating more floodplain space to reduce water levels during peak discharges. In addition, water retention and water storage areas have been created. The new measures are generally being combined with nature development strategies. Up till now, however, ecological targets of broadscale floodplain wetland restoration including sedge marshes, species-rich floodplain forests and carrs, riparian mesotrophic grasslands and other biodiverse riverine ecosystems, have hardly developed in these areas. Most studies on the conditions needed for sustainable ecological development of floodplains have focused on hydrological and geomorphological rather than biogeochemical issues (including nutrient availability and limitation). There are, however, large differences in the composition of river water and groundwater and in sediment quality between rivers in densely populated areas and those in more pristine areas, which serve as a reference. It is very likely that these factors, in combination with heavily altered hydrological regimes and the narrow areas confined between the dykes on both sides of the rivers, impose major constraints on sustainable ecological development of riverine areas. Another issue is that existing wetlands are generally considered to be very appropriate for water retention and conservation, although recent research has shown that this may pose a serious threat to their biodiversity. The present paper reviews the biogeochemical constraints on the combination of floodplain rehabilitation, water conservation and the conservation and development of wetlands. It is concluded that biogeochemical problems (mainly related to eutrophication) predominantly arise in less dynamic parts of the river system, to which the flood-pulse concept applies less. Sound knowledge of the biogeochemical processes involved will contribute to greater efficiency and a better prediction of the opportunities for restoration and development of riverine wetlands. This information can be directly applied in nature management, water management, policy-making and consultancy.

Changing concepts, changing rivers

During peak discharge periods, regulated river systems in densely populated areas such as the Netherlands have to cope with excessive volumes of water in a narrow area confined by dykes. In the past, the only strategy was to avoid flooding by raising and reinforcing dykes (levees). Since this approach has proved to be insufficient and concepts of river management have changed, new strategies have been designed which allow the development of a more natural riverine system (Admiraal et al., 1993; De Waal et al., 1995; Harper et al., 1995; Sparks, 1995; Tockner et al., 1998). Additional space is created in floodplains to avoid the heading up of water between the embankments, thereby reducing flood risks ('Space for the river' or 'Room for the river' policy; Van Stokkom et al., 2005). In this approach, floodplains are mechanically lowered (by excavation) and widened (by dyke displacement), secondary channels are excavated and large obstacles removed, so that rivers are able to expand into side channels and wetland areas (Mostert, 1998; Nienhuis et al., 1998; Leuven et al., 2000; Smits et al., 2000). In strongly regulated rivers lacking space for natural hydrogeomorphological processes, the concept of 'cyclic rejuvenation', mimicking the effects of natural channel migration, is being applied as a compromise between flood protection and ecological rehabilitation (Simons et al., 2001; Baptist et al., 2004; Geerling et al., 2006). In addition, inland areas are adapted to retain and store excess water temporarily, both in natural wetlands and in agricultural areas.

These new safety strategies are generally being combined with habitat creation strategies (Cals et al., 1998; Buijse et al., 2002; Nienhuis et al., 2002). The ecological objective is to create floodplain forests, riparian grasslands, sedge fens and other biodiverse ecosystem types (Postma et al., 1996; Bal et al., 2001; Buijse et al., 2005; Nienhuis, 2006). Up till now, however, high-biodiversity vegetation targets have hardly been achieved, especially for the permanently flooded or waterlogged locations (Pedroli et al., 2001; Buijse et al., 2002, 2005). Increased connectivity to the river has even been shown to decrease species diversity for macrophytes, macrozoobenthos, phytoplankton and zooplankton in floodplain lakes along the lower Rhine and Meuse rivers (Van den Brink et al., 1993; Van den Brink & Van der Velde, 1994).

The present paper will focus on biogeochemical problems that may impede ecological rehabilitation of floodplain vegetation, and on their relation with the flood-pulse concept. The rehabilitation discussed includes the development of ecosystem types that have become rare along rivers in densely-populated areas, such as sedge fens and alder carrs. After the discussion of constraints related to surface water, groundwater and soil quality, their possible effects on plant biodiversity will be treated. Next, implications for river management and nature conservation will be outlined.

River water: quantity and quality

The hydrological changes described above will allow rivers greater access to their floodplains (increased connectivity), and increase water level fluctuations in floodplains. In other words, the flood pulse, regulating floodplain functioning, is restored (Junk et al., 1989; Bayley, 1991; Heiler et al., 1995; Middleton, 1999; Van Geest et al., 2005). In the river deltas, more estuarine processes (tidal action and brackish water intrusion) will be allowed (Smits et al., 2000, 2006). As a result of this, large areas will be temporarily or permanently flooded, or waterlogged. Such water table fluctuations lead to alterations between oxidized and reduced soil conditions. In addition, flooding will become more frequent in many parts of the floodplains. It is evident that these changes in the water table regimes will lead to changes in vegetation composition, as a result of the susceptibility of plants to waterlogging and flooding. The duration of the flooding periods, as well as the seasonality and frequency of flooding, have major effects on species distribution and succession. This may be the result of low oxygen concentrations affecting root growth (Van der Valk, 1981; Jackson & Drew, 1984; Brock et al., 1987; Crawford, 1987, 2003; Smits et al., 1990b; Van den Brink et al., 1991; 1995; Armstrong et al., 1994; Blom et al., 1994; Lenssen et al., 1999; Visser et al., 2003; Van Eck et al., 2006) or germination (Smits et al., 1990a, 1995), decreased photosynthetic rates (Vervuren et al., 1999; 2003; Mommer et al., 2004), erosion and sedimentation, or mechanical stress by current and wave exposure (Coops & Van der Velde, 1996; Coops et al., 1996a). In addition the water depth and the fluctuation of the water table strongly control vegetation growth and zonation (Coops & Van der Velde, 1995; Coops et al., 1996a; Van Geest et al., 2005). Lastly, flooding may well increase dispersal of propagules via the surface water (Brock et al., 1987; Coops & Van der Velde, 1995; Amoros, 2001; Boedeltje et al., 2004).

Increased connectivity between rivers and their floodplains will inevitably lead to a stronger influence of the river water quality on floodplain communities, such as demonstrated for the relation between the flooding frequency and duration, and their water quality (Van den Brink & Van der Velde, 1994). The magnitude and location of sedimentation and erosion directly influence biodiversity while, conversely, the vegetation type has a significant effect on hydrogeomorphological characteristics such as hydraulic conductivity and sedimentation (Hupp, 1992; Bornette & Amoros, 1996; Coops et al., 1996b, 1999; Henry et al., 1996; Steiger & Gurnell, 2003). High connectivity seems to be a prerequisite for some aquatic species such as as *Nymphoides peltata* through the prevention of calcium deficiency (Smits et al., 1992). Surface water pollution by phosphate and nitrate increases the influx of these nutrients, both in the dissolved and the particulate fraction, in floodplain soils (Van den Brink et al., 1993; Van den Brink & Van der Velde, 1994; Knowlton & Jones, 1997; Kronvang et al., 1998; Spink et al., 1998; Darke & Walbridge, 2000). This direct pollution is known to be able to decrease biodiversity in floodplain streams and ponds through eutrophication (Van den Brink et al., 1993; 1994; Trèmolieres et al., 1994; Bornette et al., 2001; Bij de Vaate et al., 2006). Floodplain waters become dominated by algae and cyanobacteria, causing the deterioration of underwater communities.

On the other hand, nutrient retention and removal by denitrification, uptake by plants, periphyton and microorganisms, sedimentation and binding to the soil are regarded as important regulators of nutrient concentrations in the river water (Kok et al., 1990; Sjodin et al., 1997; Reddy et al., 1999; Olde Venterink et al., 2003b; Hogan et al., 2004). Van der Lee et al. (2004) studied the eutrophic river Rhine and found that nitrogen (N) retention was insignificant (<3% of the annual load), while phosphorus (P) retention amounted to 18%. For both nutrients, sedimentation was the most important retention mechanism. As a result of differences in connectivity, flow velocity and vegetation composition, nutrient accretion rates show great variability within floodplains. During floods, however, large amounts of phosphate may be released from heavily fertilized floodplain soils as they dissolve and diffuse to the water layer, as explained below. This is expected to reduce the retention capacity for this nutrient, or may even lead to a net efflux of P from the floodplains to the river. In addition, flooding is known to stimulate the loss of C, N and P as a result of increased decomposition and mineralization in soil litter, for instance, in floodplain forests (Lockaby et al., 1996).

Even if river water is eutrophic, mesotrophic vegetation types such as sedge fens may develop on floodplains. The occurrence of these often biodiverse vegetation types depends on a low total influx of nutrients. This can be determined by a lower level of connectivity (by distance), but may also result from permanent influx of nutrient-poor groundwater like in floodplains surrounded by calcareous massifs (Kohler et al., 1973; Carbiener et al., 1990; Bornette et al., 1996), or from seepage of iron-rich groundwater (see below). It can, however, also be caused by the presence of highly productive, eutrophic or hypertrophic vegetation types between the biodiverse vegetation types and the river. Nutrient-rich, species-poor stands of helophytes such as *Phragmites australis*, *Typha latifolia*, *Phalaris arundinacea* and *Glyceria maxima*, and of species-poor brushwood and forests adjacent to rivers, act as natural helophyte filters, removing nutrients and particles from the surface water and enabling biodiverse wetland types to develop farther from the river (Wassen et al., 2002; 2003). However, vast floodplains, ranging in width from several hundreds of metres to several kilometres (e.g., up to 10 km for the river Danube), are absent in densely populated lowland areas such as the Netherlands (Buijse et al., 2002). The actual space for the river will, even after the implementation of the new plans for riverine areas, in most cases be too narrow to allow for these lengthy gradients from nutrient-rich to nutrient-poor conditions, including both mineral and more organic soils. All desired wetland types of the target situation including marshes, natural grasslands and floodplain forests (Postma et al., 1996; Buijse et al., 2005), have to develop within a 'straitjacket', confined by the embankments on both sides. The better the water quality, the smaller the influence of this spatial problem, which is very difficult, if not impossible, to tackle in densely populated areas. This fact underlines the need for cleaner water and cleaner floodplain soils.

In addition to the direct eutrophication effects caused by the influx of nutrients, high sulphate and chloride concentrations in the inflowing surface water may cause adverse biogeochemical changes in many Dutch wetlands with organic soils, including former river meanders (Roelofs, 1991; Koerselman et al., 1993; Smolders & Roelofs, 1993; Lamers et al., 1998b; Beltman et al., 2000; Lamers et al., 2001; Verhoeven et al., 2001; Lucassen et al., 2004a). Concentrations of both anions in wetlands have greatly increased as a result of anthropogenic input into rivers (industrial, agricultural and mining activities) and the increased use of river water to compensate for water shortage in agricultural areas and nature reserves (Roelofs, 1991; Smolders & Roelofs, 1993; Lamers et al., 1998b). Increased chloride loading from chloride-polluted river Rhine water may negatively influence the vegetation composition of floodplain lakes (Van den Brink & Van der Velde, 1993). Large amounts of sulphate are being mobilized by the oxidation of geological iron sulphide (FeS_x) deposits by oxidation as a result of drainage (Lamers et al., 1998a; Schuurkes et al., 1988) and, indirectly, by nitrate pollution of aquifers from the intensive fertilization of arable land (Nordstrom, 1982; Kölle et al., 1985; Paul & Clark, 1989; Appelo & Postma, 1993; Hoffmann et al., 1998; Lamers et al., 2002b). In the latter case, FeS_x is oxidized anaerobically through chemolithotrophic denitrification, resulting in even stronger increases in sulphate concentrations in groundwater and, after discharge, surface water. Increased sulphate concentrations, between 1 and 4 mmol l^{-1} for Dutch surface waters, may well lead to considerable phosphate mobilization under anaerobic conditions (waterlogging and flooding; Fig. 1). The increased production of sulphide in sulphur (S)-polluted wetlands by increased microbial sulphate reduction rates interacts with iron phosphate binding, leading to the formation of iron sulphides (including pyrite), the release of phosphate and potential iron deficiency (Boström et al., 1982; Caraco et al., 1989, 1993; Roelofs, 1991; Smolders & Roelofs, 1993; Smolders et al., 1994; Lamers et al., 2002a). In addition, mineralization rates may be stimulated by increased alkalinity (Kok et al., 1990; Roelofs, 1991; Smolders & Roelofs, 1995) as a result of the reduction of electron acceptors such as sulphate or nitrate. The effect of the increased sulphate concentrations is, however, strongly dependent on the soil type, and on the concentrations (Lamers et al., 2001). One of the significant factors is the concentration of free iron that is available for phosphate binding. For wetland soils, the ratio between the pore water concentrations of iron and phosphate has a high indicator value. Phosphate mobilization to the water layer is low at values above 10 (mol mol^{-1}), and moderate between 1 and 10 (Smolders et al., 2001; Geurts et al., unpublished data). At ratios below 1, however, iron levels proved to be too low to provide a functional 'iron trap'. Although it is known that the microbial community responsible for sulphate reduction strongly changes over a salinity gradient (Laanbroek & Pfennig, 1981), information on the composition of assemblages of sulphate-reducing prokaryotes in S-polluted freshwater systems is still limited (Lomans et al., 1997; Castro et al., 2002). Potential sulphate reduction rates appear to be high within the first week after the water table is raised in drained soils, showing the presence and fast response of a viable community of sulphate-reducing prokaryotes (Miletto, unpublished data).

Groundwater changes

Floodplains show a hydrological mosaic, caused by differences in lateral fluxes between uplands, floodplains and the river, often including discharge locations. In more pristine areas, discharged groundwater will generally be nutrient-poor (De Mars et al., 1997). This discharge of hypoxic, iron- or calcium-rich groundwater may lead to increased phosphate retention, decreasing phosphate availability under natural wetland conditions. The extent to which this occurs depends on the hydrological characteristics and on soil and groundwater quality (Stumm & Morgan, 1981). As a result of large-scale water table draw-down in the catchment, groundwater fluxes have greatly changed in floodplains. This has led not only to quantitative changes, but also to qualitative changes. Discharge of iron has decreased, and groundwater may contain high concentrations of

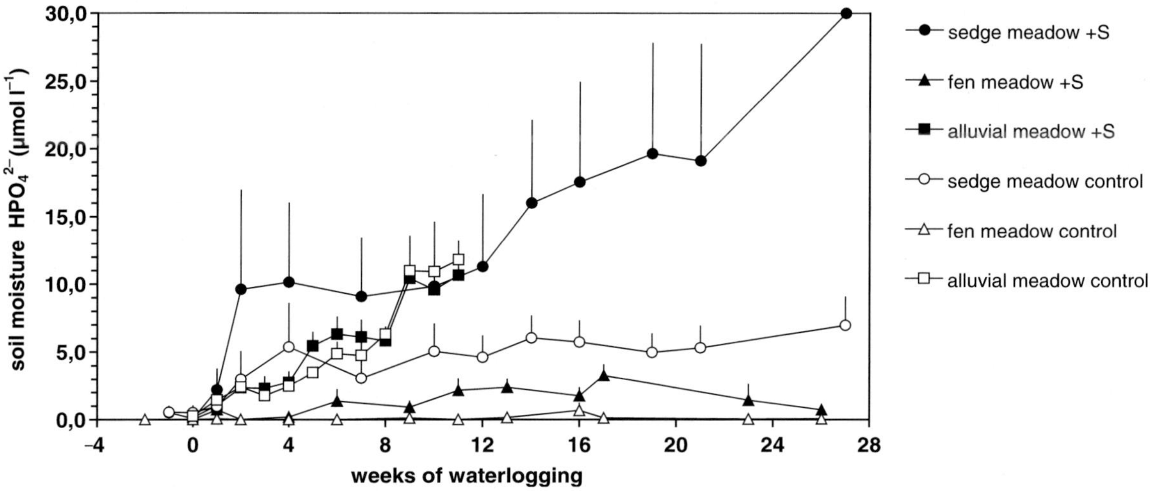

Figure 1. Phosphate mobilization (pore water) in mesocosm experiments by waterlogging with (S) or without (control) sulphate pollution (2 mmol l^{-1}) for three different sediment types. Sedge meadow: grassland abandoned by agriculture approx. 10 years before the onset of the experiment; organic sediment; Caricion nigrae vegetation. Fen meadow: organic sediment; Junco-Molinion vegetation. Alluvial meadow: clay sediment; Alopecurion pratensis vegetation. After 11 weeks (end of the experiment), phosphate mobilization in the alluvial meadow was just as high as for sulphate-treated sedge-meadows, although its initial rate was slower. In contrast to the other two types, however, sulphate did not fortify the eutrophication process. Means are given, with their standard errors ($n=6$, 5, 5, respectively). (From: Lamers et al., 2001; Loeb & Lamers, 2003).

nitrate and sulphate (up to the millimolar range; Lucassen et al., 2004a; Baker & Vervier, 2004). As a result of the high P sorption capacity of soils, phosphate concentrations in groundwater are considerably lower. Depending on the constitution of the subsoil (including water table and organic matter concentration), a large part of the nitrate may be reduced and lost from the soil by denitrification (Kester et al., 1997; Baker & Vervier, 2004; Hefting et al., 2004), although this strongly depends on the actual hydrology of the floodplain (Burt et al., 1999) and soil texture (Pinay et al., 2000). Higher water tables and concomitant absence of oxygen may, however, promote nitrate reduction to ammonium, leading to ammonium accumulation in the topsoil. On drained sites, nitrification rates are high and nitrate may accumulate (Hefting et al., 2004).

In addition to organotrophic denitrification, nitrate may also be reduced by lithotrophic denitrifiers, as explained above. Although the passage of nitrate-rich groundwater through aquifers rich in iron seems beneficial because it means that nitrate is stripped from the water, a significant new problem is created because sulphide is oxidized to sulphate. In many agricultural areas, nitrate intrusion has led to sulphate 'charging' of the groundwater in this way (Pauwels et al., 1998; Lamers et al., 2002b). In addition to this source, increased sulphate concentrations in groundwater originate from sulphide oxidation by oxygen in drained wetland soils and sulphate efflux from farmlands (Heathwaite, 1990; Freeman et al., 1996; Lamers et al., 1998a,b). Values of up to 8 mmol l^{-1} have been reported for groundwater-feeding carr woods in oxbow lakes (Boxman et al., 2003). In many Dutch riverine wetlands that still show seepage, groundwater has become the primary source of sulphate pollution (Lucassen et al., 2004a).

Sulphate pollution is especially detrimental when the efflux of water and dissolved nutrients is blocked. The creation of high water tables throughout the year as a measure against desiccation was found to be disastrous for alder carrs in former river meanders (oxbow lakes) of the river Meuse in the Netherlands. This (for this wetland type) artificial situation differs strongly from the natural conditions, which involve water table fluctuations during the year with lower values in the summer. Even within the first growing season after the introduction of this measure, the surface water had become completely covered by *Lemna* spp., while characteristic species such as *Caltha palustris*, *Calla palustris* and several *Carex* spp. had been literally drowned. Phosphate concentrations had increased severalfold as a result of prolonged anoxic conditions and accumulation (Boxman et al., 2003; Lucassen et al., 2004a, 2005). *Alnus glutinosa* even started to die at these locations. It may be clear that this artificial hydrological regime ('over-rewetting') is very detrimental and undesirable. The same holds for other types of wetlands, such as fen meadows. Aerobic conditions are needed in summer, presumably also to stimulate nitrification. Plants characteristic of this type of habitat show a strong preference for nitrate as an N source, while the accumulation of ammonium leads to toxic conditions (Smolders et al., 1996; De Graaf et al., 1998; Lucassen et al., 2003; Van den Berg et al., 2005). The re-establishment of a more natural water regime, with higher water tables during winter and tables below the hydraulic head of the seepage groundwater (allowing adequate nutrient efflux), have been shown to offer better prospects (Lucassen et al., 2004a, 2005). A positive effect of water table fluctuation on biodiversity was also found for floodplain lakes along the Lower Rhine (Van Geest et al., 2005). For fen types that need a constant water level at soil surface, however, this type of water table fluctuation does not provide a solution.

Interestingly, high nitrate concentrations in the groundwater, originating from intensive fertilization of arable land in the catchment, are able to reduce or even counteract the adverse effects of stagnating, sulphate-rich conditions. Since nitrate is a more favourable electron acceptor than sulphate, it appears to be able to prevent sulphate-induced phosphate mobilization and sulphide accumulation (Lucassen et al., 2004b). On the other hand, accumulation of ammonium (Hefting et al., 2004) may well lead to undesirable dominance of highly competitive graminoid species (Lucassen et al., unpublished data).

In addition to the plans for development of freshwater wetlands, plans have also been announced to restore more saline conditions to regions that had formerly been brackish. This is expected to lead to restored tidal water level fluctuations, affecting soil biogeochemistry (including redox reactions). Additionally, chloride and sulphate concentrations will increase. Increased sulphate and chloride concentrations in freshwater wetlands are expected to reduce biodiversity by eutrophication (nutrient-controlled systems), as described above. Higher values of both anions under brackish conditions, however, will probably lead to the rehabilitation of characteristic (eutrophic) brackish wetlands with a high biodiversity. Unlike freshwater wetlands, which are generally nutrient-controlled, these are expected to be controlled by high, fluctuating chloride concentrations (salinity controlled), or by the interaction between salinity and nutrient availability (Den Hartog, 1974; Bloemendaal & Roelofs, 1988; Moss, 1994). Higher salinity will be directly toxic to most freshwater species, but it will also influence them indirectly, through the modification of biogeochemical processes like decomposition, mineralization, redox reactions and nutrient release (Christian et al., 1990; Van den Brink & Van der Velde, 1993; Hall & Anderson, 1995; Mendelssohn et al., 1999; Hyacinthe & Van Cappellen, 2004).

Soil quality: nutrients

Research in floodplains has generally focused on hydrology and river water quality. Decomposition rates are, for instance, strongly determined by soil moisture content. In peatlands, water tables are a major factor controlling decomposition rates, with higher rates at drained sites (Bridgham & Richardson, 2003). However, in addition to hydrological changes and surface water quality, several other factors have been drastically changed in riverine areas. Although river water quality in the Netherlands has improved in recent decades, and nutrient concentrations have decreased substantially (except for N in the Meuse; Admiraal et al., 1993; Bij de Vaate et al., 2006; Fig. 2), most floodplains are still highly eutrophic, because they have received massive amounts of nutrients by fertilization (manure and fertilizer) for agricultural purposes. In the Netherlands, influxes may amount to 500 kg N and 75 kg P per hectare per year (CBS, 2004). Additionally, approximately 20–30 kg of airborne N per hectare per year is deposited in non-forest vegetation (Boxman, 2005). Due to high concentrations of amorphous

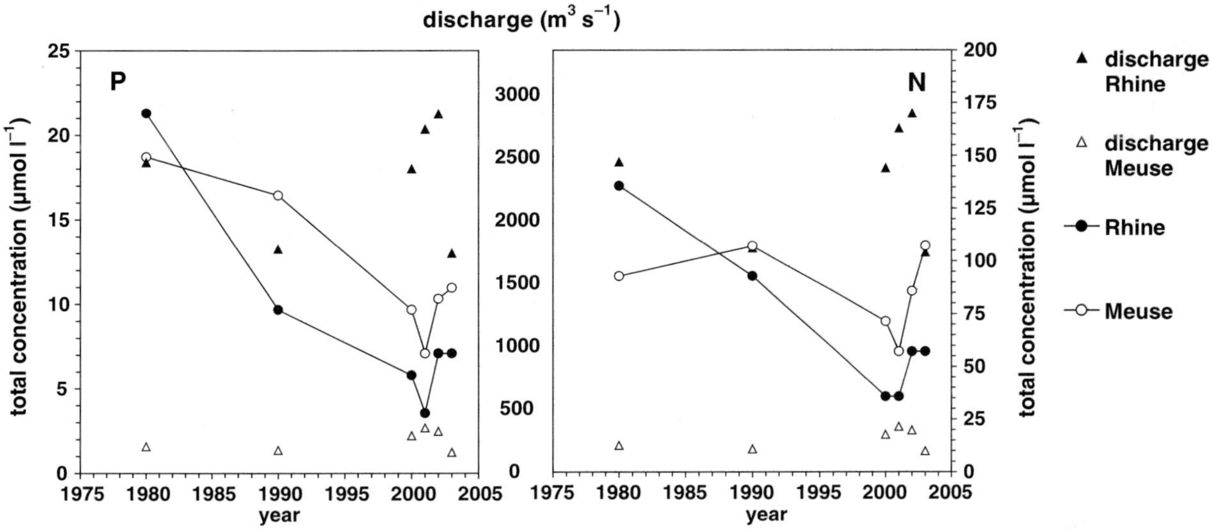

Figure 2. Total concentrations of phosphorus (P) and nitrogen (N) in the rivers Rhine and Meuse, and their discharges, between 1980 and 2003. (Data from: RIVM, 2004).

(i.e. non-crystalline) iron and aluminium (the latter especially in clay), the phosphate-binding capacity of wetlands is generally very high. Clay soils in particular are known to have an extremely high potential for P retention. Phosphate retention appears to be strongly correlated to the concentration of both compounds, especially aluminium (Richardson, 1985; Darke & Walbridge, 2000). Furthermore, pollution of the groundwater with nitrate may also lead to the accumulation of nitrate and ammonium in floodplain soils (as explained above). This soil eutrophication (P and N) appears to impose a more imminent constraint on the ecological rehabilitation of floodplains than the threat from the river water quality, especially now that this quality is improving.

It is known that flooding of former agricultural land leads to excessive P eutrophication by oxygen depletion and concomitant phosphate mobilization (Lamers et al., 1998b; Lamers et al., 2001; Young & Ross, 2001; Van Dijk et al., 2004). This is a result of limited access of oxygen to soils and the concomitant shift to other electron acceptors available for soil microorganisms, decreasing the overall redox potential and generating alkalinity (Ponnamperuma, 1984; Laanbroek, 1990). Simultaneously, the concentration of dissolved free iron increases (due to iron reduction) and phosphate is mobilized (Patrick et al., 1973; Patrick & Khalid, 1974; Khalid et al., 1977; Lamers et al., 1998a, b; Baldwin & Mitchell, 2000; Loeb et al., unpublished data). Phosphate bound to aluminium and aluminium-organic compounds will, in contrast, not be released, because the metal is not redox-sensitive (Darke & Walbridge, 2000). This means that after retention, the greater part of this fraction will stay in the soil. The same holds for phosphate bound to calcium carbonate (Golterman, 1998; Reddy et al., 1999). For calcium-rich soils, the relatively high proportion of P may be sequestered by calcium, preventing phosphate from dissolving during flooding.

Recent research has revealed that many floodplains along Dutch rivers show high concentrations of free iron in the soil moisture (up to 6000 μmol l^{-1}), preventing the accumulation of free sulphide (Loeb et al., unpublished data). Total soil iron concentrations range from 0.6 to 6% (100–1000 μmol g^{-1} DW; Loeb, unpublished data), as a result of sedimentation from river surface water (the iron concentration in suspended matter being 3% for the Rhine; ICPR, 2005). Higher concentrations of iron measured in floodplain soils are mainly the result of the discharge of anaerobic, iron-rich groundwater in the present or past. Upon reaching the (temporarily or permanently) aerobic top layer, iron is oxidized to iron hydroxides, and accumulates. This may even lead to the deposition of iron stone (marsh ore, with Fe concentrations up to 50%), for instance in river valleys whose soils comprise a mixture of sand and peat (called *dalgronden* in Dutch).

At first glance, these high iron concentrations seem to be beneficial because of their high phosphate-binding capacity. Unfortunately, this high binding capacity has, in combination with the heavy fertilization of floodplains, led to excessive loading of the soil's phosphate binding sites. As a result, the phosphate concentration in pore water may increase to 20–200 times the original concentration during flooding, leading to concentrations up to 225 μmol l^{-1} (Young & Ross, 2001; Loeb et al., unpublished data). The phosphate saturation of iron-based binding sites has proved to be a powerful diagnostic tool to predict phosphate mobilization during waterlogging or flooding (Young & Ross, 2001; Hogan et al., 2004; Loeb et al., unpublished data). In combination with ammonium loading, this phosphate 'time bomb' provides a serious biogeochemical pitfall for the creation of biodiverse wetlands, whose production is generally limited by the availability of N, phosphate, potassium or a combination of these nutrients (Aerts & Chapin, 2000; Olde Venterink et al., 2003a; Güsewell, 2004). In iron-rich, mineral floodplain soils, sulphate pollution does not lead to extra phosphate mobilization. The high availability of iron appears to be sufficient to sequester all sulphide produced without phosphate being mobilized from iron-phosphate complexes (Fig. 1; Loeb & Lamers, 2003). We expect that floodplain soils higher in organic matter and lower in iron concentration will react like fen soils when polluted with sulphate, and show both extra phosphate mobilization and sulphide accumulation (Lamers et al., 1998b; 2001). This is currently being tested. Peat-forming systems, similar to fens outside river floodplains, have become very rare in regulated riverine landscapes. This may be attributed to the lack of space that would allow for less dynamic aquatic wetland types farther from the river, showing lower sedimentation and erosion rates, and to the young age of most 'rejuvenated' floodplains, in which the accumulation of organic matter has only just started.

The extent to which phosphate mobilization occurs strongly depends on the seasonality of flooding. This has been clearly demonstrated by a mesocosm study using mineral monoliths with their vegetation. While summer floods resulted in phosphate mobilization, especially with sulphate-rich water, winter floods did not produce significant changes (Fig. 3; Bruijnzeels & Lamers, unpublished data). It was striking that this soil showed a clear response to sulphate pollution, in contrast to other mineral soils tested, as explained above. After the water table had been lowered, phosphate concentrations dropped again, as expected. In addition, sulphide hardly accumulated during the winter, while sulphide production peaked in spring, both in the sulphate and control conditions of the experiment. During water table draw-down, sulphate reduction ceased and sulphide became oxidized. Although the seasonality of flooding is clearly one of the most important variables controlling both floodplain biogeochemistry and vegetation responses, it is remarkable that river management options involving combined water storage and ecological rehabilitation often ignore this fact. Most plants will survive winter flooding, and nutrient cycling rates are modest. While water storage as a flood prevention measure during winter is relatively harmless to most vegetation types of mesotrophic wetlands (see also Van Eck et al., 2006), storage during the growing season will be disastrous.

Soil quality: toxicants

Next to nutrients and chloride, floodplains in densely populated areas have also been loaded with a large range of contaminants, including potentially toxic metals and persistent organic micropollutants (Leuven et al., 2005). There is considerable variation in concentrations between different floodplain areas along the same river, as well as along horizontal and vertical gradients within the same floodplain (Middelkoop, 2000). Concentrations of many compounds exceed quality standards at many locations (Japenga et al., 1990; Hendriks et al., 1995; Gocht et al., 2001). Even if pollutant concentrations in river water decrease, there is still accumulation in the soil (Japenga & Salomons, 1993; Leuven et al., 2005; Hollander et al., 2006; Wijnhoven et al., 2006). Although the concentrations of most contaminants in the river water have substantially

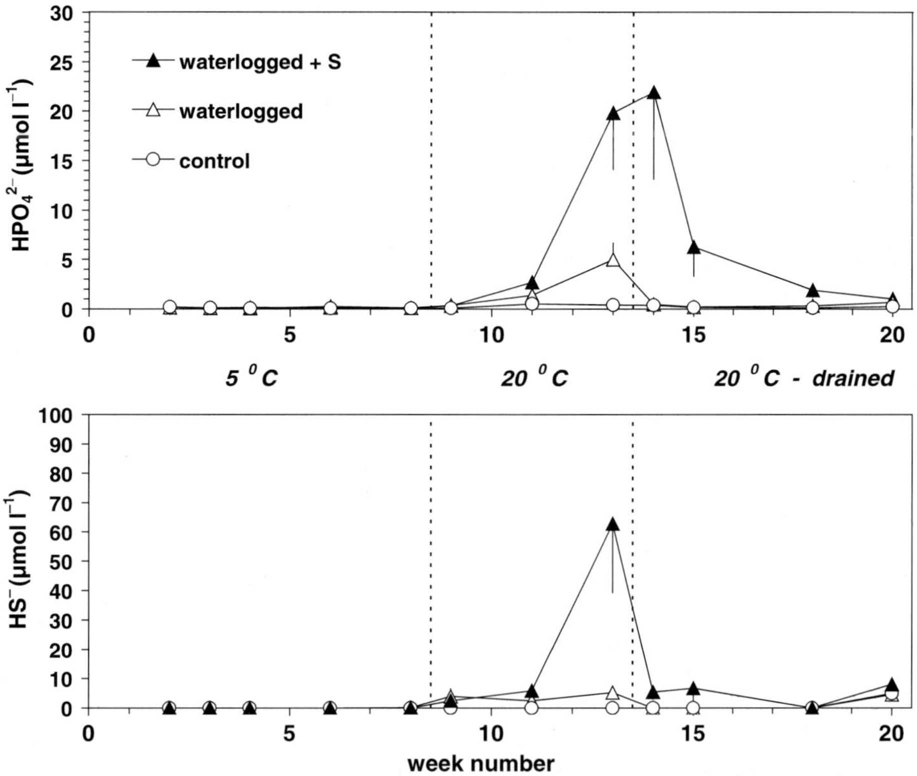

Figure 3. Effect of temperature on the mobilization of phosphate and accumulation of sulphide in pore water as a result of waterlogging without or with (S) 2 mmol l^{-1} sulphate. Mesocosm experiment with calcareous, sandy monoliths (4% organic matter) including their Junco-Molinion vegetation (Bruijnzeels & Lamers, unpublished data).

decreased during the last decades, floodplain soils may still contain high concentrations of persistent contaminants that can accumulate at different trophic levels (e.g., Hendriks & Pieters, 1993; Goodyear & McNeill, 1999). The soil profile may even provide a proxy for the estimation of historic water qualities in terms of both P and contaminants, since the age of the different layers can be determined by Cs-137 dating (Middelkoop, 2000; Winter et al., 2001). For metals, however, the actual toxicity at a particular temperature is closely related to metal speciation (Florence et al., 1992; Forstner, 1993; Tessier & Turner, 1995; Vink, 2002; Schröder et al., 2005). While total concentrations in soils may be high (suggesting great toxicity risks), bioaccumulation may be adventitiously low (Hobbelen et al., 2004). Metal speciation is determined by redox conditions, pH and the concentration and type of dissolved organic matter. Low redox potentials (waterlogging or flooding) or acidic conditions favour the dissolving of a number of metals. High total concentrations will have a much smaller effect in calcareous soils than in more acidic soils, because the metals are hardly mobile. Since desiccation of soils generally leads to acidification through oxidation processes, including the oxidation of Fe and S and nitrification (Lamers et al., 1998a; Lucassen et al, 2002), the acid neutralizing capacity (ANC) is a significant regulator of metal toxiciy. If the ANC is sufficiently high, acidification (i.e. loss of ANC; Van Breemen et al., 1983) does not lead to a drop in pH because all protons are neutralized. For S-rich wetlands, therefore, the ratio between the total concentration of S (the acidifying component) and that of calcium + magnesium (representing ANC) has proved to be a useful diagnostic tool to assess toxicity problems during drainage (Lucassen et al., 2002). Values exceeding 0.7 (mol mol^{-1}) indicate that desiccation would lead to severe acidification and metal mobilization. In conclusion, this means that data on the total metal concentrations (which are generally used in policy-making programmes) have to be combined with biogeochemical knowledge of the regulatory processes to allow toxicity to be properly predicted (Van der Welle, in press). For a specific location, current toxicity, bioaccumulation and biomagnification can be determined by measuring contaminant concentrations and physiological variables in selected organisms (Kooistra et al., 2001, 2004), or by estimating the fraction that is readily taken up by organisms (Vink, 2002).

In addition to eutrophication and pollution by metals and other toxicants, the accumulation of natural, potentially toxic substances such as sulphide and ammonium may provide an additional problem for wetland development in polluted river systems. For both substances, even low concentrations are known to be phytotoxic to helophyte and aquatic macrophyte communities (Roelofs, 1991; Van Wijck et al., 1992; Armstrong et al., 1996; Smolders & Roelofs, 1996; Smolders et al., 1996; Lamers et al., 1998b). High concentrations of iron are also known to be toxic to certain plants (Smolders et al., 1994; Lucassen et al., 2000). For eutrophic floodplain vegetation, the accumulation of organic acids such as acetic acid might be an additional problem (Laanbroek et al., 1984; Armstrong et al., 1996). The higher sensitivity to flooding of plants on organic, highly reductive substrates as compared to mineral substrates (Van den Brink et al., 1995) can be explained by the higher oxygen consumption rate in the soil and the accumulation of toxic substances mentioned above.

Biogeochemistry and biodiversity

At the highly dynamic locations near the river, increased connectivity is expected to lead to greater biodiversity as a result of increased disturbance by flooding, by sedimentation and erosion, and by increased dispersal potential (Bornette et al., 1998; Galat et al., 1998; Demars & Harper, 2005). In this situation, highly competitive species are not able to become dominant, and niches for regeneration of ruderal species (Grime, 1974) are constantly created, thereby 'resetting' succession. In addition, the dispersal of plant propagules is promoted, which is relevant in relation to habitat fragmentation. In this way, biodiversity is warranted by disturbance caused by hydrological fluctuations. For the upper Rhone river, it was indeed demonstrated that the most frequently flooded channel had the greatest macrophyte biodiversity (Bornette et al., 1998). However, from extensive field studies it is known that plant biodiversity does not necessarily increase linearly with flooding intensity. Some studies even found a negative effect, because the regional species pool of flooding-tolerant plants appeared to be relatively small (Grace & Pugesek, 1997; Grace & Jutila, 1999; Olde Venterink et al., 2001). Other studies showed that species richness was highest at intermediate levels of flooding (Day et al., 1988; Pollock et al., 1998). In a study along the lower Rhine covering 215 floodplain lakes the effect of inundation appeared to be modest, probably because flooding during the growing season is rare in this region (Van Geest et al., 2003). A stable water table in these lakes, however, appeared to reduce biodiversity and to promote the undesired dominance of *Elodea nuttallii* (Van Geest et al., 2005). Incidental temporal lowering of the water level is expected to enhance the ecological status of these lakes, like in peat lakes (Lamers et al., 2002b). In an earlier study on floodplain lakes along the lower Rhine, it was also concluded that in-lake processes (including soil resuspension by benthivorous fish) appeared to be more important than river dynamics (Roozen et al., 2003). This seems to hold particularly for less dynamic river systems, in contrast to highly dynamic systems. A complicating factor, not related to connectivity, is the fact that biodiversity can be very low in less dynamic backwaters (connected through backflows) as a result of high local nutrient loading (Bornette et al., 1998; Boedeltje et al., 2005).

In order to predict the above effects of changed nutrient biogeochemistry, it is vital to know whether the target vegetation types are limited by nutrient availability, and which of the elements is actually limiting biomass production for the total vegetation, or for potentially invading fast growing species.

In aquatic habitats, phosphate availability initially controls biomass production, except in those characterized by extremely low N availability (like softwater lakes). Increasing concentrations of phosphate strongly promote the growth of algae, decreasing light availability. Submerged macrophytes and their associated fauna start to disappear at phosphate concentrations around 5 μmol l^{-1}, and plants possessing floating leaves, such as *Lemna*, take over. At even higher phosphate concentrations (>10 μmol l^{-1}) high phytoplankton (green algae and cyanobacteria) concentrations develop, or the water becomes covered by *Azolla filiculoides*, a fern capable of N fixation thanks to its symbiosis with cyanobacteria. This means that these waters have changed from P-limited to N-limited, providing a competitive advantage for N-fixing organisms. Eutrophic floodplain lakes show clear phytoplankton dominance, suppressing submerged vegetation (Van den Brink et al., 1993; Van den Brink & Van der Velde, 1994). The growth of the vegetation in biodiverse wet meadows, poor fens and rich fens, now almost completely absent from Dutch riverine areas, has been shown to be limited by either P or N, or both (Koerselman & Verhoeven, 1995; Wassen et al., 1995; Van Oorschot et al., 1998; Bedford et al., 1999; Wheeler & Proctor, 2000). In the floodplains of the Allier and Loire (France) vegetation was found to be N-limited, as indicated by the positive correlation with N mineralization (Van Oorschot et al., 1998). Species-rich fen grasslands in the floodplain of the less polluted river Shannon were found to be co-limited by N and P (Spink et al., 1998).

The nature of nutrient limitation is generally tested by conducting fertilization experiments, by correlation analysis, or by analyzing the N:P ratio (reflecting the relative availability for a particular species). However, the results of fertilization experiments mostly refer to the biomass response of the actual total vegetation to nutrient addition, and not to the competitive advantage of fast-growing species in the longer term. The differential effects of nutrient addition can partly be explained by differences in interspecific variation in the uptake of N and P, or in the need for these two nutrients. In conditions of low P availability, small *Carex* spp. have a competitive advantage because they show higher nutrient-use efficiency than species adapted to eutrophic conditions (Chapin, 1980; Aerts & Chapin, 2000). The fact that individual species can be limited by different resources, as well as the small-scale heterogeneity in resource availability, enable species coexistence and biodiversity (Tilman, 1994; Güsewell, 2004). In the Biebrza Valley (Poland), P fertilization of floodplain fens resulted in increased growth of large sedges, grasses and non-graminoid herbs. Small sedges were shown to be outcompeted by large sedges because of the low light availability (Kotowski, 2002). A review of the literature on European fens showed that light availability was closely related to fertility, rather than to the hydrological regime (Kotowski & Van Diggelen, 2004). Contradictory results may also be explained by changes in the relative need for nutrients over time. This was demonstrated by Van der Hoek et al. (2004). After N addition, the total biomass of a fen meadow vegetation increased only during the first year of fertilization, without any effect of P fertilization. During the second year, total biomass was unexpectedly boosted by P addition, while the vegetation became dominated by the grass species *Holcus lanatus*. At higher N availability, P uptake by the vegetation as a whole had increased, leading to P shortage.

The type of nutrient limitation is strongly regulated by the extent of drainage and the mowing regime. Annual mowing of floodplains along the Polish river Jegrznia has resulted in a shift from no limitation to (periodical) N and P limitation at slightly drained riparian locations, and to P (and probably also K) limitation at the heavily drained locations (De Mars et al., 1996). Mowing generally leads to a net export of P and K (Koerselman & Verhoeven, 1995).

The extremely high nutrient concentrations (both N and P) in floodplain soils along the rivers Rhine and Meuse, due to heavy fertilization, are expected to lead to increased aboveground biomass production, and thereby to loss of species. Highly competitive, tall and fast growing species may outcompete characteristic, more slowly growing species (Day et al., 1988; Taylor et al., 1990). Vegetation types with aboveground biomass values exceeding 500 g m^{-2} always show low plant species richness. The conservation of many threatened wetland plant species requires low productive P (co)-limited conditions (Olde Venterink et al., 2001, 2003a; Wassen et al., 2005). In Dutch floodplains, the average biomass was found to be 1000 g m^{-2}, with values up to

8000 g m^{-2} for *Phragmites australis* stands. Less than one quarter of the locations monitored showed a standing stock of 500 g m^{-2} or less (Antheunisse et al., unpublished data).

Highly productive species-poor vegetation types such as those dominated by *Glyceria maxima* (Europe) or *Scirpus fluviatilis* (USA) appear not to be limited by nutrients (Wassen et al., 1995, 1998, 2003; Spink et al., 1998). In eutrophic Dutch and American floodplains, nutrient availability was found to be a less important regulator of plant production than other factors such as flooding and drought (Spink et al., 1998). In this case, nutrient availability as the controlling factor of plant growth is superseded by hydrological variables (e.g. flooding, current, sedimentation and drought) and genetic constraints (including flood tolerance and maximum growth rate). High biomass production rates reinforce nutrient accumulation, due to increased accumulation of nutrients extracted from the soil and enhanced sedimentation. This stabilizes the existing state and impedes the transition to a state controlled by nutrient availability. For eutrophic floodplains, we expect that the rehabilitation of vegetation types dominated by slow-growing species, which show high nutrient-use efficiency and a low physiological ability to respond to eutrophication, will only be possible after additional measures, as explained in the next section. In slightly acidic grassland abandoned by agriculture, the fast growing common rush, *Juncus effusus*, represents a major frustration to conservationists who are trying to convert arable land to biodiverse wetland by raising water tables. High phosphate availability in the topsoil impedes competition by more slowly growing species (Smolders et al., unpublished data).

In addition to nutrients, high concentrations of toxic organic and inorganic pollutants are also expected to impact on floodplain biodiversity. Although most studies refer to effects on fauna communities, vegetation development may also be regulated by toxicity. Potentially phytotoxic compounds include iron (Snowden & Wheeler, 1995; Lucassen et al., 2000; Batty et al., 2002), sulphide (Roelofs, 1991; Armstrong et al., 1996; Smolders et al., 1996; Lamers et al., 1998b), ammonium (e.g. Roelofs et al., 1985; Boxman & Roelofs, 1988; Smolders et al., 1996; De Graaf et al., 1998; Lucassen et al., 2003), chloride (Van den Brink & Van der Velde, 1993) and aluminium (Rengel, 1996; De Graaf et al., 1997; Rout et al., 2001; Batty et al., 2002). As ammonium and aluminium appear to be far more toxic at low pH and low Ca concentration, the toxicity of these compounds will be more pronounced at locations with a low ANC. In fen meadows, ammonium and aluminium toxicity have been shown to be quite detrimental to a number of characteristic, vulnerable species (De Graaf et al., 1998; Lucassen et al., 2003; Van den Berg et al., 2005). However, high ammonium concentrations in anaerobic soils (in which nitrification rates are low) may also be toxic to aquatic macrophytes in calcareous waters (Onaindia et al., 1996; Smolders et al., 1996; Pezeshki, 2001). Therefore, species biodiversity is expected to be regulated by concentrations of both nutrients and toxicants. Depending on speciation (regulated by pH, dissolved organic matter and redox state), high metal concentrations may accumulate in both riparian and aquatic macrophytes.

From research to policy and measures

As discussed above, we expect several biogeochemical problems to arise when Dutch rivers are allocated more space, and new hydrological regimes are established. We think that it will not be possible to simply extrapolate the ecological developments in the more pristine river systems chosen as a reference to greatly disturbed river systems such as those in the Netherlands.

Phosphate-poor conditions for the development of biodiverse wetland vegetation can only be created by either sod cutting or soil removal. To estimate the biological availability of P, extraction procedures (e.g. Olsen et al., 1954) provide a better estimate than the total concentrations. Olsen-P concentrations are very high in Dutch floodplains along the rivers Rhine, Meuse and IJssel, comparable to nature development regions outside river areas (Fig. 4a). Even for Polish floodplains along the Bug and Narew, P concentrations have been found to be quite high, while those for the Odra floodplains were extremely high, due to fertilization or pollution from the river. Efforts to deplete the strongly eutrophicated soils by mowing and biomass removal are inadequate. Total P concentrations in the soils are, on average, 10–20 times the Olsen-P values. This means that if one wanted to achieve the phosphate concentrations that are characteristic of biodiverse reference areas (amounting to 250 μmol Olsen-P kg^{-1} DW; 2500 μmol total P kg^{-1} DW), it might take 100–500 years to create the desired conditions. This is generally well beyond the scope of policy-makers, nature managers and water managers. A depth profile for P will indicate the required removal depth, which is often not equivalent to the depth of ploughing. For many habitat creation areas, however, phosphate saturation is too deep and soil removal therefore too costly (Fig. 4b). This means that the above-mentioned development of P-limited, biodiverse vegetation types is impossible. In this case, regular sod cutting, removing highly competitive species, may favour early successional stages. As deeper soil layers in floodplains may still contain high concentrations of phosphate due to historic loading (Winter et al., 2001), it is vital to analyze soil profiles before starting rehabilitation measures like excavation. In combination with this analysis, rejuvenation measures may meet extra targets by eliminating the eutrophication constraint imposed by P enrichment of floodplain soils.

Nitrate is generally a less important problem than phosphate, due to the relatively high mobility of this nutrient (Fig. 5). N concentrations in soil moisture of Dutch floodplains are generally between 0 and 100 μmol l^{-1} for ammonium, and 0 and 50 μmol l^{-1} for nitrate, which is moderate. Both in Poland and in the Netherlands, however, extremely high values, up to the millimolar range, for nitrate (drier soils) and ammonium (wetter soils) have been measured (Loeb et al., unpublished results). At locations with seepage of nitrate-rich groundwater, soil removal will not exhaust the available nitrate. At these high N concentrations, N-based eutrophication is very likely. As discussed, the relative availability of one nutrient will influence the response to another.

The ecological rehabilitation of floodplains may also be frustrated by soil contamination. As

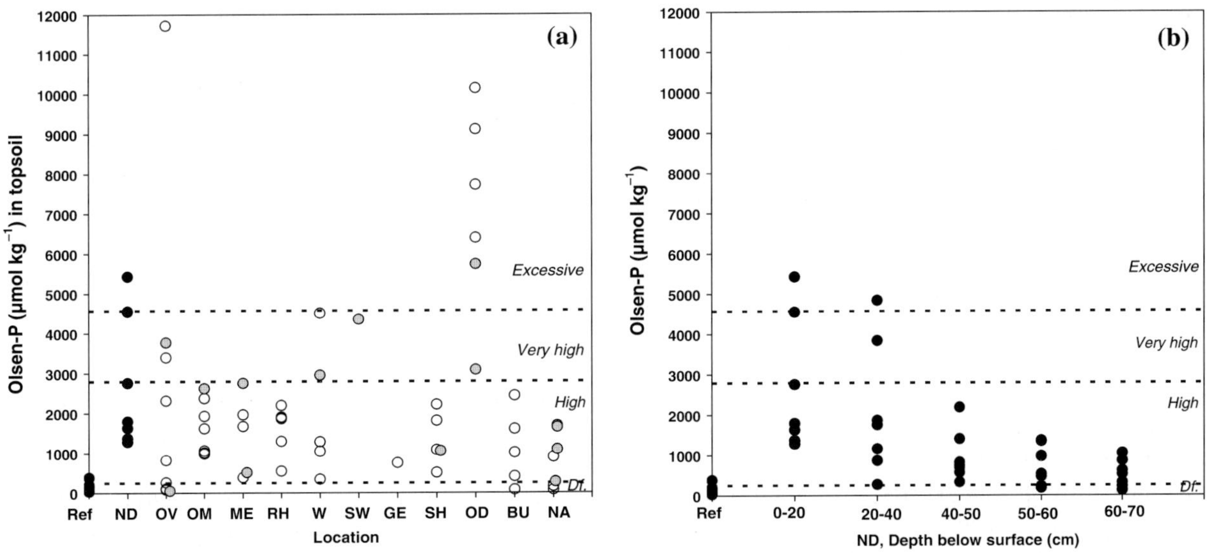

Figure 4. (a) Estimation of phosphorus availability in the upper 10–20 cm of sediments and sediments by bicarbonate extraction (Olsen et al., 1954), for various locations in the Netherlands, Poland and Ireland for riverine soils, grey dots represent organic soils, white dots mineral soils. Many locations show high P availability, some even excessive. Df.: P deficiency plausible. Ref: reference biodiverse wetlands; ND: non-alluvial, former agricultural areas destined for nature development (Smolders, unpubl. data). All other locations are alluvial wetlands, most of which were formerly or are currently in agricultural use (Antheunisse, unpublished data; Loeb, unpublished data). River codes: OV: Overijsselse Vecht, the Netherlands (NL); OM: Oude Maas (NL); ME: Meuse (NL); RH: Rhine (NL); W: Waal (NL); SW: Swalm (NL); GE: Geulle (NL); SH: Shannon, Ireland; OD: Odra, Poland (PL); BU: Bug (PL); NA: Narew (PL). (b) Olsen P values for the different depth classes at the ND locations of panel A (Smolders, unpublished data). Defic.: P deficiency plausible. Ref: reference biodiverse wetlands. Based on these estimates, decisions can be made on the extent of sediment removal needed to restore nutrient-poor conditions in ecological rehabilitation programmes.

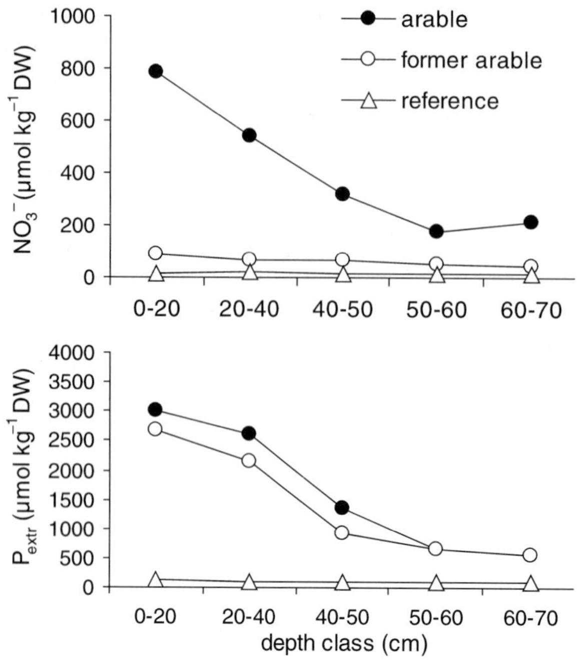

Figure 5. Concentrations of plant-available nitrate (NO_3^-; H_2O extraction) and phosphate (HPO_4^{2-}; lactate/acetate extraction) in sediments, for different depth classes; comparison between an arable land location, former arable land and reference (biodiverse nature reserve). All three are recharge areas in terms of hydrology. NO_3^- concentrations have decreased to reference values after cessation of agricultural activites 10 years ago while HPO_4^{2-} concentrations remain high. (From: Smolders et al., unpublished data).

concentrations of a large number of contaminants were extremely high during the 1970s and have since decreased in the surface water, excavation may well uncover contaminated layers (Winter et al., 2001).

The targets of ecological rehabilitation are not only related to plant species diversity, but also to ecosystem diversity and riverine landscape diversity (Amoros, 2001; Ward et al., 2002). This means that measures should be taken at these different levels. The biogeochemical constraints we have discussed operate not only at the species and ecosystem level, but also at the landscape level. For instance, vegetation development interferes with geohydrology at the landscape level, urging the need for preservation and rehabilitation of biodiversity at higher scales (Sparks, 1995; Tockner et al., 1998; Godreau et al., 1999).

The identification of possible constraints and the prediction of the outcome of restoration measures require an understanding of biogeochemical and biological key processes in riverine areas (Buijse et al., 2002) and other wetland types (Bedford, 1999). Unlike a trial and error approach, this will enable managers and policymakers to understand the factors controlling degradation or successful restoration (Bedford, 1999; Lamers et al., 2002b; Buijse et al., 2005). Field observations providing correlative information need to be coupled to experimental data, in order to find causal relationships. Once the key variables and processes have been assessed, diagnoses can be established and results for one location can be extrapolated to other areas. In addition to biogeochemical assessments, vegetation can also be used for an ecological diagnosis of riverine wetlands, because the occurrence and abundance of wetland plant species are highly indicative of their habitat characteristics (Bloemendaal & Roelofs, 1988; Demars & Harper, 1998; Dawson & Szoszkiewicz, 1999; Amoros et al., 2000; Buijse et al., 2005). Moreover, it should be possible to predict the degree of success for a given location, and to make sound choices between potential new areas combining water safety and ecological targets (risk analysis). In order to be able to include the spatial variability within riverine systems, geographic information systems applications are indispensable (Poiani & Bedford, 1995; Kooistra et al., 2001). Rehabilitation projects based on trial and error show a very unfavourable cost-benefit ratio, which is undesirable given the high costs of ecological rehabilitation.

Conclusions: how much ecological space does the 'straitjacket' provide?

Highly regulated rivers only provide a narrow area to accommodate an elaborate mosaic of wetland types, including sedge fens and carr woods. In the past, high inputs of nutrients and contaminants caused a gradually declining nutrient gradient, depending on connectivity. The heavy fertilization of floodplain soils and of groundwater feeding floodplain wetlands with N and P impose a serious constraint on the development of wetland types controlled by nutrient availability. For these areas, managers have to be satisfied with lower ecological targets, or take additional measures such as top layer removal. The need for this may be partly obviated by removing soil as a part of rejuvenation measures, provided that newly exposed soil layers do not contain high concentrations of phosphate or contaminants as a legacy from past eutrophication and pollution. For floodplain lakes, water level fluctuation seems to be vital to reduce eutrophication and increase biodiversity. In order to tackle nitrate and sulphate pollution via groundwater seepage, measures have to be taken at the landscape scale. For those communities primarily controlled by flooding regime (alternating flooded and desiccated conditions), nutrient biogeochemistry will be of minor importance, because flooding regimes overrule all other factors. Their biodiversity is mainly controlled by disturbance, either positively or negatively. Next to eutrophication, intoxication by (micro)pollutants imposes an additional constraint. After removal of the top layer, soils may even show higher concentrations as a heritage from a more polluted past.

We have argued that a trial and error approach will not be able to predict the ecological development of particular floodplain locations, nor will it be possible to select those locations that are most likely to be successful in terms of ecological rehabilitation. Biogeochemical knowledge is a prerequisite for successful ecological rehabilitation, ensuring that space for the river also implies space for ecologically sound floodplains.

Acknowledgements

The Netherlands Organisation for Scientific Research (NWO-TRIAS) and the Limburgs Landschap Foundation funded part of the research presented in this paper. Jan Klerkx provided linguistic advice. This is CWE publication 422.

References

Admiraal, W., G. van der Velde, H. Smit & W. G. Cazemier, 1993. The rivers Rhine and Meuse in the Netherlands – Present state and signs of ecological recovery. Hydrobiologia 265: 97–128.

Aerts, R. & F. S. Chapin, 2000. The mineral nutrition of wild plants revisited: A re-evaluation of processes and patterns. Advances in Ecological Research 30: 1–67.

Amoros, C., 2001. The concept of habitat diversity between and within ecosystems applied to river side-arm restoration. Environmental Management 28: 805–817.

Amoros, C., G. Bornette & C. P. Henry, 2000. A vegetation-based method for ecological diagnosis of riverine wetlands. Environmental Management 25: 211–227.

Appelo, C. A. J. & D. Postma, 1993. Geochemistry, Groundwater and Pollution. A.A. Balkema Publishers, Rotterdam.

Armstrong, J., F. Afreen-Zobayed & W. Armstrong, 1996. *Phragmites* die-back: sulphide- and acetic acid induced bud and root death, lignifications, and blockages within aeration and vascular systems. New Phytologist 134: 601–614.

Armstrong, W., R. Brändle & M. B. Jackson, 1994. Mechanisms of flood tolerance in plants. Acta Botanica Neerlandica 43: 307–358.

Baker, M. A. & P. Vervier, 2004. Hydrological variability, matter supply and denitrification in the Garonne River ecosystem. Freshwater Biology 49: 181–190.

Bal, D., H. M. Beije, M. Fellinger, R. Haveman, A. J. F. M. van Opstal & F. J. van Zadelhoff, 2001. Handboek Natuurdoeltypen. Tweede, Geheel Herziene Druk. EC-LNV, Wageningen.

Baldwin, D. S. & A. M. Mitchell, 2000. The effects of drying and re-flooding on the sediment and soil nutrient dynamics of lowland river-floodplain systems: a synthesis. Regulated Rivers: Research and Management 16: 457–467.

Baptist, M. J., W. E. Penning, H. Duel, A. J. M. Smits, G. W. Geerling, G. E. M. van der Lee & J. S. L. van Alphen, 2004. Assessment of the effects of cyclic floodplain rejuvenation on flood levels and biodiversity along the Rhine river. River Research and Applications 20: 285–297.

Batty, L. C., A. J. M. Baker & B. D. Wheeler, 2002. Aluminium and phosphate uptake by *Phragmites australis*: the role of Fe, Mn and Al root plaques. Annals of Botany 89: 443–449.

Bayley, P. B., 1991. The flood pulse advantage and the restoration of river-floodplain systems. Regulated Rivers: Research and Management 47: 75–86.

Bedford, B. L., 1999. Cumulative effects on wetland landscapes: links to wetland restoration in the United States and southern Canada. Wetlands 19: 775–788.

Bedford, B. L., M. R. Walbridge & A. Aldous, 1999. Patterns in nutrient availability and plant diversity of temperate North American wetlands. Ecology 80: 2151–2169.

Beltman, B., T. G. Rouwenhorst, M. B. van Kerkhoven, T. van der Krift & J. T. A. Verhoeven, 2000. Internal eutrophication in peat soils through competition between chloride and sulphate with phosphate for binding sites. Biogeochemistry 50: 183–194.

Bij de Vaate, A., R. Breukel & G. van der Velde, 2006. Long-term developments in the ecological rehabilitation of the main distributaries in the Rhine Delta: fish and macroinvertebrates. Hydrobiologia 565: 229–242.

Bloemendaal, F. H. J. L. & J. G. M. Roelofs, (eds), 1988. Waterplanten en Waterkwaliteit. Stichting Uitgeverij

Koninklijke Nederlandse Natuurhistorische Vereniging, Utrecht (in Dutch).

Blom, C. W. P. M., L. A. C. J. Voesenek, M. Banga, W. M. H. G. Engelaar, J. H. G. M. Rijnders, H. M. van de Steeg & E. J. W. Visser, 1994. Physiological ecology of riverside species – adaptive responses of plants to submergence. Annals of Botany 74: 253–263.

Boedeltje, G., J. P. Bakker, A. ten Brinke, J. M. van Groenendael & M. Soesbergen, 2004. Dispersal phenology of hydrochorous plants in relation to discharge, seed release time and buoyancy of seeds: the flood pulse concept supported. Journal of Ecology 92: 786–796.

Boedeltje, G., A. J. P. Smolders, L. P. M. Lamers & J. G. M. Roelofs, 2005. Interactions between sediment propagule banks and sediment nutrient fluxes explain floating plant dominance in stagnant shallow waters. Archiv für Hydrobiologie 162: 349–362.

Bornette, G. & C. Amoros, 1996. Disturbance regimes and vegetation dynamics: Role of floods in riverine wetlands. Journal of Vegetation Science 7: 615–622.

Bornette, G., C. Amoros & N. Lamouroux, 1998. Aquatic plant diversity in riverine wetlands: the role of connectivity. Freshwater Biology 39: 267–283.

Bornette, G., M. Guerlesquin & C. P. Henry, 1996. Are the Characeae able to indicate the origin of groundwater in former river channels? Vegetatio 125: 207–222.

Bornette, G., H. Piegay, A. Citterio, C. Amoros & V. Godreau, 2001. Aquatic plant diversity in four river floodplains: a comparison at two hierarchical levels. Biodiversity and Conservation 10: 1683–1701.

Boström, B., M. Jansson & C. Forsberg, 1982. Phosphorus release from lake sediments. Archiv für Hydrobiologie 18: 5–59.

Boxman, A. W., 2005. Effecten van Verminderde Stikstof-Depositie op een Grove Dennen Opstand in Natuurgebied 'De Rouwkuilen', Limburg. Report by order of the Ministry of Agriculture, Nature & Food Quality. Radboud University Nijmegen, Nijmegen (in Dutch).

Boxman, A. W., H. H. Bartelink, Ph. Bossenbroek, R. H. Kemmers & A. H. F. Stortelder, 2003. Uitvoering van Vernattingsmaatregelen op Praktijkschaal 1997–2003; Referentieproject Koelbroek. EC-LNV, Ede (in Dutch).

Boxman, A. W. & J. G. M. Roelofs, 1988. Some effects of nitrate versus ammonium nutrition on the nutrient fluxes in *Pinus sylvestris* seedlings. Effects of mycorrhizal infection. Canadian Journal of Botany 66: 1091–1097.

Bridgham, S. D. & C. J. Richardson, 2003. Endogenous versus exogenous nutrient control over decomposition and mineralization in North Carolina peatlands. Biogeochemistry 65: 151–178.

Brock, T. C. M., G. van der Velde & H. M. van de Steeg, 1987. The effects of extreme water level fluctuations on the wetland vegetation of a nymphaeid-dominated oxbow lake in the Netherlands. Archiv für Hydrobiologie Supplement 27: 57–73.

Buijse, A. D., H. Coops, M. Staras, L. H. Jans, G. J. van Geest, R. E. Grift, B. W. Ibelings, W. Oosterberg & F. C. J. M. Roozen, 2002. Restoration strategies for river floodplains along large lowland rivers in Europe. Freshwater Biology 47: 889–907.

Buijse, A. D., F. Klijn, R. S. E. W. Leuven, H. Middelkoop, F. Schiemer, J. H. Thorp & H. P. Wolfert, 2005. Rehabilitation of large rivers: references, achievements and integration into river management. Archiv für Hydrobiologie Supplement 155 – Large Rivers 15: 715–738.

Burt, T. P., L. S. Matchett, K. W. T. Goulding, C. P. Webster & N. E. Haycock, 1999. Denitrification in riparian buffer zones: the role of floodplain hydrology. Hydrological Processes 13: 1451–1463.

Cals, M. J. R., R. Postma, A. D. Buijse & E. C. L. Marteijn, 1998. Habitat restoration along the River Rhine in The Netherlands: Putting ideas into practice. Aquatic Conservation – Marine and Freshwater Ecosystems 8: 61–70.

Caraco, N. F., J. J. Cole & G. E. Likens, 1989. Evidence for sulphate-controlled phosphorus release from sediments of aquatic systems. Nature 341: 156–158.

Caraco, N. F., J. J. Cole & G. E. Likens, 1993. Sulfate control of phosphorus availability in lakes – a test and reevaluation of Hasler and Einsele model. Hydrobiologia 253: 275–280.

Carbiener, R., M. Trèmolieres, J. L. Mercier & A. Ortscheit, 1990. Aquatic macrophyte communities as bioindicators of eutrophication in calcareous oligosaprobe stream waters (Upper Rhine plain, Alsace). Vegetatio 86: 71–88.

Castro, H., K. R. Reddy & A. Ogram, 2002. Composition and function of sulfate-reducing prokaryotes in eutrophic and pristine areas of the Florida Everglades. Applied and Environmental Microbiology 68: 6129–6137.

CBS, 2004. Productie van mest en mineralen. http://www.statline.cbs.nl. CBS, Voorburg/Heerlen.

Chapin, F. S. I., 1980. The mineral nutrition of wild plants. Annual Review of Ecology and Systematics 11: 233–260.

Christian, R. R., W. L. Bryant & M. M. Brinson, 1990. *Juncus roemerianus* production and decomposition along gradients of salinity and hydroperiod. Marine Ecology – Progress Series 68: 137–145.

Coops, H., N. Geilen & G. van der Velde, 1999. Helophyte zonation in two regulated estuarine areas in the Netherlands: vegetation analysis and relationships with hydrological factors. Estuaries 22: 657–668.

Coops, H., N. Geilen, H. J. Verheij, R. Boeters & G. van der Velde, 1996b. Interactions between waves, bank erosion and emergent vegetation: an experimental study in a wave tank. Aquatic Botany 53: 187–198.

Coops, H., F. W. B. van den Brink & G. van der Velde, 1996a. Growth and morphological responses of four helophyte species in an experimental water-depth gradient. Aquatic Botany 54: 11–24.

Coops, H. & G. van der Velde, 1995. Seed dispersal, germination and seedling growth of six helophyte species in relation to water-level zonation. Freshwater Biology 34: 13–20.

Coops, H. & G. van der Velde, 1996. Effects of waves on helophyte stands: mechanical characteristics of stems of *Phragmites australis* and *Scirpus lacustris*. Aquatic Botany 53: 175–185.

Crawford, R. M. M., 1987. Plant Life in Aquatic and Amphibious Habitats. Blackwell Science Publication, Oxford.

Crawford, R. M. M., 2003. Seasonal differences in plant responses to flooding and anoxia. Canadian Journal of Botany 81: 1224–1246.

Darke, A. K. & M. R. Walbridge, 2000. Al and Fe biogeochemistry in a floodplain forest: implications for P retention. Biogeochemistry 51: 1–32.

Dawson, F. H. & K. Szoszkiewicz, 1999. Relationships of some ecological factors with the associations of vegetation in British rivers. Hydrobiologia 415: 117–122.

Day, R. T., P. A. Keddy, J. McNeill & T. Carleton, 1988. Fertility and disturbance gradients: a summary model for riverine marsh vegetation. Ecology 69: 1044–1054.

De Graaf, M. C. C., R. Bobbink, J. G. M. Roelofs & P. J. M. Verbeek, 1998. Differential effects of ammonium and nitrate on three heathland species. Plant Ecology 135: 185–196.

De Graaf, M. C. C., R. Bobbink, P. J. M. Verbeek & J. G. M. Roelofs, 1997. Aluminium toxicity and tolerance in three heathland species. Water, Air and Soil Pollution 98: 229–239.

Demars, B. O. L. & D. M. Harper, 1998. The aquatic macrophytes of an English lowland river system: assessing response to nutrient enrichment. Hydrobiologia 384: 75–88.

Demars, B. O. L. & D. M. Harper, 2005. Distribution of aquatic vascular plants in lowland rivers: separating the effects of local environmental conditions, longitudinal connectivity and river basin isolation. Freshwater Biology 50: 418–437.

De Mars, H., M. J. Wassen & W. H. M. Peeters, 1996. The effect of drainage and management on peat chemistry and nutrient deficiency in the former Jegrznia-floodplain (NE-Poland). Vegetatio 126: 59–72.

De Mars, H., M. J. Wassen & H. O. Venterink, 1997. Flooding and groundwater dynamics in fens in eastern Poland. Journal of Vegetation Science 8: 319–328.

Den Hartog, C., 1974. Brackish-water classification, its development and problems. Hydrobiological Bulletin 8: 15–28.

De Waal, L. C., A. R. G. Large, C. J. Grippel & P. M. Wade, 1995. River and floodplain rehabilitation in Western Europe: opportunities and constraints. Archiv für Hydrobiologie Supplement 101 – Large Rivers 9: 679–693.

Florence, T. M., G. M. Morrison & J. L. Stauber, 1992. Determination of trace-element speciation and the role of speciation in aquatic toxicity. Science of the Total Environment 125: 1–13.

Forstner, U., 1993. Metal speciation – General concepts and applications. International Journal of Environmental Analytical Chemistry 51: 5–23.

Freeman, C., G. Liska, N. J. Ostle, M. A. Lock, B. Reynolds & J. Hudson, 1996. Microbial activity and enzymic decomposition processes following peatland water table drawdown. Plant and Soil 180: 121–127.

Galat, D. L., L. H. Fredrickson, D. D. Humburg, K. J. Bataille, J. R. Bodie, J. Dohrenwend, G. T. Gelwicks, J. E. Havel, D. L. Helmers, J. B. Hooker, J. R. Jones, M. F. Knowlton, J. Kubisiak, J. Mazourek, A. C. McColpin, R. B. Renken & R. D. Semlitsch, 1998. Flooding to restore connectivity of regulated, large-river wetlands – natural and controlled flooding as complementary processes along the lower Missouri River. Bioscience 48: 721–733.

Geerling, G. W., A. M. J. Ragas, R. S. E. W. Leuven, J. H. van den Berg, M. Breedveld, D. Liefhebber & A. J. M. Smits, 2006. Succession and rejuvenation in floodplains along the River Allier (France). Hydrobiologia 565: 71–86.

Gocht, T., K. M. Moldenhauer & W. Puttmann, 2001. Historical record of polycyclic aromatic hydrocarbons (PAH) and heavy metals in floodplain sediments from the Rhine River (Hessisches Ried, Germany). Applied Geochemistry 16: 1707–1721.

Godreau, V., G. Bornette, B. Frochot, C. Amoros, E. Castella, B. Oertli, F. Chambaud, D. Oberti & E. Craney, 1999. Biodiversity in the floodplain of Saône: a global approach. Biodiversity and Conservation 8: 839–864.

Golterman, H. L., 1998. The distribution of phosphate over iron-bound and calcium-bound phosphate in stratified sediments. Hydrobiologia 364: 75–81.

Goodyear, K. L. & S. McNeill, 1999. Bioaccumulation of heavy metals by aquatic macro-invertebrates of different feeding guilds: a review. Science of the Total Environment 229: 1–19.

Grace, J. B. & H. Jutila, 1999. The relationship between species density and community biomass in grazed and ungrazed coastal meadows. Oikos 85: 398–408.

Grace, J. B. & B. H. Pugesek, 1997. A structural equation model of plant species richness and its application to a coastal wetland. The American Naturalist 149: 436–460.

Grime, J. P., 1974. Vegetation classification by reference to strategies. Nature 250: 26–31.

Güsewell, S., 2004. (Tansley review) N : P ratios in terrestrial plants: variation and functional significance. New Phytologist 164: 243–266.

Hall, L. W. & R. D. Anderson, 1995. The influence of salinity on the toxicity of various classes of chemicals to aquatic biota. Critical Reviews in Toxicology 25: 281–346.

Harper, D., C. Smith, P. Barham & R. Howell, 1995. The ecological basis for the management of the natural river environment. In Harper, D. M. & A. J. D. Ferguson (eds), The Ecological Basis for River Management. Wiley, Chichester, 219–238.

Heathwaite, A. L., 1990. The effect of drainage on nutrient release from fen peat and its implications for water-quality – a laboratory simulation. Water, Air and Soil Pollution 49: 159–173.

Hefting, M., J. C. Clement, D. Dowrick, A. C. Cosandey, S. Bernal, C. Cimpian, A. Tatur, T. P. Burt & G. Pinay, 2004. Water table elevation controls on soil nitrogen cycling in riparian wetlands along a European climatic gradient. Biogeochemistry 67: 113–134.

Heiler, G., T. Hein, F. Schiemer & G. Bornette, 1995. Hydrological connectivity and flood pulses as the central aspects for the integrity of a river-floodplain system. Regulated Rivers: Research and Management 11: 351–361.

Hendriks, A. J., W. C. Ma, J. J. Brouns, E. M. De Ruiter-Dijkman & R. Gast, 1995. Modeling and monitoring organochlorine and heavy-metal accumulation in soils, earthworms, and shrews in Rhine-delta floodplains. Archives of Environmental Contamination and Toxicology 29: 115–127.

Hendriks, A. J. & H. Pieters, 1993. Monitoring concentrations of microcontaminants in aquatic organisms in the Rhine delta – a comparison with reference values. Chemosphere 26: 817–836.

Henry, C. P., C. Amoros & G. Bornette, 1996. Species traits and recolonization processes after flood disturbances in riverine macrophytes. Vegetatio 122: 13–27.

Hobbelen, P. H. F., J. E. Koolhaas & C. A. M. van Gestel, 2004. Risk assessment of heavy metal pollution for detritivores in floodplain soils in the Biesbosch, the Netherlands, taking bioavailability into account. Environmental Pollution 129: 409–419.

Hoffmann, C. C., M. L. Pedersen, B. Kronvang & L. Ovig, 1998. Restoration of the rivers Brede, Cole and Skerne: a joint Danish and British EU-LIFE demonstration project, IV – Implications for nitrate and iron transformation. Aquatic Conservation – Marine and Freshwater Ecosystems 8: 223–240.

Hogan, D. M., T. E. Jordan & M. R. Walbridge, 2004. Phosphorus retention and soil organic carbon in restored and natural freshwater wetlands. Wetlands 24: 573–585.

Hollander, A., M. A. J. Huijbregts, A. M. J. Ragas & D. van de Meent, 2006. BasinBox: a generic multimedia fate model for predicting the fate of chemicals in river catchments. Hydrobiologia 565: 21–38.

Hupp, C. R., 1992. Riparian vegetation recovery patterns following stream channelization – a geomorphic perspective. Ecology 73: 1209–1226.

Hyacinthe, C. & P. van Cappellen, 2004. An authigenic iron phosphate phase in estuarine sediments: composition, formation and chemical reactivity. Marine Chemistry 91: 227–251.

ICPR, 2005. International Commission for the Protection of the Rhine, http://www.iksr.org/.

Jackson, M. B. & M. C. Drew, 1984. Effects of flooding on growth and metabolism of herbaceous plants. In Kozlowski, T. T. (ed.), Flooding and Plant Growth. Academic Press Inc, New York, 47–128.

Japenga, J. & W. Salomons, 1993. Dyke-protected floodplains – a possible chemical time bomb. Land Degradation and Rehabilitation 4: 373–380.

Japenga, J., K. H. Zschuppe, A. J. de Groot & W. Salomons, 1990. Heavy-metals and organic micropollutants in floodplains of the River Waal, a distributary of the River Rhine, 1958–1981. Netherlands Journal of Agricultural Science 38: 381–397.

Junk, W. J., P. B. Bayley & R. E. Sparks, 1989. The floodpulse concept in river floodplain systems. Canadian Journal of Fisheries and Aquatic Sciences 106: 110–127.

Kester, R. A., M. E. Meijer, J. A. Libochant, W. de Boer & H. J. Laanbroek, 1997. Contribution of nitrification and denitrification to the NO and N_2O emissions of an acid forest soil, a river sediment and a fertilized grassland soil. Soil Biology and Biochemistry 29: 1655–1664.

Khalid, R. A., W. H. Patrick & R. D. DeLaune, 1977. Phosphorus sorption characteristics of flooded soils. Soil Science Society of America Journal 41: 305–310.

Knowlton, M. F. & J. R. Jones, 1997. Trophic status of Missouri River floodplain lakes in relation to basin type and connectivity. Wetlands 17: 468–475.

Koerselman, W., M. B. van Kerkhoven & J. T. A. Verhoeven, 1993. Release of inorganic N, P and K in peat soils – effect of temperature, water chemistry and water-level. Biogeochemistry 20: 63–81.

Koerselman, W. & J. T. A. Verhoeven, 1995. Eutrophication of fen ecosystems: external and internal nutrient sources and restoration strategies. In Wheeler, B. D., S. C. Shaw, W. J. Fojt & R. A. Robertson (eds), Restoration of Temperate Wetlands. John Wiley and Sons Ltd, Chichester, 91–112.

Kohler, A., R. Wonneber & G. Zeltner, 1973. Chemical data and aquatic vascular plants as indicators for pollution in Moosach River system near Munich. Archiv für Hydrobiologie 72: 533–549.

Kok, C. J., G. van der Velde & K. M. Landsbergen, 1990. Production, nutrient dynamics and initial decomposition of floating leaves of *Nymphaea alba* L and *Nuphar lutea* (L) Sm (Nymphaeaceae) in alkaline and acid waters. Biogeochemistry 11: 235–250.

Kölle, W., O. Strebel & J. Böttcher, 1985. Formation of sulfate by microbial denitrification in a reducing aquifer. Water Supply 3: 33–40.

Kooistra, L., R. S. E. W. Leuven, P. H. Nienhuis, R. Wehrens & L. M. C. Buydens, 2001. A procedure for incorporating spatial variability in ecological risk assessment of Dutch river floodplains. Environmental Management 28: 359–373.

Kooistra, L., E. A. L. Salas, J. G. P. W. Clevers, R. Wehrens, R. S. E. W. Leuven, P. H. Nienhuis & L. M. C. Buydens, 2004. Exploring field vegetation reflectance as an indicator of soil contamination in river floodplains. Environmental Pollution 127: 281–290.

Kotowski W., 2002. Fen communities. Ecological mechanisms and conservation strategies. Ph.D Thesis, University of Groningen, Groningen.

Kotowski, W. & R. van Diggelen, 2004. Light as an environmental filter in fen vegetation. Journal of Vegetation Science 15: 583–594.

Kronvang, B., L. M. Svendsen, A. Brookes, K. Fisher, B. Moller, O. Ottosen, M. Newson & D. Sear, 1998. Restoration of the rivers Brede, Cole and Skerne: a joint Danish and British EU-LIFE demonstration project, III – channel morphology, hydrodynamics and transport of sediment and nutrients. Aquatic Conservervation – Marine and Freshwater Ecosystems 8: 209–222.

Laanbroek, H. J., 1990. Bacterial cycling of minerals that affect plant-growth in waterlogged soils – A review. Aquatic Botany 38: 109–125.

Laanbroek, H. J., H. J. Geerligs, L. Sijtsma & H. Veldkamp, 1984. Competition for sulfate and ethanol among *Desulfobacter*, *Desulfobulbus*, and *Desulfovibrio* species isolated from intertidal sediments. Applied and Environmental Microbiology 47: 329–334.

Laanbroek, H. J. & N. Pfennig, 1981. Oxidation of short-chain fatty-acids by sulfate-reducing bacteria in fresh-water and in marine-sediments. Archives of Microbiology 128: 330–335.

Lamers, L. P. M., S. J. Falla, E. M. Samborska, L. A. R. van Dulken, G. van Hengstum & J. G. M. Roelofs, 2002a. Factors controlling the extent of eutrophication and toxicity in sulfate-polluted freshwater wetlands. Limnology and Oceanography 47: 585–593.

Lamers, L. P. M., S. M. E. van Roozendaal & J. G. M. Roelofs, 1998a. Acidification of freshwater wetlands: combined effects of non-airborne sulfur pollution and desiccation. Water, Air and Soil Pollution 105: 95–106.

Lamers, L. P. M., A. J. P. Smolders & J. G. M. Roelofs, 2002b. The restoration of fens in the Netherlands. Hydrobiologia 478: 107–130.

Lamers, L. P. M., G. E. ten Dolle, S. T. G. van den Berg, S. P. J. van Delft & J. G. M. Roelofs, 2001. Differential responses of freshwater wetland soils to sulphate pollution. Biogeochemistry 55: 87–102.

Lamers, L. P. M., H. B. M. Tomassen & J. G. M. Roelofs, 1998b. Sulfate-induced eutrophication and phytotoxicity in freshwater wetlands. Environmental Science and Technology 32: 199–205.

Lenssen, J., F. Menting, W. van der Putten & K. Blom, 1999. Control of plant species richness and zonation of functional groups along a freshwater flooding gradient. Oikos 86: 523–534.

Leuven, R. S. E. W., A. J. M. Smits & P. H. Nienhuis, 2000. From integrated approaches to sustainable river management. In Smits, A. J. M., P. H. Nienhuis & R. S. E. W. Leuven (eds), New Approaches to Sustainable River Management. Backhuys Publishers, Leiden, 329–347.

Leuven, R. S. E. W., S. Wijnhoven, L. Kooistra, R. J. W. de Nooij & M. A. J. Huijbregts, 2005. Toxicological constraints for rehabilitation of riverine habitats along lowland rivers. Archiv für Hydrobiologie Supplement 155, Large Rivers 15: 657–676.

Lockaby, B. G., A. L. Murphy & G. L. Somers, 1996. Hydroperiod influences on nutrient dynamics in decomposing litter of a floodplain forest. Soil Science Society of America Journal 60: 1267–1272.

Loeb, R., & L. P. M. Lamers, 2003. The effects of river water quality on the development of wet floodplain vegetation types in the Netherlands. Proceedings of the International Ecoflood Conference 'Towards natural flood reduction strategies', Warsaw, Poland. http://levis.sggw.waw.pl/ecoflood/.

Lomans, B. P., A. J. P. Smolders, L. M. Intven, A. Pol, H. J. M. den op Camp & C. van der Drift, 1997. Formation of dimethyl sulfide and methanethiol in anoxic freshwater sediments. Applied and Environmental Microbiology 63: 4741–4747.

Lucassen, E. C. H. E. T., R. Bobbink, A. J. P. Smolders, P. J. M. van der Ven, L. P. M. Lamers & J. G. M. Roelofs, 2003. Interactive effects of low pH and high ammonium levels responsible for the decline of *Cirsium dissectum* (L.) Hill. Plant Ecology 165: 45–52.

Lucassen, E. C. H. E. T., A. J. P. Smolders, L. P. M. Lamers & J. G. M. Roelofs, 2005. Water table fluctuations and groundwater supply are important in preventing phosphate-eutrophication in sulphate-rich fens: consequences for wetland restoration. Plant and Soil 269: 109–115.

Lucassen, E. C. H. E. T., A. J. P. Smolders & J. G. M. Roelofs, 2000. Increased groundwater levels cause iron toxicity in *Glyceria fluitans* (L.). Aquatic Botany 66: 321–327.

Lucassen, E. C. H. E. T., A. J. P. Smolders & J. G. M. Roelofs, 2002. Potential sensitivity of mires to drought, acidification and mobilisation of heavy metals: the sediment S/(Ca + Mg) ratio as diagnostic tool. Environmental Pollution 120: 635–646.

Lucassen, E. C. H. E. T., A. J. P. Smolders, J. van de Crommenacker & J. G. M. Roelofs, 2004a. Effects of stagnating sulphate-rich groundwater on the mobility of phosphate in freshwater wetlands: a field experiment. Archiv für Hydrobiologie 160: 117–131.

Lucassen, E. C. H. E. T., A. J. P. Smolders, A. L. van der Salm & J. G. M. Roelofs, 2004b. High groundwater nitrate concentrations inhibit eutrophication of sulphate-rich freshwater wetlands. Biogeochemistry 67: 249–267.

Mendelssohn, I. A., B. K. Sorrell, H. Brix, H. H. Schierup, B. Lorenzen & E. Maltby, 1999. Controls on soil cellulose decomposition along a salinity gradient in a *Phragmites australis* wetland in Denmark. Aquatic Botany 64: 381–398.

Middelkoop, H., 2000. Heavy-metal pollution of the river Rhine and Meuse floodplains in the Netherlands. Netherlands Journal of Geosciences – Geologie en Mijnbouw 79: 411–427.

Middleton, B. A., 1999. Wetland Restoration, Flood Pulsing and Disturbance Dynamics. John Wiley and Sons Inc., New York.

Mommer, L., O. Pedersen & E. J. W. Visser, 2004. Acclimation of a terrestrial plant to submergence facilitates gas exchange under water. Plant Cell and Environment 27: 1281–1287.

Moss, B., 1994. Brackish and fresh-water shallow lakes – Different systems or variations on the same theme. Hydrobiologia 276: 1–14.

Mostert, E., 1998. River basin management in the European Union. How it is done and how it should be done. European Water Management 1: 26–35.

Nienhuis, P.H., 2006. Water and values: ecological research as the basis for water management and nature management. Hydrobiologia 565: 261–275.

Nienhuis, P. H., A. D. Buijse, R. S. E. W. Leuven, A. J. M. Smits, R. J. W. de Nooij & E. M. Samborska, 2002. Ecological rehabilitation of the lowland basin of the river Rhine (NW Europe). Hydrobiologia 478: 53–72.

Nienhuis P. H., R. S. E. W. Leuven, & A. M. J. Ragas (eds), 1998. New Concepts for Sustainable Management of River Basins. Backhuys Publishers, Leiden.

Nordstrom, D. K., 1982. Aqueous pyrite oxidation and consequent formation of secondary iron minerals in acid sulphate weathering. Soil Science Society of America Journal 10: 37–56.

Olde Venterink, H., M. J. Wassen, J. D. M. Belgers & J. T. A. Verhoeven, 2001. Control of environmental variables on species density in fens and meadows: importance of direct effects and effects through community biomass. Journal of Ecology 89: 1033–1040.

Olde Venterink, H., M. J. Wassen, A. W. M. Verkroost & P. C. de Ruiter, 2003a. Species richness-productivity patterns differ between N-, P-, and K-limited wetlands. Ecology 84: 2191–2199.

Olde Venterink, H., F. Wiegman, G. E. M. van der Lee & J. E. Vermaat, 2003b. Role of active floodplains for nutrient retention in the river Rhine. Journal of Environmental Quality 32: 1430–1435.

Olsen, S. R., C. W. Cole, R. Watanabe & L. A. Dean, 1954. Estimation of Available Phosphorus in Soils by Extraction with Sodium Bicarbonate. US Dpt. of Agriculture circular 939.

Onaindia, M., B. G. Debikuna & I. Benito, 1996. Aquatic plants in relation to environmental factors in northern Spain. Journal of Environmental Management 47: 123–137.

Patrick, W. H., S. Gotoh & B. G. Williams, 1973. Strengite dissolution in flooded soils and sediments. Science 179: 564–565.

Patrick, W. H. & R. A. Khalid, 1974. Phosphate release and sorption by soils and sediments – effect of aerobic and anaerobic conditions. Science 186: 53–55.

Paul, E. A. & F. E. Clark, 1989. Soil Biology and Biochemistry. 2nd edn. Academic Press, San Diego, London.

Pauwels, H., W. Kloppmann, J. C. Foucher, A. Martelat & V. Fritsche, 1998. Field tracer test for denitrification in a pyrite-bearing aquifer. Applied Geochemistry 13: 767–778.

Pedroli, B., G. de Blust, K. van Looy & S. van Rooij, 2001. Setting targets in strategies for river restoration. Landscape Ecology 17: 5–18.

Pezeshki, S. R., 2001. Wetland plant responses to soil flooding. Environmental and Experimental Botany 46: 299–312.

Pinay, G., V. J. Black, A. M. Planty-Tabacchi, B. Gumiero & H. Decamps, 2000. Geomorphic control of denitrification in large river floodplain soils. Biogeochemistry 50: 163–182.

Poiani, K. A. & B. L. Bedford, 1995. GIS based nonpoint-source pollution modeling – considerations for wetlands. Journal of Soil and Water Conservation 50: 613–619.

Pollock, M. M., R. J. Naiman & T. A. Hanley, 1998. Plant species richness in riparian wetlands – A test of biodiversity theory. Ecology 79: 94–105.

Ponnamperuma, F. N., 1984. Effects of flooding on soils. In Kozlowski, T. T. (ed.), Flooding and Plant Growth. Academic Press Inc., Orlando: 9–45.

Postma, R., M. J. J. Kerkhofs, G. B. M. Pedroli & J. G. M. Rademakers, 1996. Een stroom natuur. Natuurstreefbeelden voor Rijn en Maas. Watersysteemverkenningen 1996:1–102. RIZA nota 95.060 (in Dutch).

Reddy, K. R., R. H. Kadlec, E. Flaig & P. M. Gale, 1999. Phosphorus retention in streams and wetlands: a review. Critical Reviews in Environmental Science and Technology 29: 83–146.

Rengel, Z., 1996. (Tansley review) Uptake of aluminium by plant cells. New Phytologist 134: 389–406.

Richardson, C. J., 1985. Mechanisms controlling phosphorus retention capacity in fresh-water wetlands. Science 228: 1424–1427.

RIVM, 2004. Water quality Rhine and Meuse, 1980–2003 (figures from RIZA monitoring programme, various years), http: //www.rivm.nl/milieuennatuurcompendium/.

Roelofs, J. G. M., 1991. Inlet of alkaline river water into peaty lowlands: effects on water quality and *Stratiotes aloides* L. stands. Aquatic Botany 39: 267–293.

Roelofs, J. G. M., A. J. Kempers, A. L. F. M. Houdijk & J. Jansen, 1985. The effect of ar-borne ammonium-sulfate on *Pinus nigra* var *maritima* in the Netherlands. Plant and Soil 84: 45–56.

Roozen, F. C. J. M., G. J. van Geest, B. W. Ibelings R. Rooijackers, M. Scheffer & A. D. Buijse, 2003. Lake age and water level affect the turbidity of floodplain lakes along the lower Rhine. Freshwater Biology 48: 519–531.

Rout, G. R., S. Samantaray & P. Das, 2001. Aluminium toxicity in plants: a review. Agronomie 21: 3–21.

Schröder, T. J., T. Hiemstra, J. P. M. Vink & S. E. A. T. M. van der Zee, 2005. Modeling of the solid-solution partitioning of heavy metals and arsenic in embanked flood plain soils of the rivers Rhine and Meuse. Environmental Science and Technology 39: 7176–7184.

Schuurkes, J. A. A. R., A. J. Kempers & C. J. Kok, 1988. Aspects of biochemical sulfur conversions in sediments of a shallow soft-water lake. Journal of Freshwater Ecology 4: 369–381.

Simons, J. H. E. J., C. Bakker, M. H. I. Schropp, L. H. Jans, F. R. Kok & R. E. Grift, 2001. Man-made secondary channels along the River Rhine (the Netherlands); results of post-project monitoring. Regulated Rivers: Research and Management 17: 473–491.

Sjodin, A. L., W. M. Lewis & J. F. Saunders, 1997. Denitrification as a component of the nitrogen budget for a large plains river. Biogeochemistry 39: 327–342.

Smit, H., G. van der Velde, R. Smits & H. Coops, 1997. Ecosystem responses in the Rhine-Meuse delta during two decades after enclosure and steps toward estuary restoration. Estuaries 20: 504–520.

Smits, A. J. M., R. M. J. C. Kleukers, C. J. Kok & G. van der Velde, 1990b. Alcohol dehydrogenase isozymes in the roots of some nymphaeid and isoetid macrophytes. Adaptations to hypoxic sediment conditions?. Aquatic Botany 38: 19–27.

Smits, A. J. M., P. H. Nienhuis & R. S. E. W. Leuven (eds), 2000. New Approaches to River Management. Bakhuys Publishers, Leiden.

Smits, A. J. M., P. H. Nienhuis & H. L. F. Saeijs, 2006. Changing estuaries, changing views. Hydrobiololgia 565: 339–355.

Smits, A. J. M., P. H. van Avesaath & G. van der Velde, 1990a. Germination requirements and seed banks of some nymphaeid macrophytes – *Nymphaea alba* L, *Nuphar lutea* (L) Sm and *Nymphoides peltata* (Gmel.) O. Kuntze. Freshwater Biology 24: 315–326.

Smits, A. J. M., G. W. H. Schmitz & G. van der Velde, 1992. Calcium-dependent lamina production of *Nymphoides peltata* (Gmel.) O. Kuntze (Menyanthaceae) – implications for distribution. Journal of Experimental Botany 43: 1273–1281.

Smits, A. J. M., G. H. W. Schmitz, G. van der Velde & L. A. C. J. Voesenek, 1995. Influence of ethanol and ethylene on the seed-germination of three nymphaeid water plants. Freshwater Biology 34: 39–46.

Smolders, A. J. P., C. den Hartog, C. B. L. van Gestel & J. G. M. Roelofs, 1996. The effects of ammonium on growth, accumulation of free amino acids and nutritional status of young phosphorus deficient *Stratiotes aloides* plants. Aquatic Botany 53: 85–96.

Smolders, A. J. P., L. P. M. Lamers, M. Moonen, K. Zwaga & J. G. M. Roelofs, 2001. Controlling phosphate release from phosphate-enriched sediments by adding various iron compounds. Biogeochemistry 54: 219–228.

Smolders, A. J. P. & J. G. M. Roelofs, 1993. Sulphate mediated iron limitation and eutrophication in aquatic systems. Aquatic Botany 46: 247–253.

Smolders, A. J. P. & J. G. M. Roelofs, 1995. Internal eutrophication, iron limitation and sulphide accumulation due to the inlet of river Rhine water in peaty shallow waters in the Netherlands. Archiv für Hydrobiologie 133: 349–365.

Smolders, A. J. P. & J. G. M. Roelofs, 1996. The roles of internal iron hydroxide precipitation, sulphide toxicity and oxidizing ability in the survival of *Stratiotes aloides* roots at different iron concentrations in sediment pore water. New Phytologist 133: 253–260.

Smolders, A., J. G. M. Roelofs & G. van der Velde, 1994. Iron deficiency in *Nymphoides peltata* owing to the exhaustion of dissolved iron in anaerobic sediments. Aquatic Botany 47: 349–353.

Snowden, R. E. D. & B. D. Wheeler, 1995. Chemical changes in selected wetland plant species with increasing Fe supply, with specific reference to root precipitates and Fe tolerance. New Phytologist 131: 503–520.

Sparks, R. E., 1995. Need for ecosystem management of large rivers and their floodplains. BioScience 45: 168–182.

Spink, A., R. E. Sparks, M. van Oorschot & J. T. A. Verhoeven, 1998. Nutrient dynamics of large river floodplains. Regulated Rivers: Research and Management 14: 203–216.

Steiger, J. & A. M. Gurnell, 2003. Spatial hydrogeomorphological influences on sediment and nutrient deposition in riparian zones: observations from the Garonne River, France. Geomorphology 49: 1–23.

Stumm, W. & J. J. Morgan, 1981. Aquatic Chemistry, an Introduction Emphasizing Chemical Equilibria in Natural Waters (2nd ed.). John Wiley and Sons Inc., New York.

Taylor, D. R., L. W. Aarssen & C. Loehle, 1990. On the relationship between R/K selection and environmental carrying-capacity – A new habitat templet for plant life-history strategies. Oikos 58: 239–250.

Tessier, A. & D. R. Turner (eds), 1995. Metal Speciation and Bioavailability in Aquatic Systems. John Wiley & Sons Inc., Chichester.

Tilman, D., 1994. Competition and biodiversity in spatially structured habitats. Ecology 75: 2–16.

Tockner, K., F. Schiemer & J. V. Ward, 1998. Conservation by restoration: the management concept for a river-floodplain system on the Danube River in Austria. Aquatic Conservation – Marine and Freshwater Ecosystems 8: 71–86.

Trèmolieres, M., R. Carbiener, A. Ortscheit & J. P. Klein, 1994. Changes in aquatic vegetation in Rhine floodplain streams in Alsace in relation to disturbance. Journal of Vegetation Science 5: 169–178.

Van Breemen, N., J. Mulder & C. T. Driscoll, 1983. Acidification and alkalinization of soils. Plant and Soil 75: 283–308.

Van den Berg, L. J. L., E. Dorland, P. Vergeer, M. A. C. Hart, R. Bobbink & J. G. M. Roelofs, 2005. Decline of acid-sensitive plant species in heathland can be attributed to ammonium toxicity in combination with low pH. New Phytologist 166: 551–564.

Van den Brink, F. W. B., J. P. H. M. de Leeuw, G. van der Velde & G. M. Verheggen, 1993. Impact of hydrology on the chemistry and phytoplankton development in floodplain lakes along the lower Rhine and Meuse. Biogeochemistry 19: 103–128.

Van den Brink, F. W. B., M. M. J. Maenen, G. van der Velde & A. bij de Vaate, 1991. The (semi-)aquatic vegetation of still waters within the floodplains of the rivers Rhine and Meuse in The Netherlands: historical changes and the role of inundation. Verhandlungen der Internationale Vereinigung für theoretische und angewandte Limnologie 24: 2693–2699.

Van den Brink, F. W. B. & G. van der Velde, 1993. Growth and morphology of four fresh-water macrophytes under the impact of raised salinity level of the Lower Rhine. Aquatic Botany 45: 285–297.

Van den Brink, F. W. B. & G. van der Velde, 1994. Impact of hydrology and water chemistry on floodplain lake communities along the lower Rhine and Meuse. Water Science and Technology 29: 57–60.

Van den Brink, F. W. B., G. van der Velde, W. W. Bosman & H. Coops, 1995. Effects of substrate parameters on growth-responses of eight helophyte species in relation to flooding. Aquatic Botany 50: 79–97.

Van den Brink, F. W. B., M. M. van Katwijk & G. van der Velde, 1994. Impact of hydrology on phytoplankton and zooplankton community composition in floodplain lakes along the lower Rhine and Meuse. Journal of Plankton Research 16: 351–373.

Van der Hoek, D., A. J. E. M. van Mierlo & J. M. Van Groenendael, 2004. Nutrient limitation and nutrient-driven shifts in plant species composition in a species-rich fen meadow. Journal of Vegetation Science 15: 389–396.

Van der Lee, G. E. M., H. O. Venterink & N. E. M. Asselman, 2004. Nutrient retention in floodplains of the Rhine distributaries in The Netherlands. River Research and Applications 20: 315–325.

Van der Valk, A. G., 1981. Succession in wetlands – a Gleasonian approach. Ecology 62: 688–696.

Van der Welle, M. E. W., M. L. C. Cuppens, L. P. M. Lamers & J. G. M. Roelofs, (in press). Detoxifying toxicants: interactions between iron and sulfide toxicity in freshwater wetlands. Environmental Toxicology and Chemistry.

Van Dijk, J., M. Stroetenga, L. Bos, P. M. van Bodegom, H. A. Verhoef & R. Aerts, 2004. Restoring natural seepage conditions on former agricultural grasslands does not lead to reduction of organic matter decomposition and soil nutrient dynamics. Biogeochemistry 71: 317–337.

Van Eck, W. H. J. M., J. P. M. Lenssen, H. M. van de Steeg, C. W. P. M. Blom & H. de Kroon, 2006. Seasonal dependent effects of flooding on plant species survival and zonation: a comparative study of 10 terrestrial grassland species. Hydrobiologia 565: 59–69.

Van Geest, G. J., H. Coops, R. M. M. Roijackers, A. D. Buijse & M. M. Scheffer, 2005. Succession of aquatic vegetation driven by reduced water-level fluctuations in floodplain lakes. Journal of Applied Ecology 42: 251–260.

Van Geest, G. J., F. C. J. M. Roozen, H. Coops, R. M. M. Roijackers, A. D. Buijse, E. T. H. M. Peeters & M. Scheffer, 2003. Vegetation abundance in lowland flood plan lakes determined by surface area, age and connectivity. Freshwater Biology 48: 440–454.

Van Oorschot, M., C. Hayes & I. van Strien, 1998. The influence of soil desiccation on plant production, nutrient uptake and plant nutrient availability in two French floodplain grasslands. Regulated Rivers: Research and Management 14: 313–327.

Van Stokkom, H. T. C., A. J. M. Smits & R. S. E. W. Leuven, 2005. Flood defence in the Netherlands: a new era, a new approach. Water International 30: 76–87.

Van Wijck, C., C. J. de Groot & P. Grillas, 1992. The effect of anaerobic sediment on the growth of *Potamogeton pectinatus* l – The role of organic-matter, sulfide and ferrous iron. Aquatic Botany 44: 31–49.

Verhoeven, J. T. A., D. F. Whigham, R. van Logtestijn & J. O'Neill, 2001. A comparative study of nitrogen and phosphorus cycling in tidal and non-tidal riverine wetlands. Wetlands 21: 210–222.

Vervuren, P. J. A., C. W. P. M. Blom & H. de Kroon, 2003. Extreme flooding events on the Rhine and the survival and distribution of riparian plant species. Journal of Ecology 91: 135–146.

Vervuren, P. J. A., S. M. J. H. Beurskens & C. W. P. M. Blom, 1999. Light acclimation, CO_2 response and long-term capacity of underwater photosynthesis in three terrestrial plant species. Plant Cell and Environment 22: 959–968.

Vink, J. P. M., 2002. Measurement of heavy metal speciation over redox gradients in natural water-sediment interfaces and implications for uptake by benthic organisms. Environmental Science and Technology 36: 5130–5138.

Visser, E. J. W., L. A. C. J. Voesenek, B. B. Vartapetian & M. B. Jackson, 2003. Flooding and plant growth. Annals of Botany 91: 107–109.

Van der Hoek, D., A. J. E. M. van Mierlo & J. M. van Groenendael, 2004. Nutrient limitation and nutrient-driven shifts in plant species composition in a species-rich fen meadow. Journal of Vegetation Science 15: 389–396.

Ward, J. V., K. Tockner, D. B. Arscott & C. Claret, 2002. Riverine landscape diversity. Freshwater Biology 47: 517–539.

Wassen, M. J., H. G. M. O. Veterink & E. O. A. M. de Swart, 1995. Nutrient concentrations in mire vegetation as a measure of nutrient limitation in mire ecosystems. Journal of Vegetation Science 6: 5–16.

Wassen, M. J., R. E. van der Vliet & J. T. A. Verhoeven, 1998. Nutrient limitation in the Biebrza fens and floodplain (Poland). Acta Botanica Neerlandica 47: 241–253.

Wassen, M. J., H. Olde Venterink, E. D. Lapshina & F. Tanneberger, 2005. Endangered plants persist under phosphorus limitation. Nature 437: 547–550.

Wassen, M. J., W. H. M. Peeters & H. O. Venterink, 2002. Patterns in vegetation, hydrology, and nutrient availability in an undisturbed river floodplain in Poland. Plant Ecology 165: 27–43.

Wassen, M. J., W. H. M. Peeters & H. Olde Venterink, 2003. Patterns in vegetation, hydrology, and nutrient availability in an undisturbed river floodplain in Poland. Plant Ecology 165: 27–43.

Wheeler, B. D. & M. C. F. Proctor, 2000. Ecological gradients, subdivisions and terminology of north–west European mires. Journal of Ecology 88: 187–203.

Wijnhoven, S., G. van der Velde, R.S.E.W. Leuven & A. J. M. Smits, 2006. Modelling recolonisation of heterogeneous river floodplains by small mammals. Hydrobiologia 565: 135–152.

Winter, L. T., I. D. L. Foster, S. M. Charlesworth & J. A. Lees, 2001. Floodplain lakes as sinks for sediment-associated contaminants - A new source of proxy hydrological data? Science of the Total Environment 266: 187–194.

Young, E. O. & D. S. Ross, 2001. Phosphate release from seasonally flooded soils: a laboratory microcosm study. Journal of Environmental Quality 30: 91–101.

Hydrobiologia (2006) 565:187–200
R.S.E.W. Leuven, A.M.J. Ragas, A.J.M. Smits & G. van der Velde (eds), Living Rivers: Trends and Challenges in Science and Management
DOI 10.1007/s10750-005-1913-7

Effects of rewetting measures in Dutch raised bog remnants on assemblages of aquatic Rotifera and microcrustaceans

G.A. van Duinen[1,2,*], Y. Zhuge[1,4], W.C.E.P. Verberk[1,3], A.M.T. Brock[1], H.H. van Kleef[1,2], R.S.E.W. Leuven[2], G. van der Velde[3] & H. Esselink[1,3]
[1]*Bargerveen Foundation, Radboud University Nijmegen, P.O. Box 9010, 6500 GL Nijmegen, The Netherlands*
[2]*Department of Environmental Science, Institute for Water and Wetland Research, Radboud University Nijmegen, P.O. Box 9010, 6500 GL Nijmegen, The Netherlands*
[3]*Department of Animal Ecology and Ecophysiology, Institute for Water and Wetland Research, Radboud University Nijmegen, P.O. Box 9010, 6500 GL Nijmegen, The Netherlands*
[4]*Institute of Hydrobiology, Chinese Academy of Sciences, Donghu Experimental Station, Wuhan, Hubei 430072, P.R. China*
*(*Author for correspondence: Tel.: +31-24-3653296; Fax: +31-24-3653030; E-mail: G.vanDuinen@science.ru.nl)*

Key words: Cladocera, Copepoda, colonisation, dispersal, multivariate analysis, raised bog, peatland, Rotifera, restoration, species diversity

Abstract

Species differ in their life cycle, habitat demands and dispersal capacity. Consequently different species or species groups may respond differently to restoration measures. To evaluate effects of restoration measures in raised bog remnants on aquatic microinvertebrates, species assemblages of Rotifera and microcrustaceans were sampled in 10 rewetted and 10 non-rewetted sites, situated in 7 Dutch raised bog remnants. A total of 129 species (Rotifera 108, Cladocera 15, Copepoda 6 species) were found. The species assemblages, total numbers of species and numbers of characteristic raised bog species did not differ between the 10 rewetted and 10 non-rewetted sites. The dominant pattern in the variation in microinvertebrate assemblages could be explained by the presence or absence of open water and variation in physico-chemical variables of surface water and organic matter. Furthermore, the species assemblages of water bodies situated in the same area were on average more similar to each other than to assemblages from other areas. These differences between areas may be due to differences in environmental conditions of water bodies, and possibly also to differences in the local species pool and the subsequent immigration sequence of species. We conclude that, in contrast to earlier findings on aquatic macroinvertebrates, populations of microinvertebrate species, including characteristic species, can either persist in the raised bog remnants during the process of rewetting or (re-)establish within a relatively short period of time (less than about 5 years).

Introduction

In degraded landscapes, restoration measures are taken to rehabilitate the ecosystems and their characteristic biodiversity. These measures focus mainly on creating suitable abiotic conditions for the recovery of key plant species or characteristic vegetation types. Recovery of characteristic fauna is often assumed to follow automatically in course of time. Although animals, and especially invertebrates, make up an important part of the total species diversity, relatively little attention has been paid to how restoration measures affect the fauna (Longcore, 2003; Van Duinen et al., 2003; Van Kleef et al., 2006).

In raised bog landscapes *Sphagnum* species are key-stone species. Therefore, restoration measures in degraded raised bogs generally focus on creating suitable hydrological conditions for re-colonisation and growth of *Sphagnum* by blocking drainage ditches and building dams to retain rain water (e.g., Smolders et al., 2003; Rochefort et al., 2003; Vasander et al., 2003). These rewetting measures have been applied both in large-scaled cut-over bog remnants with little or no *Sphagnum* growth and in bog remnants degraded by drainage and small-scaled hand peat cuttings, where *Sphagnum* was still growing in parts of the area. Rewetting measures frequently resulted in a fast rise of the water table and shallow to deep flooding of large areas. The subsequent vegetation differed considerably between different rewetted sites. Floating rafts with *Sphagnum* vegetation developed if poorly humified peat was still present and swelled up or became buoyant. Submerged or floating *Sphagnum* vegetation developed if sufficient light penetrated into the (shallow) water layer and sufficient carbon dioxide was available. Inundated areas of humified peat remained almost devoid of *Sphagnum* (Smolders et al., 2003).

A comparative study on aquatic macroinvertebrates showed that rewetted sites in Dutch raised bog remnants are inhabited by only a part of the species spectra of both pristine raised bogs in Estonia and non-rewetted sites in Dutch bog remnants (Van Duinen et al., 2002, 2003). These non-rewetted sites are water bodies that have not been subject to large-scale restoration measures, but are remnants of the former use of bogs, e.g., abandoned hand peat cuttings and trenches used in buckwheat culture. The species assemblages differ considerably between rewetted and non-rewetted sites and the cumulative species richness is much higher in the non-rewetted sites. Moreover, several characteristic and rare species were found only at non-rewetted sites, irrespective of the development of a characteristic vegetation. These differences between rewetted and non-rewetted sites could not be explained by differences in the surface water quality and vegetation composition of the water bodies. The differences in species assemblage were attributed to fast and large-scaled changes in the water table caused by large-scaled rewetting measures and to lower habitat diversity in rewetted areas. Populations of rare and characteristic macroinvertebrate species, that were able to persist in degraded bog remnants, can respond negatively to rewetting of raised bog remnants (Van Duinen et al., 2003).

Contrary to these results on aquatic macroinvertebrates, Buttler et al. (1996) showed that the testate amoebae fauna of raised bogs can recover rapidly and fully, regardless of the initial condition of the cutover surface. Apparently, different invertebrate groups respond differently to restoration measures. This differential response may be explained by differences in dispersal capacity and habitat demands between invertebrate groups. Most aquatic macroinvertebrate species disperse actively. However, species may lack flight ability (e.g., Jackson, 1952) or exhibit high site fidelity, which may especially be true for species living in stable and permanent environments (Southwood, 1962; Wagner & Liebherr, 1992), such as raised bogs. As a result, for macroinvertebrates re-colonisation may be very slow. Aquatic microinvertebrates, like Rotifera and small crustaceans (Copepoda and Cladocera), on the other hand, are known to be easily spread by wind and animal vectors (Cáceres & Soluk, 2002; Cohen & Shurin, 2003). Cohen & Shurin (2003) found that zooplankton (Rotifera, Copepoda and Cladocera) disperse rapidly over short distances (10 m–1 km) and that dispersal only limits the diversity of very young communities (<1 year). Furthermore, Rotifera, Copepoda and Cladocera species often reproduce parthenogenetically and have a short life cycle that can be completed in one water body (Nogrady et al., 1993; Wetzel, 2001). Both these life-history characteristics potentially enable them to rapidly establish a population after colonisation. In contrast, most macroinvertebrates reproduce sexually and have a more complex life cycle than microinvertebrates that include larval, pupal, and adult stages. Each of these life-history stages may pose different demands on their environment. Therefore, a combination of various habitat elements may be needed to complete the life cycle (e.g., Galewski, 1971; Fairchild et al., 2003). The various habitat elements (heterogeneity) required by the species to complete their life cycle have to be present at the proper time and proper spatial scale (Verberk et al., 2001, 2005). In rewetted sites these requirements have apparently not been met for several macroinvertebrate species, contrary to the non-rewetted sites that were included in the macroinvertebrate study. This can be due to a more radical degradation (e.g., large-scaled peat extraction) in the rewetted sites, or to the abrupt process of rewetting, resulting in the disappearance of relic populations of rare and characteristic macroinvertebrate species (Van Duinen et al., 2003).

As Rotifera and microcrustaceans generally have a high dispersal capacity and a low dependence on habitat heterogeneity for the completion of their life cycle, compared to macroinvertebrates, we hypothesised that the occurrence of these microinvertebrate species, including characteristic raised bog species, do not differ between rewetted and non-rewetted sites in bog remnants. Contrary to the findings of the study on macroinvertebrates, we expected that physico-chemical conditions and vegetation structure of the water bodies are more discriminating factors between the microinvertebrate assemblages than conditions at a larger spatial scale or events in the past, including the degradation or rewetting of raised bog remnants. To test this hypothesis, a comparative study was carried out, with a similar procedure as in the earlier study on macroinvertebrates (Van Duinen et al., 2003). We sampled Rotifera, Cladocera, and Copepoda in 10 water bodies created by rewetting measures and 10 water bodies in degraded raised bog remnants, which have not been subject to large-scale rewetting measures. These two groups of sampling sites will be referred to as 'rewetted sites' and 'remnant sites', respectively. To include variation in vegetation structure, each group consisted of 5 pools with a substantial area of open water and a littoral zone and 5 puddles with little or no open water. In this paper we will answer the following questions:

1. Do remnant sites differ from rewetted sites with respect to species assemblage and (cumulative) species richness?
2. Is there a different response for characteristic raised bog species?
3. What is the relative importance of rewetting, vegetation structure and physico-chemical variables for the species assemblages of Rotifera, Cladocera and Copepoda?

Materials and methods

Sampling sites

Microinvertebrates were sampled in 20 water bodies, situated in 7 different raised bog remnants in the Netherlands (Fig. 1). Ten of these water bodies sampled were created by large-scale rewetting measures 5–30 years ago. The 10 remnant sites were pools and puddles created by peat cutting and trenches used in buckwheat culture. These remnant sites were abandoned more than 50 years ago and had not been subject to large-scale rewetting measures. The water bodies sampled differed in age, size and water and substrate quality. Characteristics of the sampling sites (year of rewetting, dimensions, and physico-chemical variables) are given in the appendix. Both the remnant and rewetted sites included five pools with a substantial area of open water in the centre and *Sphagnum cuspidatum* and other plants in the littoral zone, and five water bodies with little open water, viz. relatively small puddles and a trench used in buckwheat culture.

Microinvertebrate sampling and identification

To avoid large changes in species assemblages during the sampling period, all sites were sampled in the period of April 1 to May 6, 2002. At each sampling site two sets of subsamples were taken. Each set consisted of pooled subsamples collected at different spots in the water body. One set of subsamples was taken by means of two plankton nets with 45 and 115 μm mesh size, respectively, and was fixed immediately by adding formaldehyde to a final concentration of 2%. This concentration was later raised to 4% for storage. The other set of subsamples was taken using only a plankton net with 45 μm mesh size, was not fixed and was examined under the microscope to identify illoricate Rotifera. For the 10 pools, subsamples taken in the littoral zone were kept separately from those taken in the open water. Species were identified using the keys of Donner (1965), Koste (1978), Einsle (1993, 1996), Nogrady & Pourriot (1995), Segers (1995), De Smet (1996, 1997), and Flößner (2000). Hypochlorite solution was used for rotifer trophi analysis. Species lists are published by Zhuge & Van Duinen (2005).

Physico-chemical variables

At each sampling site, surface water was sampled without air-bubbles in iodated polyethylene bottles. Organic matter (top layer of sedimented

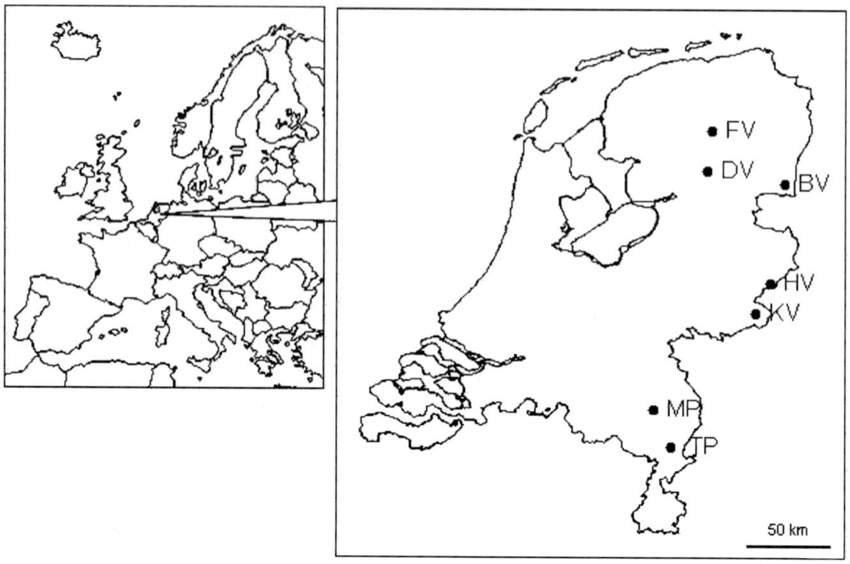

Figure 1. Locations of raised bog areas in the Netherlands at which samples were taken. FV, Fochteloërveen (0/2); DV, Dwingelerveld (2/0); BV, Bargerveen (1/3); HV, Haaksbergerveen (0/3); KV, Korenburgerveen (5/0); MP, Mariapeel (0/2); TP, Tuspeel (2/0). Figures in brackets are the numbers of remnant/rewetted sites sampled at each of the areas.

detritus and accumulated peat) was collected using the plankton net with 115 μm mesh size. Both were sampled on the same day as the microinvertebrates. Electric conductivity (EC) and dimensions of the water bodies (length, width and depth) were assessed at the sites. The pH, alkalinity and turbidity of surface water samples were measured within 24 h after sample collection. After adding 1 mg citric acid per 5 ml of water, water samples were stored at −20 °C in iodated polyethylene bottles until further analysis. Organic matter was dried (48 h at 70 °C) and ground up in liquid nitrogen. Nitrogen and carbon concentrations were measured in dried and ground organic matter with a CNS analyser (type NA 1500, Carlo Erba Instruments, Milan). Dried and ground organic matter was digested in sealed Teflon vessels in a Milestone microwave oven adding nitric acid and hydrogen peroxide. Digestates were analysed for total-P, Na, K, Ca, Cl, Mg, Mn, Fe, Si, Zn, Al, and S. The surface water samples were analysed for colour (extinction at 450 nm; Schimadzu spectrophotometer UV-120-01) and the concentrations of total inorganic carbon, NO_3, NH_4, PO_4, total-P, Na, K, Ca, Cl, Mg, Mn, Fe, Si, Zn, Al, and S. The following were determined colorimetrically; NO_3 according to Kamphake et al. (1967), NH_4 according to Grasshoff & Johanssen (1977), *o*-PO_4 according to Henriksen (1965), Cl according to O'Brien (1962). Na and K were determined with a Technicon flame photometer. Ca, Mg, Mn, Fe, Si, Zn, Al, P, and S were measured by inductively coupled plasma emission spectrophotometry. Total inorganic carbon in surface water was measured with an infrared carbon analyser. Selected water and organic matter quality data are presented in the Appendix.

Data analysis

Cumulative curves for species richness were compiled for both remnant and rewetted sites by calculating the averages and standard deviation of the total species richness of all possible combinations for one to 10 sampling sites. Significance of differences in average species richness between rewetted and remnant sites was tested with a one-way ANOVA, in which the factor 'rewetting' was nested within the factor 'area' to account for possible differences in diversity between areas. Similarity in microinvertebrate assemblages of site pairs was calculated using the Sørensen coefficient (Sørensen, 1948) $S = 2j/(a + b)$, with j being the number of species occurring in both sites, a and b

the total species numbers of the two respective sites. Average similarity coefficients were compared within and between areas and between remnant and rewetted sites. Significance of differences in averages between categories of sites was tested using a Mann–Whitney *U*-test. Significance of differences in environmental variables between rewetted and remnant sites was tested with a one-way ANOVA.

Species were considered to be characteristic for raised bogs when they are listed as typical of bogs or acidophilous species (preferring acid water bodies) in literature (data in Zhuge & Van Duinen, 2005). These data were taken from Myers (1931), Donner (1965), Koste (1978), Bērziņš & Pejler (1987), Pejler & Bērziņš (1993a), Nogrady & Pourriot (1995), Segers (1995) and De Smet (1996, 1997).

Multivariate analyses of individual species presence/absence data and environmental variables were performed in Canoco for Windows version 4.0 (Ter Braak & Šmilauer, 1998), to study the relative importance of rewetting, open water and the physico-chemical variables mentioned in the previous paragraph. A first analysis was performed with the samples taken in puddles, in open water and in littoral zones of pools separately (30 samples). A second analysis was performed in which the species presence data from the open water and littoral zone of pools were pooled (10 samples taken in pools and 10 samples in puddles). From preliminary detrended correspondence analyses (DCA) it was concluded that most of the species response curves could be best described by a linear response model (following Ter Braak, 1995). Therefore, principal components analyses (PCA) were performed for analysis of species presence data and environmental variables. Significance of the effect of environmental variables was tested using a Monte Carlo resampling procedure with 500 permutations in redundancy analyses (RDA). To test whether rewetting could significantly explain variation that remains after fitting the most important environmental variables, or the areas in which the sampling sites were situated, two partial RDAs were performed, with the binomial variable 'rewetting vs. remnant' as the only environmental variable and either the most important environmental variables, or 7 binomial dummy variables for the areas in which the sampling sites were situated, as covariables. This analysis was performed on the data set with littoral zone and open water samples pooled.

Results

Species richness and frequency

A total of 129 microinvertebrate species was found at the 20 sites sampled: 108 Rotifera, 15 Cladocera and 6 Copepoda species. The remnant sites comprised 107 species and rewetted sites 95 species. The higher number of species in remnant sites mainly concerned non-characteristic species encountered in only one of the sampling sites (Fig. 2). Most of the more frequently found species were present in both remnant and rewetted sites (non-lined parts in Fig. 2). Fifteen species were encountered in 15 or more sites. These included the characteristic species *Acantholeberis curvirostris, Lecane galeata, Trichocerca parvula, Diacyclops nanus, Alonella exisa* and *Keratella serrulata.* Seventy-three species were found at both rewetted sites and remnant sites, whereas 56 species were encountered at either the rewetted sites, or the remnant sites only. Of these 56 species, 34 were only found in the remnant sites, of which 7 species were classified as characteristic species. For the rewetted sites these numbers were 22 and 8, respectively. Eighty-five percent of these 56 species were encountered at only one or two sampling sites. Few species were encountered considerably more frequently at either remnant or rewetted sites. These were the characteristic Rotifera species *Pleurotrocha robusta* (5 remnant sites vs. no rewetted sites), *Monommata phoxa* (6 vs. 1), and *Euchlanis triquetra* (2 vs. 7), the rotifer *Encentrum martes* (7 vs. 0) and the copepod *Acanthocyclops robustus* (2 vs. 9).

The proportions of Rotifera, Copepoda and Cladocera in the total number of species were fairly similar between rewetted and remnant sites (Fig. 2). The proportion of Cladocera was higher in the frequently encountered species than in the less frequently encountered species. In total, 38 species were classified as characteristic to acid water or bogs, of which 30 were recorded at the remnant sites and 31 at the rewetted sites. The average number of species and characteristic species did not differ significantly between remnant

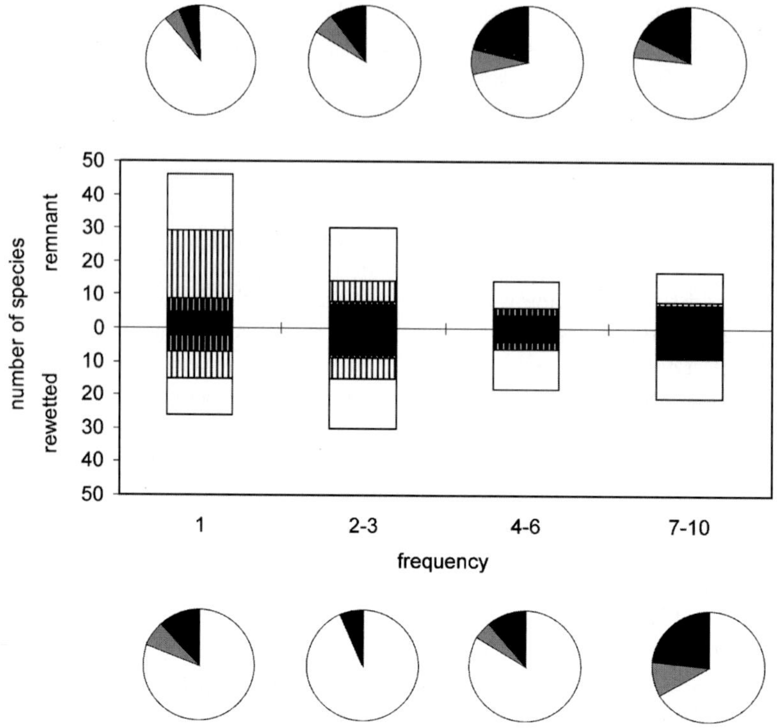

Figure 2. Numbers of species found at the 10 remnant sites (above horizontal axis) and the 10 rewetted sites (under horizontal axis) for different frequency classes. Characteristic species are shown with the black parts and non-characteristic species with the white parts. Species encountered in one of both groups of sites exclusively are shown with the lined parts of the black and white parts of the bars. The pie charts indicate the proportion of Rotifera (white), Copepoda (grey) and Cladocera (black) in the total number of species of the respective bars.

sites and rewetted sites (Table 1). The cumulative species richness of rewetted sites and remnant sites was similar, especially for the characteristic species (Fig. 3).

Species assemblages and correlations

In the first PCA (Fig. 4a) samples from rewetted and remnant sites were not separated along the first four PCA-axes. The distribution of samples along the first PCA-axis was positively or negatively correlated (with correlation coefficient <-0.5 or >0.5) to the Na and Cl concentration, electric conductivity, and pH of the surface water and to the C:N quotient of the organic matter. The open water samples were separated from the samples taken in the littoral zone of pools and in puddles along the second PCA-axis. The first four PCA-axes explained 15.1, 10.5, 8.6 and 7.7% of the variation in species data. After selecting all environmental variables that had a correlation coefficient <-0.5 or >0.5 with one or more of the first four PCA-axes, the variables Na and *o*-PO_4 concentration and pH of the surface water, open water and 'rewetting vs. remnant' significantly (Monte Carlo procedure: $p < 0.05$) explained the variation in species assemblages (Table 2). The binomial variable 'rewetting vs. remnant' explained 6.0% of the variation in species data,

Table 1. Average (±SE) species richness and numbers of characteristic species in the remnant and rewetted sites

	Remnant sites	Rewetted sites	ANOVA
	n = 10	n = 10	
Species richness	42.5 ± 6.1	44.6 ± 6.3	n.s.
Characteristic species	15.2 ± 1.5	16.3 ± 2.1	n.s.

n, number of sampling sites; n.s., not significant.

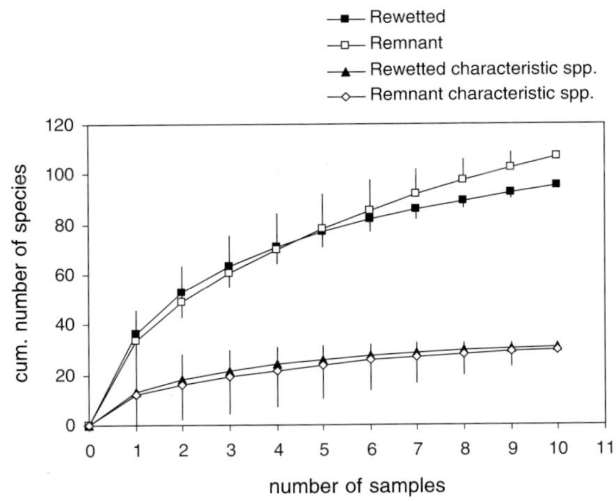

Figure 3. Cumulative species richness curves (±SD) for all species and characteristic species sampled in rewetted and remnant sites.

which is less than the variables pH, Na, *o*-PO_4 and open water, that explained 10.3, 9.6, 6.8 and 7.2%, respectively.

In the second PCA – with pooled species data of open water and littoral zone of pools – (Fig. 4b) rewetted sites scored negatively and most remnant sites positively on the second PCA-axis, whereas pools and puddles were not separated along the first four PCA-axes. The first four PCA-axes explained 18.3, 10.3, 9.8 and 8.9% of the variation in species data. After selecting all environmental variables having a correlation coefficient <−0.5 or >0.5 with one or more of the first four PCA-axes, the pH and the concentration of *o*-PO_4 in the surface water significantly (Monte Carlo procedure: $p < 0.05$) explained the variation in species assemblages (Table 3). These variables explained 12.4 and 8.2% of the variation in species data, which is again more than the (non-significant) variable 'rewetting vs. remnant' that explained 7.8%.

The average concentrations of K, Ca, Fe and Si were significantly (one-way ANOVA: $p < 0.05$) higher at remnant sites. For the other environmental variables no significant differences were found. Rewetting could not significantly explain (Monte Carlo procedure: $p = 0.144$) the variation in species data that remained after fitting the environmental variables: open water, pH, *o*-PO_4, Cl and Ca of surface water, and C:N quotient of organic matter.

Table 2. Eigenvalues (Lambda), *p*, and *F*-values of environmental variables having a correlation coefficient with one or more of the first four PCA-axes of less than −0.5 or higher than 0.5 in the PCA of puddles, open water, and littoral zone samples (Fig. 4a). *p*-values <0.05 are in italics

	Marginal effects	Conditional effects		
	Lambda 1	Lambda A	*p*	*F*
pH surface water	0.10	0.10	*0.002*	3.21
Na surface water	0.10	0.04	*0.010*	1.72
Cl surface water	0.09	0.03	0.112	1.32
EC surface water	0.08	0.03	0.475	1.01
Open water	0.07	0.07	*0.002*	2.34
o-PO4 surface water	0.07	0.07	*0.002*	2.15
P organic matter	0.07	0.04	0.174	1.25
P surface water	0.07	0.03	0.331	1.09
Rewetting	0.06	0.05	*0.010*	1.67
Al surface water	0.06	0.03	0.074	1.43
C:N organic matter	0.06	0.03	0.293	1.13
Turbidity surface water	0.04	0.04	0.104	1.32
Ca organic matter	0.04	0.03	0.287	1.12

Table 3. Eigenvalues (Lambda), *p*, and *F*-values of environmental variables having a correlation coefficient with one or more of the first four PCA-axes of less than −0.5 or higher than 0.5 in the PCA of puddles and pools, with open water and littoral zone samples pooled (Fig. 4b). *p*-values <0.05 are in italics

	Marginal effects	Conditional effects		
	Lambda 1	Lambda A	*p*	*F*
pH surface water	0.12	0.12	*0.002*	2.55
Na surface water	0.11	0.06	0.086	1.37
Cl surface water	0.10	0.04	0.607	0.85
P organic matter	0.10	0.04	0.601	0.93
EC surface water	0.10	0.03	0.679	0.78
Zn surface water	0.09	0.06	0.112	1.35
o-PO4 surface water	0.08	0.09	*0.044*	1.78
Rewetting	0.08	0.06	0.128	1.30
C:N organic matter	0.08	0.05	0.421	1.04
P surface water	0.08	0.04	0.593	0.86
C:P organic matter	0.07	0.03	0.737	0.81
Depth surface water	0.06	0.05	0.323	1.13

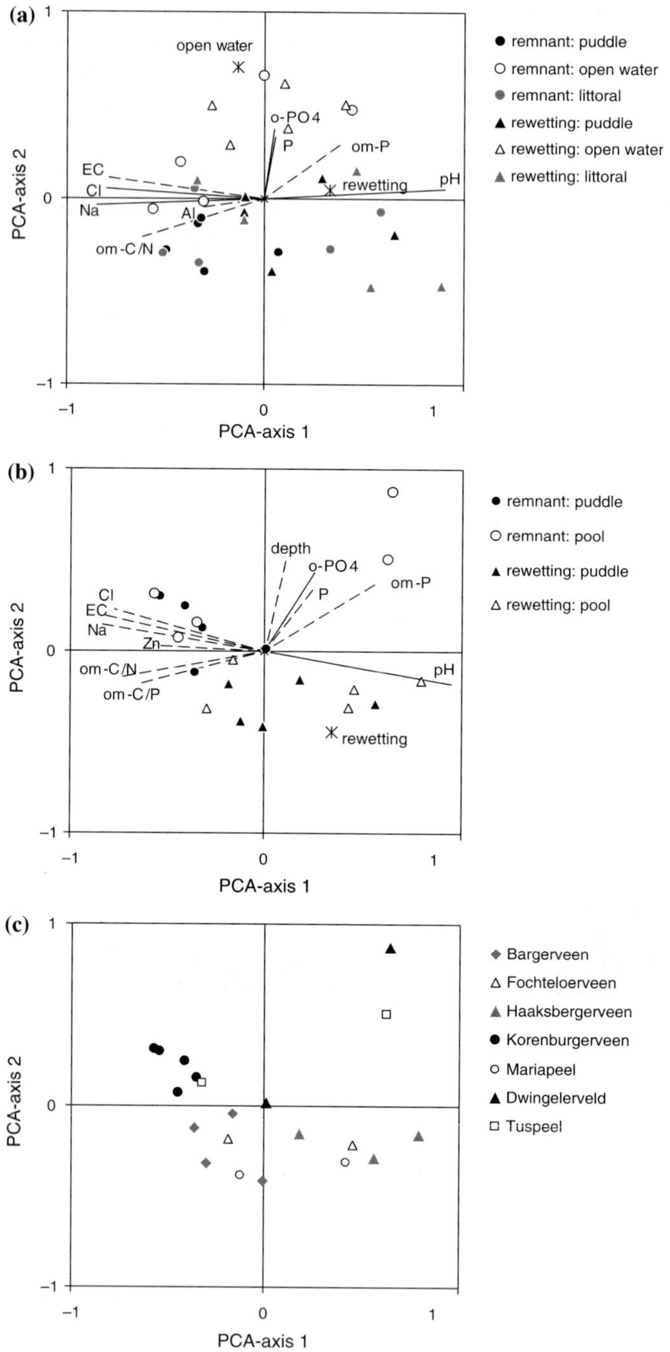

Figure 4. PCA-plots of samples based on species presence data. In figure (a) samples from open water and the littoral zone of pools are plotted separately, whereas they are pooled in (b) and (c). Samples are categorised in remnant and rewetted sites (a and b) in the area in which the sampled water bodies are situated (c). Significant explaining variables are shown with uninterrupted lines for continuous variables (score multiplied by 3) and an asterisk for centroids of binomial variables (score multiplied by 2). Interrupted lines are used for non-significant environmental variables (score multiplied by 3) having a correlation coefficient under −0.5 or above 0.5 with one or more of the first four PCA-axes. 'om', organic matter.

The samples taken in the same raised bog remnant lie relatively close to each other in the PCA-plots (Fig. 4c). This was most clear for the Korenburgerveen, Bargerveen and Haaksbergerveen reserves. The average Sørensen similarity coefficients differed strongly (13%) between pairs of sites situated in the same area and pairs of sites in different areas ($p < 0.001$; Table 4). The difference was much smaller (4%), but still significant ($p < 0.05$), between the average similarity coefficient for pairs of rewetted vs. remnant sites and for pairs within the groups of rewetted sites and remnant sites. The average similarity coefficients did not differ significantly between pairs of remnant sites and pairs of rewetted sites. Rewetting could not significantly explain (Monte Carlo procedure: $p = 0.960$) the variation in species data that remained after fitting the 7 areas in which the sampling sites were situated.

Discussion

The average and cumulative number of characteristic Rotifera and microcrustacean species and species richness did not differ between remnant sites and rewetted sites. Differences in species assemblages between remnant sites and rewetted sites could be explained by physico-chemical variables and rewetting did not significantly explain remaining variation in the species data. All but eight of the species found in more than two sampling sites were found in both remnant and rewetted sites (For comparison, this number was 21 for macroinvertebrates, with 18 of these species only found in remnant sites and 3 only in rewetted sites). Only five species were encountered considerably more frequently at either remnant or rewetted sites. Eighty-two percent of the characteristic microinvertebrate species were found in the rewetted sites and 79% in the remnant sites (These figures were 67 and 94%, respectively, for the macroinvertebrates).

Table 4. Average Sørensen similarity index (±SD) for pairs of sites situated in the same area and for pairs of sites in different areas; for pairs of rewetted sites and pairs of remnant sites; and for pairs of rewetted vs. remnant sites

	n Pairs	Sørensen-index	Mann–Whitney-*U*	
In same area	23	0.64 ± 0.10	<0.001	
In different areas	167	0.51 ± 0.07		
Remnant	45	0.56 ± 0.12	n.s.	
Rewetted	45	0.54 ± 0.08		<0.05
Rewetted vs. remnant	100	0.51 ± 0.07		

Classification of aquatic microinvertebrates, especially Rotifera, as characteristic raised bog species is more disputable than for macroinvertebrates (Van Duinen et al., 2003), as less data on the distribution of microinvertebrate species are available. This can for example be illustrated by the many new Rotifera species records for New England presented by Błędzki & Ellison (2003) and by the 36 rotifer species encountered in our studies in Dutch bog remnants – from a total number of 129 species – that were not reported earlier from the Netherlands (Zhuge & Van Duinen, 2005). Omitting species from or adding species to the list of characteristic species is not expected to considerably change the pattern described, as 89% of the species found in more than two sampling sites were found in both remnant and rewetted sites. Therefore, the conclusion that aquatic microinvertebrates species richness and number of characteristic species did not differ significantly between remnant and rewetted sites is not affected by the classification of characteristic species.

The relatively small difference in the microinvertebrate assemblage, compared to macroinvertebrates (cf. Van Duinen et al., 2003), between rewetted and remnant sites can be explained by both the less specific habitat demands and the high passive dispersal rate of microinvertebrates (Cáceres & Soluk, 2002; Cohen & Shurin, 2003). Regarding habitat demands, many Rotifera species – including many of those inhabiting raised bogs – are called ubiquitous or eurytopic, as they occur in a broad range of pH values (Bērziņš & Pejler, 1987) and are not very specific in their substrate choice (Pejler & Bērziņš, 1993b, c, 1994). In total 46 of the species found in the Dutch raised bog remnants were classified as characteristic species, but within bogs most species can be found in various ecotopes (Pejler & Bērziņš, 1993a; Jersabek, 1995). Eight of the characteristic species, as well as 14 non-characteristic species, were found in more than 50% of the water bodies sampled in the present study. However, some species prefer more wet ecotopes, whereas others can tolerate the dryer parts (cf. Bateman & Davis, 1980; Pejler & Bērziņš, 1993a).

The difference in species assemblages between open water vs. puddles and the littoral zone of pools found by us (Fig. 4a), is in accordance with results of the study by De Goeij (1987) who found the presence or absence of open water to be a main factor for differences in the species assemblages in Irish bog pools. About 80% of the Rotifera species found in our sampling sites were sessile, benthic or periphytic and thus in need of macrophytes or other substrates, whereas about 10% of the Rotifera species encountered were planktonic (data not shown).

Next to the presence or absence of open water, the Na, Cl, Zn, o-PO_4 and total-P concentration, pH, and electric conductivity of surface water and the P content and C:N and C:P quotient of organic matter were correlated to the variation in species assemblages (Fig. 4). Variation in Rotifera and microcrustacean assemblages is commonly found to be correlated to the trophic state of water bodies (Pejler & Bērziņš, 1989; Duggan et al., 2002). Epiphytic, sedimented or suspended algae, decomposing organic matter and microbes are important food sources for the groups of microinvertebrates studied (Pejler, 1983). Dissolved P and N are well known as important factors in the production of algae (Wetzel, 2001), whereas the N and P contents of organic matter and pH are important factors in its decomposition rate and the microbial activity (Kok & Van de Laar, 1991; Belyea, 1996; Smolders et al., 2002), besides the phenolic content of the organic matter (Kok et al., 1992; Kok & Van der Velde, 1994). Nutrient availability and organic matter quality could also explain the occurrence of algivorous and detritivorous aquatic oligochaete species in raised bogs (Van Duinen et al., 2006). Differences in the availability of minerals may also play a role in food quality. The higher average concentration of K, Ca, Fe and Si in remnant sites may indicate a larger influence of minerotrophic ground water in several of the remnant sites. It was beyond the scope of this study to completely unravel causal relations between species assemblages and environmental conditions. This requires more direct measurements on food quality and availability, information on nutritional requirements and food preferences of the species and probably data on food web structure. The present study showed that environmental variables related to vegetation structure, and food quality and availability could explain the dominant pattern in the variation in microinvertebrate assemblages.

Species assemblages of water bodies situated in the same area were found to be more similar to each other than to assemblages from other areas (Table 4). This was most clear for the Korenburgerveen, Bargerveen, and Haaksbergerveen reserves where three to five water bodies were sampled (Fig. 4c). Apart from the Bargerveen reserve, the sites sampled in one area were all either rewetted or not rewetted. It is unlikely that rewetting is the key factor in these differences between areas, as the Sørensen similarity index differed much stronger between areas than between rewetted vs. remnant. The differences in species assemblages between areas were correlated to several environmental variables (Fig. 4b) and might be explained by local factors affecting water quality and food availability, such as minerotrophic influence, quality of the peat substrate and atmospheric nitrogen deposition. Another explanation for the strong differences in the species assemblages between areas may be the differences in the local species pool and interspecific interactions (Shurin et al., 2000). For instance, 20 species were found in all or most of the four and five sites sampled in Bargerveen and Korenburgerveen, but the rotifers *Monommata phoxa, Pleurotrocha robusta* and *Lecane hamata* and the copepod *Acanthocyclops robustus* were found to be frequent in one of both areas and absent in the other area (data in Zhuge & Van Duinen, 2005), whereas no clear differences were found in the environmental variables between these areas (Appendix). Species already abundant in a bog remnant will be among the first to colonise adjacent, new water bodies created by peat-cutting in the past and more recently by rewetting measures. These 'early' species have a high chance to successfully establish a population. Species arriving later may generally have a lower chance to establish a population, as interactions with resident species may prevent them from increasing (Jenkins & Buikema, 1998; Shurin, 2000; Rundle et al., 2002; Cohen & Shurin, 2003).

In conclusion, the species richness, occurrence of characteristic species, and species assemblages of Rotifera, Copepoda and Cladocera did not differ between rewetted and non-rewetted sites in raised bog remnants. The variation in species assemblages could be explained by variation in physico-chemical variables and the presence of open water or vegetation in the water bodies. Contrasting to macroinvertebrate populations, that may experience a severe drawback from large-scaled rewetting measures (Van Duinen et al., 2003), populations of microinvertebrate species, including characteristic species, were not negatively affected by large-scale rewetting. Populations of microinvertebrate species were able to persist during the abrupt process of rewetting or to (re-)establish within a short period of time. At a site rewetted 5 years before sampling, 17 characteristic species were found, which is higher than the averages for remnant and rewetted sites (Table 1). Although we did not study causal relations between species occurrence and environmental variables in the raised bog remnants, the microinvertebrate assemblage is likely structured by the environmental conditions at the scale of the water body and possibly also by the local species pool and interspecific interactions. Of course, also macroinvertebrate assemblages are structured by environmental conditions. However, many macroinvertebrates complete their life cycles on larger scales in space and time. They may require different habitat elements in different life-history stages (Verberk et al., 2005). Differences in the life cycles of species, as well as differences in the dispersal capacity could explain their differential response to restoration measures in raised bog remnants and other landscape types as well.

Acknowledgements

The second author gratefully acknowledges Dr W. Koste (Germany), Dr H. Segers (Belgium) and Dr J.F.M. Geelen (The Netherlands) for the useful discussions about taxonomic difficulties and M. Orbons for her assistance in the laboratory. Mr J. Eygensteyn provided assistance during the chemical analyses and performed the C and N analysis. In addition, Vereniging Natuurmonumenten, Staatsbosbeheer and Stichting het Limburgs Landschap are acknowledged for giving us permission to enter their reserves and to take samples. Dr M.M. van Katwijk, Prof. Dr P.H. Nienhuis, Dr A.M.J. Ragas and two anonymous referees are acknowledged for their comments on an earlier version of the manuscript. This research project is part of the national research programme 'Survival Plan for Woodland and Nature,' funded by the Dutch Ministry of Agriculture, Nature and Food Quality. This is CWE-publication 423.

References

Bateman, L. E. & C. C. Davis, 1980. The rotifera of hummock-hollow formations in a poor fen in Newfoundland. Internationale Revue der gesamten Hydrobiologie 65: 127–153.

Belyea, L. R., 1996. Separating the effects of litter quality and microenvironment on decomposition rates in patterned peatland. Oikos 77: 529–539.

Bērziņš, B. & B. Pejler, 1987. Rotifer occurrence in relation to pH. Hydrobiologia 147: 107–116.

Błędzki, L. A. & A. M. Ellison, 2003. Diversity of rotifers from northeastern U.S.A. bogs with new species records for North America and New England. Hydrobiologia 497: 53–62.

Buttler, A., B. G. Warner, P. Grosvernier & Y. Matthey, 1996. Vertical patterns of testate amoebae (Protozoa: Rhizopoda) and peat-forming vegetation on cutover bogs in the Jura, Switzerland. New Phytologist 134: 371–382.

Cáceres, C. E. & D. A. Soluk, 2002. Blowing in the wind: a field test of overland dispersal and colonization by aquatic invertebrates. Oecologia 131: 402–408.

Cohen, G. M. & J. B. Shurin, 2003. Scale-dependence and mechanisms of dispersal in freshwater zooplankton. Oikos 103: 603–617.

De Goeij, A. A. M., 1987. Een onderzoek naar het voorkomen en de verspreiding van aquatische macro- en microfauna in de Ierse hoogvenen. Deel 2, Microfauna. Doctoraalverslag 221. Laboratorium voor Aquatische Oecologie, Laboratorium voor Botanie II. Katholieke Universiteit, Nijmegen (in Dutch).

De Smet, W. H., 1996. The Proalidae (Rotifera, Monogononta). Guides to the Identification of the Microinvertebrates of the Continental Waters of the World, Vol. 9. SPB Academic Publishing, The Hague.

De Smet, W. H., 1997. The Dicranophoridae (Rotifera, Monogononta). Guides to the Identification of the Microinvertebrates of the Continental Waters of the World, Vol. 12. SPB Academic Publishing, The Hague.

Donner, J., 1965. Ordnung Bdelloidea (Rotatoria, Rädertiere). Akademie Verlag, Berlin (in German).

Duggan, I. C., J. D. Green & R. J. Shiel, 2002. Distribution of rotifer assemblages in North Island, New Zealand, lakes: relationships to environmental and historical factors. Freshwater Biology 47: 195–206.

Einsle, U., 1993. Calanoida und Cyclopoida (Crustacea, Copepoda), Süßwasserfauna von Mitteleuropa. Gustav Fischer Verlag, Stuttgart (in German).

Einsle, U., 1996. Copepoda: Cyclopoida, Genera *Cyclops*, *Megacyclops*, *Acanthocyclops*. Guides to the Identification of the Microinvertebrates of the Continental Waters of the World, Vol. 10. SPB Academic Publishing, The Hague.

Fairchild, G. W., J. Cruz & A. M. Faulds, 2003. Microhabitat and landscape influences on aquatic beetle assemblages in a cluster of temporary and permanent ponds. Journal of the North American Benthological Society 22: 224–240.

Flößner, D., 2000. Die Haplopoda und Cladocera Mitteleuropas. Backhuys Publishers, Leiden (In German).

Galewski, K., 1971. A study on morphobiotic adaptations of European species of the Dytiscidae (Coleoptera). Polski Pismo entomologiczne 41: 487–702.

Grasshoff, H. & H. Johanssen, 1977. A new sensitive method for determination of ammonia in seawater. Water Research 2: 516.

Henriksen, A., 1965. An automated method for determining low-level concentrations of phosphate in fresh and saline waters. Analyst 90: 29–34.

Jackson, D., 1952. Observations on the capacity for flight of water beetles. Proceedings of the Linnean Society of London/Zoology 43: 18–42.

Jenkins, D. G. & A. L. Buikema, 1998. Do similar communities develop in similar sites? A test with zooplankton structure and function. Ecological Monographs 68: 421–443.

Jersabek, C., 1995. Distribution and ecology of rotifer communities in high-altitude alpine sites – a multivariate approach. Hydrobiologia 313/314: 75–89.

Kamphake, L. H., S. A. Hannah & J. M. Cohen, 1967. Automated analysis for nitrate by hydrazine reduction. Water Research 1: 206.

Kok, C. J. & B. J. van der Laar, 1991. Influence of pH and buffering capacity on the decomposition of *Nymphaea alba* L. detritus in laboratory experiments: a possible explanation for the inhibition of decomposition at low alkalinity. Verhandlungen der Internationalen Vereinigung für Theoretische und Angewandte Limnologie 24: 2689–2692.

Kok, C. J., C. H. J. Hof, J. P. M. Lenssen & G. van der Velde, 1992. The influence of pH on concentrations of protein and phenolics and resource quality of decomposing floating leaf material of *Nymphaea alba* L. (Nymphaeaceae) for the detritivore *Asellus aquaticus* (L.). Oecologia 91: 229–234.

Kok, C. J. & G. van der Velde, 1994. Decomposition and invertebrate colonization of aquatic and terrestrial leaf material in alkaline and acid still water. Freshwater Biology 31: 65–75.

Koste, W., 1978. Rotatoria. Die Rädertiere, Mitteleuropa. Überordnung Monogononta. Gebrüder Borntraeger Berlin, Stuttgart (in German).

Longcore, T., 2003. Terrestrial Arthropods as indicators of ecological restoration success in coastal sage scrub (California, U.S.A.). Restoration Ecology 11: 397–409.

Myers, F. J., 1931. The distribution of Rotifera on Mt. Desert Island. American Museum Novitates 494: 1–12.

Nogrady, T., R. Wallace & T. Snell, 1993. Rotifera, Biology, Ecology and Systematics. Guides to the Identification of the Microinvertebrates of the Continental Waters of the World, Vol. 4. SPB Academic Publishing, The Hague.

Nogrady, T. & R. Pourriot, 1995. The Notommatidae (Rotifera, Monogononta). Guides to the Identification of the Microinvertebrates of the Continental Waters of the World, Vol. 6. SPB Academic Publishing, The Hague.

O'Brien, J., 1962. Automatic analysis of chlorides in sewage wastes. Engineering 33: 670–672.

Pejler, B., 1983. Zoopanktonic indicators of trophy and their food. Hydrobiologia 101: 111–114.

Pejler, B. & B. Bērziņš, 1989. Rotifer occurrence and trophic degree. Hydrobiologia 182: 171–180.

Pejler, B. & B. Bērziņš, 1993a. On the ecology of mire rotifers. Limnologica 23: 295–300.

Pejler, B. & B. Bērziņš, 1993b. On relation to substrate in sessile rotifers. Hydrobiologia 259: 121–124.

Pejler, B. & B. Bērziņš, 1993c. On the ecology of Trichocercidae (Rotifera). Hydrobiologia 263: 55–59.

Pejler, B. & B. Bērziņš, 1994. On the ecology of *Lecane* (Rotifera). Hydrobiologia 273: 77–80.

Rundle, S. D., A. Foggo, V. Choiseul & D. T. Bilton, 2002. Are distribution patterns linked to dispersal mechanism? An investigation using pond invertebrate assemblages. Freshwater Biology 47: 1571–1581.

Rochefort, L., F. Quinty, S. Campeau, K. Johnson & T. Malterer, 2003. North American approach to the restoration of *Sphagnum* dominated peatlands. Wetlands Ecology and Management 11: 3–20.

Segers, H., 1995. The Lecanidae (Rotifera, Monogononta). Guides to the Identification of the Microinvertebrates of the Continental Waters of the World, Vol. 8. SPB Academic Publishing, The Hague.

Shurin, J. B., 2000. Dispersal limitation, invasion resistance, and the structure of pond zooplankton communities. Ecology 81: 3074–3086.

Shurin, J. B., J. E. Havel, M. A. Leibold & B. Pinel-Alloul, 2000. Local and regional zooplankton species richness: a scale-independent test for saturation. Ecology 81: 3062–3073.

Smolders, A. J. P., H. B. M. Tomassen, L. P. M. Lamers, B. P. Lomans & J. G. M. Roelofs, 2002. Peat bog restoration by floating raft formation: the effects of groundwater and peat quality. Journal of Applied Ecology 39: 391–401.

Smolders, A. J. P., H. B. M. Tomassen, M. van Mullekom, L. P. M. Lamers & J. G. M. Roelofs, 2003. Mechanisms involved in the re-establishment of *Sphagnum*-dominated vegetation in rewetted bog remnants. Wetlands Ecology and Management 11: 403–418.

Sørensen, T., 1948. A method of establishing groups of equal amplitude in plant society based on similarity of species content. K Danske Vidensk Selsk 5: 1–34.

Southwood, T. R. E., 1962. Migration of terrestrial arthropods in relation to habitat. Biological Reviews of the Cambridge Philosophical Society 37: 171–214.

Ter Braak, C. J. F., 1995. Ordination. In Jongman, R. H. G., C. J. F. ter Braak & O. F. R. van Tongeren (eds), Data Analysis in Community and Landscape Ecology. Cambridge University Press, Cambridge: 91–173.

Ter Braak, C. J. F. & P. Šmilauer, 1998. CANOCO Reference Manual and User's Guide to Canoco for Windows: Software for Canonical Community Ordination (version 4). Microcomputer power, Ithaca.

Van Duinen, G. A., A. M. T. Brock, J. T. Kuper, T. M. J. Peeters, M. J. A. Smits, W. C. E. P. Verberk & H. Esselink, 2002. Important keys to successful restoration of characteristic aquatic macroinvertebrate fauna of raised bogs. In Schmilewski, G. & L. Rochefort (eds), Proceedings of the International Peat Symposium – Peat in Horticulture – Quality and Environmental Challenges. International Peat Society, Finland: 292–302.

Van Duinen, G. A., A. M. T. Brock, J. T. Kuper, R. S. E. W. Leuven, T. M. J. Peeters, J. G. M. Roelofs, G. van der Velde, W. C. E. P. Verberk & H. Esselink, 2003. Do restoration measures rehabilitate fauna diversity in raised bogs? A comparative study on aquatic macroinvertebrates. Wetlands Ecology and Management 11: 447–459.

Van Duinen, G. A., T. Timm, A. J. P. Smolders & A. M. T. Brock, W. C. E. P. Verberk & H. Esselink, 2006. Differential response of aquatic oligochaete species to increased nutrient availability – a comparative study between Estonian and Dutch raised bogs. Hydrobiologia 564: 143–155.

Van Kleef, H. H., W. C. E. P. Verberk, R. S. E. W. Leuven, H. Esselink, G. van der Velde & G. A. van Duinen, 2006. Biological traits successfully predict the effects of restoration management on macroinvertebrates in shallow softwater lakes. Hydrobiologia 565: 201–216.

Vasander, H., E. -S. Tuittila, E. Lode, L. Lundin, M. Ilomets, T. Sallantaus, R. Heikkilä, M. -L. Pitkänen & J. Laine, 2003. Status and restoration of peatlands in northern Europe. Wetlands Ecology and Management 11: 51–63.

Verberk, W. C. E. P., G. A. van Duinen, T. M. J. Peeters & H. Esselink, 2001. Importance of variation in water types for water beetle fauna (Coleoptera) in Korenburgerveen, a bog remnant in the Netherlands. Proceedings of Experimental and Applied Entomology (N.E.V, Amsterdam) 12: 121–128.

Verberk, W. C. E. P., H. H. van Kleef, M. Dijkman, P. van Hoek, P. Spierenburg & H. Esselink, 2005. Seasonal changes on two different spatial scales: response of aquatic invertebrates to water body and microhabitat. Insect Science 12: 263–280.

Wagner, D. L. & J. K. Liebherr, 1992. Flightlessness in insects. TREE 7: 216–219.

Wetzel, R. G., 2001. Limnology. Lake and River Ecosystems. Academic Press, San Diego.

Zhuge, Y. & G. A. van Duinen, 2005. Rotifera, Copepoda and Cladocera in rewetted and non-rewetted raised bog remnants in The Netherlands. Report Bargerveen Foundation/Radboud University, Nijmegen.

Appendix. Selected environmental variables for the sampling sites.

	Bargerveen				Fochteloër-veen		Haaksbergerveen			Korenburgerveen					Mariapeel		Dwinge-lerveld		Tuspeel	
	1	2	3	4	1	2	1	2	3	1	2	3	4	5	1	2	1	2	1	2
Rewetting	–	1970	1975	1995	1986	1990	1970	1970	1970	–	–	–	–	–	1990	1996	–	–	–	–
Pool/puddle	Pu	Pu	Po	Po	Po	Pu	Pu	Pu	Po	Pu	Pu	Po	Po	Po	Pu	Po	Po	Pu	Po	Pu
Depth (m)	0.15	0.2	0.2	0.5	0.25	0.2	0.2	0.9	0.9	0.33	0.35	1.5	1.5	1.5	0.5	1.3	1	1	1.3	0.05
Width (m)	0.3	1	12	100	100	1	1.5	1	30	3.5	1	10	2	8	0.4	50	5	3	10	0.3
Length (m)	100	4	30	100	300	20	50	10	80	5	5	10	5	8	1.5	100	10	15	15	0.8
Surface water																				
Turbidity (ppm)	10	24	9	17	23	13	12	10	11	25	15	13	17	13	23	20	8	10	10	12
pH	3.9	4.1	3.8	3.9	4.1	4.2	4.1	4.1	4.0	3.8	3.8	3.7	3.7	3.9	4.4	4.3	4.7	4.1	4.3	4.4
EC (μS cm^{-1})	75	40	75	93	58	65	50	50	53	80	75	124	94	62	75	65	65	57	57	85
Cl (μmol l^{-1})	288	185	279	337	282	255	162	158	181	269	301	361	333	328	165	223	273	225	145	187
Na (μmol l^{-1})	278	307	312	327	302	268	161	146	166	285	324	341	327	370	224	263	263	241	207	222
o-PO_4 (μmol l^{-1})	1.5	1.7	1.2	3.9	0.9	0.8	1.1	0.5	0.6	0.9	1.2	1.5	0.9	1.1	1.0	1.1	6.5	0.5	1.4	0.4
NO_3 (μmol l^{-1})	3.2	2.5	3.0	3.0	1.8	2.4	3.2	3.4	5.2	19.5	6.0	12.6	4.7	5.4	2.4	3.0	1.1	2.6	3.6	3.5
NH_4 (μmol l^{-1})	92.4	4.9	8.7	7.6	6.9	8.2	1.2	2.1	27.2	44.4	37.4	5.1	9.6	6.9	5.4	12.6	4.4	6.1	6.7	8.2
Ca (μmol l^{-1})	26	21	29	26	30	35	26	20	28	58	47	59	46	43	17	43	33	53	30	48
Zn (μmol l^{-1})	218	427	214	206	10	32	18	20	24	265	203	202	243	267	164	215	8	16	329	190
Total-P (μmol l^{-1})	1.6	4.7	1.4	5.6	1.5	1.2	1.8	0.9	0.8	1.3	1.6	1.8	1.3	1.7	1.6	2.8	8.9	0.9	2.3	1.3
Al (μmol l^{-1})	17.5	4.9	9.7	6.9	4.1	4.9	9.3	8.5	9.3	10.9	10.3	11.1	8.2	8.5	5.7	9.1	2.1	4.4	8.1	8.5
Organic matter																				
Ca (μmol g^{-1} DW)	46	34	52	79	78	60	145	51	61	29	5	58	62	58	77	37	16	64	44	47
Total-P (μmol g^{-1} DW)	7	10	17	7	19	5	32	16	26	9	17	20	21	20	35	29	63	12	14	13
C:N quotient (g g^{-1})	48	38	24	40	23	28	18	17	17	37	26	26	21	26	19	15	14	14	21	47
C:P quotient (g g^{-1})	4514	3352	1774	4514	1591	1081	962	1063	1214	3605	1929	1986	1526	1986	903	979	473	558	1621	2269

–, No rewetting measures.

Hydrobiologia (2006) 565:201–216

R.S.E.W. Leuven, A.M.J. Ragas, A.J.M. Smits & G. van der Velde (eds), Living Rivers: Trends and Challenges in Science and Management
DOI 10.1007/s10750-005-1914-6

Biological traits successfully predict the effects of restoration management on macroinvertebrates in shallow softwater lakes

H.H. van Kleef[1,2,*], W.C.E.P. Verberk[1,3], R.S.E.W. Leuven[2], H. Esselink[1,3], G. van der Velde[3] & G.A. van Duinen[1,2]

[1]*Bargerveen Foundation, P.O. Box 9010, 6500 GL Nijmegen, The Netherlands*
[2]*Department of Environmental Science, Institute for Wetland and Water Research, Radboud University Nijmegen, P.O. Box 9010, 6500 GL Nijmegen, The Netherlands*
[3]*Department of Animal Ecology and Ecophysiology, Institute for Wetland and Water Research, Radboud University Nijmegen, P.O. Box 9010, 6500 GL Nijmegen, The Netherlands*
*(*Author for correspondence: E-mail: H.vankleef@science.ru.nl)*

Key words: acidification, bottleneck, degradation, eutrophication, life cycle, restoration ecology

Abstract

Many shallow softwater lakes are being affected by eutrophication and acidification. In these small lakes decaying organic material usually accumulates and characteristic plant and animal species disappear. In many degraded lakes organic matter and macrophytes are being removed in order to restore the lakes to their original state. To assess the effects of restoration management in softwater lakes on aquatic macroinvertebrates, changes in the species assemblages were studied in four degraded lakes in the Netherlands undergoing restoration measures. The degraded lakes still harboured species characteristic of pristine softwaters. However, most of these species were not recorded after restoration measures were taken. Species' densities declined dramatically during the execution of restoration measures. Swimming and abundant species were more likely to survive the restoration measures than other species. The first years after restoration, the lakes did not meet the habitat requirements for a number of species. Species requiring vegetation for ovipositioning, animal food sources and swards of vegetation as habitat declined. Because recolonization is expected to be restricted, it is recommended to ensure the survival of relict populations when taking measures to restore degraded softwater lakes. This may be achieved by phasing restoration measures in space and time, hereby minimizing mortality during the execution of restoration measures and by preserving habitat conditions required by characteristic species.

Introduction

In western Europe, eutrophication, acidification and desiccation have had a tremendous impact on oligotrophic ecosystems. As a result, plant and animal species characteristic of these systems have shown a strong decrease during the last century (Haines, 1981; Roelofs, 1983; Leuven et al., 1986; Ravera, 1986; Arts & Leuven, 1988; Leuven, 1988; Lamers et al., 1998). In order to prevent these species from going extinct, various restoration measures have been developed (Brouwer & Roelofs, 2001; Lamers et al., 2002; Nienhuis et al., 2002). Restoration efforts often focus on restoring the physico-chemical conditions required for the development of communities dominated by characteristic plant and animal species and usually result in the creation of early successional stages. Research on the effects of restoration mostly focuses on physico-chemical conditions and the recovery of vegetation, whereas studies on the effects on fauna are very scarce.

We studied how restoration measures in Dutch shallow softwater lakes affected fauna communities. Historically, these lakes were formed by peat cutting, branching of rivers and brooks and depressions in cover sands (e.g. coastal and ancient river dunes and glacial deposits). Nowadays, Dutch rivers and brooks are strongly regulated, cover sands planted with trees and the cutting of peat has stopped. As a result no softwater lakes are formed any more (Brouwer, 2001), emphasizing the need to preserve the existing lakes as these ecosystems harbour a large number of characteristic species. Pristine softwater lakes have a low alkalinity and are nutrient limited, with respect to inorganic carbon, nitrogen and phosphorus. These characteristics make the lakes very sensitive to both acidification and eutrophication. Acidification and eutrophication lift the nutrient limitation of the lakes and increase primary production, because fast-growing plant species are able to become dominant (Bloemendaal & Roelofs, 1988). Increased primary production leads to accumulation of organic matter, increased decomposition and anaerobic conditions. Characteristic plant species decline or disappear, because they are out competed for light and unable to germinate under anaerobic conditions (Bloemendaal & Roelofs, 1988). Characteristic fauna species may disappear as a result of oxygen deficiencies and heavy metal toxicity (Henrikson et al., 1982).

Restoration measures in deteriorated softwater lakes aim at restoring nutrient limitation, low alkalinity and aerobic sediments. Generally, these measures are carried out by removal of accumulated organic matter and macrophytes. In order to facilitate restoration activities the water table in these lakes is often lowered until only small pools of water remain. Restoration measures are devised in a way that the recovery of characteristic plant species is ensured, i.e. these species have formed persistent seed banks that are preserved during removal of organic matter. The extent to which recovery of characteristic fauna communities will occur, depends on the ability of species to persist during the draining of the lake and removal of the organic matter and macrophytes, during the recovery of the required habitat characteristics and on the species' ability to colonize the restored lake.

Due to habitat destruction and deterioration as well as the limited dispersal capacity of some species, the importance of colonization is expected to be limited (Van Kleef & Esselink, 2004a). Therefore, this study focuses on the persistence of species during and after the implementation of restoration measures. In four degraded softwaters changes in macroinvertebrate assemblages resulting from restoration efforts are studied. Observed changes in macroinvertebrate community composition are analysed using biological traits of species in order to (1) understand the observed effects of the restoration measures, (2) predict the effects in future restoration projects to anticipate unwanted side effects and (3) provide information on the effects of restoration in similar ecosystems and of similar measures. The persistence of species is studied in two different time periods: (1) when restoration measures are being taken (execution phase) and (2) during the recovery of the ecosystem after measures are taken (recovery phase). During these phases different factors will determine species' performance and species will require a set of different attributes to optimally perform during each phase.

Southwood (1977) hypothesized that the habitat acts as a templet onto which evolution has forged biological traits (the habitat templet concept). Species are adapted to live under certain conditions and can only survive in habitats that meet their requirements. Many studies have tested the habitat templet concept in river systems (Scarsbrook & Townsend, 1993; Juget & Lafont, 1994; Resh et al., 1994; Richoux, 1994; Tachet et al., 1994; Usseglio-Polatera, 1994; and others) and used biological traits to understand the distribution of species throughout the catchment. In the present study changes in the species assemblages resulting from restoration measures are related to the biological traits of the species involved. This approach provides us with a tool to predict bottlenecks for species in future restoration projects and to avoid local extinction of characteristic species. Furthermore, it can be used for assessing the performance of species that were not encountered in this study.

We hypothesize that only a limited number of species will be able to survive restoration measures. It is assumed that during the execution phase mobile species will be better able to reach the relative safety of the remaining small water bodies, when the waters are drained. Furthermore, species occurring in large numbers are more likely to persist during this phase. Part of the species that have survived the execution phase will be unable to complete their life cycles during the recovery phase, because the environment does not meet their habitat requirements, yet.

This paper addresses the following questions:

1. Do deteriorated softwater lakes still harbour species characteristic of pristine softwaters?
2. Are characteristic aquatic macroinvertebrates able to survive the process of restoration?
3. Does the abundance of species prior to restoration determine the chance of survival during the execution phase?
4. Do the changes in species assemblages during the execution and recovery phases correlate with particular biological traits?

Materials and methods

Study sites and restoration measures

Aquatic macroinvertebrates were sampled in four degraded shallow softwater lakes in the Netherlands i.e. lakes IJsbaan, Verwolde, Grenspoel and Wijnjeterperschar (Fig. 1). The surface area of these lakes was 0.67, 0.12, 3.74 and 0.67 ha and the maximum depth 0.5, 0.5, 1.0 and 0.5 m, respectively. Detailed data on water chemistry, vegetation and restoration measures of the lakes are described in Van Kleef & Esselink (2004b). Restoration measures in lakes IJsbaan, Verwolde and Grenspoel consisted of removal of accumulated organic matter and macrophytes. For these restoration activities, the water table in the lakes was lowered without completely draining the lakes, resulting in shallow pools of approximately 10 cm depth. In lake Wijnjeterperschar restoration measures were carried out without lowering the water table and were phased in space and time: half of the lake was cleared of detritus and vegetation in July 2001 and the other part in spring 2002.

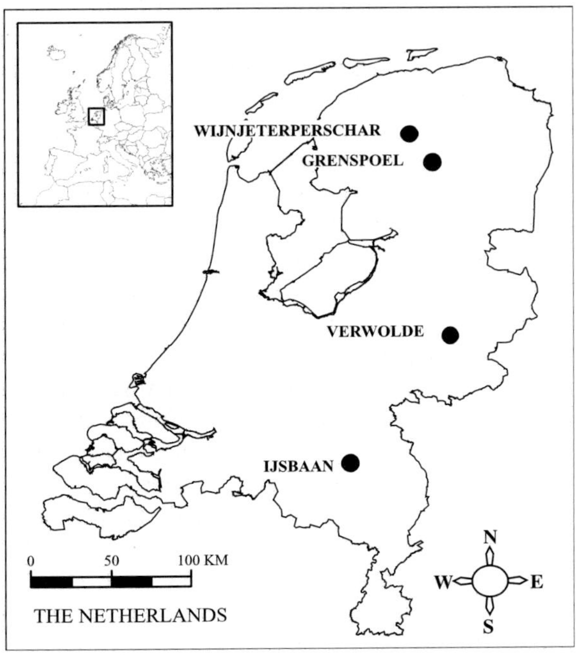

Figure 1. Locations of the study sites in the Netherlands.

Sampling

For determining changes in species assemblages resulting directly from restoration (during the execution phase), lake IJsbaan was sampled 1 week before and 1 week after restoration measures were taken. Data on the execution phase for the other lakes are unavailable. In order to determine changes in species assemblages resulting from restoration and changes in habitat (execution and recovery phases combined), each site was sampled twice before and twice after restoration measures were taken, except lake Verwolde which was only sampled once before and once after measures were taken.

Macroinvertebrates were sampled using a 20×30 cm pond net with ½ mm mesh size. Different habitat elements, such as reedbeds, submerged sphagnum vegetation and mud soils were sampled proportionally to their occurrence in the study site. Total sampling effort was kept constant at 1.5 m^2. This ensures that the list of taxa and their numbers reflects the community composition and abundance of the dominant species in the lakes (Van der Hammen et al., 1984).

Samples were transported to the laboratory and stored at 4 °C to minimize mortality. Samples were washed over three sieves with 2, 1, and ½ mm mesh size, respectively and sorted in white trays within 2 days after sample collection and preserved until identification. All macroinvertebrates were identified to the lowest taxonomical level possible (mostly species level).

Biological traits and coding of information

Biological traits were selected if they were expected to be of importance in determining the performance of species in the restored softwaters. The type of locomotion was expected to determine the ability of a species to reach the remaining patches of open water and to evade the machinery, hereby influencing the survival chances during the execution phase. Species' performance during the recovery phase was expected to be determined by a number of biological traits (Table 1). Specific demands on the environment for completing life cycles (carnivory, ovipositioning in aquatic or terrestrial vegetation, inhabiting vegetation and parasitism) were expected to decrease the abundance of species. Delicate respiratory organs (gills and lungs) that may have been damaged during restoration were expected to decrease species' performance, whereas species breeding air (terrestrial respiration) were expected to perform better than those with aquatic respiration. Success of species during the recovery phase was also expected to be influenced by their capacity for colonization (flight capability) and rapid reproduction (number of generations per year and reproduction).

In total eight different biological traits were selected of which three (locomotion, diet and respiration) were defined for juvenile as well as adult stages. Each trait had several classes to which species could be assigned (Table 1). Biological traits of different taxa were obtained from a large number of scientific papers, handbooks, taxonomic literature and information provided by experts (Dr. H.K.M. Moller Pillot and Dr. L.W.G. Higler pers. comm.). For every trait class a score of '0' or '1' was assigned to each taxon (Van Kleef & Esselink, 2005). A '0' indicating low or no affinity and a '1' indicating a high affinity with the trait class. Taxa with an equal affinity for multiple classes within one biological trait were coded a '1' for each of these classes.

Data analyses

Because species occurring in low numbers are easily missed in field studies (Foggo et al., 2003), it is nearly impossible to determine whether a species has newly colonised or disappeared from a study site. Therefore, instead of presence-absence data, the performance of species at individual sites was used to analyse changes in species assemblages during and after the execution of restoration measures.

Execution phase

Changes in species assemblages resulting from restoration were analysed by comparing samples from lake IJsbaan 1 week before and 1 week after restoration measures were carried out. The relation between locomotion type and survival of macroinvertebrates was tested with a student *t*-test (unequal variance, two-tailed). The Spearman correlation coefficient was calculated between different classes of the density of species before and the chance of recovering of these species after measures were taken.

Recovery phase

For the analysis of species' performance during the recovery phase three cross tables were compiled (Fig. 2). Cross table A consists of the species and their biological traits. Species occurring in multiple lakes were included multiple times, because on different locations they were affected by restoration measures in different ways. All taxa identified to species level were included, except oligochaetes, which were omitted because of incomplete knowledge on their biological traits. Cross table A consists of 250 species-site combinations and 54 traits. Next, for each of the lakes, all data collected before measures were carried out were pooled, as well as all data collected after restoration measures were taken. Subsequently, for every site-species combination the change in species density was calculated and assigned to five different categories of density change: (1) not recovered: species found only before restoration measures were taken; (2) decreased: species found before and after restoration measures are taken, exhibiting a decrease in density of at least 25%; (3) relatively stable: species found before and after restoration measures are taken, exhibiting a change in density less than 25%; (4) increased: species found before and after restoration measures are taken, exhibiting an increase in density of

Table 1. Biological traits associated with changes in species density

Biological trait	Classes	Code	Not recovered	Decreased	Relatively stable	Increased	New	Prediction	
Flight	Flight	A1	X					>	–
	Flightless	A2					X	<	–
Preferred life history	Semivoltine	B1		X				<	+
	Univoltine	B2		X				<	+
	Multivoltine	B3				X		>	+
Diet juveniles	Detritus	C1				X		>	+
	Plants/algae	C2				X		>	+
	Animal	C3		X				<	+
Diet adults	Detritus	D1					X	>	+
	Plants	D2				X		>	+
	Animal	D3	X					<	+
	Irrelevant	D4					X	>	+
Reproduction	Sexually	E1	X					<	+
	Asexually	E2					X	>	+
Parental care	Eggs in/on vegetation	F1		X				<	+
	Eggs on solid substrate	F2					X	>	+
	Eggs fixed terrestrial	F3	X					<	+
	Eggs on adult	F4			X			>	–
	Eggs on surface	F5				X		>	+
	Ovovivipary	F6					X	>	+
Respiration juveniles	Lung	G1					X	<	–
	Gills	G2			X			<	–
	Tegument	G3				X		>	+
	Spiracles	G4	X					>	–
	Ventilation	G5		X				>	–
	Respiration water	G6					X	<	–
	Respiration air	G7				X		>	+
Respiration adults	Lung	H1					X	<	–
	Gills	H2			X			<	–
	Tegument	H3					X	>	+
	Spiracles	H4	X					>	–
	Respiration water	H5					X	<	–
	Respiration air	H6	X					>	–
Habitat	Indifferent	I1					X	>	+
	Anaerobic sediment	I2				X		>	+
	Vegetation	I3	X					<	+
	Host	I4					X	<	–
	Aerobic organic matter	I5					X	>	+
	Bare sand	I6				X		>	+

<: decrease predicted; >: increase predicted; +: predictions confirmed; –: predictions not confirmed.

at least 25% and (5) new: species found only after restoration measures are taken. Cross table B consists of 250 species-site combinations and 5 categories of density change.

Cross table C contains data for each species-site combination on the variables determining survival during the execution phase: locomotion of juvenile and adult stages and the density of the

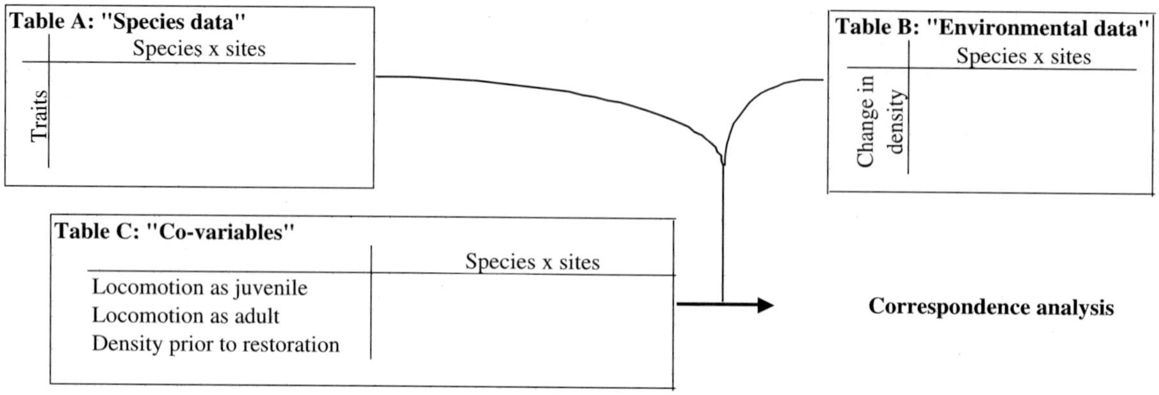

Figure 2. Schematic representation of the analysis of changes in species assemblages taking place during the recovery phase.

species prior to the execution of restoration measures (Figs. 3 and 4). The weight of locomotion type was derived from the effect analysis during the execution phase (Fig. 3). Locomotion types related to a low survival of individuals (sprawler, burrower, clinger and crawler) were given a low value (1) and locomotion types related to an higher survival (swimmer and skater) a high value (3). The locomotion type "Skater" was assumed to affect survival of individuals in a similar way as the type "Swimmer". Furthermore, species with a terrestrial adult stage during the execution of restoration measures were assumed to be the least influenced and were given the highest value (4). For each species the locomotion values for juveniles and adults were calculated as the average weight of the locomotion types. Densities of species in samples taken before measures were carried out were pooled per site and log-transformed ($10\times\log$ (average density prior to restoration $+1$)) in order to normalize the data. Cross table C consists of 250 species-site combinations, their log-transformed starting density and their locomotion type as adult as well as juvenile.

The regular use of ordination techniques is to analyse variation in species composition in a set of samples and identify correlations with environmental variables. In this study ordination tech-

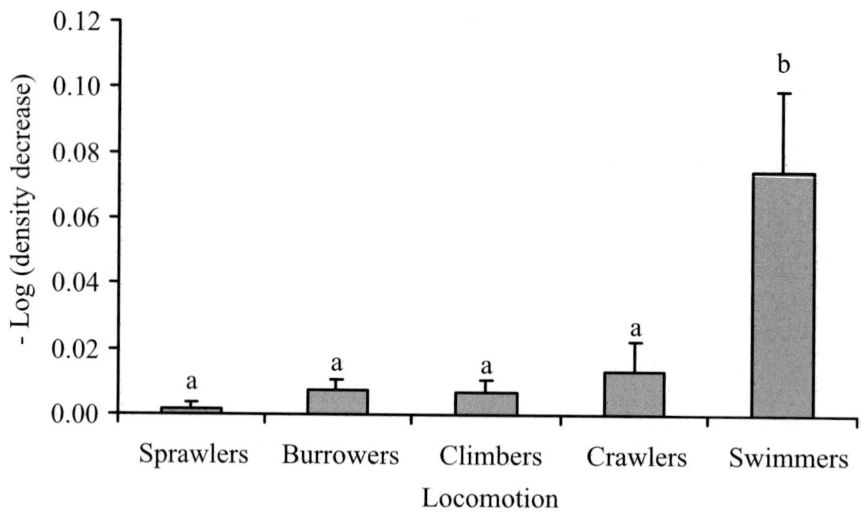

Figure 3. Chance of survival ($\pm$SE) in relation to locomotion of macroinvertebrates in lake IJsbaan. Chance of survival is calculated as the -log ((density 1 week before restoration – density 1 week after restoration)/density before restoration). Different letters (a and b) indicate significant differences (Student *t*-test, $p < 0.05$) in the response of locomotion types on restoration measures. Total macroinvertebrate density decreased from 12,436 to 163 individuals m^{-2}.

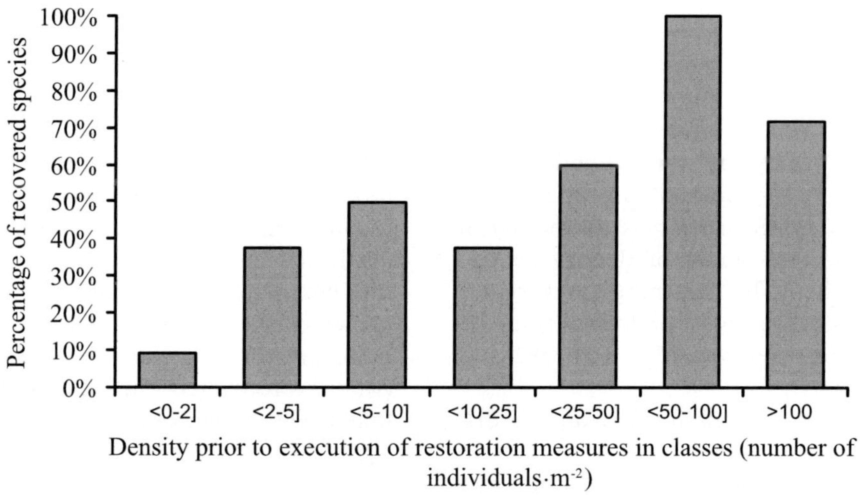

Figure 4. Relation between the number of recovered species in lake IJsbaan and their density prior to restoration. Number of recovered species is expressed in different densities classes (individuals·m^{-2}) and is calculated as the percentage of species recorded before restoration that was also recorded after restoration.

niques are used for a different purpose, i.e. to identify correlations between biological traits and the performance of species. Instead of the normal "species data", listing the distribution of species in a set of samples, cross table A was used, listing the distribution of traits in a set of species. Cross table B containing data on the performance of the species was used instead of the normal "environmental data", describing the environmental conditions at the study sites. This part of the analysis requires data on species' performance during the recovery phase. However, the observed changes in species density (cross table B) as measured before and after restoration measures are the combined effect of the execution and the recovery phase. Therefore it is necessary to include a step that eliminates the variation in the data resulting from the execution phase and allows the analysis of the species' performance during the recovery phase separately. For this reason variables explaining survival during the execution phase (cross table C) were included as co-variables in the analysis.

In order to identify correlations between the variation in traits and the performance of species an indirect analysis (principal components analysis PCA or correspondence analysis CA) was chosen, because it does not constrain the variation data to the measured species' performance as a direct analysis (redundancy analysis RDA or canonical correspondence analysis CCA) would. In addition, a direct analysis was performed, because it allows these correlations to be tested for their significance (Monte Carlo resampling procedure (500 permutations)).

Because of the large number of species-site combinations and biological traits, the variation in the dataset was expected to be large suggesting a unimodal response curve. This was confirmed in a preliminary detrended correspondence analysis performed in Canoco for Windows version 4.0 (Ter Braak & Smilauer, 1998) which revealed a length of gradient larger than 2.5, indicating that a CA was suitable for data analysis. Therefore, co-occurring biological traits were distinguished and correlated with species' performance using a CA and their significance was tested in a CCA.

In the correspondence analysis the variation in the data is described by Canoco on four ordination axes, with the first axis describing most variation, followed by the second, third and fourth. In order to determine the affinity of the biological traits with the species' performance, classes of biological traits were assigned to the nearest category of density change in the four-dimensional ordination space. Distances between ordination scores of classes of biological traits and scores of categories of density changes were calculated, where the scores were weighted for the amount of variance explained by the four CA-axes calculated by Canoco (Equation 1) with: D_{X-Y} = Distance

$$D_{X-Y}=\sqrt{(V_1*(T_{1X}-D_{1Y}))^2+(V_2*(T_{2X}-D_{2Y}))^2+(V_3*(T_{3X}-D_{3Y}))^2+(V_4*(T_{4X}-D_{4Y}))^2} \quad (1)$$

between class of biological trait X and density change category Y, V_i=Amount of variance explained by axis i, T_{iX}=Score of class of biological trait X on axis i, D_{iY}=Score of density change category Y on axis i. The percentage of variance explained in the distribution of biological traits by the four ordination axes was 12.9, 11.4, 9.2 and 8.2%, respectively.

A cluster analysis was performed on the distribution of biological traits over the species in order to visualise the co-occurrence of traits and facilitate the interpretation of the previous analysis. The cluster analysis was performed using Twinspan (Hill, 1979).

Execution and recovery phases combined

A list of macroinvertebrate species characteristic of shallow softwater lakes in the Netherlands was obtained from Arts (2000). Arts (2000) describes a characteristic species as a species which occurs more often in shallow softwater lakes than in other water types. A Wilcoxon signed ranks test was used to test if the number of observed characteristic species changed significantly. Relative differences in species' performance between characteristic and non-characteristic species were tested with a χ^2-test.

Results

One hundred and sixty nine species were identified in this study (Oligochaeta: 6, Hirudinea: 6, Mollusca: 9, Crustacea: 1, Araneae: 1, Hydracarina: 20, Chironomidae: 24, Chaoboridae: 4, Dixidae: 2, Cylindrotomidae: 1, Ephemeroptera: 4, Trichoptera: 8, Neuroptera: 1, Megaloptera: 1, Heteroptera: 25, Coleoptera: 40 and Odonata: 16 species) of which fourteen were characteristic (Table 3).

Execution phase

During the execution phase, abundance of aquatic macroinvertebrates decreased with 99.7% in Lake IJsbaan. Species in lake IJsbaan were not equally affected by the restoration activities. Species that were able to swim had a significantly higher chance to survive compared to species with other types of locomotion (Fig. 3), indicating that the ability to evade the restoration machinery is important for the survival of macroinvertebrates during this phase. Of the 50 species that were recorded 1 week before restoration measures were taken, only 20 could be recovered one week after measures were taken. The chance of recovering a species showed a significant positive correlation (Spearman correlation, $p=0.013$) with the density of the species prior to the execution of restoration measures (Fig. 4), indicating that abundant species have a higher survival chance than scarce species.

Recovery phase

The categories of changes in species densities "Not recovered" and "New" as well as "Decreased" and "Increased" were plotted opposite of each other and both groups (not recovered/new and decreased/increased) were plotted perpendicular to each other on the first and second CA-axis (Fig. 5). The categories of species densities changes "Not recovered" and "Decreased" correlated significantly (Monte Carlo resampling procedure, $p=0.016$ and 0.048, respectively) with the variation in biological traits. After omission of these two variables the category "New" also correlated significantly with the first CCA-axis (Monte Carlo resampling procedure, $p=0.012$).

Species that were not recovered or decreased in abundance after restoration measures were taken, are mainly semi- or univoltine and reproduce sexually, preventing them from a rapid build up of their numbers (Table 1). They require special habitat features for ovipositioning (aquatic or terrestrial vegetation) and nutrients (animal food sources). They also prefer to inhabit swards of vegetation. Other biological traits associated with a negative reaction of species to restoration measures are the ability to fly and respiration by spiracles and ventilation. Biological traits dominating in species that increased in abundance or newly appeared in the

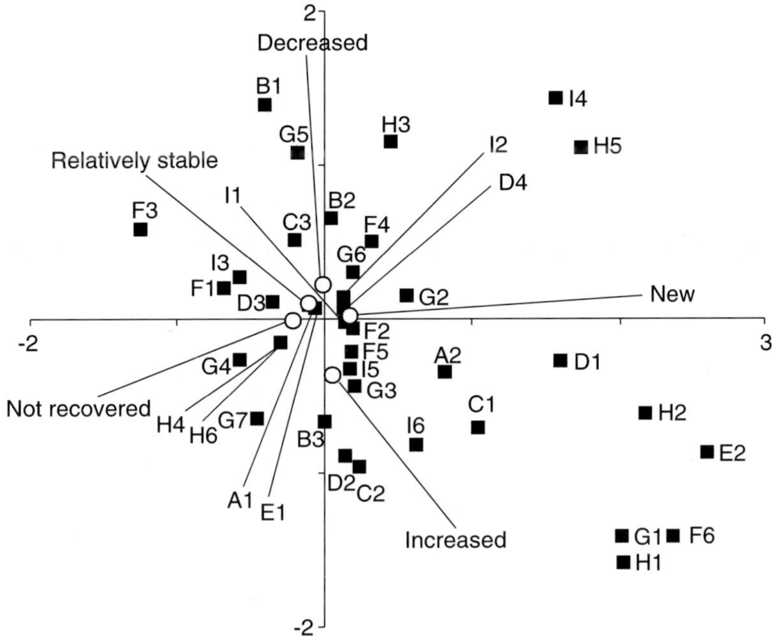

Figure 5. Plot of the first two axes of the correspondence analysis of biological traits. Squares and codes represent classes of biological traits (Table 2): flight capability (A), preferred life history (B), diet of juveniles (C), diet of adults (D), reproduction (E), parental care (F), respiration of juveniles (G), respiration of adults (H) and preferred habitat (I). Changes in the density of species are plotted as open circles: species found only before restoration ("Not recovered"), species with a density decrease of more than 25% ("Decreased"), species with a density change of less than 25% ("Relatively stable"), species with a density increase of more than 25% ("Increased") and species found only after restoration ("New").

treated lakes enable them to rapidly increase their numbers (multiple generations per year, asexual reproduction). Classes of parental care that make a species independent on special habitat features for ovipositioning (ovovivipary, depositioning of eggs on water surface and solid substrates) are associated with species increasing in abundance. Species profiting from restoration measures inhabit organic sediments, bare sand, host species or have no particular habitat preference. Furthermore, they utilize readily available food such as detritus, algae and terrestrial food sources. Other biological traits associated with a positive reaction of species to restoration measures are the absence of the ability to fly and respiration by lungs and tegument.

Nearly all predictions for changes in species densities for the biological traits diet, habitat, parental care, life history and reproduction were confirmed (Table 1). Mainly species with different classes for flight capacity and respiration did not react to restoration measures as predicted. The reason that not all predictions were confirmed is because classes of biological traits were not randomly distributed across species. Species with similar biological traits were grouped together in clusters (Table 2). Flight ability and respiration through spiracles, both expected to induce an increase in abundance, co-occurred with sexual reproduction, which was expected to decrease. The decrease in abundance of species capable of flight may also be induced by carnivory and ovipositioning in vegetation, which both occurred solely in flight capable species. Whereas species, unable to fly and breathing by gills or lungs, often featured traits that were expected to induce an increase in abundance (detritivory and asexually reproduction).

Execution and recovery phases combined

Species characteristic of softwater lakes were found in all study sites prior to restoration (Table 3). After restoration measures were taken the number of characteristic species did not increase significantly (Wilcoxon signed ranks test,

Table 2. Co-occurrence of biological traits

Biological trait	Classes	Code	Cluster							
			I	II	III	IV	V	VI	VII	VIII
Preferred life history	Semivoltine	B1	0.4							
Diet adults	Irrelevant (adults terrestrial)	D4	1.0	1.0						
Respiration juveniles	Ventilation	G5	0.5	0.1						
Preferred habitat	Aerobic organic Matter	I5	0.4	0.3			0.03			
Parental care	Eggs on surface	F5	0.3	0.8		0.04				
Preferred habitat	Anaerobic sediment	I2		0.1						
Preferred habitat	Bare sand	I6		0.1						
Respiration juveniles	Tegument	G3		0.6	0.8			0.2		
Diet juveniles	Detritus	C1		0.6					1.0	0.7
Respiration juveniles	Gills	G2	0.6	0.2	0.2		0.1			1.0
Preferred life history	Multivoltine	B3		0.8	1.0	1.0	0.03			0.4
Diet juveniles	Plants	C2	0.1	0.6	1.0	0.4	0.1		0.7	0.7
Preferred habitat	Vegetation	I3	0.7	0.2			0.3			
Flight capability	Flight	A1	1.0	1.0	1.0	1.0	0.9	0.8		
Reproduction	Sexually	E1	1.0	1.0	1.0	1.0	1.0	1.0	0.7	0.1
Respiration adults	Spiracles	H4	1.0	1.0	1.0	1.0	1.0			
Respiration adults	Respiration air	H6	1.0	1.0	1.0	1.0	1.0			
Parental care	Ovoviviparous	F6		0.03					0.3	
Diet adults	Plants	D2			1.0	0.4	0.1	0.1	0.7	0.7
Respiration juveniles	Respiration air	G7		0.1		0.6	0.3			
Parental care	Eggs in/on vegetation	F1	0.5		0.8	0.1	0.3			
Parental care	Eggs on solid substrate	F2	0.3	0.8	0.2	0.9	0.6	0.9	0.7	0.9
Respiration juveniles	Respiration aquatic	G6	1.0	0.9	1.0	0.4	0.7	1.0	1.0	1.0
Diet juveniles	Animals	C3	1.0	0.4		0.6	1.0	1.0		
Preferred habitat	Indifferent	I1	0.2	0.4	1.0	1.0	0.7	1.0	1.0	1.0
Preferred life history	Univoltine	B2	0.7	0.2			1.0	1.0	1.0	0.6
Diet adults	Animals	D3				1.0	0.9	0.8		
Respiration juveniles	Spiracles	G4		0.1		1.0	0.9			
Parental care	Eggs fixed terrestrial	F3					0.1			
Parental care	Eggs on adult	F4					0.03	0.1		0.1
Respiration adults	Tegument	H3						0.2		
Preferred habitat	Host	I4						0.8		
Respiration adults	Respiration aquatic	H5						1.0	1.0	1.0
Flight capability	Flightless	A2				0.04	0.1	0.2	1.0	1.0
Diet adults	Detritus	D1				0.04	0.1	0.1	1.0	0.7
Reproduction	Asexually	E2							0.3	0.9
Respiration juveniles	Lung	G1							1.0	
Respiration adults	Lung	H1							1.0	
Respiration adults	Gills	H2								1.0
	Number of species		20	38	6	25	34	25	3	7

The clusters are groups of species exhibiting a high similarity in biological traits. Clusters and traits are ordered in a way that clusters with a high similarity and frequently co-occurring traits are close together in the table. Numbers indicate the relative number of species within the cluster that features the biological trait.

Table 3. Changes in presence of aquatic invertebrates characteristic for softwater lakes

	IJsbaan	Verwolde	Grenspoel	Wijnjeterperschar
Heteroptera				
Arctocorisa germari (Fieber)			–	+
Cymatia bonsdorffi (Sahlberg)	+	=	+	+
Glaenocorisa propinqua (Fieber)			–	+
Notonecta obliqua Thunberg		+	+	
Notonecta viridis Delcourt	+		+	
Sigara scotti (Douglas & Scott)	+		=	+
Coleoptera				
Agabus labiatus (Brahm)			+	
Helochares punctatus Sharp	–		=	–
Ranthus suturellus (Harris)			+	
Odonata				
Coenagrion lunulatum (Charpentier)				+
Trichoptera				
Holocentropus dubius (Rambur)				–
Chironomidae				
Polypedilum uncinatum (Goetghebuer)	+	–	–	–
Telmatopelopia nemorum (Goetghebuer)				–
Zalutschia humphresiae Dowling & Murray			+	

–, = and +: species was recorded only before, before and after or only after restoration measures were carried out, respectively.

$p = 0.102$). Many characteristic species, recorded after restoration measures were carried out (16 out of 19), were not encountered before restoration, indicating an ability to rapidly increase in abundance either by reproduction or by colonisation. However, characteristic species present before restoration appear to be negatively affected by restoration measures as most of them (9 out of 12) were not recorded after measures were taken.

Relative numbers of species recorded only before, before and after or only after restoration measures were carried out differed for characteristic and non-characteristic species (Fig. 6). Species composition of non-characteristic species

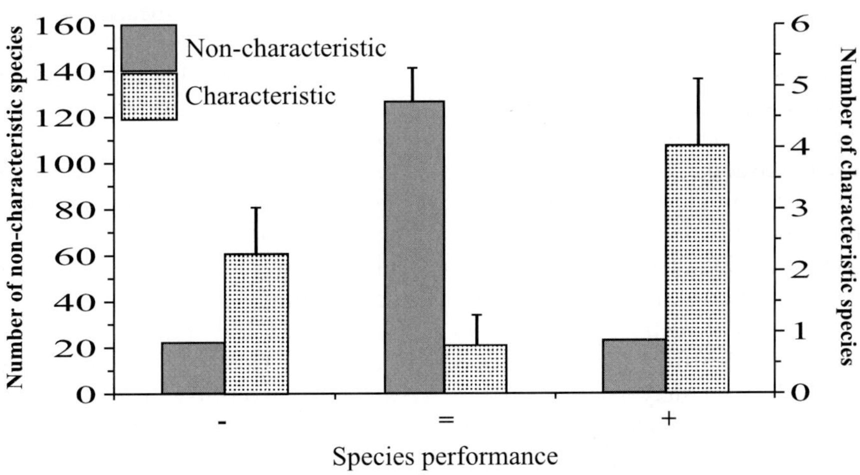

Figure 6. Number of characteristic and non-characteristic species (± SE) that were recorded only before (–), before and after (=) or only after restoration measures were carried out (+).

was more stable than that of characteristic species. I.e. characteristic species, recorded before restoration measures were taken, had a significantly higher chance of not being recorded after restoration, compared to non-characteristic species (χ^2-test, $p<0.005$). The relative number of characteristic species observed only after restoration, on the other hand, was higher than that of non-characteristic species (χ^2-test, $p<0.01$).

Discussion

Biological traits were used to analyse changes in species abundance and community composition. Results suggest that this approach is suitable for understanding changes in species assemblages during different phases of softwater lake restoration (Figs. 3 and 5 and Table 1). Even though the studied lakes differed in morphology, vegetation composition, water quality and in the way the restoration measures were carried out (Van Kleef & Esselink, 2004b), the analysis still produced significant results supporting our hypothesis, indicating that this method is very robust. The same approach has also been applied in order to explain and predict the distribution of aquatic macroinvertebrates in the Rhone basin (Juget & Lafont, 1994; Richoux, 1994; Tachet et al., 1994; Usseglio-Polatera, 1994). These studies tested the habitat templet concept, that postulates that species are adapted through biological traits to the spatial and temporal variability in the habitat. Their results, however, only partly confirmed their predictions. Understanding differences in invertebrate community structure between different habitats requires detailed knowledge of the habitats and the adaptations of species to physical, chemical and biological conditions. For example, distasteful secretions from skin glands in water mites are an adaptation against predation (Kerfoot, 1982). However, the functionality of this adaptation depends on the predators present as it only protects against fish but not against water bugs, that feed on body fluids. Furthermore, a restriction imposed by the habitat may be tackled by different adaptations (e.g. some dragonflies avoid predation by burrowing in the sediment, while water mites excrete toxins) or prevent a positive response of a trait that otherwise would have been successful in the habitat. The latter possibly occurred in this study, where flight ability and respiration did not react as predicted. They were probably overruled by traits that induced a different response in species.

The analysis of biological traits in this study yielded satisfactory results for two reasons. (1) Species assemblages before and after restoration measures were compared within water bodies instead of comparing differences between study sites. Hereby, the number of factors involved in structuring the communities (e.g. local species pool, site area, depth) is strongly reduced. This also allowed for the analysis of different effects caused by the same disturbance, i.e. it was possible to separate processes in the execution as well as in the recovery phase. (2) The habitat changes resulting from large scale restoration measures are so pronounced and influential in structuring the species assemblage that the level of detail required from the ecological knowledge is limited. I.e. in this study data on morphology and life history were sufficient for interpretation and no data on physiology, behaviour and species interactions were needed. Understanding of the changes in species assemblages resulting from restoration measures would have been more complicated when the measures had induced a more gradual change in habitat or when the lakes would have had more time to recover.

This study shows that the macroinvertebrate assemblage is selectively impoverished by the way restoration measures are being carried out nowadays. Restoration measures erect multiple filters (Tonn, 1990; Poff, 1997) through which species have to pass in order to maintain their population. These filters can be separated in two categories, each posing restrictions on a different set of biological traits. The first set of filters is in operation during the execution phase, at the moment that restoration measures are taken. It is applied when the water table is lowered and animals are removed along with the accumulated organic matter and macrophytes. As a result, the abundance of all species seriously decreases. However, species that are able to swim are less susceptible to restoration measures than species with a low mobility in the water layer. The lower mortality of swimming species is probably the result of their ability to reach the relative safety of the small pools of waters that remain after lowering of the water table. High mortality may result in the complete disappearance of species. Especially species occurring in low densities have a high chance of being removed along with the organic matter entirely. These species often include relict populations of characteristic species, the very same species that the restoration efforts are meant to preserve.

Although the effects of a strongly reduced population size could not be determined in this study, it theoretically may decrease the persistence of the species in a number of ways. Isolated populations consisting of a small number of individuals have an increased vulnerability to inbreeding depression (Charlesworth & Charlesworth, 1987), demographic factors (Menges, 1991), such as mortality by chance (Reed et al., 2003), dispersal (Rohani et al., 1996) and the Allee-effect (Boukal & Berec, 2002; Engen et al., 2003).

The second set of filters applies to species that have survived the execution phase and have to complete their life cycles under the newly created conditions (recovery phase). We hypothesized that species will experience problems completing their life cycles when they depend on specific conditions to be present in the lake. Our predictions were confirmed in 25 out of 39 cases. Species that require specific habitat features for ovipositioning (aquatic or terrestrial vegetation), nutrients (animal food sources) and habitats (swards of vegetation) decline or disappear after restoration measures have been taken. Classes of biological traits are not randomly distributed (Table 2). I.e. biological traits that were expected to influence the performance of species in opposite ways coincided and clustered together in the ordination analysis. This may have been the reason why species with different classes for the biological traits respiration and flight capability did not react to restoration measures as predicted. Negative responses induced by delicate respiratory systems (e.g. lungs and gills in water snails) were compensated through asexual reproduction and detritivory. Positive responses resulting from flight ability were negated by food (carnivory) and parental care bottlenecks (ovipositioning in vegetation).

Due to the restrictions put on the performance of specific biological traits during the process of restoration, the abundance of many species declined and most characteristic species were not recorded again after restoration. It cannot be excluded that these species were still present in the lakes and have been missed during sampling as a result of their reduced numbers. And as many species depend on aquatic vegetation, they may recover in time as the vegetation recovers. However, due to wind action and restored nutrient limitation the lakes can remain in a pioneer stage for hundreds of years (Brouwer, 2001) and development of later successional vegetations and the recovery of accompanying fauna species may take a long time. The disappearance of species as a consequence of restoration efforts along this path has been recorded in the Netherlands for a number of rare and characteristic dragonflies, e.g. *Coenagrion hastulatum* (Charpentier) and *Aeshna subarctica* (Walker) (pers. comm. R. Ketelaar).

Even when species are not able to pass through the multiple filters imposed on them by restoration measures, their survival on the long term is not necessarily endangered if they are able to recolonize the restored lakes. In fact the number of observed characteristic species increased relatively faster than the number of other species (Fig. 6), indicating a good colonisation capacity and/or ability to increase their numbers through reproduction. However, shallow softwater lakes in the Netherlands have a long history of degradation and habitat fragmentation resulting from land reclamation, acidification, eutrophication and the loss of natural and anthropogenic dynamics (Arts & Leuven, 1988). At present, there are no intact softwater lakes left and many characteristic species have become very rare and their remaining populations small (Leuven et al., 1987; Leuven, 1988). Furthermore, in some characteristic macroinvertebrate species, most of the specimens have a poorly developed flight musculature, e.g. the water beetle *Hygrotus novemlineatus* (Stephens) (Jackson, 1956). These species are unlikely to cover the large distances required to colonize the restored waters. Restoration efforts in fragmented landscapes should therefore not only focus on restoring ecosystem functions, but also on preserving remaining relict populations and (re)creating suitable habitats.

The problem of declining and disappearing characteristic fauna species as a result of restoration measures most likely is not limited to softwater lakes. In many other ecosystems similar restoration measures are being taken (e.g. restoring meanders in streams, sod cutting of heather, rewetting of raised bogs, etc.). These measures are often carried out on a large scale and may result in strong disturbances and the (temporary) disappearance of advanced successional stages. Van Duinen et al. (2003) demonstrated that large scale rewetting measures in raised bog remnants benefit only a limited number of characteristic species. Degraded bog remnants still harboured many characteristic bog species. They attributed the limited species spectrum of treated locations to a decrease in habitat diversity resulting from the large scale at which the measures were taken. They argue that restoration measures should not only focus on restoring a single aspect of the ecosystem (in their case hummock-hollow vegetation types). The preservation and restoration of the whole array of habitats associated with an ecosystem must become the objective of restoration efforts, as many animal species depend on these habitats or on a combination of them (Verberk et al., 2001, 2002; Smits et al., 2002; Van Duinen et al., 2003). Their argumentation also holds true for the restoration of softwater lakes, which at the moment mainly focuses on restoring physico-chemical conditions for early pioneer communities dominated by isoetid plants.

What do the results of this study indicate for restoration management in shallow softwater lakes? Restoration measures are necessary in order to prevent the extinction of a large number of characteristic species. In fragmented landscapes, this goal is not reached the way measures are being carried out nowadays. Therefore, the measures have to be adjusted in order to ensure the survival of relict populations during the process of restoration. Adjustments of restoration measures should decrease the number of individuals that are being removed along with the organic matter and should not lead to an environment in which species are unable to complete their life cycles. Eventhough there is no actual experience in optimizing restoration measures with respect to the preservation of relict populations, there are several intuitively feasible alternatives. Temporary storage of removed organic matter on the shores of the lake facilitates the return to the water of species that are large and mobile, hereby reducing their initial mortality. By preserving parts of the original vegetation, refuges are created for less mobile species and it enables species that rely on vegetation to complete their life cycles. The best chances of survival for species will be provided by phasing the restoration measures in space and time. Verberk et al. (2001) and Van Duinen et al. (2003) also recommended phasing of restoration measures for raised bog remnants and complex landscapes. One part of the lake is restored, while the other part remains untreated until the characteristic species have colonized the restored part. It is important to divide the lake in compartments in order prevent the redistribution of excess nutrients. In order to counteract habitat fragmentation, lakes can be created close to populations of characteristic species or on locations where they have been lost to land reclamation.

Acknowledgements

This project is part of the Survival Plan Forest and Nature and financed by the Dutch Ministry of Agriculture, Nature and Food Quality. Many thanks go to D. Bal, J. Holtland, P. van den Munckhof, F. Bink, A. Barendregt, H. Siepel, B. Higler, H. Sierdsma and H. Moller Pillot, who provided us with lots of useful comments and data on biological traits. J. Kuper, T. Peeters, A. Brock, T. Timm and H. Vallenduuk made a major contribution in the processing of large numbers of macroinvertebrates and water chemistry samples. Staatsbosbeheer (State Forestry) and the Unie van Bosgroepen (Union of Forest groups) were so kind to allow research in their areas. STOWA (Foundation for Applied Water Research) supplied data on lake Grenspoel.

References

Arts, G. H. P., 2000. Natuurlijke levensgemeenschappen van de Nederlandse binnenwateren deel 13, Vennen. Achtergronddocument bij het "Handboek Natuurdoeltypen in Nederland". Expertisecentrum LNV, Ministerie van Landbouw, Natuurbeheer en Visserij, Ede-Wageningen (in Dutch).

Arts, G. H. P. & R. S. E. W. Leuven, 1988. Floristic changes in shallow softwaters in relation to underlying environmental factors. Freshwater Biology 20: 97–111.

Bloemendaal, F. H. J. L. & J. G. M. Roelofs (eds), 1988. Waterplanten en waterkwaliteit. Stichting uitgeverij Koninklijke Nederlandse Natuurhistorische Vereniging, Utrecht (in Dutch).

Boukal, D. S. & L. Berec, 2002. Single-species models of the Allee effect: Extinction boundaries, sex ratios and mate encounters. Journal of Theoretical Biology 218: 375–394.

Brouwer, E., 2001. Restoration of atlantic softwater lakes and perspectives for characteristic macrophytes. PhD thesis, University of Nijmegen, Nijmegen.

Brouwer, E. & J. G. M. Roelofs, 2001. Degraded softwater lakes: possibilities for restoration. Restoration Ecology 9: 155–166.

Charlesworth, D. & B. Charlesworth, 1987. Inbreeding depression and its evolutionary consequences. Annual Review of Ecology and Systematics 18: 237–368.

Engen, S., R. Lande & B. E. Saether, 2003. Demographic stochasticity and allee effects in populations with two sexes. Ecology 84: 2378–2386.

Foggo, A., S. D. Rundle & D. T. Bilton, 2003. The net result: evaluating species richness extrapolation techniques for littoral pond invertebrates. Freshwater Biology 48: 1756–1764.

Haines, T. A., 1981. Acidic precipitation and its consequences for ecosystems: a review. Transactions of the American Fisheries Society 110: 669–707.

Henrikson, L., J. B. Olofsson & H. G. Oscarson, 1982. The impact of acidification on Chironomidae (Diptera) as indicated by subfossil stratification. Hydrobiologia 86: 223–229.

Hill, M. O., 1979. TWINSPAN – a FORTRAN program for arranging multivariate data in an ordered two-way table by classification of the individuals and attributes. Cornell University, Ithaca, New York.

Jackson, D. J., 1956. The capacity for flight of certain water beetles and its bearing on their origin on the western Scottish Isles. Proceedings of the Linnean Society of London, London 167: 76–96.

Juget, J. & M. Lafont, 1994. Theoretical habitat templets, species traits, and species richness: aquatic oligochaetes in the upper Rhone River and its floodplain. Freshwater Biology 31: 327–340.

Kerfoot, W. C., 1982. A question of taste: crypsis and warning coloration in freshwater zooplankton communities. Ecology 63: 538–554.

Lamers, L. P. M., S. M. E. van Roozendaal & J. G. M. Roelofs, 1998. Acidification of freshwater wetlands: combined effects of non-airborne sulfur pollution and desiccation. Water, Air and Soil Pollution 105: 95–106.

Lamers, L. P. M., A. J. P. Smolders & J. G. M. Roelofs, 2002. The restoration of fens in the Netherlands. Hydrobiologia 478: 107–130.

Leuven, R. S. E. W., 1988. Impact of acidification on aquatic ecosystems in the Netherlands. PhD Thesis, University of Nijmegen, Nijmegen.

Leuven, R. S. E. W., G. van der Velde, J. A. M. Vanhemelrijk & R. L. E. Eeken, 1986. Impact of acidification on the distribution of aquatic insects in lentic softwaters. Proceedings of the 3rd European Congress of Entomology, Dutch Entomological Society, Amsterdam, 103–106.

Leuven, R. S. E. W., J. A. van der Velden, J. A. M. Vanhemelrijk & G. van der Velde , 1987. Impact of acidification on chironomid communities in poorly buffered waters in the Netherlands. Entomologica Scandinavica Supplement 29: 269–280.

Menges, E. S., 1991. The application of minimum viable population theory to plants. In Falk, D. & K. Holsinger (eds) Genetics and Conservation of Rare Plants. Oxford University Press, New York, 45–61.

Nienhuis, P. H., J. P. Bakker, A. P. Grootjans, R. D. Gulati & V. N. de Jonge, 2002. The state of the art of aquatic and semi-aquatic ecological restoration projects in the Netherlands. Hydrobiologia 478: 219–233.

Poff, N. L., 1997. Landscape filters and species traits: towards a mechanistic understanding and prediction in stream ecology. Journal of the North American Benthological Society 16: 391–409.

Ravera, O., 1986. The ecological aspects of acid deposition. Experientia 42: 329–588.

Reed, D. H., J. J. O'Grady, B. W. Brook, J. D. Ballou & R. Frankham, 2003. Estimates of minimum viable population sizes for vertebrates and factors influencing those estimates. Biological Conservation 113: 23–34.

Resh, V. H., A. G. Hildrew, B. Statzner & C. R. Townsend, 1994. Theoretical habitat templets, species traits and species richness: a synthesis of long-term ecological research on the upper Rhone River in the context of concurrently developed theory. Freshwater Biology 31: 539–554.

Richoux, P., 1994. Theoretical habitat templets, species traits, and species richness: aquatic Coleoptera in the upper Rhone River and its floodplain. Freshwater Biology 31: 377–395.

Roelofs, J. G. M., 1983. Impact of acidification and eutrophication on macrophyte communities in softwaters in the Netherlands. I. Field observations. Aquatic Botany 17: 139–155.

Rohani, P., R. M. May & M. P. Hassell, 1996. Metapopulations and equilibrium stability: the effects of spatial structure. Journal of Theoretical Biology 181: 97–109.

Scarsbrook, M. R. & C. R. Townsend, 1993. Stream community structure in relation to spatial and temporal variation: a habitat templet study of two contrasting New Zealand streams. Freshwater Biology 29: 395–410.

Smits, M. J. A., G. A. van Duinen, J. G. Bosman, A. M. T. Brock, J. Javois, J. T. Kuper, T. M. P. Peeters, M. A. J. Peeters & H. Esselink, 2002. Species richness in a species poor system: aquatic macroinvertebrates of Nigula Raba, an intact raised bog system in Estonia. Proceedings of the international Peat Symposium, Pärnu, 283–291.

Southwood, T. R. E., 1977. Habitat, the templet for ecological strategies? Journal of Animal Ecology 46: 337–365.

Tachet, H., P. Usseglio-Polatera & C. Roux, 1994. Theoretical habitat templets, species traits, and species richness: Trichoptera in the upper Rhone River and its floodplain. Freshwater Biology 31: 397–415.

Ter Braak, C. J. F. & P. Smilauer, 1998. CANOCO Reference Manual and User's Guide to Canoco for Windows: Software for Canonical Community Ordination (Version 4). Microcomputer power, Ithaca, New York.

Tonn, W. M., 1990. Climate change and fish communities: a conceptual framework. Transactions of the American Fisheries Society 119: 337–352.

Usseglio-Polatera, P., 1994. Theoretical habitat templets, species traits, and species richness: aquatic insects in the upper Rhone River and its floodplain. Freshwater Biology 31: 417–437.

Van der Hammen, H., T. H. L. Claassen & P. F. M. Verdonschot, 1984. Handleiding voor hydrobiologische milieu-inventarisatie. Interprovinciale Ambtelijke Werkgroep Milieu-inventarisatie, subwerkgroep Hydrobiologie, Haarlem (in Dutch).

Van Duinen, G. A., A. M. T. Brock, J. T. Kuper, R. S. E. W. Leuven, T. M. P. Peeters, J. G. M. Roelofs, G. van der Velde, W. C. E. P. Verberk & H. Esselink, 2003. Do restauration measures rehabilitate fauna diversity in raised bogs? A comparative study on aquatic macroinvertebrates. Wetlands Ecology and Management 11: 447–459.

Van Kleef, H. & H. Esselink, 2004a. Monitoring van abiotiek, vegetatie, dansmuggen en kokerjuffers in gerestaureerde zwakgebufferde oppervlaktewateren. Kolonisatie van diersoorten. Expertisecentrum LNV, Ministerie van Landbouw, Natuur en Voedselkwaliteit, Ede-Wageningen (in Dutch).

Van Kleef, H. & H. Esselink, 2004b. Analyse van de effecten van herstelmaatregelen op watermacrofauna in zwakgebufferde oppervlaktewateren. Een vergelijkend onderzoek in vier vennen waar herstelmaatregelen zijn uitgevoerd. Expertisecentrum LNV, Ministerie van Landbouw, Natuur en Voedselkwaliteit, Ede-Wageningen (in Dutch).

Van Kleef, H. & H. Esselink, 2005. Aquatic Macroinvertebrates in Deteriorated Shallow Softwater Lakes Under Going Restoration. Their Abundances and Biological Traits. Bargerveen Foundation, Department of Environmental Science, Radboud University Nijmegen, Nijmegen.

Verberk, W. C. E. P., G. A. van Duinen, T. M. J. Peeters & H. Esselink, 2001. Importance of variation in watertypes for water beetle fauna (Coleoptera) in Korenburgerveen, a bog remnant in the Netherlands. Proceedings Experimental and Applied Entomology, Dutch Entomological Society, Amsterdam 12: 121–128.

Verberk, W. C. E. P., A. M. T. Brock, G. A. van Duinen M. van Es, J. T. Kuper, T. M. J. Peeters, M. J. A. Smits L. Timan & H. Esselink, 2002. Seasonal and spatial patterns in macroinvertebrate assemblage in a heterogeneous landscape. Proceedings Experimental and Applied Entomology, Dutch Entomological Society, Amsterdam 13: 35–43.

Hydrobiologia (2006) 565:217–228

R.S.E.W. Leuven, A.M.J. Ragas, A.J.M. Smits & G. van der Velde (eds), Living Rivers: Trends and Challenges in Science and Management
DOI 10.1007/s10750-005-1915-5

Analysis of naturalization alternatives for the recovery of moist-soil plants in the floodplain of the Illinois River

C. Ahn[1,*], D.M. Johnston[2], R.E. Sparks[3] & D.C. White[4]

[1]*Department of Environmental Science and Policy, George Mason University, 4400 University Dr., MS 5F2, Fairfax, VA 22030, USA*

[2]*Department of Landscape Architecture, University of Illinois at Urbana-Champaign, 611 East Lorado Taft Dr., Champaign, IL 61820, USA*

[3]*National Great Rivers Research and Education Center, 110 Olin Science Building, 5800 Godfrey Road, Godfrey, IL 62035-2466, USA*

[4]*Illinois Water Resources Center, University of Illinois at Urbana-Champaign, 1101 West Peabody Dr., Urbana, IL 61801, USA*

*(*Author for correspondence: E-mail: Cahn@gmu.edu)*

Key words: floodplain restoration, floodplain-river naturalization, moist-soil plants, levee removal, navigation dam, ecohydrology

Abstract

The hydrologic regime of the Illinois River has been substantially altered by floodplain levees, navigation dams, and water diversion. Unnaturally frequent and untimely water level fluctuations, large and small, have decreased the productivity of many floodplain vegetation communities that provide important ecological services, including the moist-soil plant community. We simulated three scenarios, including two that were expected to benefit moist-soil plants: (1) existing conditions, with levees and navigation dams closed during the summer growing season; (2) levees opened to reconnect the river and its floodplain during the growing season; and (3) both the downstream navigation dam and the levees opened during the growing season. A 1-dimensional hydraulic model generated daily hydrographs of the river at three positions in the 135 km study reach: (1) near the downstream dam, (2) in the middle of the reach, and (3) near the upstream dam. These hydrographs then were used to run a model that predicts the growth of moist-soil plants at a range of floodplain elevations. As expected, the model predicted that plants would grow over a larger area with levees open during the growing season than under the existing conditions, but the outcomes showed a strong location dependency. Moist-soil plant production would increase in the upper and mid-reach locations, but there would be no change near the downstream dam despite opening the levees. Modelling revealed that the existing operation of the navigation dam permanently floods most of the floodplain zone where moist soil plants might grow for at least 15 km upstream of the dam. Trees currently grow all the way to the low water line and are likely to exclude moist soil plants from any restored portion of the floodplain. Sites for reconnecting the river with its floodplain should be carefully chosen to maximize the chances of recovering the important moist-soil plant community in this regulated river.

Introduction

During the past 100 years, the hydrologic regime of the Illinois River has been substantially altered by locks and dams, floodplain levees, and water diversion (Sparks et al., 1998, 2000; Schneider, 2000; Koel & Sparks, 2002). The generally higher water levels and more frequent fluctuations (including fluctuations far smaller than flood stage) due to the hydrologic alterations profoundly affect many plants that thrive in the Illinois floodplain-river ecosystem, including moist-soil plants.

Moist-soil plants are typically annual grasses and forbs (though some species may live for several years under favorable conditions) that grow on mud flats exposed as seasonal floods recede. One member of this group, the decurrent false aster (*Boltonia decurrens*) is endemic to the Illinois River and is listed as an endangered species by the U.S. Fish & Wildlife Service. Productivity of moist-soil plants along the Illinois River has declined because of unnaturally frequent fluctuations that inundate these mud flats and drown the plants during the critical summer growing season. In common with other types of floodplain plant communities, including forests and submergent vegetation, the moist-soil plants perform several ecological functions. Their roots stabilize the sediments deposited along shorelines of the channels and floodplain lakes where wave action would otherwise resuspend the sediment and thereby increase the turbidity of the water. Even in death, at the end of the summer growing season, the stems and leaves of moist-soil plants contribute organic matter to detritus-based food webs. The seeds, rhizomes and tubers of these plants are an important food source for resident beaver (*Castor canadensis*) and muskrat (*Ondatra zibethicus*) and for waterfowl during their spring and fall migrations in the Mississippi Flyway, of which the Illinois River is an important part (Bellrose et al., 1979; Fredrickson & Taylor, 1982; Bellrose et al., 1983; Havera, 1999).

Currently, there are seven locks and dams on the Illinois River that create the Illinois Waterway (Demissie & Knapp, 2000). Two of these dams, on the lower 370 km of the Illinois River, are wicket dams. Wickets are lowered to the river bottom during periods of high flow; during periods of low flow the wickets are raised to maintain the required navigation depth of 2.74 m (Sparks et al., 1998).

There are also 36 levee and drainage districts, mostly on the lower Illinois River (Thompson, 2002). The levees isolate about 73,000 ha, or 57% of the floodplain, from the river (Bellrose et al., 1983). Levees that protect industrial and urban areas occupy a very small proportion of the 73,000 ha, including two small districts that provide flood protection for the towns of Liverpool and Beardstown in the La Grange Reach. By isolating large areas of the floodplain, primarily for row crop agriculture, levees reduce the area of the floodplain available to store and convey floods and the extent of the biologically rich floodplain, a biological richness and productivity that had evolved under the influence of the natural flood pulse (Junk et al., 1989). The remaining 43% of the floodplain remains "active", in the sense that it is still subject to inundation by the river.

Recently, public interest in the conservation and recovery of natural services has prompted major public and private investments in conversion of some levee and drainage districts back to functional floodplains along some rivers (Galat, 1998; Sparks et al., 2000). Restoration of native moist-soil plants is important in these conversions because they provide food for wildlife and other valuable ecological services; they are also indicators of successful restoration of the flood pulse (Ahn et al., 2004a,b). However, there are very few precedents for this type of naturalization to guide current efforts and there is a controversy regarding whether the converted levee districts should be opened to the river or isolated from the river, for reasons explained next.

State, federal and non-governmental agencies are currently investing significant resources to naturalize areas within the Illinois floodplain-river ecosystem (Sparks et al., 2000; Clancy, 2001). There are two distinctly different approaches to re-establishing a more natural hydrologic regime on the floodplain (Sparks et al., 1998). One is to keep the existing levees and manage the water levels on project sites independently from the river in order to produce wetland vegetation, including moist-soil plants, primarily to attract and feed migrating waterfowl. This water level management in impoundments maintained by private duck hunting clubs and government wildlife agencies mimics the natural water regime, with the mud flats intentionally flooded from late fall through winter and late spring, then exposed during the summer low flow season. Pumps are often used to pump water in or out of the impoundments, if river levels do not permit the filling or emptying of the impoundments through gates at the desired times. One former levee district that is just south of the dam at Peoria is currently not connected to the river at all. Local rainfall, groundwater and river water that seeps through or under the levees are allowed to raise water levels and pumps are used to lower the water levels. Water is pumped out into the river, but river water is not allowed directly in.

However, this option does not provide a direct connection between the river and its floodplain that is essential to many fish species that spawn and feed on the floodplains (e.g. basses and sunfishes, Family *Centrarchidae*, which are sought by sport fishermen) and to many important ecological processes (dispersal of floating seeds, nutrient cycling).

A second, more systemic approach is to reconnect the floodplain to the river either by breaching the levees or (more likely) installing gates in the levees – in either case, the levee is "opened" to the river, the term that will be used throughout this paper. The approach is systemic in the sense that the river itself provides the water regime in all the reconnected floodplains, instead of management staff who operate pumps and gates in each isolated floodplain compartment. Such an approach also allows fish access and promotes exchanges of nutrients, organic matter, and plant propagules (seeds, shoots, and winter buds) between the river and its floodplain, but it exposes the floodplain to the unnatural water fluctuations of the regulated river.

The Nature Conservancy, a non-governmental conservation organization, proposes to reconnect the Illinois River to a recently acquired 21.3 km^2 agricultural levee district in the middle section of the La Grange reach of the Illinois River (Fig. 1). Called "Emiquon," the area to be naturalized once contained two large backwater lakes connected to the Illinois River during rises in the river level. The area was drained and leveed in the 1920s and was farmed for 80 years (Clancy, 2001). The model described in this paper was developed to assist in planning the Emiquon reconnection and in the selection of other potential reconnection sites. The Nature Conservancy regards Emiquon as a demonstration and learning project. If reconnection to the river promotes recovery here, then other private and government agencies might be persuaded to reconnect the floodplain compartments they manage. Rather than simply breaching the levee, The Nature Conservancy proposes to install gates, so that the reconnection can be tested and even stopped, if excessive sedimentation and water level fluctuations impede recovery.

Another systemic approach, besides opening the levees, would be to "open" the dams. In the case of the navigation dams at Peoria and La Grange, "opening" could be accomplished by lowering the wickets at the dams during the low water, growing season, to expose the floodplain that is now permanently inundated. Such an approach is unlikely because it would shorten the navigation season or require that barges be lightly loaded so that they could operate in shallow water, so we did not even model this scenario initially. However, after the simulation model predicted no moist soil plants would grow in the 15 km upstream of the dam, even if the levees were opened (Table 1), we did model the additional effect of opening the downstream dam (lowering the wickets) at La Grange during the growing season. Our interest was in better understanding the effects of the dam, after this rather surprising prediction.

Naturalization planning for the entire Illinois River is currently in progress, guided by both hydrologic and ecological models which predict the impacts of various naturalization and management scenarios on water flow, sedimentation, and vegetation. We have developed a moist-soil plant growth simulator (Ahn et al., 2004a) as part of an inter-disciplinary modelling approach linking the timing, duration and extent of flooding to the annual success and areal extent of moist-soil plants (Ahn et al., 2004b).

Here we describe how the plant model was used to assess naturalization scenarios along a representative section of the lower Illinois River, between the La Grange dam and Peoria dam (Fig. 1). We examined rather extreme scenarios involving opening the levees and the La Grange Dam. Our approach and analysis provide information and insights that are useful in selecting restoration sites and in addressing the controversy over isolating vs. connecting the converted levee districts to the river.

Materials and methods

Site description: the La Grange Reach of the Illinois River

The 135km long La Grange Reach of the Illinois River is bounded by dams at La Grange and Peoria (Fig. 1). The wickets are lowered during high flows, so that commercial barges and migratory fishes can pass directly over the structures without going

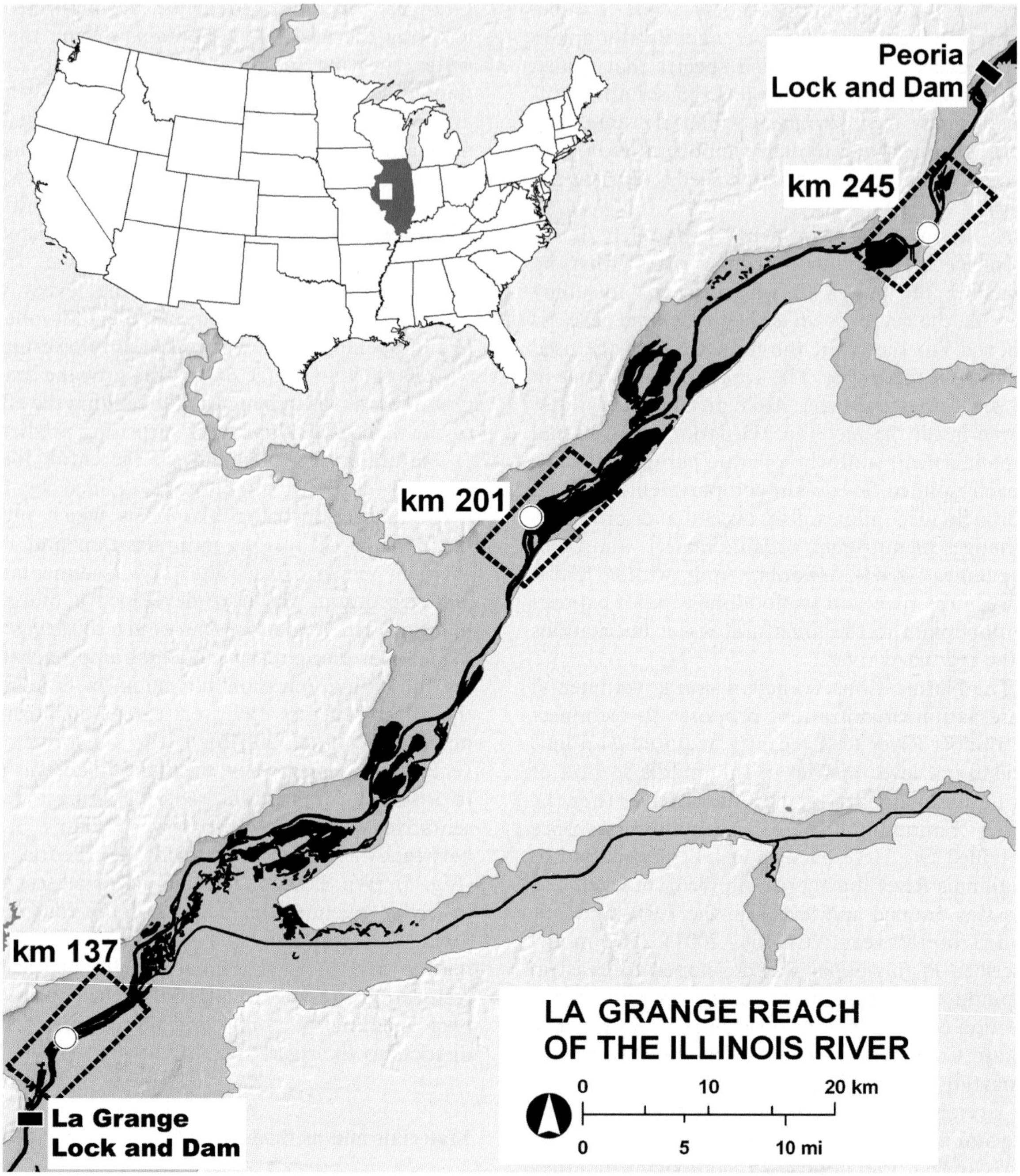

Figure 1. The La Grange Reach of the Illinois River. The growth of moist soil vegetation was modelled in three 15 km zones: near the upstream dam at Peoria (the center point of the zone is Illinois River km 245, measured upstream from the confluence with the Mississippi River); in mid-reach (km 201); and near the downstream dam at La Grange (km 137). The 21.3 km^2 Emiquon Floodplain Restoration Site, an agricultural drainage and levee district purchased by The Nature Conservancy, is in the middle zone.

through the locks. As a result, these dams have a much smaller effect on the river and its biota than high storage dams do on rivers elsewhere.

Several levee and drainage districts are located along the La Grange reach. These districts separate a total of about 33,000 ha of formerly active

Table 1. Variation in water levels (coefficient of variation, CV) and predicted floodplain area (km^2) of moist-soil plants under three scenarios at three locations in the La Grange reach of the Illinois River

Location within navigation reach	Scenario	Variation (CV) in water levels during growing season	Area available for moist soil plants[a] (km^2)	Gain[b] (km^2)
Upper 15 km	Existing	0.44	0.58	
	Levees open	0.37	5.34	4.76
	Levees and dam open	0.46	5.23	4.65
Middle 15 km	Existing	0.33	2.09	
	Levees open	0.28	2.44	0.35
	Levees and dam open	0.40	2.44	0.35
Lower 15 km	Existing	0.22	0.00	
	Levees open	0.19	0.00	0.00
	Levees and dam open	0.55	1.17	1.17

[a]The area above the low water line and below the elevation of the tree line where moist soil plants can germinate, grow and set seed in four out of ten years, on average (i.e. probability of success in any given year = 40%).
[b]Difference between the area currently available for moist soil plants and what would be available under two scenarios: (1) if levees were opened, and (2) if levees and the downstream navigation dam were opened.

floodplain from the hydrologic influence of the river, and are potential floodplain restoration sites, if the landowners are willing to sell the land or accept payments in return for flood or conservation easements.

Hydraulic model

We have calibrated a hydraulic model for the La Grange Reach of the river, using 61 years of daily water level elevations at three gages within the reach. The model used is UNET (HEC, 1993; United States Army Corps of Engineers, 1995). UNET is an acronym for *Un*steady Flow through a Full *Net*work of Open Channels. The 61 years encompass the period from 1940, when the modern navigation dams became operational, to the most recently available gage readings. UNET simulates one-dimensional unsteady flow through a defined, or bounded, network of open channels. We used the model to estimate river stages within the upper, middle, and lower segments of the La Grange Reach (Fig. 1) during the summer growing season under three conditions: (1) current conditions (with levees and dams in place); (2) levees opened; and (3) levees and La Grange dam opened (wickets down). The model was calibrated using historic records under current conditions (levees and dams in place) using standard engineering practices. Modeled results allow estimation of stages at locations along the reach not having recorded observations. As there are no recorded observations under the other two conditions, we must assume the model simulates river stages under these conditions.

The levees open scenario is represented in the model by removing the levees from the elevation cross-sections across the channel and floodplain. In actuality, it would be prohibitively expensive to physically remove the entire levee system; instead, levees would be opened at the upstream and downstream ends, so that flood waters could flow through the former levee and drainage districts in the old natural floodways and floodplain lakes that existed before the levees. Because the hydraulic volume of the floodplain is large in comparison to the volume of earth in the levee, the portion of the levee that would be left after breaching or installing gates at the upstream and downstream ends would have little effect on water levels. Moreover, the main levee parallels the main channel, so flood waters would pass on both the channel and floodplain side of the breached or gated levee, with little loss in conveyance capacity due to flow restriction.

The hydraulic model assumes that the topography remains the same after opening the levees or the dam. This is a reasonable assumption over a time span of at least 30 years, based on modelling the Emiquon site under five "opening" scenarios that included full connection and various operating rules for either one or two gates in the levee (Illinois State Water Survey, 2004). The most sedimentation occurs with full connection, with 3.30 cm of fill in the restoration site in 30 years. The other four scenarios, where one or two gates are opened and closed under various operating rules, lead to much less sedimentation, down to a minimum of 0.25 cm in 30 years.

Plant growth model

The resulting simulated hydrographs from UNET were used as input to a moist-soil plant growth model developed by Ahn et al. (2004a), which quantifies moist-soil plant "success" in terms of germination and growth. The model uses physiological parameters for millets (*Echinochloa* sp.), one group of moist soil plants. Millets are summer annual plants that grow up to 1.5 m tall and prefer wet (but not inundated) soils and warm temperatures to germinate. Although individual species of moist soil plants undoubtedly vary in their tolerances for inundation, little published information is available on species other than those, such as millets, that are important in waterfowl management.

The plants require that mudflats be exposed during the period from 10 July to 1 October (Bellrose et al., 1983). Inundation during germination or early seedling development kills the plants. They cannot tolerate flooding until the plants are at least 15 cm tall (Fredrickson & Taylor, 1982). If moist-soil plants can grow for at least 70 days without being overtopped by water, they will mature and produce seeds. The timing of the flood is critical for a successful seed crop.

The upper land elevation boundary for moist soil plants is the tree line. Above this land elevation, the annual moist-soil plants are excluded by the perennial, but less flood tolerant woody species. The land elevation of the tree line along the La Grange reach was determined from aerial photographs and topographic maps. The vertical range of the potential zone for moist-soil plants is only about 2–3 m, extending from the tree line down to the permanent water level, either the low stage in the river and its connecting channels or the low stage in floodplain lakes that are seasonally disconnected from the river.

As reported elsewhere (Ahn et al., 2004a), the plant model was validated by comparing model predictions to 19 vegetation inventories conducted in the 1930s and 1940s. The historical record included years of success and years of failure of the moist soil plants. Although the model predictions matched the historical observations in all 19 cases, the validation must be considered qualitative, because the agreement was limited to whether the plants were absent or present. Presence/absence was the only common response measure we could use because the historical observations reported plant cover (as a percentage of the total area available for moist soil plants) while the model predicts biomass density and height of the plants.

Daily water level fluctuations during the growing season determine the success of the plants in a given year at a specified land elevation. The plant model is run at a series of land elevations, starting at the tree line and decreasing at 10 cm intervals until the lowest recorded water level is reached. This step-wise process allows us to determine, for each year, the *lowest* land elevation at which moist-soil plants are successful, with success defined as producing 90% of the potential maximum biomass (Ahn et al., 2004a). The *lowest successful elevation* is an annual hydrologic parameter (analogous to "maximum daily flow," or "7-day low flow"; Ahn et al., 2004b). For example, at River km 245 the lowest land elevation in 1998 where moist soil plants germinated and grew to at least 90% of their maximum potential biomass was 132.5 m above mean sea level.

Repeating this process for the 61 years of daily river stage records since the modern dams became operational provides us with a historical distribution of the *lowest successful elevation* parameter. This distribution enables us to employ frequency analysis, a commonly used technique in hydrology (Chow, 1964), to make statements about the probability of future plant success at a given land elevation, assuming that the underlying source of variability (such as year-to-year differences in weather) remains the same. For example, a probability of 0.4 means that moist soil plants would be successful 4 out of 10 years. The purpose of this modelling is to allow us to compare alternative naturalization strategies, which requires computing a probability distribution for each scenario. We reduced the computational requirements of scenario assessment by about two-thirds while retaining the full range of historical variability, by selecting the first, last and every 5th percentile observation (21 years in all) from the historical distribution and modelling only those years.

Scenario testing

Using the procedure described above, we modelled three scenarios: existing conditions (neither the levees nor the dam are typically open during the growing season); levees opened; and both levees and the downstream dam open (wickets down) during the growing season. The scenarios are admittedly extreme because they assume that the entire floodplain along the study reach would be reconnected to the river. Examining such unlikely cases provides a useful check of effectiveness of the naturalization strategy. If wholesale reconnection produces no reduction in flood heights and improvements in moist-soil plants, then there would be little point in developing scenarios for less reconnection. If the extreme case *does* produce benefits, then there would be value in testing smaller scale efforts at locations where potential benefits are expected to be high or costs low, or where the land use change would be politically acceptable.

Results

Under existing conditions, the spring flood and smaller rises in water levels are moderated as they move down the La Grange reach, because of the well-known dampening effects of flow resistance and hydraulic storage capacity in the channel itself and in the floodplain and its associated channels and backwaters that remain open to the river (compare panels a, b and c in Fig. 2). The coefficients of variation in water levels under existing conditions are greatest upstream, intermediate in the middle segment, and lowest in the downstream segment (Table 1). Also, the wickets at the downstream dam are raised as the spring flood recedes, so the river is not allowed to get as low as it once did naturally (Fig. 2c). The effects of the dam diminish with distance upstream from the dam.

Opening the levees would reduce flood heights during the spring and early growing season (Fig. 2). The reduction would be over 0.5 m for the upper 15 km segment of the reach (Fig. 2a) and somewhat less in the mid-reach segment (Fig. 2b). However, the reduction in the water level accompanying levee removal would be negligible along the lower segment because the dam maintains the water levels (Fig. 2c).

Opening the downstream dam would have the greatest effect on the downstream segment (Fig. 2c). As expected, water levels would drop markedly close to the dam, but progressively less upstream (Fig. 2a,b).

Predicted effects of opening levees and dams on moist soil plants

The predicted response of moist soil plants varied dramatically by location within the navigation reach (Table 1, Fig. 3). In the upper 15 km of the navigation reach (in the vicinity of River km 245; Fig. 1), the model predicted that moist soil plants would do well if the floodplains were reconnected to the river by opening the levees. The area available for moist soil plants would increase substantially, from 0.58 to 5.34 km^2 (Table 1). Lowering of the wickets at the downstream navigation dam would have little additional effect on area (Fig. 3a).

In the mid-reach segment (in the vicinity of River km 201, including Emiquon, Fig. 1), the probability of moist-soil plant success would increase with levee removal, as would the predicted area covered by moist-soil plants (a gain of 0.35 km^2; Table 1 and Fig. 3b).

However, in the downstream segment (River km 137 vicinity) no moist soil plants are produced under existing conditions and the model predicted none would be produced even if more floodplain were provided by opening the levees (Table 1; Fig. 3c). At this location, it would be necessary to open the downstream (La Grange) dam to provide areas (a total of 1.17 km^2, Table 1) where moist-soil plants could potentially grow.

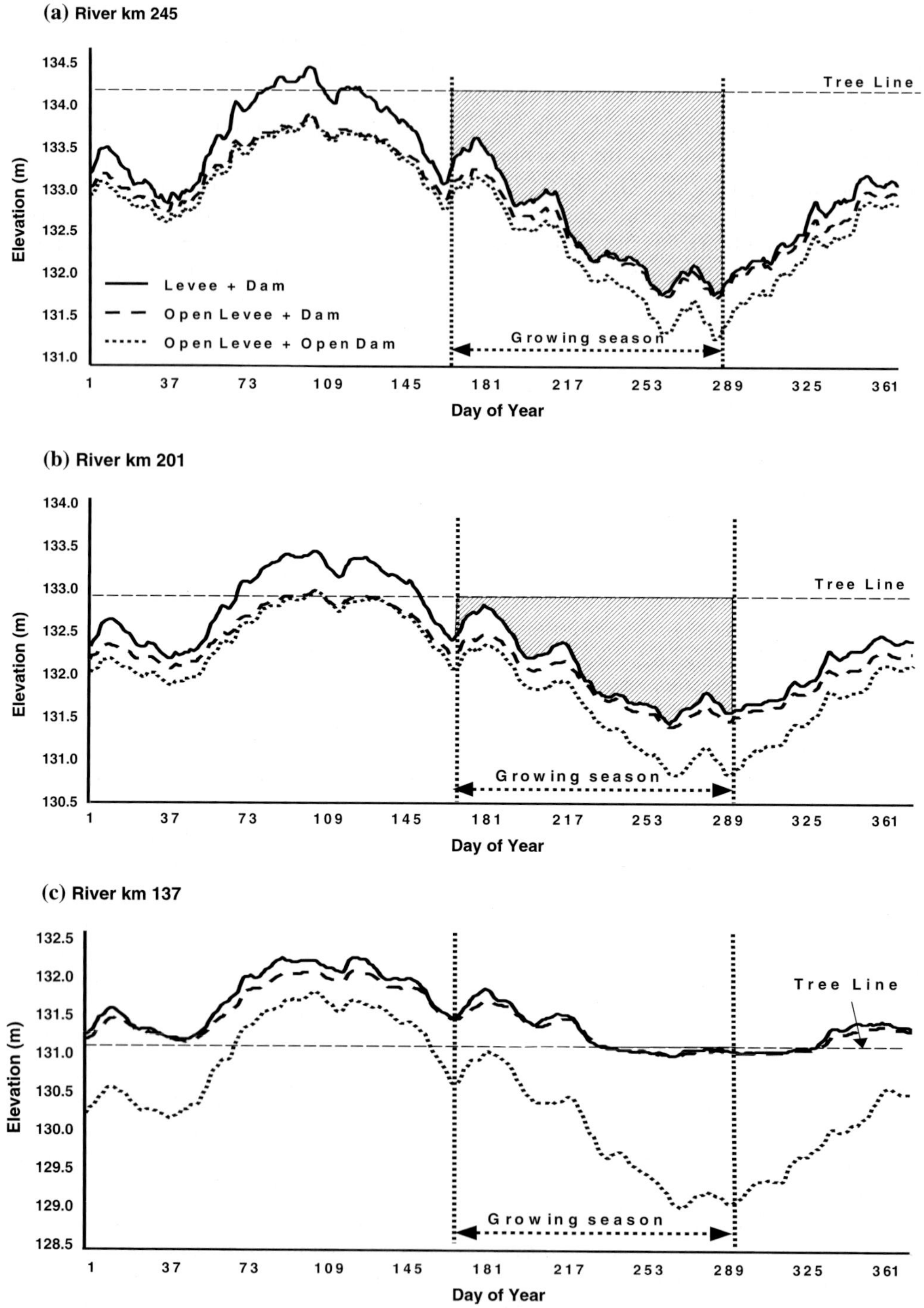

Figure 2. Simulated mean daily water levels at three locations, (a) km 245; (b) km 201; (c) km 137, representing three 15 km zones within the La Grange Reach of the Illinois River, under three scenarios: (1) existing conditions, with levees and locks and dams; (2) with levees open; and (3) with levees and the water level control wickets lowered at the downstream navigation dam (dam open). Vertical dotted lines bracket the growing season for moist-soil plants. The shaded grey area represents the elevation zone, bounded by the tree line at the upper end and by the average daily low water elevation at the low end, where moist soil plants could grow. See text for details about the determination of the tree line and low water elevations.

Discussion

The effects of levees and navigation dams on water levels vary significantly from one location to the next, depending on proximity to the dams and the characteristics of the adjoining floodplain. Under regulation, the Illinois River has: (1) greater and more frequent water level fluctuations below the upstream dam during the growing season, and (2) permanently higher, but more stable water levels near the downstream dam. This upstream–downstream pattern in the water level regime is generally characteristic of navigation reaches in the Illinois and Upper Mississippi rivers (Sparks et al., 1990). Under existing conditions, the upstream fluctuations reduce the success of and the area covered by moist soil plants in most years by preventing germination or drowning plants that do germinate. In some downstream segments, including the downstream portion of La Grange reach, there is no moist soil zone at all under existing conditions, because this zone is now permanently inundated by the navigation dam during the summer low water season and the tree line begins at the margin of the water (Fig. 2c).

Opening the levees would have the expected effect of increasing the area available for moist soil plants in the middle and upper segments of the navigation reach, although the increase is much greater in the upper than in the middle segment (Table 1 and Fig. 3a,b). The increased hydraulic storage capacity would have the additional beneficial effect of damping the harmful, mid-summer water level fluctuations. If recovery of moist soil vegetation and reconnection of the floodplains to the river are management goals, our results indicate that acquisition of flood easements or outright purchases of levee and drainage districts would be most effective in the upper segments of La Grange Reach and probably in other navigation reaches as well. Having found a positive result with the extreme scenario of opening all the levees, the next step will be to simulate the effects of opening only those levees in the upstream half of the Reach, where the moist soil plants have the greatest probabilities of success. The increased success for moist soil plants occurred despite the detrimental water level fluctuations during the 61-year record used for modelling. If the water fluctuations in the main river were reduced by altering dam operations and increasing the water retention of the tributary watersheds the gains should be much greater.

In contrast to the upstream segments, opening the levees in the downstream segment would have no beneficial effect on moist soil plants. This seemingly anomalous result is explained by examining the results from the simulation where the navigation dam was opened during the growing season. Normally, the dam is closed during the summer low flow season and the water level never drops below the tree line (Fig. 2c).

Like the moist-soil plants, many of the floodplain trees will only germinate on moist soil, not under water. Likewise, both seedling trees and moist-soil plants can be drowned by untimely floods. However, trees can become established if they have a few flood-free years in which to grow taller than the depth of the next flood. Many floodplain-adapted tree species can tolerate short-duration floods that do not overtop their upper branches and leaves but do kill plants of shorter stature. The moist-soil plants thus depend on a flood pattern that will exclude trees, but not exclude them. Since trees have life spans measured in decades or centuries and produce propagules nearly every year when mature, they have opportunities to become established at unusually low elevations in the floodplain during rare events, such as prolonged droughts, when floods may not occur at all or may recede exceptionally early. Trees may then persist in what would ordinarily be the moist soil plant zone until they are cut down, die of old age or are killed by rare floods that extend through the growing season (Sparks et al., 1998). The tree lines in our study were determined from observation of aerial photographs. While moist soil plants currently cannot grow higher than the existing tree lines, the tree lines are not a permanent natural boundary, and would change in response to deforestation by humans and/or large herbivores, droughts, protracted floods, or naturalization schemes that change the existing water level patterns. Although a germination and growth model for floodplain forests was beyond the scope of our study, it would certainly be useful to develop such a model to support site selection and planning for naturalization.

The modelling approach described in this paper can be useful in predicting responses of moist soil

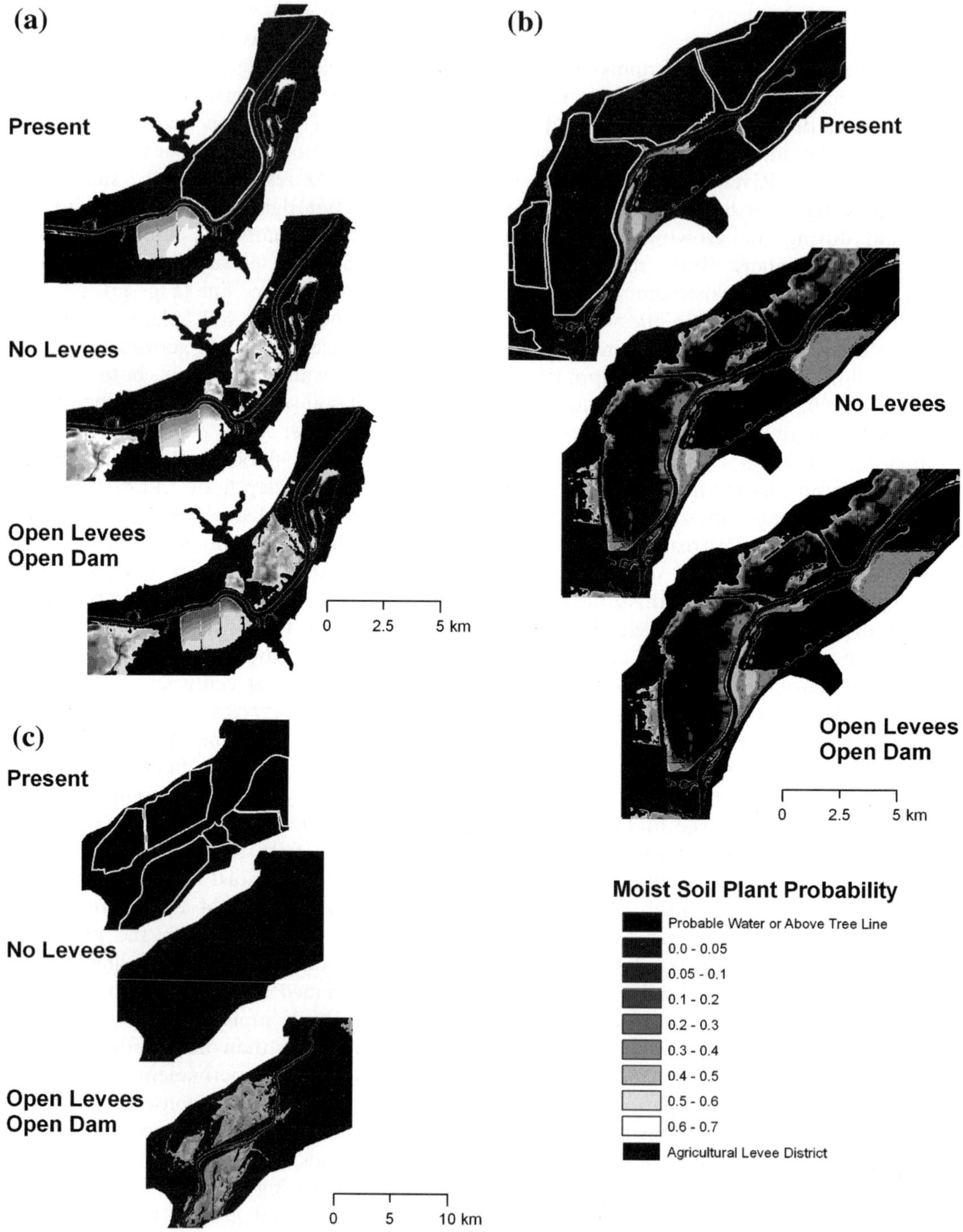

Figure 3. Probability of moist soil plant success (defined as 90% of potential maximum biomass production during the growing season) in three zones: (a) upper 15 km; (b) mid 15 km; and (c) lower 15 km in the La Grange Reach of the Illinois River, under three scenarios: (1) existing conditions, with levees and locks and dams; (2) levees open; and (3) levees and the downstream navigation dam open. At a floodplain elevation where the probability of successful germination and growth of the moist soil plants is 0.50, moist soil plants are replaced by trees. No moist soil plants grow at low elevations in permanent water (probability of success = 0.00). White lines designate levees.

plants to *any* water level fluctuations at any location proposed for naturalization, including: (1) natural fluctuations; (2) dam operations on the main-stem river that might be altered to benefit floodplain vegetation; (3) moderating effects of conservation practices in tributary watersheds; or (4) combinations of approaches. Since moist-soil plants are a good food source for muskrats, beaver, and migratory waterfowl, there is also a link between the success of these plants, used by wildlife, and value to humans who hunt, fish, or view wildlife. Moist soil vegetation provides ecological services, including stabilization of shorelines that are otherwise subject to wave erosion. These linkages could be incorporated in evaluations of probable benefits and costs of river management alternatives.

Conclusion

The framework and approach of linking hydrological and ecological models in this study enabled us to predict and compare consequences of alternative naturalization scenarios for the Illinois floodplain–river system. The models are well grounded in their respective literature and reflect the current state of knowledge of river hydraulics and moist-soil plant ecology. However, not all of the factors involved in moist-soil plant success are fully understood and factors such as nutrient levels, sedimentation rates, and competition with woody vegetation were not included in our modelling, and may prove important. In addition, the choice of indicators or key ecosystem components to model may be critical in examining each floodplain–river system, and any model must be parameterized for the particular system that is to be modelled.

Moist-soil plants are a good indicator for a naturalistic water level regime but not for river-floodplain connectivity, because the flooding pattern these plants require can be created by regulated flooding within leveed areas of the floodplain. These areas can be inundated using local water, including rainwater, ground water and water from small local tributaries, and then dewatered with pumps or gravity drains. Other organisms, such as fish species that migrate between the river and the floodplain, are better indicators for connectivity.

The analytical framework in our study should also be viewed as part of an iterative process of adaptive management, including: modelling (as part of planning), implementing, monitoring, analysis and modelling again. Collecting more experimental and field-oriented information on the response of moist-soil plants, as naturalization efforts proceed, is critical to this adaptive management process. It is also important to have fine-scale elevation data for floodplain sites proposed for restoration because plants respond to elevational differences on the scale of 10–15 cm.

A clear result of this study is that location does matter. Even if the entire floodplain were reconnected to the river near the downstream dam in our study reach, moist soil plants are unlikely to grow there because of the effects of the dam on the water levels. Success is much more likely at locations farther upstream. Modelling approaches like those described in this paper can help with the critical step of site selection for naturalization.

Acknowledgements

The study was supported by the National Science Foundation (NSF BCS-00-03208) and the Illinois Chapter of The Nature Conservancy. We gratefully acknowledge S. Havera and F. Bellrose at the Forbes Biological Station of the Illinois Natural History Survey, and R. Adams at the Chautauqua National Wildlife and Fish Refuge of the U.S. Fish and Wildlife Service, both at Havana, Illinois for information on moist-soil plants. We thank Dr. M. Demissie and Dr. Y. Lian (Director and Associate Professional Scientist, respectively, Center for Watershed Science of the Illinois State Water Survey, Champaign, Illinois) for the hydraulic modelling output that was used in the plant model. Thanks also to R. M. Sparks for improving the manuscript. The opinions expressed in this article are those of the authors, and not necessarily their agencies or sponsors. This is contribution number 7 from the National Great Rivers Research and Education Center, Alton, Illinois and 237 from the Illinois Water Resources Center, Urbana, Illinois. The paper was presented at the 7th Intecol Wetlands Conference in Utrecht, The Netherlands 2004.

References

Ahn, C., R. E. Sparks & D. C. White, 2004a. A dynamic model to predict responses of millets (*Echinochloa* sp.) to different hydrologic conditions for the Illinois floodplain-river. River Research and Applications 20: 485–498.

Ahn, C., D. C. White & R. E. Sparks, 2004b. Moist-soil plants as ecohydrologic indicators to recover the flood pulse in the Illinois River. Restoration Ecology 12: 207–213.

Bellrose, F. C., S. P. Havera, F. L. Paveglio & D. W. Steffeck, 1983. The fate of lakes in the Illinois River Valley. Illinois Natural History Survey Biological Notes 119: 1–27.

Bellrose, F. C., F. L. Paveglio & D. W. Steffeck, 1979. Waterfowl populations and the changing environment of the Illinois River Valley. Illinois Natural History Survey Bulletin 32: 1–54.

Chow, V. T., 1964. Statistical and Probability Analysis of Hydrologic Data, Section 8-I. In Chow, V. T. (ed.), Handbook of Applied Hydrology. McGraw-Hill Inc, New York: 1–42.

Clancy, P., 2001. Restoring the Illinois River. Nature Conservancy 51: 12–19.

Demissie, M. & V. Knapp, 2000. Hydrology and hydraulics of the Illinois River. Proceedings of the 2000 Joint Conference on Water Resources Engineering and Water Resources Planning and Management. In Hotchkiss, R. H. & M. Glade (eds). Environmental and Water Resources Institute of the American Society of Civil Engineers, Reston, VA. CD-ROM.

Fredrickson L. H. & T. S. Taylor, 1982. Management of Seasonally Flooded Impoundments for Wildlife. Publication 148, United States Department of the Interior, Fish and Wildlife Service, Washington, DC.

Galat, D. L., 1998. Flooding to restore connectivity of regulated, large-river wetlands. BioScience 48: 721–733.

Havera S. P., 1999. Waterfowl of Illinois. Special Publication No. 21, Illinois Natural History Survey, Champaign, IL.

HEC, 1993. One-dimensional Unsteady Flow through a Full Network of Open Channels. User's Manual. U.S. Army Corps of Engineers, Hydraulic Engineering Center, Davis CA.

Illinois State Water Survey, 2004. Predicted sedimentation in Emiquon. Report to the Emiquon science advisory council of the nature conservancy, April 2004. http://ilrdss.sws.uiuc.edu/search/search_results.asp?pr = 500&strSearch = emiquon.

Junk, W. J., P. B. Bayley & R. E. Sparks, 1989. The flood pulse concept in river-floodplain systems. In Dodge D. P. (ed.), Proceedings of the International Large River Symposium. Vol. 106. Canadian Special Publication of Fisheries and Aquatic Sciences: 110–127.

Koel, T. M. & R. E. Sparks, 2002. Historical patterns of river stage and fish communities as criteria for operations of dams on the Illinois River. River Research and Applications 18: 2–19.

Schneider, D. W., 2000. Backwater lakes: history shows their values to Illinois. Wetland Matters 5: 1–11.

Sparks, R. E., J. C. Nelson & Y. Yin, 1998. Naturalization of the flood regime in regulated rivers. BioScience 48: 706–720.

Sparks, R. E., P. B. Bayley, S. L. Kohler & L. L. Osborne, 1990. Disturbance and recovery of large floodplain rivers. Environmental Management 14: 699–709.

Sparks R. E., J. B. Braden, M. Demissie, P. Mitra, D. W. Schneider, D. C. White & R. Xia, 2000. Technical Support of Public Decisions to Restore Floodplain Ecosystems: A Status Report on the Illinois River Project, USA. In Smits A. J. M., P. H. Nienhuis & R. S. E. W. Leuven, (eds), New Approaches to River Management. Backhuys Publishers, Leiden: 225–247.

Thompson, J., 2002. Wetlands Drainage, River Modification, and Sectoral Conflict in the Lower Illinois Valley, 1890–1930. Southern Illinois University Press, Carbondale, IL.

United States Army Corps of Engineers, 1995. Floodplain Management Assessment of the Upper Mississippi River and Lower Missouri Rivers and Tributaries Main Report, United States Army Corps of Engineers, Rock Island, IL.

Hydrobiologia (2006) 565:229–242 © Springer 2006
R.S.E.W. Leuven, A.M.J. Ragas, A.J.M. Smits & G. van der Velde (eds), Living Rivers: Trends and Challenges in Science and Management
DOI 10.1007/s10750-005-1916-4

Long-term developments in ecological rehabilitation of the main distributaries in the Rhine delta: fish and macroinvertebrates

A. bij de Vaate[1,3,*], R. Breukel[1,3] & G. van der Velde[2,3]
[1]*Institute for Inland Water Management and Waste Water Treatment (RIZA), P.O. Box 17, 8200 AA Lelystad, The Netherlands*
[2]*Department of Animal Ecology and Ecophysiology, Institute for Wetland and Water Research, Radboud University Nijmegen, Toernooiveld 1, 6525 ED Nijmegen, The Netherlands*
[3]*Participant of the Netherlands Centre for River Studies (NCR), P.O. Box 177, 2600 MH Delft, The Netherlands*
*(*Author for correspondence: E-mail: b.bdvaate@riza.rws.minvenw.nl)*

Key words: water quality, river engineering, non-indigenous species, exotic species

Abstract

Significant anthropogenic pressure in the Dutch part of the River Rhine is present from the 12th century. River engineering and water pollution were the main stress factors for flora and fauna. From the middle of the 20th century measures were taken to reduce water pollution. Recently, from 1987 onwards, these activities were put into a wider context of ecological river rehabilitation. Effects of improvements on fish and macroinvertebrates in the main distributaries in the Rhine delta are reviewed. The conclusions are that (a) most of the alterations in the Rhine delta are irreversible due to hard socio-economic boundary conditions (e.g. safety, navigation); (b) chances for the development of riverine biotopes have therefore to be found in the forelands and not in the main channels of the Rhine delta; (c) further reduction of pollutants, especially thermal pollution, is needed to help original species to colonise the Rhine delta again; (d) non-indigenous species clearly leave a mark on recolonisation possibilities of original species.

Introduction

The Rhine valley, including the Rhine–Meuse delta, has been densely populated for many centuries. From the point of view of water management the river has different functions now, varying from discharge of water and waste to transport route for raw materials and products (Anonymous, 2001; Bij de Vaate, 2003). Anthropogenic influence evolved in the 12th century with the construction of levees along the main channel for the purpose of protecting inhabitants against flooding (Middelkoop, 1997). Later on, shipping asked for the adjustment of the main channel to allow larger freighters sailing. These activities led to irreversible changes, not only in the occurrence of riverine habitats caused by disruption of the natural sequence of backwaters, but also in the disappearance of aquatic-terrestrial transition zones (e.g. Gore & Shields, 1995; Sparks, 1995; Galat et al., 1998; Nienhuis & Leuven, 2001). In the Rhine delta (The Netherlands) the smaller distributaries were dammed, and the remaining larger ones (called IJssel, Nederrijn/Lek and Waal; Fig. 1) changed from meandering streams with extensive floodplains, into shipping canals surrounded by relatively low, so-called summer dikes, on the banks and major dikes at a greater distance (Fig. 2; Middelkoop, 1997; Van Urk & Smit, 1989). Moreover, from the 19th century the summer beds in the distributaries were fixed with groynes of basalt stones to promote channel bottom erosion (Van Urk, 1984; Kalweit, 1993), a measure to deepen the channel in a 'natural' way for shipping.

Effects of domestic and industrial waste water discharges into the whole catchment area

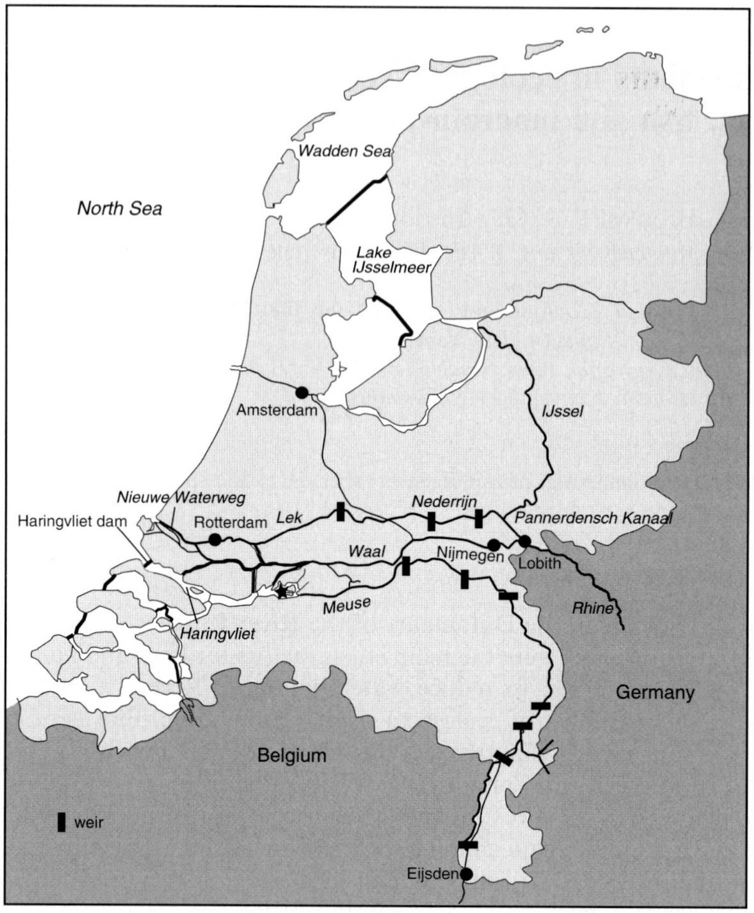

Figure 1. The main distributaries in the Rhine delta.

surfaced in the second part of the 19th and first part of the 20th century and resulted in a strong reduction of population sizes and large scale extinction of many riverine organisms (Klink, 1989; Lelek, 1989; Van den Brink et al., 1990). However, from the 1960s measures have been taken to improve water quality (Cals et al., 1996). The first signs of water quality improvement became visible in macroinvertebrate and fish communities from the second half of the 1970s (Lelek, 1989; Van Urk & Bij de Vaate, 1990; Admiraal et al., 1993).

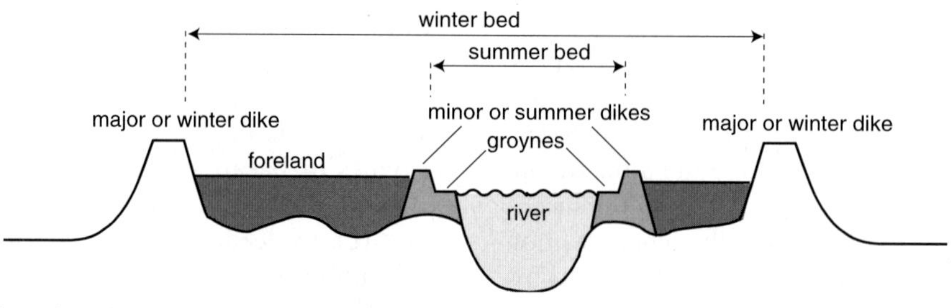

Figure 2. Cross section of the large rivers in the Netherlands. (winter bed, major or winter dike, forelands, minor or summer dike, groynes, summer bed).

An international ecological rehabilitation programme for the river Rhine was initiated in 1987 after the so-called Sandoz-accident in November 1986, an environmental disaster caused by extinguishing water used for a fire control in the Sandoz chemical concern in the vicinity of the Swiss town of Basel (Van Dijk et al., 1995). In addition to further water quality improvement, ecological rehabilitation of the river Rhine in the Netherlands has been focused from that time on the river forelands, the remaining part of the floodplain between the summer and major dikes on both sides of the river channel (Fig. 2; Schropp & Bakker, 1998). The reason for exclusion of the main channel in river rehabilitation is its function as a transport route and the need for unhampered discharge of water, ice and suspended matter.

Floodplain rehabilitation and nature development in the Dutch part of the river Rhine has been based on insights formulated by De Bruin et al. (1987) and WWF (1992). The aim is to give the river back, as much as possible, its natural dynamics in the forelands with regard to its imposed functions (Cals et al., 1996). Effects of biotope alterations on fish and macroinvertebrates in the Rhine delta are the topic of this paper. After an overview of river engineering activities and water quality developments their influence on fish and macroinvertebrate communities are discussed in view of results of investigations mainly performed from the 1980s providing water managers and decision makers a better insight into the functioning of a river ecosystem still influenced by severe anthropogenic stress.

River engineering

River regulation and damming are considered to have the most important destructive impact on biota due to terrestrialisation and fragmentation of the river floodplain system (Schiemer, 1999). Alterations in the catchment area changed runoff patterns. Its originally large water storage capacity had reduced strongly with the result that the relatively slow release of water during wet periods had disappeared. The consequence was an increase in water level fluctuations, which in turn resulted in alterations in sediment transport. The forelands between the summer and winter dikes (Fig. 2) became a trap for fine-grained sediment and in the course of time the natural relief was levelled down by the deposition of clay and by actions of farmers and river managers (Nienhuis & Leuven, 2001). Geo-morphological processes decreased drastically, and were captured in straight jackets by dikes, groynes and weirs. During and after the process of river normalisation the lateral connectivity between the main channel and the water bodies in the forelands gradually disappeared. Incision of the river bed in the remaining single channel was the main effect; loss of shallow lotic habitats the main result.

After embankments in the 12th–14th century the former floodplain was regularly exposed to flooding as the result of dike breaches, particularly in the 18th and in the first part of the 19th century. These breaches were the result of a combination of the poor discharge capacity of the river channels and the relatively poor condition of the river dikes (Middelkoop, 1997). Ice jams in spring were an important cause of the breaches (Driessen, 1994). River engineering activities in the main channel of the Rhine distributaries initially focused on the improvement of the discharge of water and ice, and thereafter on the improvement of navigation. Adjustments of the discharge distribution at the bifurcations in the Rhine delta contributed to those improvements. The first adjustment occurred in the 17th century. Main interventions in the past three centuries are listed in Table 1 (Van der Ven, 1996; Middelkoop, 1997, 1998).

Water quality development

Klink (1989) distinguished four types of river pollution covering more or less successive phases in pollution history. Pollution started with the discharge of organic substances in domestic waste waters, causing an increase of COD (chemical oxygen demand) and BOD (biological oxygen demand), resulting in a decrease of the dissolved oxygen concentration. This was followed by the pollution from heavy metals, being the combined result of mining and industrial activity. The third type is formed by pollution with organic compounds (e.g. PCBs, PAHs), and the lasting the contamination by pesticides. Pollution in both the latter types is caused by organic micropollutants, which have been produced by the chemical industry

Table 1. Main interventions in the Rhine delta in the past three centuries (Van der Ven, 1996; Middelkoop, 1997, 1998) (for geographic names see Fig. 1)

Period/year	Intervention
1707	Opening of the Pannerdensch Kanaal, constructed for the improvement of the northern distributary at the first bifurcation in the Rhine delta. Later on this bifurcation was adjusted several times
1727–1734	Damming of the connections between the rivers Meuse and Waal at Heerewaarden and Voorn. Both connections served as an overflow for the River Waal in periods with high Rhine discharge
1775	Reconstruction of the second bifurcation where the Pannerdensch Kanaal splits into the rivers Nederrijn and IJssel
1850–1870	Digging of the Nieuwe Merwede
1850–1885	Normalisation of the river IJssel
1868	Opening of the Nieuwe Waterweg. Construction of this canal was needed to improve the entrance to the harbours of Rotterdam for sea-going vessels. It forms an artificial outlet in the Rhine delta
1875–1916	Normalisation of the river Waal
1932	Completion of the Zuiderzee damming (De Jong & Bij de Vaate 1989)
1954–1967	Construction of three weirs in the Nederrijn/Lek
1970	Completion of the Haringvliet damming, the joint estuary of the rivers Rhine and Meuse
1989-present	Lowering of forelands and construction of secondary channels along the distributaries (Cals et al., 1996; Schropp & Bakker, 1998)

in the river valleys (point sources of pollution), are everywhere used, causing diffuse pollution.

Discharge of industrial waste water became a serious problem during the process of industrialisation of the river valleys in the 19th century. However, water quality data from the river Rhine are very scarce from the period before 1965. Analysis of dated sediment layers in the forelands have shown to be a helpful tool in reconstructing pollution history with heavy metals and organic micropollutants (Klink, 1989; Beurskens et al., 1993).

The implementation of national laws against pollution and protection of the environment, and the establishment (in 1950) of the International Commission for Protection of the Rhine (ICPR), which got its internationally recognised juridical basis after signing the Treaty of Bern in 1963, were important impulses for the development and realisation of chemical monitoring programmes (Dieperink, 1997). Permanent monitoring stations were erected at several sites along the distributaries. The station in the vicinity of Lobith, at the German–Dutch border, became the primary Dutch reference station for water quality developments in the river Rhine (recently this station was integrated into the two km downstream situated German monitoring station at Bimmen on the south bank).

The Rhine Action Programme, launched in 1987, initially focused on water quality improvement. The countries in the Rhine basin, united in the ICPR, agreed upon a target reduction of at least 50% of the pollution caused by priority compounds by the year 1995 (compared with the situation in 1985). Furthermore, water quality targets for the river Rhine were set for about 50 priority compounds (reduction of 70–90%), not only based on requirements for drinking water production and the protection of aquatic life, but also on human tolerance levels for fish consumption (Van Dijk et al., 1995).

Although calamities in the river Rhine caused a political reveille needed for the rehabilitation of the river Rhine (Dieperink, 1997; *viz.* the endosulphan and the so-called Sandoz calamity in 1969 and 1986, respectively), calamitous pollution accidents still occur (Institute for Inland Water Management and Waste Water Treatment, unpublished data). Long and short term impacts on the river fauna of these short lasting accidents are unknown due to the lack of suitable monitoring programmes. Furthermore, it is difficult to assign effects of specific pollutants on colonisation, growth and reproduction of autochthonous aquatic species in an environment of continuous changing water quality variables in the case of long term effects (e.g. Hellawell, 1989).

Main pollutants

Around the turn of the 19th and 20th centuries, the river was already seriously polluted with domestic and industrial waste water (Tittizer & Krebs, 1996). Phenols in the river made the commercially interesting fish species, such as Atlantic Salmon (*Salmo salar*) and Eel (*Anguilla anguilla*), unfit for consumption (Lobrecht & Van Os, 1977; Van Drimmelen, 1987). Rock bottom of the river pollution was reached in the first part of the 1970s. At that time the river water was acute toxic for water flies and trout embryos, and caused malformation and other effects in fish and insect larvae (Alink et al., 1980; Poels et al., 1980; Slooff, 1982, 1983a, b; Slooff et al., 1983; Van Urk & Kerkum, 1986, 1987; Van der Gaag, 1987). From the second half of the 1970s water quality improved considerably because of:

a. the implementation of environmental protection laws;
b. the construction of waste water treatment plants;
c. international agreements;
d. foundation of policy with monitoring results;
e. development of cleaner production methods;
f. reduction of spills (e.g. caused by calamities);
g. increased public awareness.

Firstly, focussing on priority substances, the organic load and the amount of polluting substances such as heavy metals and organic micropollutants in discharges of domestic and industrial waste water decreased dramatically within a period of approximately 10 years (De Kruijf, 1982; Van der Weijden & Middelburg, 1989; Heymen & Van der Weijden, 1991; Van der Klei et al., 1991). A significant decrease of the organic load resulted in better oxygen conditions for the river fauna (Table 2; Fig. 3). Between 1952 and 1972 the oxygen concentration, measured at the German–Dutch border, had significantly decreased. However, from 1973 onwards, a significant improvement was observed (Table 2). Of relevance here is the fact that the yearly observed

Table 2. One-way ANOVA of trends in water quality parameters measured in the river Rhine at the German–Dutch border

Parameter	Unit	Period	N^a	Slope[b]	R^2	F	p
O_2	mg l^{-1}	1952–1972	21	−0.019	0.560	26	<0.001
O_2	mg l^{-1}	1973–2003	31	0.018	0.810	124	<0.001
COD	mg l^{-1}	1968–1996	23	−0.043	0.846	122	<0.001
Chlorophyll a	μg l^{-1}	1977–2003	26	−0.063	0.582	36	<0.001
Cl^-	mg l^{-1}	1971–2003	33	−0.021	0.600	42	<0.001
Cl^-	mg l^{-1}	1990–2003	14	−0.068	0.810	51	<0.001
Cd_{total}	μg.l^{-1}	1975–2003	29	−0.149	0.761	90	<0.001
Cu_{total}	μg l^{-1}	1970–2003	34	−0.066	0.819	145	<0.001
Hg_{total}	μg l^{-1}	1971–2003	33	−0.035	0.916	350	<0.001
Pb_{total}	μg l^{-1}	1971–2003	33	−0.077	0.737	85	<0.001
Zn_{total}	μg l^{-1}	1971–2003	33	−0.091	0.908	318	<0.001
Mineral oil[g]	mg kg^{-1}	1972–1987	16	−0.270	0.888	111	<0.001
$PCB_{sum}^{c,g}$	μg kg^{-1}	1988–2003	16	−0.031	0.244	5.8	<0.05
PAHs[d]	μg l^{-1}	1979–1988	10	−0.239	0.891	66	<0.001
γ-HCH[e]	μg l^{-1}	1973–2003	30	−0.139	0.907	285	<0.001
ACEI[f]	μg l^{-1}	1973–2003	31	−0.111	0.734	84	<0.001

[a]number of yearly averages (number of observations ≥6 per year); [b]based on ln transformed averages; [c]sum of the 28, 52, 101, 118, 138, 153 and 180 PCB congeners; [d]six of Borneff; [e]linden; [f]acetyl-choline esterase inhibitors; [g]in dry suspended matter.

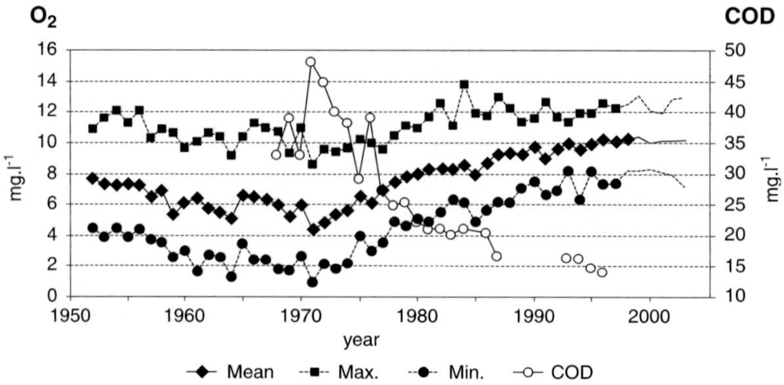

Figure 3. Yearly average, minimum and maximum observed dissolved oxygen concentrations, and yearly average of the COD in the river Rhine measured at the German–Dutch border.

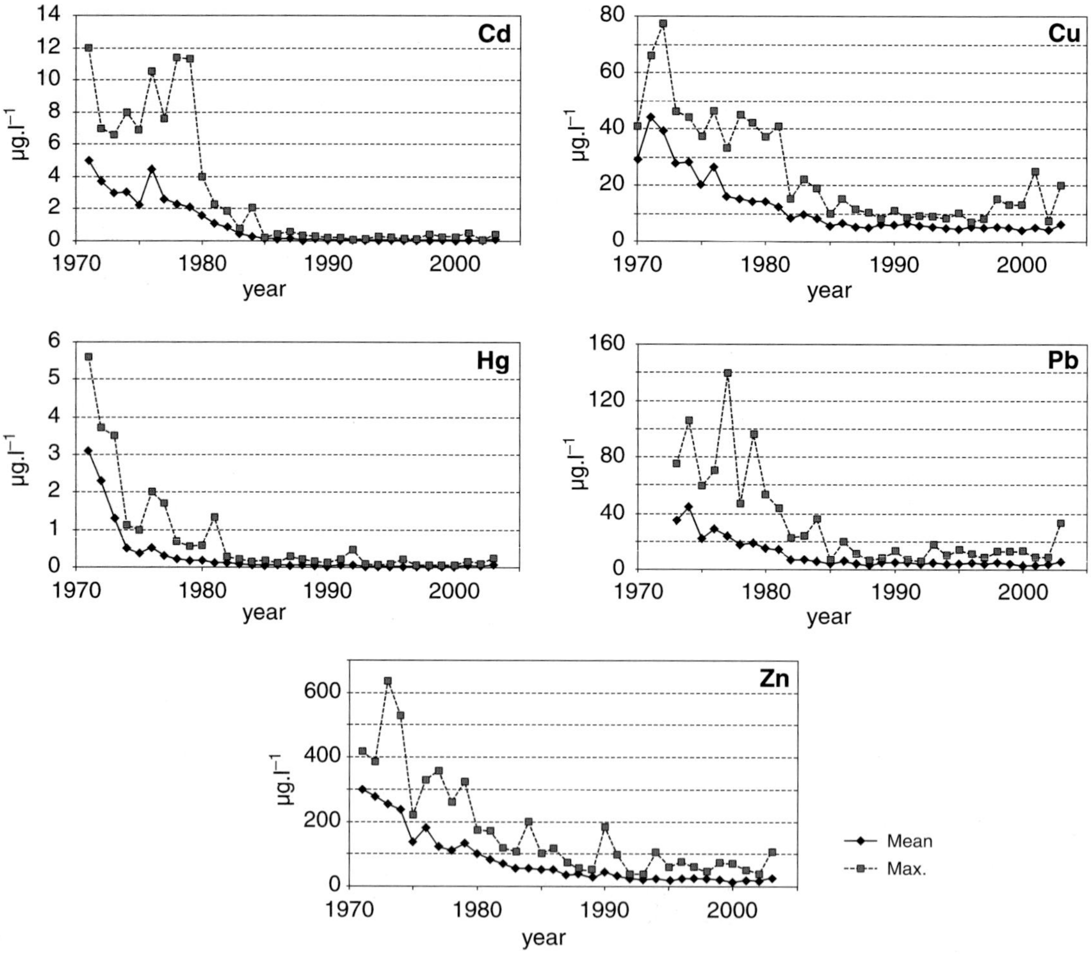

Figure 4. Yearly average and maximum observed total concentrations of Cd, Cu, Hg, Pb and Zn in the river Rhine measured at the German–Dutch border.

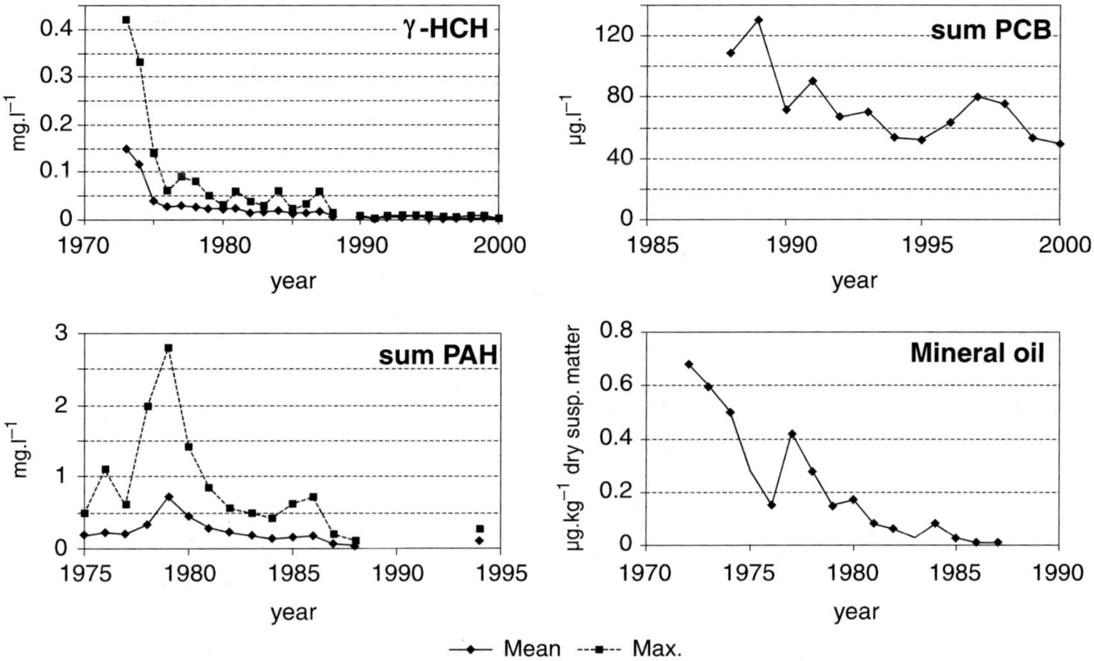

Figure 5. Yearly average concentrations of γ-HCH, sum PCB, sum PAH and mineral oil in the river Rhine measured at the German–Dutch border, including maximum observed concentrations of γ-HCH and sum PAH.

minimum concentration in the latter period increased from 2 to 8 mg l^{-1}.

Concentrations of toxic substances in the river water such as heavy metals, organic micropollutants (e.g. PAHs, γ-HCH), as well as mineral oil, significantly reduced with at least a factor 10 in the period 1970–1990 (Fig. 4 and 5; Heymen & Van der Weijden, 1991; Van der Velde et al., 1991; Admiraal et al., 1993; Van Urk et al., 1993). As a result of this decrease, the differences, in general, between the yearly average and the maximum observed concentrations were also strongly reduced (Fig. 4 and 5).

Although water quality improvement of the river Rhine became a success story, toxicity of the Rhine water did not disappear completely because of the sum-toxicity of thousands of chemical compounds present in the water, most of them in concentrations below detection level (Hendriks et al., 1994; Leuven et al., 1998; Van der Velde & Leuven, 1999). In 1993, for example, Nolan et al.

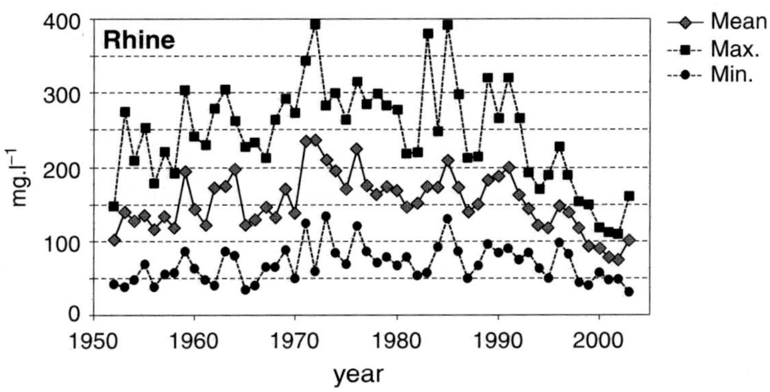

Figure 6. Yearly average, minimum and maximum observed chloride concentrations in the river Rhine measured at the German–Dutch border.

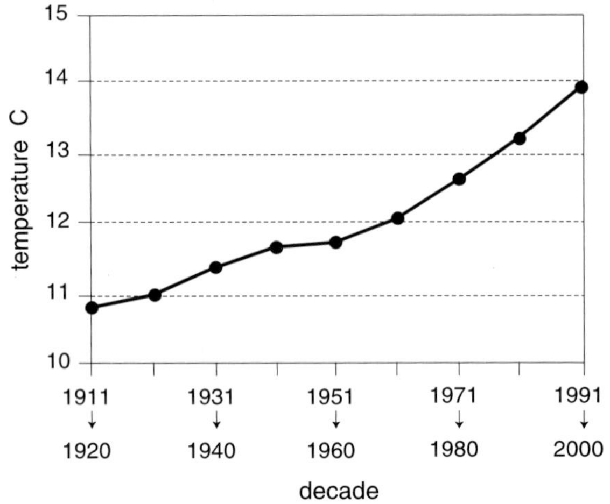

Figure 7. Decade averages of the water temperature of the river Rhine.

(1998; 2000) observed toxic effects on the skin and in the immune system of sea trout smolts (*Salmo trutta*). Other important water quality related habitat components for the river fauna, such as chloride concentration and water temperature, did not improve to the same extent (chloride, Fig. 6) or, quite the contrary, even deteriorated (water temperature, Fig. 7). Main sources of chloride enrichment are potassium mines in the Alsace (France) and brown coal mines in Germany. In the period 1960–2000 the yearly average chloride concentration varied between 90 and 240 mg l^{-1} (Fig. 6) with minimum and maximum concentrations measuring 35 and 420 mg l^{-1}, respectively (this maximum value was reported in the annual reports of the Combined Rhine and Meuse Waterworks, RIWA, Amsterdam). To put these values into the right context, it should be noted that the natural chloride concentration assessed was 12 mg l^{-1} on average (Molt, 1961). Reduction of the chloride discharge of the potassium mines was an important issue in international discussions on water quality improvement in the period 1960–1990. A final agreement with France was reached in 1991 consisting of a salt discharge reduction of 60 kg s^{-1} and temporary salt storage during periods in which the chloride concentration in the river Rhine exceeds 200 mg l^{-1} at the German–Dutch border (Dieperink, 1997). Results of all negotiations were visible in a significant decline of the average chloride concentration from 1971, noticeably after 1990 (Table 2).

Thermal pollution has been mainly caused by discharge of heated cooling water of electric power plants and industries. Recently part of the water temperature increase could also be attributed to climate changes probably due to global warming. Compared with the situation around 1900, the average water temperature had increased by 3 °C till around 1980 (Wessels, 1984), and has increased with 0.5 °C per 10 years from 1952 (Fig. 7).

Effects on the fauna

Before the initiation of the National Biological Monitoring Programme in the river Rhine in 1992, data on macroinvertebrates and fish were not collected systematically with a long-term horizon, but only in relatively short-term research projects. In 1973 Van Urk (1981, 1984) started to study changes in the epilithic littoral macroinvertebrate fauna in the river IJssel, one of the three main distributaries of the river Rhine. This study has been continued until today and has resulted in the longest macroinvertebrate time series known in larger rivers in the Netherlands. From the 1960s till 1992 fisheries independent fish stock monitoring was mainly focussed on commercially interesting species (Klinge et al., 1998).

Macroinvertebrates

As indicated above, the macroinvertebrate community on riprap in the littoral zone of the river IJssel has been monitored from 1973 on a yearly basis (Van Urk, 1984; Van Urk & Bij de Vaate, 1990; Bij de Vaate, 1994). Apart from an incidental survey made by Lauterborn in the beginning of the 20th century (Lauterborn, 1918), the IJssel dataset is the only source for data older than 15 years. To fill this gap, results of palaeolimnological investigations are used (Klink, 1989) to reconstruct macroinvertebrate communities for e.g. target and reference descriptions. However, only the remains of insects can be found in sediment layers. On the other hand, insect larvae constitute the main part of the species richness in a healthy river. Van den Brink et al. (1990) combined different sources to reconstruct macroinvertebrate development in the Rhine delta from the beginning of the 20th century. They concluded that around 1985 the number of rheophilic species had declined, while the number of euryoecious species had increased. As a result of the increased chloride concentration (Fig. 6) a number of brackish water crustaceans were found approximately 100 km upstream from their original limits of distribution (Den Hartog et al., 1989).

From recolonisation patterns of autochthonous species in the Rhine delta, two phases in water quality recovery could be distinguished. Between 1975 and 1980 some pollution stress tolerant insects (in this initial period mainly chironomid and a few caddis fly species) and molluscs began to recolonise the river. Density increase of these macroinvertebrates in the years after that period correlated well with a decrease of chemical pollutants in the river water (Van Urk, 1981; Van Urk & Bij de Vaate, 1990; Van Urk et al., 1993).

Recolonisation of the lower Rhine by the burrowing mayfly *Ephoron virgo* in 1991 (Bij de Vaate et al., 1992) and its mass development in the following years marked the start of the second phase in water quality rehabilitation of the river. Concentrations of toxicants had strongly decreased in

Table 3. Dominant (••) and subdominant (•) macroinvertebrate species in two habitats in the free flowing distributaries in the Rhine delta in the period 1975–2000 (R = riprap in the littoral zone; B = channel bottom)

Taxon	1975		1980		1990		1995		2000	
	R	B	R	B	R	B	R	B	R	B
River Waal										
Corbicula fluminalis								•		
Corbicula fluminea						•		••		••
Dreissena polymorpha							•			
Potamopyrgus antipodarum						•				
Propappus volcki								••		
Chelicorophium curvispinum					••	••	••		••	
Jaera istri									••	
Dikerogammarus villosus									••	
Gammarus tigrinus					•			••		•
Cricotopus bicinctus									•	
Kloosia pusilla								•		
Neozavriella species									•	
River IJssel										
Corbicula fluminea										•
Dreissena polymorpha	•				•					
Ancylus fluviatilis									•	
Potamopyrgus antipodarum								•		
Hypania invalida										••
Asellus aquaticus	••									
Jaera istri									••	
Chelicorophium curvispinum					••		••	••	••	••
Dikerogammarus villosus									•	••
Gammarus tigrinus					•		•	••		
Hydropsyche contubernalis			•							
Dicrotendipes nervosus			•							
Cricotopus intersectus							•			
Cricotopus species			••							

Species were considered to be dominant or subdominant if their density contributed to > 20% of the total macroinvertebrate density or >20% of the macroinvertebrate density minus the dominant species respectively. If no dominant species were present, those contributing to 10–20% of the total macroinvertebrate density were considered to be subdominant. Non-indigenous species underlined.

the past two decades, and oxygen concentration had reached a level that no longer seemed to be the limiting factor for the colonisation of many macroinvertebrate species with a higher demand for oxygen. Another example is the dragonfly *Gomphus flavipes* that recolonised the Rhine delta from 1996 (Habraken & Crombaghs, 1997; Goudsmits, 1998). Although potentially several other sensitive autochthonous macroinvertebrate species could also extend their territory in the Rhine distributaries due to water quality improvement, other factors such as the absence of physical habitat and the increased water temperature most probably prevent successful colonisation of these species.

Macroinvertebrate communities are currently (since 1985) dominated by non-indigenous species (Table 3), mainly species from the Ponto-Caspian area. The colonisation of the River Rhine by Ponto-Caspian species was accelerated by the opening of the Main-Danube Canal in September 1992 (Bij de Vaate et al., 2002; Van der Velde et al., 2002). Some Ponto-Caspian species (e.g. *Chelicorophium curvispinum* and *Dikerogammarus villosus*) have had a relatively strong negative impact on densities of other macroinvertebrates (Van den Brink et al., 1991, 1993; Van der Velde et al., 1994, 2002; Rajagopal et al., 1999; Dick & Platvoet, 2000).

Fishes

Van den Brink et al. (1990) gave a brief overview of developments in the fish fauna in the Rhine delta from the start of the 20th century till 1985. In general, densities of most species were relatively low in the period 1960–1980. The groups of the rheophilic and anadromous species suffered the most from the combination of water quality deterioration, river engineering, damming, and the closure of river outlets (e.g. Lelek, 1989). On the other hand, the fish fauna took most advantage of the ecological rehabilitation programmes that were introduced after the Sandoz-accident in 1986, especially Atlantic salmon (*Salmo salar*) and sea trout (*Salmo trutta*) which have been reintroduced on a large scale in the main tributaries. In 2000, species richness was higher than the previous century (Fig. 8), being the result of colonisation of the Rhine delta by introduced species. Nature development in the river forelands has been important from the point of view of fish stocks as well. The construction of secondary channels positively contributed to fish stocks in the main channel of the distributaries (Grift, 2001).

Rehabilitation perspectives

Rivers restoration has become an important issue from the end of the 1980s (Boon et al., 1992; Sparks, 1995; Nienhuis & Leuven, 1998; Pedroli & Postma, 1998; Nienhuis et al., 2002; Buijse et al., 2005). Important general aspects are (a) improvement of the lateral and (b) longitudinal connectivity and (c) connectivity with the groundwater. Dutch policy aim for river restoration is also to reduce habitat fragmentation since river valleys are considered to be important corridors for migration and dispersal of aquatic and terrestrial ani-

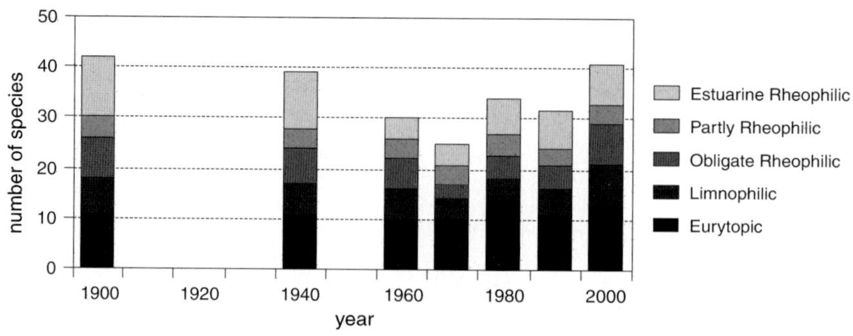

Figure 8. Number of fish species per guild in the Rhine distributaries (1900–2000). Sources: Lelek (1989), Oosterbroek (1990), Van den Brink et al. (1990), De Nie (1997), and unpublished results of the National Biological Monitoring Programme 1992–2000.

mal species, and for biodiversity conservation (Anonymous, 1998; 2001). As the consequence of this policy, structure and functioning of ecological networks need to be improved in order to create viable populations of target species (ICPR, 1998; Foppen & Reijnen, 1998; Chardon et al., 2000). However, restoration of geo-morphological processes to improve connectivity in the Rhine distributaries is only possible in a very limited way because of the functions assigned. Unhampered discharge of water and ice and the economy related functions remain more important than their ecological functions due to safety and socio-economic reasons, respectively (Anonymous, 2001). In practice, possibilities for restoration of large rivers in the Netherlands are thus mainly possible in aquatic/terrestrial transition zones in the Rhine delta (Van Dijk et al., 1995; Heiler et al., 1995; Simons et al., 2001; Buijse et al., 2002, 2005). Several large ecologically important reaches (1000–6000 ha each), with smaller areas in between, were identified along the Rhine distributaries, of which totally about 7500 ha of floodplain area have an important ecological function (Van Dijk et al., 1995). Improvement of longitudinal connectivity within the river channel was realised in 2004 by the construction of fish ladders at the weirs in the Nederrijn/Lek distributary. Further improvement will be realised by optimalisation of the discharge regime of the sluices in the barrier dams separating the Wadden Sea and Lake IJsselmeer in the northern, and the North Sea and lake Haringvliet in the south western part of the Netherlands (Fig. 1). Measures will focus on improvement of fish migration through both dams and on the creation of a brackish zone on both sides of the Haringvliet dam (Smit et al., 1997).

Conclusions

The Rhine delta had become an area with a strong anthropogenic pressure on the environment. This will not change in the future due to the resulting unacceptable social and economical impacts. Therefore the conclusions from this review are that (a) most of the alterations in the Rhine delta are irreversible; (b) chances for the development of riverine biotopes have therefore to be found in the relatively narrow forelands and not in the main channels of the Rhine delta or in the original floodplain; (c) further reduction of pollutants, especially thermal pollution is needed to help original species to colonise the Rhine delta again; (d) the ongoing process of invasions by non-indigenous species clearly leaves a mark on the existing communities and on recolonisation possibilities of original species. This is CWE publication no. 425.

References

Admiraal, W., G. van der Velde, H. Smit & W. G. Cazemier, 1993. The rivers Rhine and Meuse in The Netherlands: present state and signs of ecological recovery. Hydrobiologia 265: 97–128.

Alink, G. M., E. M. H. Frederix-Wolters, M. A. van der Gaag, J. F. J. van der Kerkhoff & C. L. M. Poels, 1980. Induction of sister-chromatid exchanges in fish exposed to Rhine water. Mutation Research 78: 369–374.

Anonymous, 1998. Vierde nota waterhuishouding. Regeringsbeslissing. Ministerie van Verkeer en Waterstaat, 's Gravenhage (in Dutch).

Anonymous, 2001. Beheersplan voor de Rijkswateren. Programma voor het beheer in de periode 2001–2004. Report Ministerie voor Verkeer & Waterstaat, 's Gravenhage (in Dutch).

Beurskens, J. E. M., G. A. J. Mol, H. L. Barreveld, B. van Munster & H. J. Winkels, 1993. Geochronology of priority pollutants in a sedimentation area of the Rhine River. Environmental Toxicology and Chemistry 12: 1549–1566.

Bij de Vaate, A., 1994. Long-term changes in the macroinvertebrate fauna of the River IJssel, The Netherlands. Verhandlungen der Internationale Vereinigung für theoretische und angewandte Limnologie 25: 1563–1567.

Bij de Vaate, A., 2003. Degradation and recovery of the freshwater fauna in the lower sections of the rivers Rhine and Meuse. Ph.D. thesis Wageningen University, Wageningen.

Bij de Vaate, A., A. Klink & F. Oosterbroek, 1992. The mayfly, *Ephoron virgo* (Olivier), back in the Dutch parts of the rivers Rhine and Meuse. Hydrobiological Bulletin 25: 237–240.

Bij de Vaate, A., K. Jazdzewski, H. Ketelaars, S. Gollasch & G. van der Velde, 2002. Geographical patterns in range extension of macroinvertebrate Ponto-Caspian species in Europe. Canadian Journal of Fisheries and Aquatic Sciences 59: 1159–1174.

Boon, P. J., P. Calow & G. E. Petts, 1992. River Conservation and Management. John Wiley, Chichester.

Buijse, A. D., H. Coops, M. Staras, L. H. Jans, G. J. van Geest, R. E. Grift, B. W. Ibelings, W. Oosterberg & F. C. J. M. Roozen, 2002. Restoration strategies for river-floodplains along large lowland rivers in Europe. Freshwater Biology 47: 889–907.

Buijse, A. D., F. Klijn, R. S. E. W. Leuven, H. Middelkoop, F. Schiermer, J. Thorp & H. Wolfert, 2005. Rehabilitation of large rivers: references, achievements and integration into river management. Archiv für Hydrobiologie, Supplement 155, Large Rivers 15: 715–738.

Cals, M. J. R., R. Postma & E. C. L. Marteijn, 1996. Ecological river restoration in The Netherlands: state of the art and strategies for the future. Aquatic Conservation: Marine and Freshwater Ecosystems 8: 61–70.

Chardon, J. P., R. P. B. Foppen & N. Geilen, 2000. LACRH-RIVER: a method to assess the functioning of rivers as ecological networks. European Water Management 3: 35–43.

De Bruin, D., D. Hamhuis, L. van Nieuwenhuijze, W. Overmars, D. Sijmons & F. Vera, 1987. Ooievaar : de toekomst van het rivierengebied. Report Stichting Gelderse Milieufederatie, Arnhem (in Dutch).

De Jong, J. & A. bij de Vaate, 1989. Dams and the environment. The Zuiderzee damming. Bulletin International Commission on Large Dams 66: 1–85.

De Kruijf, H. A. M., 1982. Progress in the application of the convention against the chemical pollution of the River Rhine. Toxicological and Environmental Chemistry 6: 41–63.

De Nie, H., 1997. Atlas van de Nederlandse zoetwatervissen. Media Publishing, Doetinchem (in Dutch).

Den Hartog, C., F. B. W. van den Brink & G. van der Velde, 1989. Brackish-water invaders in the River Rhine. A bioindication for increased salinity level over the years. Naturwissenschaften 76: 80–81.

Dick, J. T. A. & D. Platvoet, 2000. Invading predatory crustacean *Dikerogammarus villosus* eliminates both native and exotic species. Proceedings of the Royal Society of London B Biology 267: 977–983.

Dieperink, C., 1997. Tussen zout en zalm. Lessen uit de ontwikkeling van het regime inzake de Rijnvervuiling. Ph.D. thesis, University Utrecht, Utrecht (in Dutch).

Driessen, A. M. A. J., 1994. Watersnood tussen Maas en Waal. Overstromingsrampen in het rivierengebied tussen 1780 en 1810. Walburg Pers, Zutphen (in Dutch).

Foppen, R. P. B. & R. Reijnen, 1998. Ecological networks in riparian systems: Examples for Dutch floodplain rivers. In Nienhuis, P. H., R. S. E. W. Leuven & A. M. J. Ragas (eds), New Concepts for Sustainable Management of River Basins, Backhuys Publishers, Leiden, 85–93.

Galat, D. L., L. H. Fredrickson, D. D. Humburg, K. J. Bataille, J. R. Bodie, J. Dohrenwend, G. T. Gelwicks, J. E. Havel, D. L. Helmers, J. B. Hooker, J. R. Jones, M. F. Knowlton, J. Kubisiak, J. Mazourek, A. C. McColpin, R. B. Renken & R. D. Semlitsch, 1998. Flooding to restore connectivity of regulated, large-river wetlands. BioScience 48: 721–733.

Gore, J. A. & F. D. Shields Jr., 1995. Can large rivers be restored? BioScience 45: 142–152.

Goudsmits, K., 1998. De rivierrombout (*Gomphus flavipes*) terug in Nederland. Nieuwsbrief Nederlandse Vereniging voor Libellenstudie 2(3): 2 (in Dutch).

Grift, R. E., 2001. How fish benefit from floodplain restoration along the lower River Rhine. Ph.D. thesis Wageningen University, Wageningen.

Habraken, J. M. P. M. & B. H. J. M. Crombaghs, 1997. Een vondst van de rivierrombout (*Gomphus flavipes* (Carpentier)) langs de Waal. Brachytron 1(1): 3–5 (in Dutch).

Heiler, G., T. Hein, F. Schiemer & G. Bornette, 1995. Hydrological connectivity and flood pulses as the central aspects for the integrity of a river-floodplain system. Regulated Rivers: Research & Management 11: 351–361.

Hellawell, J. M., 1989. Biological Indicators of Freshwater Pollution and Environmental Management. Elsevier, London.

Hendriks, A. J., J. L. Maas-Diepeveen, A. Noordsij & M. A. van der Gaag, 1994. Monitoring response of XAD-concentrated water in the Rhine delta: a major part of the toxic compounds remains unidentified. Water Research 28: 581–598.

Heymen, R. & M. van der Weijden, 1991. Resultaten van waterkwaliteitsonderzoek in de Rijn in Nederland. Report 91.047, Institute for Inland Water Management & Waste Water Treatment, Lelystad (in Dutch).

ICPR, 1998. Bestandsaufnahme der ökologisch wertvollen Gebiete am Rhein und erste Schritte auf dem Weg zum Biotopverbund. Report International Commission for Protection of the Rhine against Pollution (ICPR), Koblenz (in German).

Kalweit, H. (eds), 1993. Der Rhein unter der Einwirkung des Menschen: Ausbau Schiffahrt, Wasserwirtschaft. International Commission for Hydrology of the River Rhine, Lelystad (in German).

Klinge, M., A. D. Buijse, W. G. Cazemier, E. H. R. R. Lammens & K. H. Prins, 1998. Biologische monitoring zoete rijkswateren: vis in de zoete rijkswateren, 1992–1996. Report 98.017. Institute for Inland Water Management & Waste Water Treatment, Lelystad (in Dutch).

Klink, A., 1989. The Lower Rhine: palaeoecological analysis. In Petts G. E. (ed.), Historical Change of Large Alluvial Rivers: Western Europe. John Wiley, Chichester, 183–201.

Lauterborn, R., 1918. Die geographische und biologische Gliederung des Rheinstroms III. Sitzungsberichte der Heidelberger Akademie der Wissenschaften. Mathematisch-Naturwissenschaftliche Klasse 9: 1–87.

Lelek, A., 1989. The Rhine River and some of its tributaries under human impact in the last two centuries. In Dodge, D.P. (ed.), Proc. Intern. Large River Symposium. Canadian Special Publication of Fisheries and Aquatic Sciences 106: 469–487.

Leuven, R. S. E. W., J. L. M. Haans, A. J. Hendriks, R. A. C. Lock & S. E. Wendelaar Bonga, 1998. Assessing cumulative impacts of multiple stressors on river systems. In Nienhuis, P. H., R. S. E. W. Leuven & A. M. J. Ragas (eds), New Concepts for Sustainable Management of River Basins. Backhuys Publishers, Leiden: 241–260.

Lobregt, P. & J. van Os, 1977. De laatste Riviervissers. De Walburg Pers, Zutphen (in Dutch).

Middelkoop, H., 1997. Embanked floodplains in the Netherlands: geomorphological evolution over various time scales. Nederlandse Geografische Studies 224: 1–341.

Middelkoop, H. (eds), 1998. Twee rivieren. Rijn en Maas in Nederland. Report 98.041 Institute for Inland Water Management & Waste Water Treatment, Arnhem (in Dutch).

Molt, E .L., 1961. Pollution of the River Rhine water. In Anonymous, The River Rhine. Report Technical University Delft, Delft (in Dutch).

Nienhuis, P. H., A. D Buijse, R. S. E. W. Leuven, A. J. M. Smits, R. J. W. de Nooij & E. M. Samborska, 2002. Ecological rehabilitation of the lowland basin of the River Rhine (NW Europe). Hydrobiologia 478: 241–260.

Nienhuis, P. H. & R. S. E. W. Leuven, 1998. Ecological concepts for the sustainable management of lowland river basins: a review. In Nienhuis P. H., R. S. E. W. Leuven & A. M. J. Ragas (eds), New Concepts for Sustainable Management of River Basins. Backhuys Publishers, Leiden: 7–33.

Nienhuis, P. H. & R. S. E. W. Leuven, 2001. River restoration and flood protection: controversy or synergism? Hydrobiologia 444: 85–99.

Nolan, D. T., R. H. Hadderingh, F. A. T. Spanings, H. A. Jenner & S. E. Wendelaar Bonga, 2000. Acute temperature elevation in tap and Rhine water affects skin and gill epithelia, hydromineral balance, and gill Na^+/K^+-ATPase activity of brown trout (*Salmo trutta*) smolts. Canadian Journal of Fisheries and Aquatic Sciences 57: 708–718.

Nolan, D. T., R. H. Hadderingh, H. A. Jenner & S. E. Wendelaar Bonga, 1998. The effects of exposure to Rhine water on sea trout smolt (*Salmo trutta trutta* L.): an ultrastructural and physiological study. In Nienhuis, P. H., R. S. E. W. Leuven & A. M. J. Ragas (eds), New Concepts for Sustainable Management of River Basins. Backhuys Publishers, Leiden: 261–271.

Oosterbroek, F. M. J., 1990. Drie jaar monitoring van de visstand in de Waal bij de EPON-centrale te Nijmegen. Report 287, Catholic University, Nijmegen, Department of Aquatic Ecology & Biogeology (in Dutch).

Pedroli, G. B. M. & R. Postma, 1998. Nature rehabilitation in European river ecosystems: three cases. In Nienhuis, P.H., R.S.E.W. Leuven & A.M.J. Ragas (eds), New Concepts for Sustainable Management of River Basins. Backhuys Publishers, Leiden: 67–84.

Poels, C. L. M., M. A. van der Gaag & J. F. J. van de Kerkhoff, 1980. An investigation into the long-term effects of Rhine water on rainbow trout. Water Research 14: 1029–1035.

Rajagopal, S., G. van der Velde, B. G. P. Paffen & A. bij de Vaate, 1999. Population densities, biomass, growth and production of *Corophium curvispinum* G.O. Sars (Amphipoda) in the Lower Rhine. In Schram, F. R. & J. C. von Vaupel Klein (eds), Crustaceans and the Biodiversity Crisis: Proceedings 4th international Crustacean Congress. Vol. 1, Brill, Leiden: 457–472.

Schiemer, F., 1999. Conservation of biodiversity in floodplain rivers. Archiv für Hydrobiologie Supplement 115: 423–438.

Schropp, M. H. I. & C. Bakker, 1998. Secondary channels as a basis for the ecological rehabilitation of Dutch rivers. Aquatic Conservation: Marine and Freshwater Ecosystems 8: 53–59.

Simons, H. E. J., C. Bakker, M. H. I. Schropp, L. H. Jans, F. R. Kok & R. E. Grift, 2001. Man-made secondary channels along the River Rhine (The Netherlands); results of post-project monitoring. Regulated Rivers: Research & Management 17: 473–491.

Slooff, W., 1982. Skeletal anomalies in fish from polluted surface waters. Aquatic Toxicology 2: 157–173.

Slooff, W., 1983a. Study on the use of feral fish as indicators for the presence of chemical carcinogens in Dutch surface waters. Aquatic Toxicology 3: 127–139.

Slooff, W., 1983b. Rijn-, Lek-, Waal- en IJsseluiterwaarden onder invloed van ingrepen en verontreinigingen. In Hekstra G. P. & W. Joenje (eds), Rijnwater in Nederland. Ecologische kring, Arnhem, 13–31.

Slooff, W., D. de Zwart & J. F. J. van de Kerkhoff, 1983. Monitoring the rivers Rhine and Meuse in The Netherlands for toxicity. Aquatic Toxicology 4: 189–198.

Smit, H., G. van der Velde, R. G. Smits & H. Coops, 1997. Ecosystem developments in the Rhine-Meuse delta during two decades after enclosure and steps towards estuary restoration. Estuaries 20: 504–520.

Sparks, R. E., 1995. Need for ecosystem management of large rivers and their floodplains. BioScience 45: 168–182.

Tittizer, T. & F. Krebs (eds), 1996. Ökosystemforschung: der Rhein und seine Auen, eine Bilanz. Springer Verlag, Berlin (in German).

Van den Brink, F. W. B., G. van der Velde & W. G. Cazemier, 1990. The faunistic composition of the freshwater section of the River Rhine in The Netherlands: present state and changes since 1900. In Kinzelbach, R. & G. Friedrich (eds), Biologie des Rheins. Vol. 1, Limnologie aktuell: 191–216.

Van den Brink, F. W. B., G. van der Velde & A. bij de Vaate, 1991. Amphipod invasion on the Rhine. Nature 352: 576.

Van den Brink, F. W. B., G. van der Velde & A. bij de Vaate, 1993. Ecological aspects, explosive range extension and impact of a mass invader, *Corophium curvispinum* Sars, 1895 (Crustacea: Amphipoda), in the Lower Rhine (The Netherlands). Oecologia 93: 224–232.

Van der Gaag, M. A., 1987. Tests for growth retardation and pathology in fishes exposed to complex mixtures: Experiences on polluted river water. In Vouk V. B., G. C. Butler, A. C. Upton, D. C. Parke & S. C. Asher (eds), Methods for Assessing the Effects of Mixtures of Chemicals. Wiley, Chichester: 775–796.

Van der Klei, W., R. H. Dekker, H. Kersten & J. A. W. de Wit, 1991. Water management of the River Rhine: past, present and future. Journal European Water Pollution Control Federation 1: 9–18.

Van der Velde, G., G. van Urk, F. W. B. van den Brink, F. Colijn, W. A. Bruggeman & R. S. E. W Leuven, 1991. Rein Rijnwater, een sleutelfactor in chemisch oecosysteemherstel. In Hekstra, G. P. & F. J. M. van der Linden (eds), Flora en fauna chemisch onder druk. Pudoc, Wageningen: 231–266 (in Dutch).

Van der Velde, G., F. W. B. van den Brink, B. G. P. Paffen, A. bij de Vaate & H. A. Jenner, 1994. Decline of zebra mussel populations in the Rhine: competition between two mass invaders (*Dreissena polymorpha* and *Corophium curvispinum*). Naturwissenschaften 81: 32–34.

Van der Velde, G. & R. S. E. W. Leuven, 1999. Polluted river systems: monitoring and assessment of ecotoxicological risks. Acta Hydrochimica et Hydrobiologica 27: 251–256.

Van der Velde, G., I. Nagelkerken, S. Rajagopal & A. bij de Vaate, 2002. Invasions by alien species in inland freshwater bodies in Western Europe: the Rhine delta. In Leppäkoski,

E., S. Gollasch & S. Olenin (ed) Aquatic Invasive Species of Europe. Distribution, Impacts and Management. Kluwer Academic Publishers, Dordrecht: 360–372.

Van der Ven, G.P. (ed.), 1996. Man-made Lowlands. History of Water Management and land reclamation in The Netherlands. Stichting Matrijs Utrecht.

Van der Weijden, C. H. & J. J. Middelburg, 1989. Hydrogeochemistry of the River Rhine: long-term and seasonal variability, elemental budgets, base levels and pollution. Water Research 23: 1247–1266.

Van Dijk, G. M., E. C. L. Marteijn & A. Schulte-Wülwer-Leidig, 1995. Ecological rehabilitation of the River Rhine: plans, progress and perspectives. Regulated Rivers: Research & Management 11: 377–388.

Van Drimmelen, D. E., 1987. Schets van de Nederlandse rivier- en binnenvisserij tot het midden van de 20ste eeuw. Organisatie ter Verbetering van de Binnenvisserij, Nieuwegein (in Dutch).

Van Urk, G., 1981. Changes in the macroinvertebrate fauna of the River IJssel. H_2O 14: 494–499.

Van Urk, G., 1984. Lower Rhine – Meuse. In Whitton, B. A. (ed.), Ecology of European Rivers. Blackwell, Oxford: 291–315.

Van Urk, G. & F. C. M. Kerkum, 1986. Misvormingen bij muggelarven uit Nederlandse oppervlaktewateren. H_2O 19: 624–627 (in Dutch with English summary).

Van Urk, G. & F. C. M. Kerkum, 1987. Chironomid mortality after the Sandoz accident and deformities in *Chironomus* larvae due to sediment pollution in the Rhine. Aqua 4: 191–196.

Van Urk, G. & H. Smit, 1989. The Lower Rhine: geomorphological changes. In Petts, G. E. (ed.), Historical Changes of Large Alluvial Rivers: Western Europe. Wiley & Sons, Chichester: 167–182.

Van Urk, G. & A. bij de Vaate, 1990. Ecological studies in the Lower Rhine in The Netherlands. In Kinzelbach, R. & G. Friedrich (eds), Biologie des Rheins. Vol. 1, Limnologie Aktuell: 131–145.

Van Urk, G., F. Kerkum & A. bij de Vaate, 1993. Insects and insecticides in the Lower Rhine. Water Research 27: 205–213.

Wessels, H. R. A., 1984. Water temperature of the River Rhine, 1911–1984. H_2O 17(18): 396–399.

WWF., 1992. Living Rivers. World Wide Fund for Nature, Zeist.

Hydrobiologia (2006) 565:243–257
R.S.E.W. Leuven, A.M.J. Ragas, A.J.M. Smits & G. van der Velde (eds), Living Rivers: Trends and Challenges in Science and Management
DOI 10.1007/s10750-005-1917-3

Flood detention, nature development and water quality along the lowland river Sava, Croatia

M.J. Baptist[1,2,*], M. Haasnoot[1], P. Cornelissen[3], J. Icke[1], G. van der Wedden[4], H.J. de Vriend[4] & G. Gugić[5]
[1]*WL|Delft Hydraulics, P.O. Box 177, 2600 MH Delft, The Netherlands*
[2]*Faculty of Civil Engineering and Geosciences, Water Resources Section, Delft University of Technology, Stevinweg 1, 2628 CN Delft, The Netherlands*
[3]*Rijkswaterstaat, RIZA, Zuiderwagenplein 2, 8284 AD Lelystad, The Netherlands*
[4]*Faculty of Civil Engineering and Geosciences, Hydraulic Engineering Section, Delft University of Technology, Stevinweg 1, 2628 CN, Delft, The Netherlands*
[5]*Lonjsko Polje Nature Park, HR-44324, Jasenovac, Croatia*
*(*Author for correspondence: E-mail: m.j.baptist@citg.tudelft.nl)*

Key words: ecohydrology, hydrologic modelling, nature management, phosphorus storage, sediment storage, vegetation succession

Abstract

The construction or designation of detention areas along lowland rivers is considered along many European rivers. Since Croatia accommodates large detention areas, both natural (e.g., Mokro Polje) and controlled (Lonjsko Polje), it serves as an excellent example for planned detention areas elsewhere in Europe. This modelling study focuses on the controlled detention area of Lonjsko Polje. The flooding characteristics of the area are assessed in combination with the vegetation development and the transport and storage of sediment and phosphorus. Results of the modelling show that it is not so much the intake capacity that determines the flood duration time of a detention area, but the drainage capacity. A too long inundation duration following a flood event is shown to lead to major shifts in the vegetation composition. The results further indicate that about 30% of the sediment and adsorbed phosphorus that enters the detention area during an extreme (1:100 years) flood is retained within the area; this is about 10% of the total sediment and adsorbed phosphorus load of the Sava. Results of this study can be used to properly design and manage detention areas along lowland rivers.

Introduction

Detention areas serve as a temporary storage of water that is diverted from the river channel. This lowers the peak level of the flood and thus alleviates the flood risk for downstream areas. The construction or designation of detention areas along lowland rivers forms part of strategies to reduce flood water levels and is considered along many European rivers (Van Stokkom et al., 2005).

Detention areas need to be large enough to be effective, while inlet and outlet structures and embankments may allow managing the timing of the flood alleviation. Flood detention may involve large areas and will also affect the ecological functioning of such areas.

The Lonjsko Polje detention area in Croatia (Fig. 1) is probably the best example in Europe to study the hydrology, ecology and water quality of a large, controlled flood detention area. Besides its role in flood control, it has very important ecological values, on a regional, national, and even global scale.

The objective of this study is to evaluate different flood control strategies for the Lonjsko Polje detention area and to assess their implications for

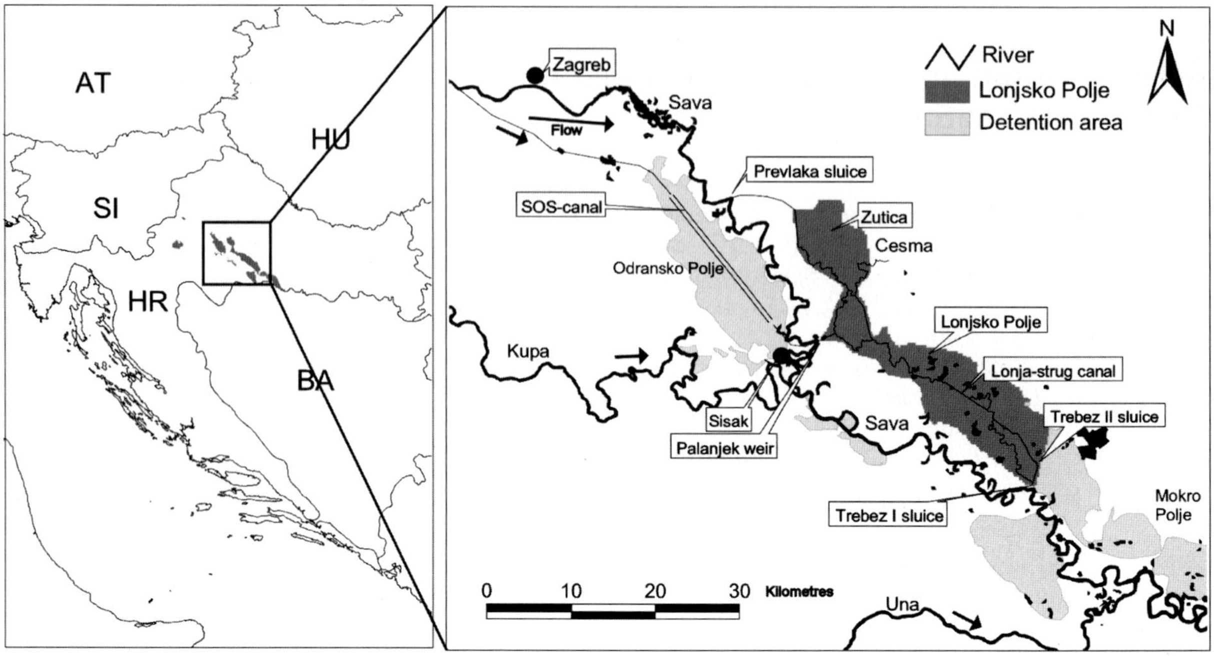

Figure 1. The Lonjsko Polje detention area and flood protection works in the Central Sava Basin, the towns of Zagreb and Sisak, and the rivers Sava, Kupa, Una and Česma. AT is Austria, HU is Hungary, SI is Slovenia, HR is Croatia and BA is Bosnia–Herzegovina.

flood safety, ecosystem development and phosphorus storage. Flood control strategies were evaluated using an overland flow model for the flooding of Lonjsko Polje. Subsequently, the effect on vegetation development was evaluated on the basis of knowledge rules describing ecohydrological relationships for vegetation. Finally, water quality modelling shows the sediment and phosphorus deposition in Lonjsko Polje.

Material and methods

Study area

Flood control in the Central Sava Basin (CSB) is accomplished through controlled flooding of (semi-)natural detention areas, transforming the flood wave of the river Sava and its tributaries. The flood control system includes artificial canals, dams and sluices for controlled distribution of the water, and detention areas for flood water storage. An overview of the various flood control works in the CSB and their history is given by Petričec et al. (2004).

The Lonjsko Polje *detention area* is the largest detention area in the CSB. It measures 237 Mm^2 and has an estimated maximum detention capacity of 634 Mm^3 (Brundić et al., 2001). It forms part of the Lonjsko Polje *Nature Park*, which has an area of 510 Mm^2 and consists of various detention areas and floodplains downstream of Zagreb.

The Lonjsko Polje Nature Park is a protected wetland under the Ramsar treaty and is an important bird area (Heath & Evans, 2000). The present vegetation distribution and composition is the result of the hydrological conditions and the land use of the past. In the past decades the vegetation composition has changed from a predominantly open landscape with pastures to a landscape with a mosaic of riparian forests and herbaceous vegetation. Due to economic and demographic reasons, the traditional grazing with indigenous breeds of cows, horses and pigs has declined. In addition, the hydrology is affected by the construction of embankments and sluices around the Lonjsko Polje detention area in the 1970s (Gugić & Ćosić-Flajsig, 2004).

Figure 1 presents an overview of Lonjsko Polje. In this paper, the name Lonjsko Polje refers to the detention area, not the Nature Park. A Sava flood wave enters Lonjsko Polje through the Prevlaka sluice, which has a maximum capacity of 600 $m^3 s^{-1}$. It then flows through the Lonja-Trebež river to the Žutica area in the northwest. In Žutica more water may be added from the smaller streams Lonja and Česma. From Žutica onwards, the flood wave progresses overland and via the Lonja river. The flood wave subsequently leaves Lonjsko Polje via the Trebež I sluice, draining into the Sava River, with a maximum capacity of 500 $m^3 s^{-1}$.

Flood management plans

Since the Lonjsko Polje detention area is located *downstream* of Zagreb it cannot directly serve as a flood protection system for Zagreb. However, as part of a large flood management study completed in 1972 by a consortium of local and foreign specialists under the auspices of the United Nations Development Office (UNDO, 1972), the Sava-Odra-Sava (SOS) canal was designed. This canal takes in water *upstream* of Zagreb and transports part of the flood wave back into the Sava River, just upstream of Sisak. On the opposite side of the river, a planned sluice at Palanjek then fills the Lonjsko Polje detention area, thus providing flood safety for Zagreb. In this plan, the size of the Lonjsko Polje detention area is increased by 8% to 256 Mm^2, and the maximum detention capacity is supposed to increase to 915 Mm^3 (Brundić et al., 2000). This plan will be further referred to as the 1972-plan.

The execution of the 1972-plan ceased in the beginning of the 1980s (Petričec et al., 2004). The SOS-canal is in its current state partly finished and now fills the Odransko Polje during floods. The Palanjek sluice has never been built and neither has the planned Trebež II sluice that connects Lonjsko Polje with the downstream Mokro Polje detention area.

More recently, an Environmental Impact Assessment for the World Bank, executed by Croatian Waters (the water management authority of Croatia), the Hydrological Project Bureau of Croatia and Euronatur, proposed an alternative flood management plan (CW, 2001). In this plan, Odransko Polje will be transformed into a detention area, which will be filled by the SOS-canal in its present, unfinished state. In Palanjek a weir is proposed, instead of a sluice, providing capacity to fill Lonjsko Polje. Furthermore, it is proposed to build the Trebež II sluice, to connect Lonjsko Polje with Mokro Polje. In this plan, the maximum detention capacity is supposed to increase to 733 Mm^3 (Brundić et al., 2001). This plan will be further referred to as the WB-plan.

Case definition

Three planning alternatives were evaluated:

1. The current situation;
2. The 1972-plan;
3. The WB-plan.

It must be noted here that the expansion of the detention area from 237 to 256 Mm^2 in the 1972-plan is not taken into account in our model, since we do not know where exactly the additional area is planned.

The hydrology of floods in the CSB is complicated. Different origins of floods for different seasons bring different flood events. Consequently, Croatian Waters has defined three different flood types, characterised by an extremely high Sava, Kupa or Una discharge, respectively. One of these flood types is applied, i.e. the SL1 event, which is the 1:100 years event for an extremely high Sava discharge at the border between Slovenia and Croatia, combined with flood hydrographs from the other major tributaries to the Sava. This event yields the highest flood levels.

In this study the three planning alternatives were considered in combination with two hydrological scenarios, the SL1 event and the year 1997 as an average hydrological year, yielding six cases in total.

Hydrological model

The numerical model Sobek is applied in this study to investigate the flow and quality of water. Moreover, it serves as a base model for flood management evaluation and for assessing the ecohydrological developments. Sobek is a flow and transport model that solves the momentum and continuity equations for surface flow, as well as the advection diffusion equation for transport, in one and two dimensions, to model shallow overland flow in, for example, floodplains, polders or detention areas (Verwey, 2001; Postma et al., 2003). The model application for this study builds upon the existing one-dimensional (1-D) model of the Sava, which has been developed for the International Commission for the Protection of the Danube River (Van Gils & Bendow, 2000).

The model application simulates the flow of water in the Sava. It has an upstream boundary in Slovenia and a downstream boundary in Belgrade at the confluence of the Sava and Danube. One-dimensional modelling elements include the Sava itself, the SOS-canal near Zagreb, as well as canals and (former) streams in the Lonjsko Polje detention area, the Česma, the Lonja and the Lonja-Strug canal. Cross sections and stage–discharge relationships were obtained from Croatian Waters and the Hydrological Project Bureau (VPB) of Croatia.

Various sluices and weirs that affect the flooding of Lonjsko Polje were included in the model. These were the existing Jankomir weir at the entrance of the SOS-canal, and the Prevlaka inlet sluice and the Trebež I outlet sluice, connecting Lonjsko Polje with the Sava. Two planned water control devices were included in the model as well, to be able to simulate the consequences of the 1972-plan and the WB-plan. These were the Palanjek inlet weir into Lonjsko Polje and the Trebež II outlet sluice, which drains into Mokro Polje. Detailed specifications were obtained from Croatian Waters.

Tributaries to the Sava, such as the Kupa and Una, were modelled as lateral inflow locations, using a stage–discharge relationship. The SOS-canal was included in the model in two different ways. In the present situation and in the World Bank-plan, the SOS-canal ends in the Odransko Polje region, thus filling this area in case of a flood. In the 1972-plan, the SOS-canal is connected to the Sava, upstream of Sisak.

The Lonjsko Polje detention area is schematised on a two-dimensional (2-D) rectilinear grid, with grid cells of 250×250 m. A Digital Elevation Model made available by the Nature Park Service describes the main part of the topography. For the remaining part (viz. the Žutica area), the topography was derived from topographic maps. The water balance in the model for Lonjsko Polje is determined by the inflow and outflow discharges through the sluices, the inflow from the Česma and Lonja, the precipitation, evaporation and infiltration rates, the drainage through small ditches, and the hydraulic resistance of the canals and the land. The inflow and outflow discharges through the sluices are dependent on the Sava water levels and their operating regime. In the model, operating rules have been defined that open and close the sluices depending on required water levels in the downstream region. The discharge of the Česma can be significant (835 $m^3 s^{-1}$ for the 1:100 year event) and is included in the scenario analysis. The discharge of the Lonja is insignificant and was defined at a constant value of 10 $m^3 s^{-1}$. Monthly precipitation and evaporation rates in Lonjsko Polje were obtained from the VPB. The infiltration rate to groundwater aquifers was estimated at 1 mm per day. The flow through small ditches plays a significant role in the drainage of Lonjsko Polje. For this purpose, additional 1-D elements, representing small canals, were added to the model schematisation. Finally, the hydraulic resistance of the canals and the land affects the flow rates and water levels. A larger hydraulic resistance results in slowing down the celerity of the flood wave and increasing water levels. Since the hydraulic resistance is not exactly known, this quantity is used as a calibration parameter.

The model was calibrated with water level data from a flood event in November 1998, obtained from Croatian Waters. The calibration aimed at simulating the flood levels and the flood wave propagation in both the Sava and Lonjsko Polje as good as possible. After calibrating the bed roughness of the Sava, the modelled results for water depth in the Sava did not differ more than 5% from the measurements which stretched over 10 days. The calibration of roughness values in the Lonjsko Polje resulted in less than 10% error between measured and modelled water levels, for three locations. However, measurements of flooding were available over a 3.5 day period only, whereas the flood duration lasted for weeks. In addition, a sensitivity analysis was carried out on the flood duration, taking into account the discharges in and out of Lonjsko Polje, the bed roughness and the evaporation. It showed that the most sensitive and also uncertain parameter is the operating regime of the sluices that determines the discharges in and out of Lonjsko Polje.

The model was validated with an independent set of data from a small flood event in the spring of 2004. The calibration coefficients for the hydraulic resistance remained unchanged and the modelled results were compared with measurements. Measurements for the Sava consisted of time-series of hourly water levels at Prevlaka and Trebež, obtained from Croatian Waters. Measurements for Lonjsko Polje consisted of recorded maximum water level at the embankment near the village of Mužilovčica, and measurements of water depths at 11 locations, conducted on May 6th 2004, by WL|Delft Hydraulics. The comparison of modelled with measured water levels in the Sava resulted in a maximum difference of 14% in water depth. The comparison of water depths in Lonjsko Polje was rather ambiguous. The maximum recorded water level at Mužilovčica was underestimated by 0.50–0.75 m by the model. On the other hand, the correspondence between our own measurements and the model results yielded an accuracy of 0.15 m in water depth.

Vegetation succession model

The vegetation succession model describes the effects of flooding and land use on the distribution of vegetation over time. Site factors influencing the presence of vegetation are flood duration, spring groundwater level, soil texture and land use. The most discriminating factor for the distribution of plant species along rivers is the flood duration (Dister, 1980; Crawford, 1992; Rademakers & Wolfert, 1994; Gurnell, 1997; Vartapetian & Jackson, 1997; Pollock et al., 1998; Van de Steeg & Blom, 1998; Vervuren et al., 2003; Van Geest, 2005). Moreover, summer floods or floods in the growing season, have a large impact on the zonation of riparian plant species (Brock et al., 1987; Van den Brink et al., 1991; Vervuren et al., 2003; Van Eck et al., 2004).

Ecohydrological relationships between vegetation types of riverine wetlands and floodplains were based on a classification for the average inundation duration, with class boundaries of 2, 20, 50, 150 and >360 days per year (Van der Meijden, 1996; Van Splunder, 1998; Löffl, 1999; Peters, 2002; Klijn et al., 2004). Table 1 presents the vegetation types, their corresponding flooding durations (in days per year) and some examples of typical species found. Knowledge rules for the succession of vegetation types were defined, dependent on the inundation duration and the land use, for mean hydrological years, based on experiences of Lonjsko Polje park managers and on data from floodplains in The Netherlands

Table 1. Vegetation types, corresponding flooding duration and typical species

Vegetation type	Flooding (days year^{-1})	Typical species
Dry hardwood forest	< 20	*Quercus robur, Carpinus betulus*
Hardwood forest	20–50	*Quercus robur, Fraxinus excelsior, Ulmus minor*
Aspen plantation	< 150	*Populus tremula*
Softwood forest	20–150	*Salix alba*
Wet hardwood forest	50–150	*Fraxinus excelsior, Alnus glutinosa*
Marsh forest	> 150	*Alnus glutinosa, Carex* spp.
Hardwood shrub	< 50	*Crataegus monogyna, Rosa canina, Prunus spinosa*
Softwood shrub	50–150	*Salix alba*
Dry herbaceous	< 20	*Bromopsis inermis, Arctium lappa, Brassica nigra*
Wet herbaceous	20–150	*Phalaris arundinacea, Epilobium hirsutum, Cirsium arvense*
Helophytes	> 150	*Scirpus* spp., *Carex* spp., *Phragmites australis*
Floodplain hayfield	< 50	*Arrhenatherum elatius, Pimpinella major, Pastinaca sativa*
Wet hayfield	50–150	*Alopecurus pratensis, Rumex* spp., *Agrostis stolonifera*
Dry meadow	< 20	*Bromopsis inermis, Cynosurus cristatus, Ranunculus repens*
Wet meadow	20–150	*Agrostis stolonifera, Potentilla anserina, Trifolium* spp.
Arable floodplain	< 20	
Lake	365	

(Peters, 2002; Van Velzen et al., 2003; Baptist et al., 2004). The stage of vegetation succession after 2, 10, 20, 50 and 100 years was predicted based on these knowledge values (Table 2, left panel).

In addition to vegetation succession under mean hydrological conditions, lethal threshold values for inundation duration under extreme conditions were defined. The most important constraint that plants have to deal with during flooding is oxygen deficiency (Crawford & Brändle, 1996; Vartapetian & Jackson, 1997). Van den Brink et al. (1995) demonstrated that this effect is strongest with respect to soil with a high organic matter content and least on mineral soil. A lethal threshold value gives the number of days a certain vegetation type can survive while flooded. If the threshold is exceeded, the vegetation type is assumed to be set back to herbaceous vegetation in our model. Most major floods occur in autumn or winter, but the inundation duration can extend into spring. Since vegetation is more vulnerable to flooding in the growing season, lethal thresholds for both winter and growing season were defined (Blom et al., 1990; Van Eck et al., 2004). The threshold values were based on expert knowledge and the experience of the Lonjsko Polje park managers.

Three types of land use management affect the vegetation composition and succession in Lonjsko Polje, i.e., forestry (replanting of forest), grazing by a low density of herbivores and mowing. Table 2 presents the knowledge rules for vegetation succession for a mean hydrological year and for an extreme flood event, without land management (no grazing, no mowing, no planting of hardwood forest), and with land management consisting of grazing, mowing and planting of hardwood forest. The latter type of land use resembles the current land use the most.

Table 2. Vegetation succession rules for mean hydrological years and following a 1:100 years flood event

Vegetation succession for mean hydrological years							Vegetation succession following a flood event threshold (days)						
Vegetation type	Code	Year					Nov–Mar	Apr–Jun	Year				
		2	10	20	50	100			2	10	20	50	100
Dry hardwood forest	11	./.	./.	./.	./.	./.	20	10	31/31	21/21	21/.	./.	./.
Hardwood forest	12	./.	./.	./.	./.	./.	50	20	31/31	21/21	21/.	./.	./.
Aspen plantation	13	./.	./.	./.	./.	15/13	100	20	31/31	21/.	21/.	12/.	12/.
Softwood forest	14	./.	./.	./.	./.	15/15	150	90	32/42	22/42	./42	./42	15/42
Wet hardwood forest	15	./.	./.	./.	./.	./.	150	90	32/32	22/22	14/14	14/14	./.
Marsh forest	16	./.	./.	./.	./.	./.	150	90	33/42	33/42	./42	./42	./42
Hardwood shrub	21	./.	./.	./.	12/12	12/12	50	20	31/41	./41	./41	12/41	12/41
Softwood shrub	22	14/14	14/14	14/14	14/14	15/15	150	90	32/42	./42	14/42	14/42	15/42
Dry herbaceous	31	./51	21/51	21/51	11/51	11/51	20	10	./51	21/51	21/51	11/51	11/51
Wet herbaceous	32	./42	22/42	14/42	14/42	15/42	150	50	./41	22/41	14/41	14/41	15/41
Helophytes	33	./.	./.	./.	16/16	16/16	150	50	./.	./.	./.	16/16	16/16
Floodplain hayfield	41	./.	./.	21/.	21/.	12/.	50	20	31/.	21/.	21/.	12/.	12/.
Wet hayfield	42	./.	32/.	32/.	22/.	14/.	150	50	31/.	22/.	14/.	14/.	15/.
Dry meadow	51	./.	./.	31/.	21/.	11/.	20	10	31/.	21/.	21/.	11/.	11/.
Wet meadow	52	./42	32/42	32/42	22/42	14/42	50	20	32/42	22/42	14/42	14/42	15/42
Arable floodplain	61	31/41	21/41	21/41	11/41	11/41	50	20	31/41	21/41	21/41	12/41	12/41
Lake	71	./.	./.	./.	./.	./.	n.a.	n.a.	./.	./.	./.	./.	./.

Each vegetation type has its code listed in the code-column. Vegetation succession is depicted by codes for successive years. Left of the slash the codes are given for succession without land management, right of the slash the codes are given for succession with land management consisting of grazing in low density, mowing and planting of hardwood forest. When the vegetation equals the initial vegetation type, a dot is used, indicating no change.

An actual vegetation map made up in the mid-nineties was compared with the result from the ecohydrological knowledge rules applied on mean hydrological years and with land management. The result differed from the observed vegetation map in that it has a different classification methodology and there were some discrepancies, possibly caused by erroneous ecohydrological model rules, hydrological model results, or the presence of vegetation that has not adapted to new flood durations. For further analysis the ecotope map that results from the application of our knowledge rules was applied.

Three planning alternatives were distinguished: the current situation, the 1972-plan and the WB-plan. For each of these alternatives, the mean hydrological conditions differ. For these reasons, adjusted maps were made up that do not correspond exactly to the observed situation, but contained the theoretical vegetation composition as it has adapted itself to the flood durations resulting from our hydrological modelling. This means that it was assumed that vegetation development would follow our model rules. This is a necessary step for further vegetation succession modelling based on the current level of knowledge. The analysis of the results can be described as follows: if the 1972-plan would have been implemented, what would then be the spatial and temporal distribution of the vegetation for mean hydrological years and following a 1:100 year flood event.

For each of the three physical planning alternatives, the inundation duration (days per year) in each grid cell was calculated by the hydrological model, for average hydrological years and for the 1:100 years flood event. These data were used to calculate the vegetation succession, based on the knowledge rules from Table 2. The extreme flood was simulated as a single event, affecting the present day vegetation and commencing on January 1st.

Water quality model

The Sobek 1-D–2-D water quality module, which is coupled to the Sobek 1-D–2-D flow module was applied in this study to assess the sediment and phosphorus balance of Lonjsko Polje. The water quality model applies the advection diffusion equation in two horizontal directions to simulate the transport, mixing, loads and processes of sediment and phosphorus. The modelling of sediment deposition is carried out in 2-D, thus differing from the approach by Asselman & Van Wijngaarden (2002), who applied Sobek in 1-D mode.

The accumulation of fine sediment in Lonjsko Polje is due to the sedimentation of suspended matter, which is transported from the Sava river. The balance between sedimentation and resuspension depends on the flow velocities inside Lonjsko Polje and is based on the Partheniades–Krone formulae (Krone, 1962; Partheniades, 1962). At low flow velocities, net sedimentation occurs, whereas at high flow velocities, net resuspension occurs. The spatial distribution of sediments in Lonjsko Polje will, therefore, be determined by the inflow of sediments and the flow patterns of the water. Together with the sediment, phosphorus is brought into Lonjsko Polje.

The fate and transport of phosphorus is highly dependent on that of the sediment. Phosphorus occurs in different forms in the environment, but the majority is adsorbed to mineral particles, such as silt and clay. During a flood event, high suspended sediment loads enter Lonjsko Polje, yielding a high load of adsorbed phosphorus. The most important loss route for adsorbed phosphorus from the water to the soil is through sedimentation (Behrendt & Opitz, 1999). Phosphorus can also be released from the soil into the water column, and it can be transformed to organic phosphorus via plant uptake.

Since little data on nitrogen is available, and the removal rate in the detention area is expected to be insignificant (Pérez et al., 1999; Van der Lee et al., 2004), nitrogen was omitted in our analysis.

The input to the water quality model is the concentration of suspended sediment and phosphorus in the Sava at the border between Croatia and Slovenia. A power function with an additive constant term gives a good estimate for a sediment rating curve (Asselman, 2000). Alternatively a non-linear deterministic approach can be applied (Sivakumar & Wallender, 2005). However, there is not enough data available to fit a reliable rating curve or other function for high discharges in the Sava. Instead, it is assumed in this study that the suspended sediment concentration is simply proportional to discharge, reaching a maximum of about 2000 g m^{-3} at the peak discharge. The total phosphorus concentration is proportional to discharge as well, reaching a maximum of about 2 g m^{-3} at the peak discharge. These values were estimated on the basis of available measurements in the Danube near Vienna during a major flood event in August 2002 (Danubs, 2004). Effects of retardation and exhaustion were disregarded. At base flow conditions, the water quality monitoring data for the Sava was applied.

Unfortunately, there is no data on sediment and phosphorus available from inside the detention area, making calibration of the water quality model impossible. The results of the water quality modelling should, therefore, be considered merely indicative.

Results

Flood protection by Lonjsko Polje

Table 3 presents the maximum discharge at Sisak, upstream of the confluence with the Kupa, following from the model simulations for the 1:100 years event. The results show that the 1972-plan does not provide enough safety for Sisak, since the discharge here should be kept below 2200 $m^3\ s^{-1}$. Upstream of Sisak, the flow from the SOS-canal adds up to the flow in the Sava. Part of the flow is redirected to Lonjsko Polje via the Palanjek weir. Although the maximum capacity is 2370 $m^3\ s^{-1}$, the computations show that only 600 $m^3\ s^{-1}$ flows through the Palanjek weir. The remaining discharge on the Sava is well above the desired discharge. Both in the current situation and the WB-plan, flood protection is more effective than in the 1972-plan.

Flooding of Lonjsko Polje

About four days after the flood wave has passed the border between Slovenia and Croatia, the upstream part of Lonjsko Polje area starts to fill. Two to three weeks later, the Lonjsko Polje area is completely filled up, with an average depth of 3.3–4.2 m, depending on the planning alternative. Maximum water depths of more than 6 m will occur. Table 3 presents the maximum water depth and detention capacity that result from the model simulations. As compared with the planned capacity (Brundić et al., 2001), our results suggest a much higher capacity in the current situation, a higher capacity for the WB-plan, and a slightly higher capacity for the 1972-plan. Note that in the latter case, an extension of the area by 8% was originally planned, but was not taken into account in our computations.

Rather surprisingly, the detention capacity in the current situation is the largest. This is due to the fact that the drainage rates of the Lonjsko Polje area differ significantly between cases. For both the 1972-plan and the WB-plan, the outlet capacity of the additional Trebež II sluice drains the area faster. In addition, the flow direction in the Palanjek weir reverses once the flood wave has passed, yielding additional drainage capacity. In the current situation, the water is trapped inside Lonjsko Polje. In case of a high Sava discharge, the area keeps filling for a longer period compared with the other plans. The peak levels in Lonjsko Polje will be reached 8 days sooner for the 1972-plan, compared with the current situation, i.e. 19 days after the onset upstream of Zagreb, instead of 27 days. For the WB-plan the peak is reached after 21 days.

Furthermore, the propagation speed of the inundation of Lonjsko Polje differs between the planning alternatives. In case of the 1972-plan, the propagation speed is highest. The low-lying middle regions of Lonjsko Polje will be flooded slightly more than 2 days sooner, compared with the current situation and the WB-plan (5.9 days after the onset upstream of Zagreb, instead of

Table 3. Computed maximum Sava discharge at Sisak, computed maximum water depth in Lonjsko Polje, computed maximum and planned maximum detention capacity for the 1:100 SL1 flood event

Alternative	Max. discharge ($m^3\ s^{-1}$)	Max. depth (m)	Max. capacity (Mm^3)	Planned capacity (Mm^3)
Current	1900	7.45	1005	634
1972-plan	2400	7.13	932	915
WB-plan	2000	6.65	790	733

8 days, or 7.8 days respectively). The difference is due to the presence of the SOS-canal, which carries the flood wave swiftly to Lonjsko Polje via Palanjek. The 1972-plan thus gives less time for evacuation than the other two alternatives.

The relative inundation areas, expressed in hectares of inundation, are presented in Figure 2. In the current situation, for a mean hydrological year, almost 60% of the area is inundated for more than 20 days per year. With the implementation of the 1972-plan or the WB-plan, Lonjsko Polje becomes dryer, especially in the low-lying southern part, with a difference from the current situation between 20 and 40 days. Around 50% of the total area will be flooded more than 20 days per year.

In case of a 1:100 years event, the whole area is inundated for more than 20 days per year, and there are locations that are inundated continuously in the first year after the event, such as near the Palanjek sluice and in the lower southern part. If a flood event occurs, the 1972-plan results in the lowest flood duration times.

Vegetation development

Figure 3 presents the results of the computations for vegetation development. To limit the number of vegetation types presented in Figure 3, some have been aggregated, for example *hardwood forest* consists of the sum of dry hardwood forest, hardwood forest and aspen plantation, *herbaceous* consists of the sum of dry and wet herbaceous vegetation. The category herbaceous swamp consists of helophytes. Note that the percentage cover differs between the physical alternatives already at $t=0$. This is due to the model methodology in which it was assumed that the vegetation has been adapted to the mean hydrological situation of the corresponding physical alternative. Since the 1972-plan shows the driest situation, the vegetation composition has been adapted accordingly. A methodological problem, however, is that the adaptation period differs between vegetation types. In the case of hardwood forest, for example, the adaptation to drier conditions is slow.

For mean hydrological years, the model rules imply that, in the current situation, the open parts of Lonjsko Polje will be gradually covered by more shrubs and, after 100 years, will change into a climax stage with hardwood forests, totalling 93% cover, in case no land use management is implemented. Typical wet types, such as helophytes (6% cover) will gradually be replaced by a higher successional stage. In the moist low-lying parts, a mixture of wet hardwood, softwood and marsh forest will be found. If land use of mowing and grazing at low animal densities is implemented, the growth of shrubs, rough herbaceous vegetation and softwood shrubs will be delayed and, therefore, a more open landscape will be maintained at

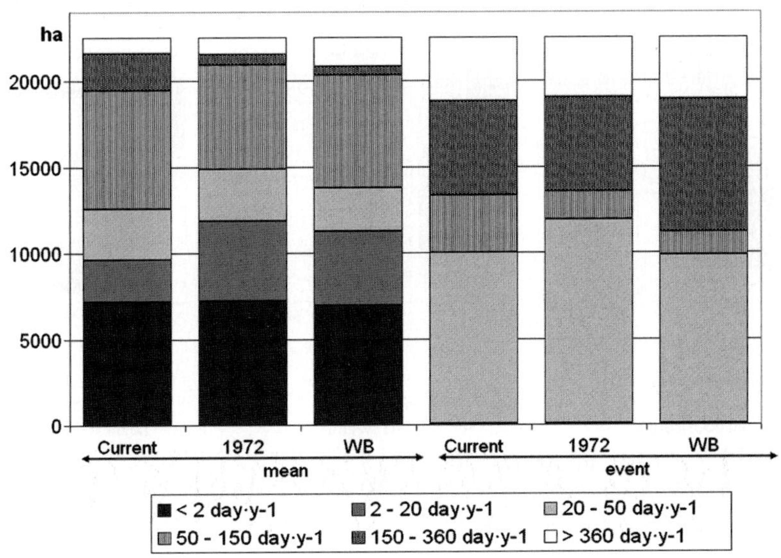

Figure 2. Relative share of areas (ha) with various flood duration times (day year^{-1}) for various cases (for mean hydrological years or the 1:100 years flood event, the current situation, the 1972-plan and the WB-plan).

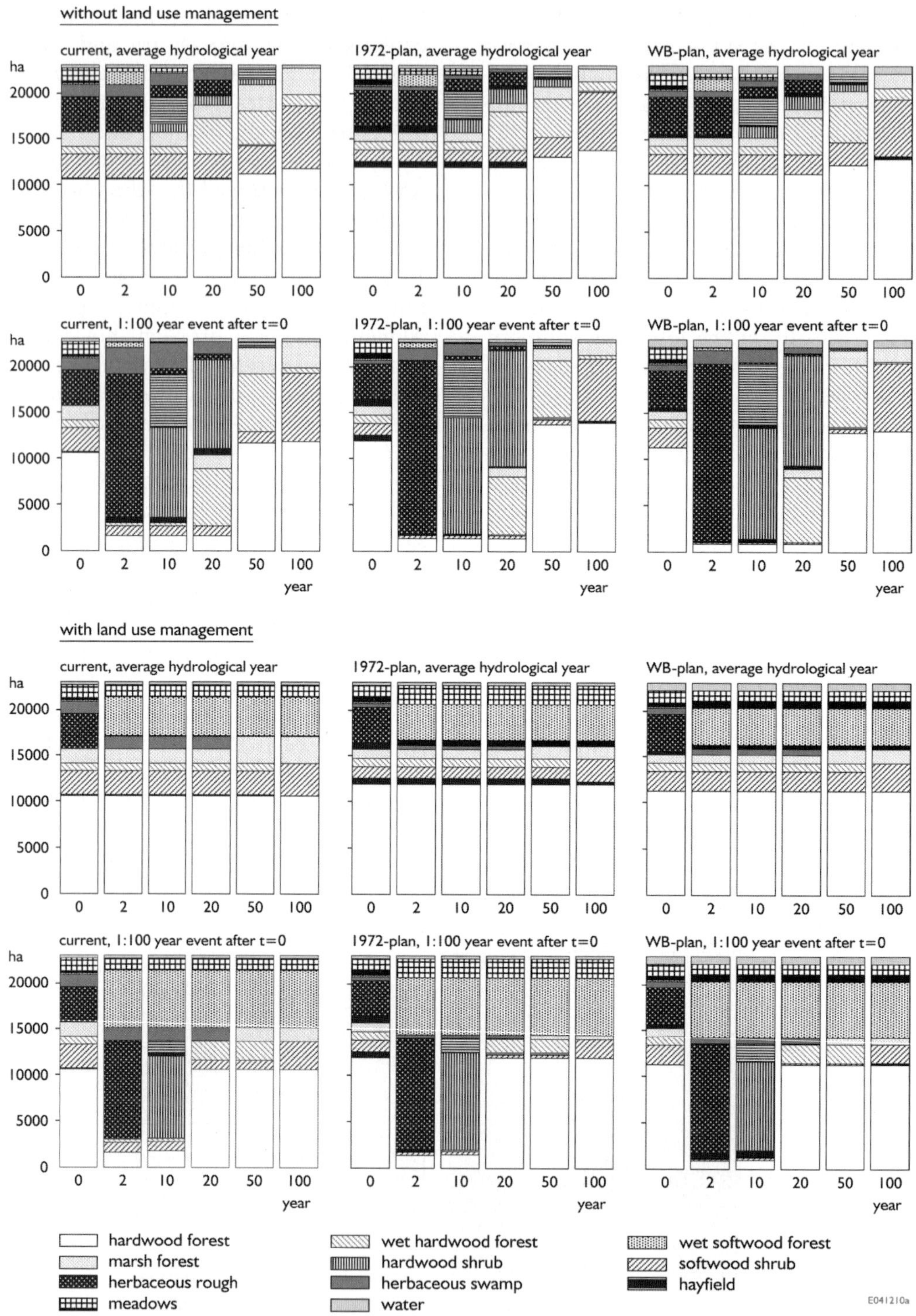

Figure 3. Results of the computations for vegetation development (ha) for the current situation, after implementation of the 1972-plan and after implementation of the WB-plan, without land use management and with land use management consisting of grazing, mowing and replanting of forest.

the cost of hardwood forest (74% cover instead of 93%). Furthermore, herbaceous vegetation will be managed and changed into hayfields to be used for the cattle.

If the 1972-plan would have been implemented, the northern part (Žutica) will not differ much from the current situation, being dominated by hardwood forest. In the remaining part of Lonjsko Polje, the dryer situation results in an increase of hardwood forest area with 6% at the cost of softwood and marsh forest. Especially in the low-lying southern part, there would be more hardwood forest compared with the current situation. Typical wet vegetation types will diminish. Part of the area covered by helophytes will change into herbaceous vegetation, initially diminishing the helophyte cover from 6 to 2%, and finally resulting in the absence of helophytes as a result of succession. The cover of marsh forest will decrease from 12.5 to 6%. Under conditions of land use with grazing and mowing the result differs from the 1972-plan without management, in that it has more hayfields and meadows (+4%) and much less wet hardwood forest (−15%).

If the WB-plan would have been implemented, the vegetation composition will be in between the current situation and the 1972-plan. There is a net increase of dry vegetation types, such as hardwood forest (+3%), at the cost of wet vegetation types, such as helophytes and marsh forest (both −3%), compared with the case for the current alternative. However, the WB-plan results in wetter conditions at specific locations; the total area permanently under water increases by 2%. The results for the land use scenario show 3% more hayfields and meadows, and a 15% decrease in wet hardwood forest compared to the WB-plan without management.

In case of a 1:100 year event, in the current situation, 2 years after an event, large areas of herbaceous vegetation appear (an increase from 17 to 70% cover) at the cost of hardwood and softwood forest. Ten years later, this will be followed by a large cover of shrubs, replaced by softwood and ultimately by hardwood forest. The area of marsh forest will be initially replaced by helophytes and subsequently succeeded by marsh forest again. Overall, the vegetation is less diverse, compared to the vegetation composition that would have been present without the event. As a result of land use management (grazing, mowing and replanting of forest), part of the herbaceous vegetation will be replaced by meadows, arable land and hayfields and ten years later, there will be much less shrubs (46% instead of 70% cover). As a result of the replanting of hardwood forest, this vegetation type is recovering faster than in the case without replanting.

If the 1972-plan would have been implemented, the hardwood forest will consist of relatively more cover of dryer types at the cost of wet hardwood forest and marsh forest, in comparison with the current situation. Whereas the cover of helophytes could reach 12% in the current situation, it is limited to only 6% in the 1972-plan. Would the 1972-plan be implemented, the vegetation composition would adapt to the corresponding flood durations. Since the flood duration for mean hydrological years is less for the 1972-plan, the vegetation consists of types that favour dryer conditions. The result of land use is comparable to that for the current situation, be it that there are less wet vegetation types.

In the case of implementation of the WB-plan, the results resemble those for the 1972-plan, with slightly less effect on the wet vegetation types.

Water quality

In the 1:100 year events, about 30% of the sediment and the adsorbed phosphorus that enters the Lonjsko Polje area deposits. This equals to 8–12% of the sediment and phosphorus loads that flow through the river Sava during this event (Table 4). In the current situation, more sediment and phosphorus is deposited than with the 1972-plan or the World Bank-plan implemented. The 1972-plan and the World Bank-plan are quite alike with respect to sediment and phosphorus entrapment.

One simulation has been made for a representative winter (1996–1997) in the current situation. It showed that 64% of the suspended sediment and 68% of the particulate phosphorus that enters Lonjsko Polje is deposited there. As compared with the sediment and phosphorus that is transported by the Sava, however, this is no more than 4 and 3%, respectively.

After a 1:100 event, for the current situation, the thickness of the freshly deposited sediment

Table 4. Deposition of sediment and phosphorus in Lonjsko Polje (LP), and storage of sediment and phosphorus, expressed as percentage of total input to LP and as percentage of total load of the Sava, for events and for a mean winter

Alternative	Sed. dep. (10^6 kg)	P dep. (10^3 kg)	Sed. trap. (% LP)	Sed. trap. (% Sava)	P trap. (%LP)	P trap. (% Sava)
Current, event	634	737	33	12	36	13
1972-plan, event	444	462	29	8	30	8
WB-plan, event	479	468	28	9	27	9
Current, mean ('96–'97)	7	14	64	4	68	3

layer is approximately 1 cm at maximum, assuming a porosity of 0.80 in the new sediment layer. The sediment layer is not evenly distributed over the Lonjsko Polje area. Most sediment settles in the deeper parts, downstream from the locations where the water enters the detention area. No accurate measurements have been made, but the Park Service confirms the presence of a layer of a few millimetres of fine sediment after a flood event.

Discussion

In retrospect, the flood control system in the CSB was ahead of its time. The plan that was made up in 1972 (UNDO, 1972) sought for a solution not only in river channel regulation and dike construction, but also in the storage of flood waters in natural inundation areas. On the other hand, the plan did have a very technocratic character. The economic and ecological values of lowland forests were ignored and the 1972-plan lacks any assessment of the biodiversity in the catchment of the Sava river and the impacts upon it from the planned construction works (CW, 2001). The latter is a major shortcoming, since the CSB is of unique ecological importance within Europe. The World Bank-plan, therefore, considered the natural values of the CSB and proposed some important changes to the original 1972-plan.

In this study a combination of a 1-D model for rivers and canals with a 2-D model for overland flow was applied, which has major advantages over existing models for the Sava (Filipovic et al., 2000). The flow model is coupled to a water quality model, making modelling of sediment deposition possible.

The knowledge rules for the vegetation model stem partly from Dutch floodplain studies and partly from expert knowledge of the Lonjsko Polje park managers. A demerit of our model approach is that we assume that the vegetation adapted itself to the corresponding new hydrological situation before we apply vegetation succession modelling. Furthermore, the knowledge rules given in Table 2 describe only the development of the dominating vegetation type. This means that not-dominating vegetation types (theoretically up to 49% cover) were not mentioned in the tables, but may form a substantial part of the vegetation of the area. Finally, we have assumed one type of land use to be executed in the entire area, whereas in reality, there is more differentiation in land use practise. For this study it was not possible to determine detailed land-use practise.

A following step would be to apply probabilistic modelling for the vegetation development. The existing vegetation composition then is the result of a spectrum of hydrological conditions and events. It might be possible to apply the hydrological model to simulate the inundation durations for a wider variety of hydrological events (1:90 years event, 1:80 years event, etc.). Subsequently, a number of discharge scenarios can be drawn, for example a series of dryer years, or wetter climate conditions, etc. For each of these scenarios the vegetation succession can be calculated yielding a probabilistic view of vegetation development. At this moment in time, however, detailed enough knowledge on the vegetation response to hydrological changes is lacking.

Another future step would be to use these model instruments to find an optimised scenario for both flood safety and ecology. It is then necessary to have a multi-criteria evaluation method. The models can quantify the parameters needed for evaluation of different scenarios. For ecological evaluation, the parameters should be extended with faunal species. The project results have proven that the models can help to find ecologically sound flood management strategies. An interesting follow-up of this project would therefore be the exploration of an ecological flood management strategy for the whole CSB.

We limited ourselves to suspended sediment and adsorbed phosphorus modelling. Suspended sediment sampling carried out by Wattendorf (2000) indicated that these suspended sediments may contain heavy metals and polycyclic aromatic hydrocarbons, responding to floodplain sedimentation as well (Walling & Owens, 2003). These substances have not been modelled in this study, but require attention in future research. The results for the removal percentages for suspended sediments and adsorbed phosphorus are of almost equal value in the 1:100 event simulations. This is partly due to the artificial boundary conditions in the model, which have been defined under the assumption that the concentrations of suspended sediments and adsorbed phosphate increase with the same ratio during a 1:100 event. The storage percentage of sediment found in our study (4% of the Sava load, for mean hydrological conditions, or maximum 12% for a 1:100 years flood event) is rather low. It is lower than that found by Walling et al. (1998), who measured 39 and 49%, or Middelkoop & Asselman (1998), who found 19% for a 1:40 years flood or Sweet et al. (2003), who determined 50–60%. The hydrological circumstances, however, differ a lot. The Lonjsko Polje area is not an active floodplain any longer and only fills via sluices under moderate flood conditions. Our results thus indicate that this controlled detention area is less efficient in storing sediment than natural floodplains. Finally, it is highly recommended to perform monitoring during flood events, to improve upon model simulations.

Conclusion

With respect to the effects of a 1:100 year flood event in Lonjsko Polje, some important general conclusions can be drawn. First, it is not so much the intake capacity that determines the flood duration time, but the drainage capacity. Second, the long inundation duration following a 1:100 flood event leads to major setbacks in the vegetation composition, even when the detention area is regularly flooded in mean hydrological years. However, the effect is less severe if the vegetation has been able to adapt itself to (small) regular flooding. Third, the vegetation distribution can be largely determined by land use management. Land use management can even be considered a prerequisite to get a wider variety in vegetation types, yielding more habitats for fauna. And fourth, about 30% of the sediment and adsorbed phosphorus that enters the detention area during an extreme flood is retained within the area. In terms of the purifying capacity of this detention area for the Sava, however, its effect is limited to maximum 10%, since the majority of the nutrients flows past the detention area through the main channel.

The evaluation of different flood control strategies for Lonjsko Polje showed that the current situation offers the highest flood safety for downstream regions, it has the highest cover of the important wet vegetation types and it has the highest nutrient storage.

An important lesson for the planning and construction of controlled detention areas is to take care of a large enough drainage capacity. The longer the inundation duration, the larger the effects of anoxia to vegetation will be. Shortening the flood duration is, therefore, advantageous to vegetation. Usually, full attention is given to the size of the intake structures, but the fast drainage of such areas can pose a problem and deserves attention, especially in low-lying regions, such as in the Netherlands. On the other hand, a shorter flood duration leads to lower sediment and nutrient storage rates. Major effects of inundation on vegetation development might be mitigated when the water management of a detention area is such that the vegetation composition is already adjusted to wet conditions. This can be achieved by regular flooding of the detention area, which also results in hydrological connectivity between the river and floodplain, enabling exchange processes and favouring migration of aquatic organisms (Hohensinner et al., 2004).

Acknowledgements

The authors wish to thank Croatian Waters, especially D. Biondić, G. Ćosić-Flajsig and

river Waal. I will end with a few interdisciplinary challenges in research on 'water and values'.

Ecology in development

Ecology is a relatively young branch of science. When I started my studies in biology at the end of the 1950s at the Free University in Amsterdam, ecology as a discipline was hardly existing in the academic curricula in the Netherlands. Ecology embraces the study of the relations between organisms and between organisms and their environment, a multi-disciplinary field of natural sciences. A closer look at the subjects I studied reveals that that there was quite a lot of 'ecology' there, but the fields of interest had other specifications like zoogeography and biosystematics. Textbooks did not exist, and the youthfully field of science had to be studied from lecture notes. Big names before 1940 were (a.o.) the German August Thienemann, and the Englishmen Charles Elton and Alfred George Tansley, who coined the concept of the ecosystem in 1935 (Tansley, 1935). After the Second World War ecology became booming business, not in the last place owing to its potential and actual role in describing and curing the increasing threat of environmental problems, getting out of societal control. A large number of ecologists are, and have been actively engaged in integrative or trans-disciplinary ecological research, including the late Eugene and Howard Odum.

The Odum brothers have had a great impact on the development of the science of ecology. Their holistic vision on the structure and the functioning of organisms, communities and ecosystems has influenced generations of ecologists all over the world. Their textbook 'Fundamentals of Ecology', of which the first print appeared in 1953, has been the only textbook on system ecology worldwide for many years. I have read the strongly revised third edition of this book (Odum, 1971a) to pieces. Odum's textbook has recently been called the book that has made the greatest impact on the careers of biologists and ecologists in English speaking countries and parts of Western Europe, particularly among population biologists and system ecologists (Barrett & Mabry, 2002). This evaluation cannot be necessarily shared by all the scientists in biological sciences, particularly those that work in evolutionary biology, but it is certain that Odum's book is paradigmatic. The holistic perspective presented has influenced generations of scientists, biologists and environmental scientists alike.

The list of ecosystem attributes, published as early as 1969 by E.P. Odum, stood many storms in ecology (except a few statements such as the relation between stability and diversity). Odum (1969) presented a tabular model of ecological succession, trends to be expected in the development of ecosystems from the early developmental stages to mature stages (Table 1). The holistic approach is perhaps best summarized by Odum's oft-quoted statement that 'the ecosystem is greater than the sum of its parts'. Odum made the ecosystem concept central to the discipline of ecology, and the principles of ecosystem ecology were translated into lessons in preserving the natural environment. In his famous papers in Science (Odum, 1969, 1977) Eugene Odum argues that knowledge of the principles of ecological succession is of great value for a better understanding of the conflicts between man and nature. Cycles of chemical matter, energy fluxes and food chains are central notions in ecosystem research, and demand a multidisciplinary approach. In his system analysis, in which the description and possible solution of environmental problems is at stake, finally also economics and sociology are involved. To that extent Eugene Odum may be coined as one of the founding fathers of environmental sciences. His brother Howard Odum went even a few steps further in his holistic way of thinking, and he introduced political, ethical and religious aspects in his systems approach (Odum, 1971b, 1981). It is for that reason that he is ignored in circles of sober (Dutch) ecologists, predominantly oriented on reductionism; also for me Howard Odum oversteps the mark.

The classic dichotomy in ecology is between structure and function, in other words which life-forms do we see in nature, and how does nature work. Starting from this split, we see that the descriptive trend of the 50th and 60th changed into causal-analytic studies on the functioning of ecosystems in the second half of the 60th and in the 70th. The famous series of handbooks of the International Biological Program appeared in those years. I remember clearly the lively discussions on the 'unifying concepts in ecology', held at the first International Ecology Congress in Den

Table 1. A tabular model of ecological succession: trends to be expected in the development of ecosystems (Odum, 1969)

Ecosystem attributes		Developmental stages	Mature stages
Community energetics			
1	Gross production/community respiration (P/R ratio)	Greater or less than 1	Approaches 1
2	Gross production/standing crop biomass (P/B ratio)	High	Low
3	Biomass supported/unit energy flow (B/E ratio)	Low	High
4	Net community production (yield)	High	Low
5	Food chains	Linear, predominantly grazing	Weblike, predominantly detritus
Community structure			
6	Total organic matter	Small	Large
7	Inorganic nutrients	Extrabiotic	Intrabiotic
8	Species diversity – variety component	Low	High
9	Species diversity – equitability component	Low	High
10	Biochemical diversity	Low	High
11	Stratification and spatial heterogeneity (pattern diversity)	Poorly organized	Well-organized
Life history			
12	Niche specialization	Broad	Narrow
13	Size of organism	Small	Large
14	Life cycles	Short, simple	Long, complex
Nutrient cycling			
15	Mineral cycles	Open	Closed
16	Nutrient exchange rate, between organisms and environment	Rapid	Slow
17	Role of detritus in nutrient regeneration	Unimportant	Important
Selection pressure			
18	Growth form	For rapid growth ("r-selection")	For feedback control ("K-selection")
19	Production	Quantity	Quality
Overall homeostasis			
20	Internal symbiosis	Undeveloped	Developed
21	Nutrient conservation	Poor	Good
22	Stability (resistance to external perturbations)	Poor	Good
23	Entropy	High	Low
24	Information	Low	High

Haag in 1975 (Van Dobben & Lowe-McConnell, 1975). Descriptive research became obsolete, and blooming university departments in botany and zoology were decimated in favour of new developments in molecular biology, population biology and systems ecology. The diversity of plants and animals – and everything in nature turns on this – became the domain of out-of-university circles. In 1992 the concept 'biodiversity' popped up (United Nations Conference on Environment and Development, 1992), and suddenly knowledge on the structure of ecosystems appeared in urgent need. The question of 'how' cannot be studied separated of the question of 'what'. Plant and animal species show up and disappear in the time span of decades, and the causes are often unknown. Natural climate changes or anthropogenic influences? Who knows?

University departments were not able to present answers to the ever growing demand on information on the distribution of plants and animals. The specialists who were supposed to know the answers, were all gone. The abolition of fields of interest such as plant- and animal taxonomy and morphology in favour of molecular biology has been a precarious development. How can the evolution epos be understood without basic knowledge of the exuberant diversity of plants and animals that surrounds us?

Some universities showed their inventiveness to solve this problem. The Radboud University Nijmegen founded the Netherlands Centre for Nature Research (NCN), a cooperation between a number of departments of the University, and a number of foundations and societies in the field of the distribution and monitoring of flora and fauna (s.c. PGO's, Private Data Collecting Organizations). Their joint aim is to do research in favour of nature management and nature policy. Here, the fundamental questions supposed to be posed and answered by university departments intertwine with the pool of knowledge of the PGO's.

Now in 2005, ecology has become a mature branch of the natural sciences. In the 1960th approximately 500 biologists gained their degree in the Netherlands, in the 1990th the numbers of graduates have increased to 4000 (NIBI, 2000). The relative increase in the number of ecologists is even larger: in the sixties some tens of people called themselves ecologist, now many hundreds of them have their jobs, distributed over universities, state departments and private bureaus. Ecological knowledge is much-needed to solve present and future problems on water resources.

The past 15 years an avalanche of papers has appeared with the central focus on a holistic way of thinking in environmental management and nature management, promoting concepts like 'ecosystem health' and 'ecosystem integrity' (e.g. Rapport, 1992; Costanza et al., 1999; Winterhalder et al., 2004; Vugteveen et al., 2006). Essentially, a large part of these concepts appears to be new wine in old bottles: Odum revisited. Quite a number of ecologists reject these 'health' and 'integrity' terms for their imprecision (e.g. Lackey, 2001; Davis & Slobodkin, 2004), whereas others wholeheartedly embrace them and seek to develop them further in theoretical and practical terms, at the interface of scientific ecology, societal values and public policy (Rapport, 1995; Carpenter et al., 2001; Vugteveen et al., 2006).

According to Winterhalder et al. (2004) the conflict in fact is linked with the debate over the term 'ecosystem' itself. Some consider the term semi-mystical and unacceptable (Davis & Slobodkin, 2004), whereas others, including myself, consider it a core concept in ecology. From its origin in 1935 (Tansley, 1935) the 'ecosystem' proved to be a strong and convincing holistic and integrative ecological concept that combined living organisms and the physical environment into a system. Golley (1993) wrote an interesting book on the history of the ecosystem concept, tracing its evolution, describing how numerous American and European researchers contributed to its evolution, and discussing the explosive growth of ecosystem studies. The ecosystem concept has long been central to the public perception of ecology and to increasing awareness of environmental degradation. More recently, ecosystem integrity and health are both defined as referring to the state or condition of an ecosystem in which its dynamic attributes are expressed, related to its ecological stage of development.

Ecology of rivers and estuaries

Rivers and estuaries under pressure

On a global scale freshwater is a rare resource. Considering the predictions on the global amount of freshwater in 2025, the growth of the world population and the consequent demand on usable freshwater is many times more important than the warming effects of climate change. Almost the entire net increase of the world population until 2025 will take place in urban areas in developing countries. In 2025 almost 60% of the world population – roughly 5 billion people – live in cities; many of these cities are situated on a river (Malmqvist & Rundle, 2002). Yet, 1.8 billion people live on earth greatly in need of freshwater resources, and the numbers will increase, mainly in Africa and Asia (Vörösmarty et al., 2000).

The relatively small amount of freshwater on earth is threatened from many sides. In the process of building large dams in rivers, meant to regulate the discharge of water, the remedy is often worse than the disease. Numerous wetlands have been lost and are lost owing to changing land use. Owing to canalisation, and urban compacting and petrifaction of the grounds surrounding the rivers, the hydrology of most rivers has been changed irreversibly, with consequent accelerated discharge of water to the sea. River floods following human interference, it is the order of the day. In general, the chemical pollution of the surface water has increased mainly in the course of the 20th century. In North America and Western Europe, however, drastic reductions have been reached over the past 25 years for heavy metals, organic toxicants and acidifying substances. The concentrations of plant nutrients in Western European river water, however, hardly decrease. Long-term data for underdeveloped countries are largely lacking; the situation for those areas is considerably worse than in rich countries (Tockner & Stanford, 2002).

On a global scale, unfortunately, the same negative image as sketched for the future of our rivers, also counts for our estuaries. Habitat loss is the most important threat for the estuarine and coastal ecosystems, mainly owing to the steadily increasing human population. By far the largest numbers of the six billion people that live on earth, are situated in coastal areas, and the numbers will only increase. Habitat destruction, e.g. the transformation of natural habitats in commercially exploitable fish- and shrimp-ponds has far-reaching ecological consequences: biodiversity decreases, the functioning of the remaining coastal ecosystems is disturbed by excessive discharges of sewage water and plant nutrients, followed by chronic eutrophication, and oxygen depletion of water and sediment. Over-fishing and chemical pollution can also be mentioned as serious threats for the estuarine and coastal environment, not only in the tropics (Kennish, 2002). The loss of estuarine habitats counts for the rich and the developing countries alike, but there is a difference: in the rich countries deterioration is ongoing for (sometimes) hundreds of years, whereas in developing countries the accelerated breakdown of coral reefs and mangroves is of a more recent date. Particularly in many tropical countries the people are engaged in other problems as the preservation of their coastal environment. In rich countries the tendency exists to spend money for the conservation and restoration of coastal wetlands during periods of economic prosperity.

Calculations of Costanza et al. (1997) showed that the values of the world's remaining wetland habitats are far more greater than we had ever realized, in terms of natural capital and services to man. This notion encompasses a strong motivation to perform ecological research as the basis for water- and nature management.

Ecological concepts in river science

In the course of time quite a number of ideas have been published on the biology of river-estuary systems. Two concepts are of particular interest to river scientists, the River Continuum Concept (RCC; Vannote et al., 1980) and the Flood Pulse Concept (FPC; Junk et al., 1989). The concepts explain the ecological continuum of a river from source to sink, a gradient without sharp interruptions (RCC), and the sudden and unpredictable impact of river floods (FPC). The division of a river basin into compartments by the construction of large dams or by amputation of tributaries, and the serious demands for navigability, cannot go unpunished. Vital communities will then disappear, and migratory behaviour of fish that migrate from the open sea to their upstream spawning grounds, is blocked. And that does not only hurt the fish, but also the fisheries. The loss of the salmon fisheries on the river Rhine, already before World War II, gives evidence of this phenomenon (De Groot, 2002).

The river flows into the sea, there is erosion in the upstream sections and sedimentation in the downstream sections. Following its course, changes occur in current velocity, light intensity, the turbidity of the water, the biotic communities, and the production and decomposition processes. The impact of the river can be noticed far into the sea. The river is everywhere, not only flowing within its bed, but also invisible for the naked eye, invisible in the PCB's in the Wadden Sea, invisible in the unwanted plant nutrients, loading more than half of the surface water of the Netherlands (Gulati & Van Donk, 2002).

The differences in the dynamics of water movements is a characteristic distinction between the estuary and the river proper. An estuary, such as the Dollard or the Westerschelde, is dominated by the tidal rhythm, the twice a day repeated inundation and emersion of the tidal flats. A river like the Rhine or Meuse, is characterized by the irregular inundation of the floodplains, once or twice a year. Along the river continuum, both phenomena grade into each other: before the execution of the Delta Plan, the tidal movements from the North Sea were noticeable to Zaltbommel

(West of Tiel, Fig. 3) on the river Waal, approximately 100 km from the North Sea.

The periodic inundation of the tidal flats and the salt marshes in the estuaries, in contrast with the irregular, episodic inundation of the river floodplains has led to completely different ecosystems and food webs. An almost fully water-bound food web has evolved on and in the tidal flats, comprising animals that filter their food from the water, with adaptations to the oxygen depleted and even toxic (sulphide) soil. The habitat is dominated by algae, molluscs and crustaceans. Insects, with some exceptions do not survive in these circumstances.

The food web of river floodplains is far less known than the food web of tidal flats, particularly in its reaction to the episodic flooding. Flooding is a catastrophe for terrestrial animals; the water mass moves fast, and within some days the floodplains are completely inundated, and the choice for the animals is to flee or to drown. Except in pools and gullies, the food web in the floodplains is an almost completely terrestrial one. Many flowering plants occur there, and numerous insects, earthworms, birds of prey and mammals such as mice, voles, shrews and moles. Each species shows its typical survival strategy and colonization pattern after a flood event (Wijnhoven et al., 2003, 2006).

Research in river basins in the Netherlands

After these general notions, I will now concentrate on some pieces of work I did in the Netherlands, the small and rich country on the North Sea. Why the Netherlands? I have to make a choice from my work during the past 40 years. My ecological research put me in touch with many (specialists on) rivers and estuaries all over the world, in Europe, the USA, Asia, and Africa. I could as well discuss the results of our fieldwork into the structure and functioning of coastal seagrass ecosystems in the Indonesian archipelago and along the coast of Mauritania, West Africa, as the results of long discussions with students from developing countries during the many lectures in tropical marine ecology I gave at the Free University of Brussels. Still, most of my work was done at home in Western Europe. Most data were gathered in those extremely manipulated, regulated and managed half-natural (if you like, not-natural) ecosystems in the basins of the rivers Rhine and Meuse, where the intense relation between science and policy gives the discussion on 'water and values' a specific tension.

1953, 1995, 2003: a career in a nutshell

My career has been determined by two natural disasters, or to put it more precisely, by a mega-disaster, 'de Ramp' of 1953, and a near-disaster, the river flood of 1995. The flooding disaster of 1953 led to the foundation of the Delta Institute for Hydrobiological Research in 1957 in Yerseke (now Netherlands Institute of Ecology). The research institute was founded to analyse the ecological consequences of the execution of the Delta Works – the construction of a system of mega-dams to close the mouths of the estuaries of the rivers Rhine and Meuse, in order to prevent future damage by storm surges from the North Sea. I took service in Yerseke in 1965. The near-disaster of 1995 was caused by extremely high floods in the rivers Rhine and Meuse. There was a real danger that the weak river dikes should be overtopped or should breach, and as a precautionary measure approximately 250,000 citizens were evacuated from the threatened area. In 1995 I worked already for several years at Radboud University Nijmegen.

A disaster evokes conflicting feelings. On the one hand there is the human suffering; the loss of lives or goods and chattels has great impact. On the other hand there is the renewed societal interest in the backgrounds of a disaster. How may these disasters be prevented in the future? Disasters and incidents lead to new research. There is a saying 'no policy without calamity', and this notion is justified. In any case, the water policy not only in the Netherlands, is strongly decided by disasters and incidents (Anonymous, 2003). The introduction of costly, unpopular measures to guarantee the safety for the population, is only successful if there is a sense of urgency. These measures cost a lot of money. The few parts per thousand of the project costs, spent on scientific research, allowed us to do our research in relative luxury, both in the Delta area after the storm flood disaster of 1953, and in the river area after the high water incident of 1995. In fact, these additional credit facilities may be considered as a necessary supplement to the meagre basic funding for which universities are supposed to do their research. But the 'memory for disasters' of society is short. Soon more urgent problems attract attention again.

The flooding disaster of 1953 induced the Delta Plan. Under the motto 'this never again' (1835 human casualties; uncountable damage to human goods and chattels) the open mouths of the estuaries in the south-western Netherlands had to be closed by massive seawalls. The Delta Act was already implemented in 1957, and the estuaries were closed, one by one in the sixties and seventies. In 1986 the construction of the storm surge barrier in the mouth of Oosterschelde estuary was finished (Fig. 1). The barrier was considered the crown on the Delta Plan. Bombastic words were written on the proud Delta Works. I quote: "The former archipelago of isolated, sometimes forgotten islands changed in a short period in a delta of new possibilities, a junction of multicoloured activities, connected to the other parts of the Netherlands by dams and bridges. Inland lakes were created for amenity purposes, industry and new employment arose" (Van der Maas, 1986).

However, each medal has two sides. The side of ecology, the disappearance of a unique tidal delta, in favour of safety for the human population: it was in fact not done to bring that up after the horrendous disaster of 1953. Ecologists, of course, did worry about the consequences of the Delta Plan, and their arguments have played a decisive role in the political choice in favour of the saline

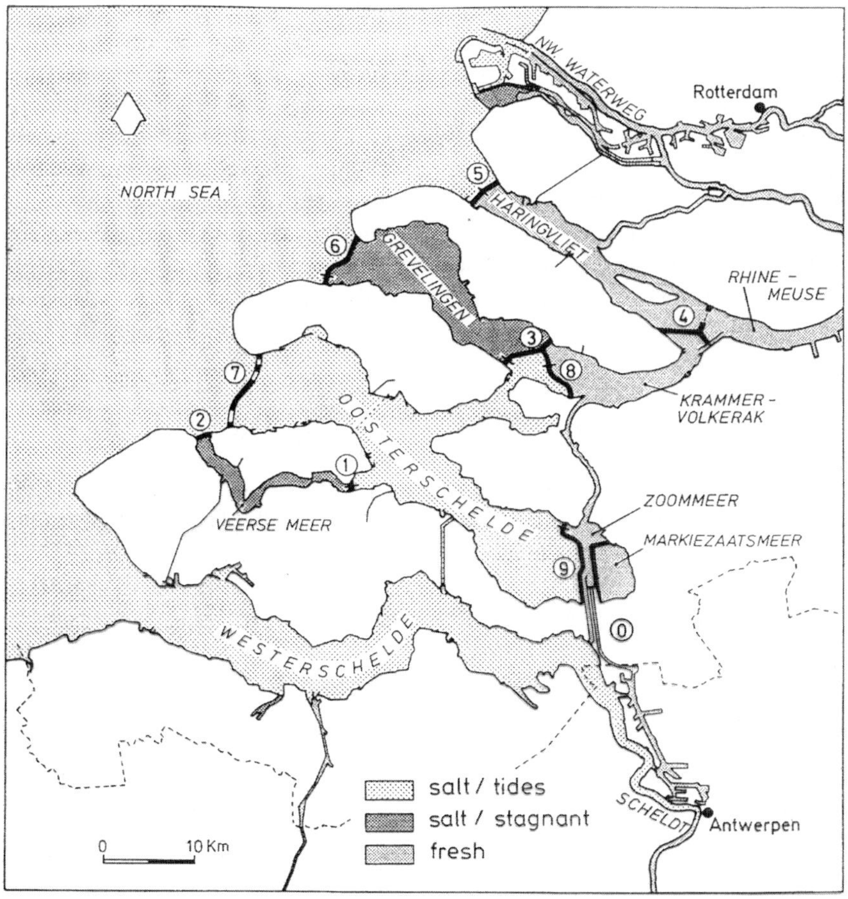

Figure 1. Delta area of the rivers Rhine, Meuse and Scheldt in the SW Netherlands, with various water bodies as resulting from the Delta project engineering scheme. 1 = Zandkreekdam, 1960; 2 = Veersegatdam, 1961; 3 = Grevelingendam, 1964; 4 = Volkerak-dam, 1969; 5 = Haringvlietdam, 1970; 6 = Brouwersdam, 1971; 7 = Oosterschelde storm surge barrier, 1986; 8 = Philipsdam, 1987; 9 = Oesterdam, 1986. Markiezaatsmeer has been closed off from Zoommeer by Markiezaatsdam in 1983. The connection between Oosterschelde and Westerschelde (0 = Zandkreekdam) was already closed in 1867.

Grevelingen, and the semi-tidal Oosterschelde. To illustrate this, I will quote a judgement in 1970 of my former director: "When we think about our own country, we may be short about the Delta Plan. This scheme does not mean management but complete annihilation" (Vaas, 1970). This statement is typical of those days. Annihilation means destruction. The statements were made, but in an academic setting, and at that time hardly anyone took note of these words. Besides my work as a scientist involved in 'water', I have made statements on the 'values' of the Delta on several occasions. In 1982 I published a very negative ecological balance sheet of the Delta Works. A negative balance from which the special features disappeared and the common features appeared (Nienhuis, 1982).

Time passes by quickly: water and values are changing entities over time. The past 15–20 years the appreciation for ecology as a science, and for ecological values has increased enormously. The understanding has grown that estuaries, with their tidal rhythm and their gradients from saline to freshwater, are ecologically much more healthy than the present separated and isolated waters. The Delta waters developed into instable aquaria, difficult to manage. There are problems caused by Cyanobacteria in the Volkerak-Zoommeer, by stinking sea lettuce (*Ulva* species) in the Veerse Meer, by eroding sandy shoals, that eventually disappear under water in the Oosterschelde, by turbid water and the appearance of annoying exotic species in the Grevelingen. Only 19 years have passed since the construction of the storm surge barrier in the Oosterschelde was finished. The crown on the Delta Plan has deteriorated during that short period to a rattling set of false teeth in the mouth of the estuary. Now, in the year 2005 even engineers of the State Department for Public Works and Water Management speak freely about the 'New Delta', about the return of tidal dynamics, about the reconstruction of the natural connections between the separated water bodies, under the prerequisite of full maintenance of safety for the human population. The near-disasters of 1993 and 1995, the river floods, has amplified these ideas (Hendriksen, 2002). Climate change leads, among other phenomena, to increased river discharges, and the Delta of Rhine and Meuse is needed, just as before 1953, for the discharge of superfluous river water to the sea. Working on living rivers also implies working on a living Delta.

The incidents of 1993 and 1995, the extremely high water levels in the large rivers, induced the same shock-effect as in 1953: this never again. The Delta Plan Large Rivers passed parliament within a few months. Hundreds of kilometres of river dikes were reinforced, widened and heightened. In contrast to the situation in the 1950th, when ecology did not play any role, now the values of rare plants and animals and their distribution were explicitly taken into account. Safety, of course, remained the prime argument, but where ever possible landscape and cultural historic values were preserved. In the meantime the statistical experts of our climate came with their alarming modelling results. The seasonal pattern of river discharges will change, and high water levels will increase in frequency in the future. The idea took post that we deal with a structural change in the discharge regime of the large rivers. In the policy document 'Space for the River (MinV&W, 2000), as part of the 'Water Policy in the 21st Century', a number of possible measures has been mentioned to cope with the high water problems of the future (Van Stokkom et al., 2005). This is food for ecologists and environmental scientists.

Research and management in the Grevelingen (ca. 1970–1990)

In 1971 the Grevelingen estuary was closed off from the sea by a large seawall, and the system became a large lake filled with clear, good quality seawater (Fig. 1). Within a period of 10 years extensive seagrass (*Zostera marina*) beds developed all over the lake. It started with patches of a few hundreds of hectares, and gradually expanded into massive underwater meadows of over 4500 ha, to 5 m water depth (Fig. 2).

This phenomenon was remarkable, and it took attention of Dutch and foreign specialists. I have directed a team of biologists, chemists, and other specialists who did quite some research into this remarkable ecosystem. The structure of the system was unravelled. The seagrass system appeared to be a habitat for numerous invertebrates and fish, looking for food and shelter. It was an

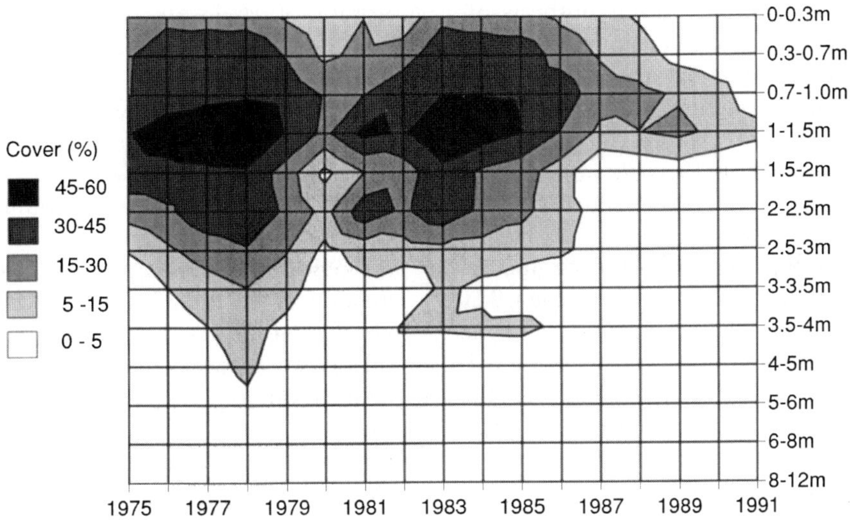

Figure 2. The relation between water depth (m) and eelgrass cover (%) over time (summer 1975–summer 1991), derived from the digitised maps and integrated for Grevelingendam lagoon, based on 400 sampling points spread over the lake. Eelgrass abundance is based on 17 annual surveys; 10 surveys contain actual data, the other surveys have been interpolated (Nienhuis et al., 1996).

inexhaustible resource for plant-eating and fish-eating birds, particularly ducks, geese, mergansers and cormorants. The second step was to study the functioning of the seagrass system, the primary production of water plants and algae, the growth and reproduction of the plants during spring, and the decomposition during winter, the consumption by animals, the carrying capacity for birds and fish, the cycle of chemical elements (carbon; nitrogen; phosphorus), the sediment–water exchange, and many other subjects. Using these data, mathematical ecological models were constructed, in order to predict the future developments of the ecosystem (e.g. Verhagen & Nienhuis, 1983). The outcome of these models was of great interest for water managers, because they wanted to know how the underwater vegetation could be restrained, in view of the interests of fisheries and recreation.

In 2005 only a few traces remained of the once flourishing seagrass beds in the past. In the 90th the decrease set in, and in 2000 it was concluded that *Zostera marina* became extinct in Lake Grevelingen. What is the reason of this extinction after 20 years of wax and wane? On the basis of our research some hypotheses were formulated. The causal factor might be the constant high salinity of the water, or the associated low concentrations of silicate and ammonium as limiting factors for growth and development. Another hypothesis could be a sequence of severe winters, eventually killing the roots and rhizome mats of the seagrass. Even a mysterious infectious disease was coined as possible cause for the deterioration of the vegetation (Nienhuis et al., 1996). We had a handful of statistical correlations but no causal relations. The most likely explanation appeared to be the constant high salinity of the water. The estuarine *Zostera marina* population was closed off from the sea in 1971, and the originally brackish water circumstances under which seagrass thrived, gradually changed into an undiluted, evaporating mass of seawater. Obviously, a constant and high salinity is recognized by the brackish water vegetation as a severe and continuous stress situation (Kamermans et al., 1999).

Our results offered input for many model parameters, still in use to formulate and calculate the rules in behavioural patters of *Zostera marina* populations all over the world. Our research, however, was carried out in a young, instable, not resilient ecosystem. The models were able to predict the wax of the seagrass vegetation, but not the wane. The reason for that is simply because the property of the population to react negatively to high salinity was not included in the model as a variable. And we know, models cannot deliver better predictions than the data allow.

Restoration of the seagrass vegetation might be expected when the estuarine dynamics in Lake

Grevelingen will be restored. This hypothesis offers the water managers a solid argument (next to other arguments) to work on the restoration of the estuarine gradient in the lagoon.

Research and management in the Afferdensche and Deestsche Waarden (ca. 1990–2005)

In 1994, I became full professor of Environmental Sciences at the Radboud University Nijmegen. My move from the NIOO in the estuarine part of the rivers Rhine and Meuse, to the University of Nijmegen on the river Waal, meant a shift to the freshwater side of the river continuum. A great difference between the research activities of roughly 30 years ago, and now, is that the number of people directly or indirectly involved in research has multiplied, and this counts both for estuarine research as well as for river research. The interference of people in research activities has increased enormously, particularly on the policy side, the 'paper' side of projects, exemplified in thorough planning, in procedures to be followed, in communication on the progress of the project, etc. Complete research teams are now involved in data collecting, monitoring and processing of data, and in computer modelling.

Our large rivers have been used – and abused – for over 1000 years. They are canalised and constricted between high levees in far too narrow beds. The nutrient-poor river basin grasslands, the unfertilised hay lands, and the clear-water river creeks, rich in species, have disappeared, mainly owing to the intensive agricultural practice. Floodplain forests, river dunes, and sand and gravel banks have systematically been removed to facilitate the fast discharge of river water. Unbridled chemical pollution before and after the Second World War until 1970, has had a devastating impact on populations of many typical river species. The river water is heated by effluents of cooling water of power plants. Connected to the rise in temperature of the water, invasions of exotic species, such as freshwater shrimps and bivalve molluscs, have turned the food network upside down (Van der Velde et al., 2002).

Since a few decades governmental and private organizations have invested quite a lot in ecological restoration of the spoiled rivers. The most comprehensive results have been booked on the terrain of the water quality: the water is now considerably cleaner than it was 25 years ago, and concentrations of non-degradable organic micro-pollutants and heavy metals have decreased spectacularly. There is much debate, however, about the results reached with regard to the ecological restoration of river-bound habitats. 'Nature development' is a new instrument in the ecological restoration process. The phenomenon is introduced in the 80th and it comprises the conversion of non-profitable agricultural land (meadows; maize fields) in wetlands, including the restoration of the characteristic natural river dynamics. The return of a large number of invertebrates and fish after the digging of secondary channels in the framework of nature development projects, is a positively valued action. On the other hand, spontaneously growing sand banks, and the natural development of floodplain forests are not allowed, however, for reasons of safety, because they are hindering the quick dispatch of water during extreme river floods.

I will focus now on Afferdensche and Deestsche Waarden, a complex of old floodplains and river creeks at the river Waal, close to Nijmegen, 300 ha of silted-up land, comprising the river bed itself, meadows lined with hawthorns, soggy reed lands, wetland marshes, and a number of deep clay pits exploited by two (now closed) brick factories (Fig. 3). A substantial part of the research of the Department of Environmental Science of Radboud University is focused on this wetland. We are carrying out projects, in cooperation with many other scientists from universities and research institutes. We participate in large projects subsidized by the government (such as Land Ocean Interactions in the Coastal Zone, and Stimuleringsprogramma Systeemgericht Ecotoxicologisch Onderzoek).

'To guarantee safety' is the buzz-word after the floods of 1995, and the government is planning a re-arrangement of the area, with the prime objective to give the river Waal more space for a quicker run off of water during extremely high discharges. A secondary motive is 'nature development' in former agricultural areas. Large areas of the floodplain will be dug off and lowered, and a secondary discharge channel will be scooped out, parallel to the main channel. When these dredging plans will be realized approximately 500,000 m^3 contaminated soil will be uncovered. The soil

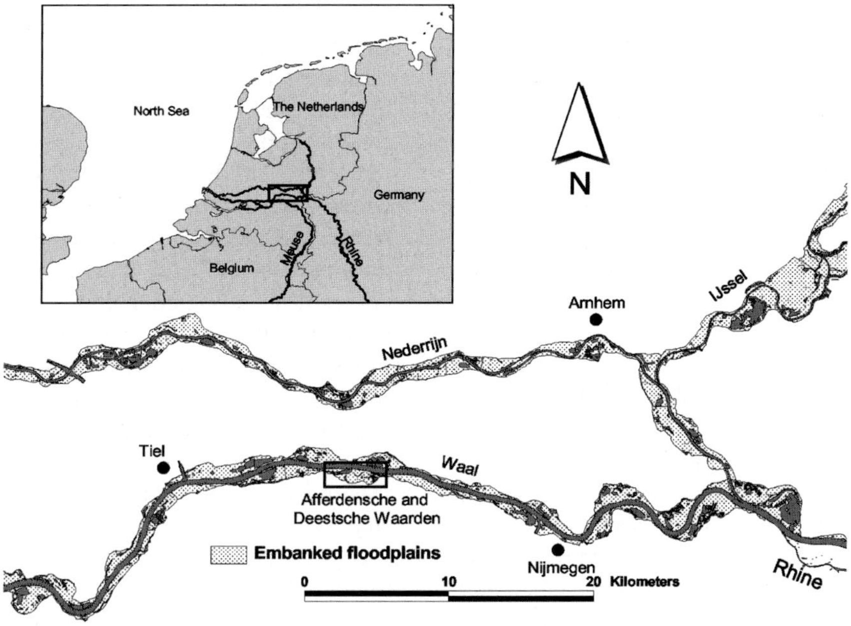

Figure 3. Location of the Afferdensche and Deestsche Waarden floodplain along the river Waal (Kooistra et al., 2002).

contains contaminants such as heavy metals and non-degradable organic micro-pollutants that have been deposited, packed in silt and clay, during the 'dirty' years after the Second World War. Because the river became cleaner in the course of time, the old polluted deposits have recently been covered with cleaner sediments. The massive dredging activities in Afferdensche and Deestsche Waarden, and in other floodplains as well, introduce the likely risk for plants and animals in the food chain of becoming seriously contaminated (Leuven et al., 2005).

Quite a lot of research activities have been devoted to the effects of water and soil pollution on individual species in the floodplain food web in the past 20 years. Important questions remained, however. The heterogeneity of soil contamination in river floodplains connected to the distribution patterns and behaviour of animals living and feeding in floodplains was poorly understood, particularly in relation to episodic flooding (Nienhuis, 2003b; Kooistra et al., 2001; Kooistra, 2004).

Animals low in the food chain, such as earth worms comprise a dominant group of invertebrates, showing a biomass up to 0.15 kg m^{-2}. Multivariate statistical analysis revealed that hydrodynamics and the moisture content of the soil are the most important steering variables for their population dynamics and distribution patterns. Species living close to the surface of the soil are almost eradicated by flooding, while worms living deeper in the soil survive flooding. It is suggested that serious soil contamination by arsenic and zinc does not negatively influence the earth-worm populations (Zorn, 2004).

For animals higher in the food chain, annual flooding has a major impact on densities, distribution and re-colonization patterns. Flooding may completely erase the populations of small mammals from the inundated parts of the floodplain. The re-colonization rate and pattern varies per species, and the process takes several months. This casts new light upon ecological risks of the displacement of sediment in the framework of flood defence measures. The exposure risks to soil pollution in cadmium and zinc contaminated environments and the mobilization and bio-availability of these heavy metals in polluted floodplain soils have to be remodelled on the basis of detailed data on population dynamics of the dominant small animals (Wijnhoven et al., 2003, 2006).

When it comes to the top predators in the floodplain food chain, Kooistra (2004) developed procedures, in the framework of ecological risk assessment, to incorporate spatial components of exposure of key species (e.g. *Athene noctuca*) to soil pollution (Fig. 4). By linking a geographical information system (GIS) with a probabilistic exposure model, spatial aspects of exposure have been taken into account (Fig. 5). The resulting site-specific risk maps of Afferdensche and Deestsche Waarden identify potential high-risk areas that can be of use to floodplain managers to make wise decisions in their planning of floodplain reconstruction (Leuven et al., 2002, 2005).

Challenges in 'water and values' research

Notwithstanding the semi-natural status of our large rivers, the landscape is appreciated as 'typical Dutch', comprising values that belong to the national heritage. A strictly ecological view on the rivers also reveals large values. River basins fit fully into the Ecological Main Structure, the prestigious international concept of interconnected nature reserves. A string of National and Regional Nature Reserves along the rivers exemplifies those values. Large areas are covered by the Habitat Directive and the Birds Directive of the European Commission, guaranteeing international protection.

Since 1995 the boundary condition for all measures in river basins is 'promotion of safety', and this forces the government to take inevitable, draconic measures. It is my concern, however, that after 40 years from now, the realization of the measures foreseen in the Physical Planning Core Decision 'Space for the River' (implementation of the National Act on Physical Planning, to be ratified by Parliament) will in ecological terms be appreciated as negatively as the Delta Works nowadays. Recently it became clear that the feared discharge of 18,000 $m^3 s^{-1}$ at the Rhine near Lobith is extremely unlikely. When this doom-scenario occurs, large parts of Nordrhein-Westfalen will be flooded, and this leads to strongly decreased discharges in the Netherlands (Duits-Nederlandse Werkgroep Hoogwater, 2004). It is true, maintenance of the riverbed is in arrears, the downstream sections of Rhine and Meuse are net collectors of sand and silt, and dredging must continue forever. But 'nature development' should not be used as a mere eyewash. Nature development suggests that 'nature' is absent, or vaguely noticeable, in the present situation, and that by dredging gullies and lowering the surface level of the floodplains, 'new nature' will arise. This thought is diametrically opposed to the current appreciation for the river landscape, moulded by man in the course of centuries. Nature development is a hardly defendable option at places where present cultural-historic and landscape values dominate. An ecologically underpinned total vision on the developments in the river basins is lacking. It is not sensible to take too big steps in a process of which the long-term ecological consequences cannot be estimated. It should be avoided that everywhere the same solution is chosen, under the motto "secondary gullies improve biodiversity, so we shall dredge a secondary gully everywhere, even though there has never been one in the past".

It must be possible to adapt a number of stakeholder interests to sustainable, ecological river management, and not the other way around. My plea: do not chose the same solution everywhere. Do not decide overhasty about the future development of ancient floodplains. Leave some floodplains, demonstrating specific natural and cultural values, untouched. As Westhoff (1999) repeatedly stated: "To manage our semi-natural landscapes, the best action is to perform everywhere a different measure, but to maintain these measures over time. The common management practice nowadays is a continual change in measures, and everywhere the same measures". Illustrative for this type of management is the dredging of secondary gullies, the grazing with large cattle, compacting and manuring the soil, and the uniform mowing regime on the dikes. The ecological plea, optimisation of habitat and ecosystem diversity, implies management in favour of communities at spots poor in plant nutrients and at sandy spots, to take advantage of seepage habitats, to manage the wetlands without the threat of hidden toxicants, and to allow natural succession the time it takes.

In the course of time the Dutch river landscape has gained values that should be counted for. These values may be called 'the fifth dimension of the landscape' (Lenders, 2003). Besides the three physical dimensions of a landscape, there is 'time' as the fourth dimension, the

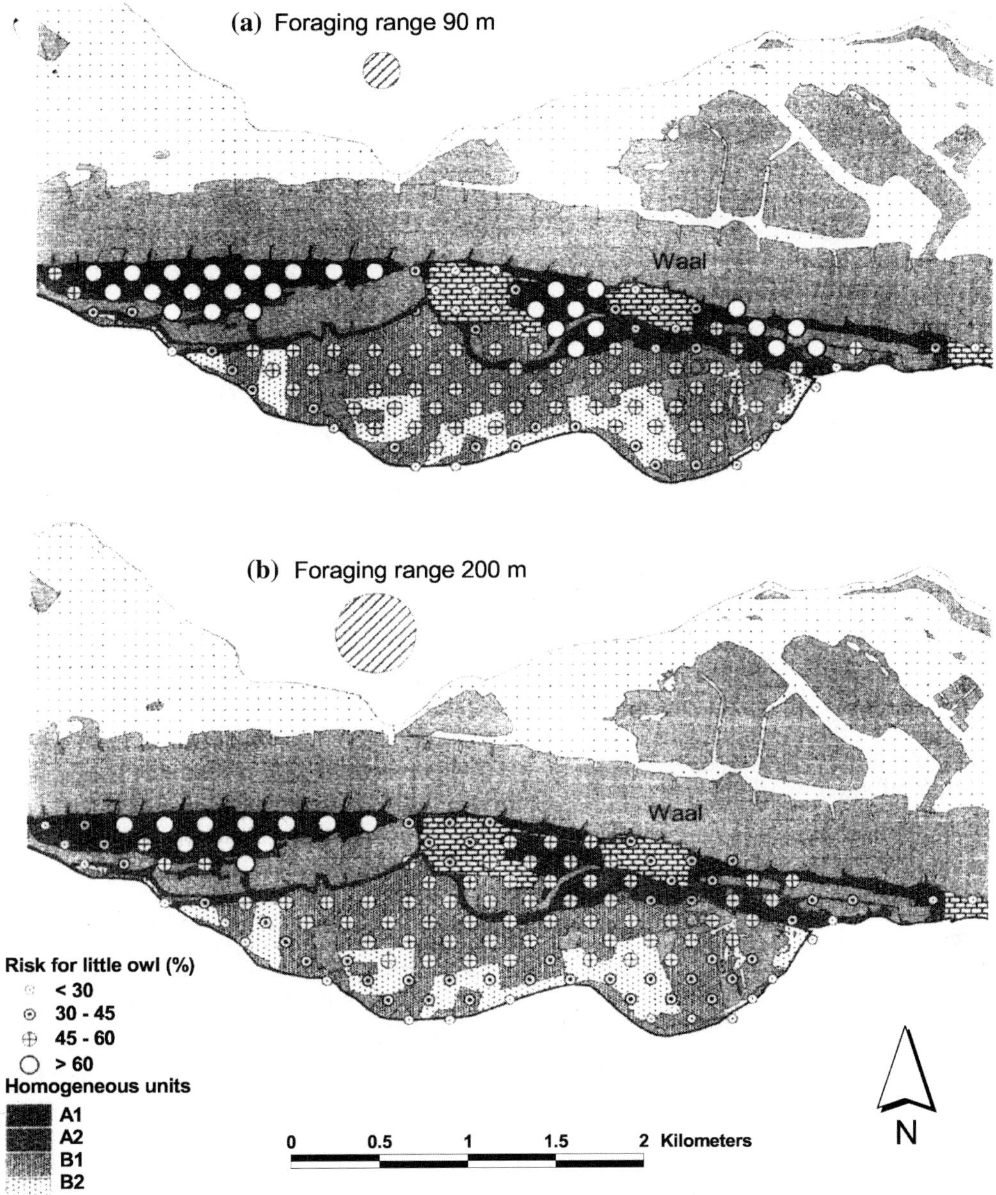

Figure 4. Site-specific cadmium exposure risk for the little owl (*Athene noctua*) with a foraging range of (a) 90 m and (b) 200 m. Exposure risk data indicate the percentage of the population struck by detrimental effects in certain stages of the life cycle, by the consumption of contaminated prey (Kooistra et al., 2002).

history of its origin and development. Then there is the fifth dimension, comprising the cultural and historical assets, shaping the landscape to what it really is. Natural succession needs time, sometimes hundreds of years, to show the full development of biodiversity. My concern is that the concept of 'cyclic floodplain rejuvenation' (take care, by management, that floodplain ecosystems stay young; prevent the full growth of floodplain forests; Duel et al., 2001; Baptist et al., 2004) is too much going down on one's knees for the river managers, having as a consequence that the yellow dragline will be visible and audible in the floodplains more frequently than the black stork. The fifth dimension of the Dutch semi-natural river landscape is anchored

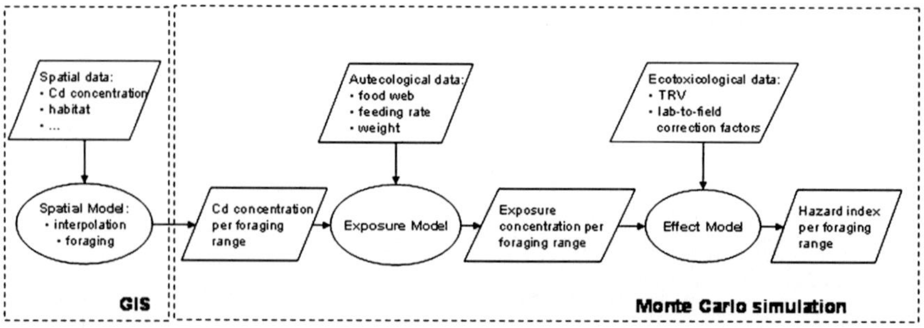

Figure 5. Schematic overview of the conceptual model for the incorporation of spatial components in the ecological risk assessment of river floodplains (Kooistra, 2004).

in the collective memory of the inhabitants of the river basins (Lenders, 2003). Part of this collective memory is formed by the floodplain pastures studded with cows, the flowering hawthorn hedges bordering the meadow lots, the unfertilised, luxuriously flowering hay lands, the endangered wetland birds, and the tens of thousands of migratory geese in winter. The cultural side of the fifth dimension comprises the archaeological remnants dating back to the Roman 'limes', when the river Rhine was the northern border of their world empire, and also fortifications smothered in weeds as reminder of the paramount strategic function of rivers in the military history.

References

Anonymous, 2003. Het waterbeleid kan niet zonder rampen. De Water 88: 8 (in Dutch).

Baptist, M. J., W. E. Penning, H. Duel, A. J. M. Smits, G. W. Geerling, G. E. M. van der Lee & J. S. L. van Alphen, 2004. Assessment of the effects of cyclic floodplain rejuvenation on flood levels and biodiversity along the river Rhine. River Research and Applications 20: 285–297.

Barrett, G. W. & K. E. Mabry, 2002. Twentieth-century classic book and benchmark publications in biology. BioScience 52: 282–285.

Carpenter, S., B. Walker, J. M. Anderies & N. Abel, 2001. From metaphor to measurement: resilience of what to what? Ecosystems 4: 765–781.

Costanza, R., R. d'Arge, R. de Groot, S. Farber, M. Grasso, B. Hannon, S. Naeem, K. Limburg, J. Paruelo, R. V. O'Neill, R. Raskin, P. Sutton & M. Van den Belt, 1997. The value of the world's ecosystem services and natural capital. Nature 387: 253–260.

Costanza, R., B. G. Norton & B. D. Haskell, 1999. Ecosystem Health: New Goals for Environmental Management. Island Press, Washington, D.C.

Davis, M. A. & L. B. Slobodkin, 2004. The Science and Values of Restoration Ecology. Restoration Ecology 12: 1–3.

De Groot, S. J., 2002. A review of the past and present status of anadromous fish species in the Netherlands: is restocking the Rhine feasible?. Hydrobiologia 478: 205–218.

Duel, H., M. J. Baptist & W. E. Penning (eds), 2001. Cyclic Floodplain Rejuvenation. A New Strategy Based on Floodplain Measures for Both Flood Risk Management and Enhancement of the Biodiversity of the River Rhine. IRMA-SPONGE Program. Publ. 14–2001, Netherlands Centre for River Studies, Delft.

Duits-Nederlandse Werkgroep Hoogwater, 2004. De Rijn kent geen grenzen – Der Rhein kennt keine Grenzen. Provincie Gelderland, Arnhem.

Golley, F. B., 1993. A History of the Ecosystem Concept in Ecology: More Than the Sum of the Parts. Yale University Press, New Haven, London.

Gulati, R. D. & E. van Donk, 2002. Lakes in the Netherlands, their origin, eutrophication and restoration: state-of-the-art review. Hydrobiologia 478: 73–106.

Hendriksen, J., 2002. De nieuwe Delta, voorwaarts naar vroeger. De Water 85: 3–5(in Dutch).

Junk, W. J., P. B. Bayley & R. A. Sparks, 1989. The flood pulse concept in river-flood plain systems. In Dodge, D.P. (ed.), Proc. Intern. Large River Symposium, Canadian Special Publication Fisheries and Aquatic Sciences 106: 110–127.

Kamermans, P., M. A. Hemminga & D. J. de Jong, 1999. Significance of salinity and silicon levels for growth of a formerly estuarine eelgrass (*Zostera marina*) population (Lake Grevelingen, the Netherlands). Marine Biology 133: 527–539.

Kennish, M. J., 2002. Environmental threats and environmental future of estuaries. Environmental Conservation 29: 78–107.

Kooistra, L., 2004. Incorporating spatial variability in ecological risk assessment of contaminated river floodplains. PhD Thesis University of Nijmegen, Nijmegen.

Kooistra, L., R. S. E. W. Leuven, R. Wehrens, L. M. C. Buydens & P. H. Nienhuis, 2001. A procedure for incorporating spatial variability in ecological risk assessment of Dutch river floodplains. Environmental Management 28: 359–373.

Kooistra, L., R. S. E. W. Leuven, R. Wehrens, L. M. C. Buydens & P. H. Nienhuis, 2002. GIS as a tool for incorporating the spatial variability of pollutants in ecological risk assessments of Dutch river floodplains. In Leuven, R. S. E. W. I. Poudevigne & R. M. Teeuw (eds) Application of Geographic Information Systems and Remote Sensing in River Studies. Backhuys Publishers, Leiden, 133–146.

Lackey, R. T., 2001. Values, policy, and ecosystem health. BioScience 51: 437–443.

Lenders, H. J. R., 2003. Environmental rehabilitation of the river landscape in the Netherlands – a blend of five dimensions. PhD thesis, University of Nijmegen, Nijmegen.

Leuven, R. S. E. W., I. Poudevigne & R. M. Teeuw, 2002. Application of Geographic Information Systems and Remote Sensing in River Studies. Backhuys Publishers, Leiden.

Leuven, R. S. E. W., S. Wijnhoven, L. Kooistra, R. J. W. de Nooij & M. A. J. Huijbregts, 2005. Toxicological constraints for rehabilitation of riverine habitats along lowland rivers. Archiv für Hydrobiologie Supplement 155 (1–4) Large Rivers 15 (1–4): 657–676.

Malmqvist, B. & S. Rundle, 2002. Threats to the running water ecosystems of the world. Environmental Conservation 29: 134–153.

MinV&W, 2000. Ministerie van Verkeer en Waterstaat. Ruimte voor de Rivier. Anders omgaan met water – Waterbeleid in de 21ste eeuw. Den Haag (in Dutch).

NIBI, 2000. Adresboek voor Biologen 2001. Nederlands Instituut voor Biologie, Utrecht (in Dutch).

Nienhuis, P. H., 1982. De oecologische consequenties van de Deltawerken. In Wolff, W.J. (ed.), Wadden, Duinen, Delta. PUDOC, Wageningen: 101–132 (in Dutch).

Nienhuis, P. H., 2003a. Water en Waarden. Ecologisch onderzoek als basis voor waterbeheer en natuurbeheer. Valedictory Address. Radboud University, Nijmegen (in Dutch).

Nienhuis, P. H., 2003b. Het laagland met de plassen, verkaveld door de sloten. De blauwe milieuwetenschappen. In Boersema J. J., T. Pulles, J. van der Straaten & J. Bertels (eds), De Oogst van Milieu. Boom, Amsterdam. Milieu 17 (4/5): 50–64 (in Dutch).

Nienhuis, P. H., B. H. H. de Bree, P. M. J. Herman, A. M. B. Holland, J. M. Verschuure & E. G. J. Wessel, 1996. Twenty five years of changes in the distribution and biomass of eelgrass, *Zostera marina*, in Grevelingen Lagoon, the Netherlands. Netherlands Journal of Aquatic Ecology 30: 107–117.

Odum, E. P., 1969. The strategy of ecosystem development. Science 164: 262–270.

Odum, E. P., 1971a. Fundamentals of Ecology (3rd edn.). Saunders, Philadelphia.

Odum, E. P., 1977. The emergence of ecology as a new integrative discipline. Science 195: 1289–1291.

Odum, H. T., 1971b. Environment, Power and Society. John Wiley & Sons, Chichester.

Odum, H. T., 1981. Energy Basis for Man and Nature. McGraw Hill, New York.

Rapport, D. J., 1992. Evaluating ecosystem health. Journal of Aquatic Ecosystem Health 1: 15–24.

Rapport, D. J., 1995. Ecosystem health: more than a metaphor?. Environmental Values 4: 287–309.

Tansley, A. G., 1935. The use and abuse of vegetational concepts and terms. Ecology 16: 284–307.

Tockner, K. & J. A. Stanford, 2002. Riverine flood plains: present status and future trends. Environmental Conservation 29: 134–153.

United Nations Conference on Environment and Development, 1992. Declaration of the United Nations Conference on Environment and development, Agenda 21. Rio de Janeiro: 6–11.

Vaas, K. F., 1970. Problemen bij het beheer van het mariene milieu. In Van de Kamer, J. C. (ed.) Het verstoorde evenwicht – Een pleidooi voor het behoud van het natuurlijk milieu. Oosthoek, Utrecht, 217–230 (in Dutch).

Van der Maas, C., 1986. De andere Delta – Stormvloed 1953 Kering 1986. Ministerie van Verkeer en Waterstaat, Den Haagin Dutch.

Van der Velde, G., I. Nagelkerken, S. Rajagopal & A. bij de Vaate, 2002. Invasion by alien species in inland freshwater bodies in Western Europe: the Rhine delta. In Leppäkoski, E., S. Gollasch & S. Olenin (eds), Invasive Aquatic Species of Europe. Kluwer, Dordrecht: 360–372.

Van Dobben, W. H. & R. H. Lowe-McConnell, 1975. Unifying Concepts in Ecology. Junk-Kluwer, Dordrecht.

Vannote, R. L., G. W. Minshall, K. W. Cummins, J. R. Sedell & C. E. Cushing, 1980. The river continuum concept. Canadian Journal of Fisheries and Aquatic Sciences 37: 130–137.

Van Stokkom, H. T. C., A. J. M. Smits & R. S. E. W. Leuven, 2005. Flood defence in the Netherlands: a new era, a new approach. Water International 30: 76–87.

Verhagen, J. H. G. & P. H. Nienhuis, 1983. A simulation model of production of eelgrass (*Zostera marina*), seasonal changes in biomass and its distribution in Lake Grevelingen. Marine Ecology Progress Series 10: 187–195.

Vörösmarty, C. J., P. Green, J. Salisbury & R. B. Lammers, 2000. Global water resources: vulnerability from climate change and population growth. Science 289: 284–288.

Vugteveen, P., R. S. E. W. Leuven, M. A. J. Huijbregts & H. J. R. Lenders, 2006. Redefinition and elaboration of river ecosystem health: perspective for river management. Hydrobiologia 565: 289–308.

Westhoff, V., 1999. Natuurbeheer, onze verantwoordelijkheid voor de biosfeer. Victor Westhoff-lezing. Nijmegen University Press, Nijmegen (in Dutch).

Wijnhoven, S., A. J. M. Smits, R. S. E. W. Leuven & G. van der Velde, 2003. Impact of flooding on small mammals along the river Waal (The Netherlands). In Leuven, R.S.E.W., A.G. van Os & P.H. Nienhuis (eds), Proceedings NCR-days 2002: Current themes in Dutch river research. NCR-publications 20–2003, Netherlands Centre for River Studies, Delft: 156–159.

Wijnhoven, S., G. van der Velde, R. S. E. W. Leuven & A. J. M. Smits, 2006. Modelling recolonisation of heterogeneous river floodplains by small mammals. Hydrobiologia 565: 135–152.

Winterhalder, K., A. F. Clewell & J. Aronson, 2004. Values and science in ecological restoration – a response to Davis and Slobodkin. Restoration Ecology 12: 4–7.

Zorn, M. I., 2004. The floodplain upside down: interactions between earthworm bioturbation, flooding and pollution. PhD thesis, Free University, Amsterdam.

Hydrobiologia (2006) 565:277–288
R.S.E.W. Leuven, A.M.J. Ragas, A.J.M. Smits & G. van der Velde (eds), Living Rivers: Trends and Challenges in Science and Management
DOI 10.1007/s10750-005-1919-1

Opinion Paper

Advancing science for water resources management

G.E. Petts[1,]*, J. Nestler[2] & R. Kennedy[3]
[1]*School of Geography, Earth and Environmental Sciences, University of Birmingham, Edgbaston B15 2TT, UK*
[2]*US Army Engineer Research and Development Centre, Vicksburg, Mississippi, USA*
[3]*US Army Engineer Research and Development Centre, European Research Office, London, UK*
*(*Author for correspondence: E-mail: g.e.petts@bham.ac.uk)*

Key words: flow variability, integrated physical-biological models, long-term datasets, science-based management, water resources

Abstract

Despite the major advances in science to underpin water resources and river management that have taken place over the past two decades, a need remains to establish a unifying framework that will lead to new, appropriate tools for water resources management. In Europe, this need has been highlighted by the promotion of the Water Framework Directive. From a scientific perspective, key questions focus on the ecological significance of flow variability over a range of timescales and the linkage between flow variability, habitat variability and biological population responses, and the biological interactions among these populations. Creation of scientifically sound tools requires development of knowledge at the level of first principles to realize sustainable developments within the context of adaptive management. Similitude analyses provide a mechanism for upscaling from fine 'research' scales to the coarser scales of water resource managers. Lack of appropriate data is the major obstacle to the development of these tools, especially those concerned with large rivers.

Introduction

Water level and discharge control and landscape transformation in the modern period has attracted considerable attention (Cosgrove & Petts, 1990). Water security and flood control have been at the centre of the development of civilizations and remain key issues today (Nienhuis & Leuven, 2001; Van Stokkom et al., 2005). The impressive engineering solutions that dominated water control and river management through to the late 20th century are often environmentally unsustainable and the cause of dramatic degradation of many aquatic systems (Boon et al., 2000; Smits et al., 2006). In response to the growing pressures for sustainable approaches to water management, the last two decades of the 20th century witnessed a paradigm shift from 'control by construction' to 'stewardship and dynamic/adaptive management', in which environmental concerns are balanced with the economic benefits of water resources development (Leuven et al., 2000). Within a political framework of escalating expectation and regulation, and growing recognition of the enormous complexity of environmental systems, managers have embraced adaptive management as an approach to coping with uncertainty (Clark, 2002). Adaptive management focuses on responses and scenario building, and requires informed and aware stakeholder and public participation. However, river managers still lack appropriate science-based tools to realize the integration of human and environmental water needs. Those managers concerned with water resources planning have a tradition based in human needs driven by the economic risks of drought founded upon flow duration statistics. Engineers struggle to design deterministic and probabilistic solutions for flood control and navigation improvement that have sustainable environmental benefits. Scientists strive to develop the understanding of ecosystem response to flow changes necessary for advancing tools to analyze the sustainability of alternative water resource management plans.

In the UK, one of the main goals of the Environment Agency's Vision (Environment Agency, 2001) is to achieve integrated and sustainable management of inland waters that includes protection and enhancement of wildlife. In achieving this goal the Agency must consider all users, the environment, and the views of nature conservation agencies and organizations. The requirements of the European Union's (EU) Water Framework Directive that came into force in December 2000 (Logan & Furse, 2002; Irvine, 2004) are exposing the inadequacies of our scientific understanding that prevent development of planning models to describe, forecast, and manage impacts of different designs or operations.

This paper addresses issues for establishing a unifying framework to advance the scientific knowledge necessary to develop appropriate tools for water resources management. The paper is based upon discussions at a workshop sponsored by the U.S. Army Corps of Engineers as part of the Water Operations and Technical Support Program. The Workshop brought together a small group of physical and biological scientists, engineers, planners and policy makers. Most offered a western European perspective, with a range of traditions and experiences but all founded in the long history of basin development and river regulation, in nearly four decades of practical experience in river rehabilitation, and in an evolving legislative framework that will lead to the full implementation of the EU Water Framework Directive. The paper addresses two key themes: (i) the need for scientists to focus on environmental variability at different scales and (ii) the need for a scientific framework that is shared by both scientists and managers.

Scientific questions and approaches

Issues that deal with scales of space and time, and hierarchy remain an overarching concern. River basin management must deal with a multitude of scales that range from patches and habitats for individual species, some of which may be very small, to the catchment landscape. The EU Water Framework Directive requires catchment-level assessment, placing emphasis on ecological status and integrity. But is ecological integrity scale-dependent? If so, how should it be measured? To develop an integrated measure describing the quality or status of a river basin, methods for assessing characteristics of a stream reach, the local area draining into the reach, and the catchment up-stream from the reach, are needed. For example, a common assumption is that the width of the riparian zone is an important criterion for estimating ecosystem integrity, especially along the middle and lower reaches of natural rivers where the river corridor is characterized by a dynamic mosaic of habitat patches. Here, successional gradients are steepest and habitat turnover is greatest, but high connectivity (Tabacchi et al., 1998; Andersson et al., 2000) and rich propagule banks (Goodson et al., 2002) sustain high inter-habitat dispersal rates and promote resilience. In regulated rivers, seed banks and biotic dispersal mechanisms, such as active migration or drift, cannot maintain viable metapopulations if habitat fragments are too small, if habitats are insufficiently interconnected, or if certain successional stages are underrepresented because water control has reduced riparian zone width, reduced the complexity of the habitat mosaic, or altered the geomorphology of the river channel in other ways. The scale thresholds at which these effects emerge are unknown (Geerling et al., 2006). Scientists do not have adequate understanding of the importance of river corridor architecture across a range of spatial scales nor do they know the timing and duration of disturbance needed to maintain this dynamic architecture.

Indicators of condition

A key element of adaptive management involves monitoring of carefully defined indicators of river condition. It is axiomatic that any analysis of river corridors should consider both functional and structural components. However, because of the lack of appropriate process-based models, managers continue to depend upon structural indicators that provide surrogates for process. This is quite difficult to do in practice because the relationship between function and structure is often neither clear nor appreciated. For example, critical biogeochemical processes may be "invisible" at scales used to describe and manage river morphology. To allow managers to incorporate process in their evaluations, readily apparent indicators of processes and functions will need to be discovered and developed from improved scientific understanding.

The way forward requires that the fine scales at which process and function are understood and described by scientists must be reconciled with the coarse structural scale at which managers make environmental management decisions. Reconciliation can be achieved by merging together the two approaches into a single integrated framework that spans the needs of scientists and resource managers. In this framework, scientists study process and function at a scale determined by experimental accuracy, but with an understanding of the scale at which managers make decisions. Once scientists have quantitatively described process and function at a scale appropriate for understanding and peer review confirmation, then the quantitative description can be simplified by recursively dropping terms, reducing dimensions, and coarsening time and space scales until the answer provided by the detailed quantitative description diverges from the answers provided by the simplified surrogate. The step just prior to divergence represents the most parsimonious assessment tool having a scientific basis that can be used by managers for water resources decision-making. Alternatively, spatial or structural surrogates could be evaluated against fine scale descriptions of process and function to quantitatively determine which measures of structure best correlate to patterns in process and function. This systematic "ecological model similitude analysis" is the only sure way to defensibly span the gulf between the fine-scale process and function world of scientists with the coarser-scale structure world of managers.

New structure-based surrogate indicators of integrity, especially applicable to sector- and reach-scale classifications should clearly be relatable to process and function (Table 1). At the sector scale, information is needed on feedback

Table 1. Proposed indicators of (structural) integrity within fluvial hydrosystems that could be linked to process and function for large floodplain rivers (after K. Tockner, personal communication)

Structural measures	Measures of variability with discharge
Proximity to tributaries	Shoreline length
Width of active corridor	Connectivity between habitats
Riparian zone width	Area or number of channels
Number of islands	Components of detrital organic matter budget
Area of exposed sediments	Degree of surface – ground water exchange
Number of ponds	
Volume of large wood	
Relative number of species:	
colonizing opportunists	
competing species	
ubiquitous species	

effects between reaches to evaluate the importance of continuity for exchange of material, biota, and propagules. These will require new quantitative knowledge of the dynamics of key processes such as surface water and groundwater interactions; sediment routing (input, transport, dispersion, deposition, erosion); organic matter production (and changing composition related to the importance of the different autogenic, allogenic and anthropogenic sources) and utilization through food chains; and seed/propagule dispersal and vegetation succession. It will also be important to "scale-up" knowledge developed at the smaller, reach and site scales using the similitude concept described above. That is, parsimonious tools based on similitude analysis do not exhibit the computational and mathematical challenges of detailed process tools and therefore are more amenable to up scaling by aggregation from smaller spatial scales to larger spatial scales.

Focus on flow variability

The introduction of structure-based surrogates that are sensitive to process and function associated with water control such as proposed in Table 1, would represent a major step forward in assessing the integrity of riverine ecosystems. However, this in no way masks the need for fundamental research on process variability and how process variability is preserved in structure-based tools. Three key abiotic factors, flow, temperature and substratum stability, are assumed to be the primary drivers of lotic and floodplain ecosystems over a range of scales if such a similitude analysis were to be done (e.g. Poff & Ward, 1990; Petts, 2000; Tockner et al., 2000; Milner et al., 2001; Church, 2002; Gurnell & Petts, 2002; Richards et al., 2002). Furthermore, the degree of predictability of flow variations, year on year, has been hypothesized as a major determinant of the importance of abiotic versus biotic factors in regulating population and community processes and patterns (see Poff & Ward, 1989). An evolutionary perspective suggests that a predictable disturbance regime will offer more opportunity for a functional assemblage to develop than an unpredictable one. Ward et al. (2001) showed that it is the natural disturbance regime that drives habitat turnover, the expansion and contraction of resource gradients and the potential for competitive exclusion among species. Clearly, understanding the complex effects of abiotic variations upon habitat structure and the dynamic interrelated responses of species and biotic communities over ecosystem-relevant timescales is a major challenge for the future.

Despite considerable progress over the past two decades (see Special Issue of *River Research and Applications* 19, 5–6, 2003) the management of flows remains embedded in expert judgement. This is supported by tools such as the Instream Flow Incremental Methodology (IFIM), implemented within river basins using local applications of 'Hands-off flows' (HOF) applied to abstractions and constrained by the legacy of historic water allocations (e.g. Petts, 1996; Petts et al., 1999). HOF is a term used by water resource managers in the UK to define the flow (or river level) threshold at which abstractions are switched off or cut back; equivalent to 'flow reservations' in the US. New science is required to better understand the behaviour of ecosystems, not individual abiotic or biotic components. Unfortunately, there is a lack of long-term data at the appropriate scale on most key components to understand system dynamics. Increasingly, scientists have relied on the assumption that return to a more natural flow regime would help to restore natural river functions. However, without direct incorporation or analysis of process-based information such flow-based approaches must typically be done in an adaptive management framework (Walters, 1986) to compensate for inadequate process-based knowledge. The whole flow regime perspective of Poff et al. (1997) and Richter et al. (1997) has been adopted in many management situations, as discussed in Poff et al. (2003), often as collaborative efforts with the US Corps of Engineers.

One important question is how to translate carrying capacity into population status. This requires an understanding of the linkages between habitat and key biological processes, such as reproduction, mortality and energetics. The challenge is to determine how habitat variability affects population dynamics not only for aquatic organisms (e.g. Whittaker & Shelby, 2000; Cattaneo et al., 2002; Railsback & Harvey, 2003) but also for terrestrial taxa that utilize seasonally exposed habitats (Loreau et al., 2001; Sadler et al., 2004). The impact of extreme high- and low-flow events is likely to relate to the timing of the event in relation to species traits and life cycles. In this context, changes in temperature become important in determining the timing of key biological 'windows'. The biota of river corridors have a wide range of generation times, from one or more generations per year for many aquatic insects to one generation in 3–8 years for salmon, and many are particularly vulnerable at key times in their life cycle. For example, terrestrial invertebrates on exposed-riverine-sediments exhibit classic ecotonal characteristics: high productivity and high species richness coupled with high levels of species fidelity and rarity (Sadler et al., 2004) but many are most vulnerable in mid-late summer when larvae are active on the sediments. Similarly, adult trout have been shown to suffer habitat limitation in summer low-flow periods (Gourand et al., 2001) and their young can suffer severe losses if high flows occur between emergence and their first summer (Latterell et al., 1998; Cattaneo et al., 2002; Capra et al., 2003).

It is important to understand the effects of both disturbance events and subsequent recovery processes, especially ways in which landforms and communities age and change in their sensitivity to disturbance. Major floods can reset entire systems; moderate floods disturb and rejuvenate systems and frequent, low-magnitude floods can enhance recovery/succession. However, within each river reach, the frequencies associated with these three types of event are likely to vary with time since the last major flood and with the level of human impact on the river corridor (e.g. Gurnell & Petts, 2002).

In the past scientists and engineers have attempted to model rivers as steady-state systems. Although rates of colonization by plants and animals may be rapid, key geomorphological changes can require decades and in some cases centuries. Channel form and the complex habitat-mosaic of river corridors respond to the flow regime but in many cases such as along regulated rivers below dams, channel change and the restructuring of channel-floodplain architecture can require more than a century to establish a new quasi-equilibrium condition (Petts & Gurnell, 2005). Petts (1987) suggested that research and management should be directed to modelling the sequences of transient ecosystem states that evolve in response to the actual series of flows and sediment loads in each reach over management timescales, rather than focusing, as in the past, on predicting 'regime' changes. Petts & Gurnell (2005) further suggest that the modelling of channel – and physical habitat – changes, and the resultant ecological changes, should focus on the spatial scale of the reach over time periods of 10–30 years. It is important to understand, if not isolate, natural variations over this timescale and describe underlying trends in response to changes in the driving hydrological controls. It is clear that better data are needed if scientists are to develop conceptual models and then the modelling tools and functional indicators for achieving integrated management for human and environmental water needs.

Developing a common science framework

The development of a common science-based framework for managing river systems requires new science/knowledge to address the emerging issue of variability within fluvial ecosystems and the integration of physical and biological processes from first principles. First principles means that a problem is decomposed into component parts until it is irreducible. In a modern context, first principles can be considered to mean the description of process and function in a way that avoids unfounded over-simplification or approximation. This will require new experimental and empirical studies to provide new knowledge of mechanisms and new data to develop new models.

The data dilemma

The plea by scientists for more data conflicts with the claim of many managers that they have data overload! From a practical perspective, too much data hinders wise management decisions when managers really require consistent, simple, and defensible tools and approaches. The dilemma is that scientists need detailed process data across a range of temporal and spatial scales to develop conceptual models, but managers typically only need trends in time and space of transformed data to make informed decisions. Scientists trained to decompose processes are perceived as being unwilling or unable to 'coarsen' their view of the world to predict the qualitative trends that managers claim they need. The data dilemma can be solved by considering the two groups to be at ends of a spectrum and that the gap between them can be spanned through a systematic, incremental process of simplification using the concept of similitude analysis described earlier.

Undoubtedly, a large amount of data exists in many countries (see Boon et al., 2000) but can it be mined to close the gap between scientists and managers? Many rivers are characterized by datasets collected as part of long-term monitoring programmes comprising regular or periodic surveys and environmental monitoring to meet regulatory requirements and focused on describing trends in time or space. Long-term monitoring data are abundant but mostly for variables that are easy to measure. However, often there is a disconnection between the scale at which scientists need to work and the scale of the information available; there is little synoptic data for inferring component interactions, functions and processes, and thereby facilitating the development of surrogate variables. All assessments of river system integrity must be considered in the context of normal, natural variability. There is limited data on the scale of natural variations of habitats, populations or communities; there is little coupled data to advance improved models of abiotic–biotic interactions; and even less data to define threshold conditions for biota. It may be possible to extrapolate knowledge about ecological response data from one system to another by using habitat template studies as done by Poff & Allan (1995) for fish, showing that the 'functional' groups of fish species across sites were sensitive to differences in natural flow variability using limited biological data.

The collection of data over regular time and space intervals as part of river management frameworks, while valuable, is often of limited utility for describing environmental processes at the resolution required to develop forecasting tools and predictive models capable of assessing management scenarios. Event-associated processes often elude the sampling designs of regular monitoring programs so that synoptic, flow-associated sampling necessary to describe river processes is generally not available, particularly for extreme high flows or droughts and especially for large rivers. Therefore, some findings about rivers are "black box" in the sense that ecological processes cannot be monitored in "real time" and must instead be inferred from "post-event" forensic evaluations of a river. These attributes make the response of rivers to alternative management actions, particularly at flow extremes, difficult to forecast. The information needed to predict the results of a particular management scenario on the integrity of the basin often requires experimental 'cause and effect' sampling. Monitoring is at the heart of the EU Water Framework Directive and there is an opportunity for advancing data monitoring designs that could interface with scientific research objectives as well as would meet management-information needs.

Problems of large rivers

Particular problems arise in advancing knowledge of large, natural rivers. Large rivers are complex systems that reflect many different hierarchically organized processes occurring over a diverse range of temporal and spatial scales. It is inherently difficult to accurately associate changes in the characteristics of a large river with the blend of variables responsible for a particular change. In addition, deployment of instruments and development of comprehensive monitoring programs for large rivers can be challenging because large spatial scales discourage high-resolution sampling. Furthermore, particularly in natural rivers, high energy environments and mobile channels defy many sampling technologies, unstable flow and bed conditions make maintaining regular sampling stations difficult, and in long deep, slow-flowing lowland rivers, high turbidity limits direct observation and the application of some optical techniques.

A large river is hierarchically organized so that tools to forecast its response must integrate across the scales used by the different disciplines that build tools to manage rivers. If these tools can be successfully developed then the disciplines can be broadly separated into three tiers: (i) basic scientists who work at the level of "first principles" within their discipline, (ii) applied scientists who attempt to build management tools using good science, and (iii) resource managers who attempt to use the tools to manage or restore large rivers. Hierarchically structured monitoring programs could be designed to collect information for basic scientists to improve their understanding of large rivers, to allow applied scientists to build better management tools, and, in turn, resource managers to better manage large rivers. These tiers can be considered in the context of the similitude analysis described earlier: process-oriented basic scientists develop tools that are scientifically valuable, but of little or no use to managers. Applied scientists take the tools developed by basic scientists and, using the principle of similitude analysis, build structural- or surrogate-based tools that output results of a scale appropriate to decision-making. Managers have the responsibility of learning the tools and the assumptions and simplifications made in their development and applying them appropriately. Most importantly, managers should have the responsibility of providing feedback to the scientists to make sure that applied science can support wise and equitable water resources decision-making.

Issues in 'modern' landscapes

In 'modern' landscapes, the regulated nature of flows within 'trained' river channels may make monitoring more practicable. However, an additional constraint to the progress of knowledge development is the artificial nature of catchments with complex and continually changing land uses, and the artificially influenced water quality that is typically more turbid and nutrient-rich than natural rivers. In consequence, much of our understanding of the dynamics of large rivers is qualitative, or at best, semi-quantitative, and often based upon partially regulated systems. Greater research effort is needed on 'reference rivers, reaches and sites' to better understand the links between process, function, and structure (Ward et al., 2001) and the nature of feedback processes that sustain the important goods and services of natural riverine ecosystems (Postel & Carpenter, 1997).

Advances through new tools for monitoring

New concepts can advance understanding, scientifically informed monitoring can provide new insights to system behaviour, and detailed empirical field research and increasingly sophisticated experiments are necessary to test hypotheses about system interactions and to reveal fundamental mechanisms. However, the development of new tools can have the most dramatic impact in advancing knowledge of ecosystem dynamics. This can be illustrated by recent advances in understanding dissolved organic matter (DOM) quality. DOM is a key element of the detritus-bacteria-microinvertebrate loop that may be responsible for processing the majority of allochthonous carbon (Sinsaburgh & Findley, 2003) within fluvial system food-webs, although many elements of these remain to be resolved. New advances in technology for measuring DOM fluorescence (Baker, 2001; 2002) have shown that DOM, once seen as an inert residue of biological activity, can be labile and more variable. DOM fluorescence is fast, requires only small samples, provides detection limits in the ppb to ppm range, depending on DOM type, and is non-invasive and potentially automatable (Amon et al., 2003). The technique has advanced knowledge on DOM quality, sources and bioavailability and led to important new questions being raised, including: does DOM quality determine the bacterial community or vice versa? What is more important to ecosystems: small amounts of labile DOM or large quantities of relatively inert DOM?

Conflicting traditions

The varying traditions and conventions used by the different disciplines that study rivers, as suggested above, exacerbate the difficulties of developing a common science framework to forecast the response of rivers to management actions. The contrast in approach is reflected by advances in aquatic ecology during the last three decades that can be broadly separated into two approaches, synthetic versus engineering, each with its own distinct origin and history of development.

In the synthetic approach, scientists have attempted to understand the ecology of rivers at a holistic level and to describe how important riverine processes vary over time and space. Noteworthy conceptual advances include the River Continuum Concept (Vannote et al., 1980), the Flood Pulse Concept (Junk et al., 1989), the Patch Dynamics Concept (Townsend, 1989), the Natural Disturbance concept (Resh et al., 1988), and the Natural Flow Paradigm (Poff et al., 1997).

In the engineering approach, researchers have tried to develop suites of tools that could be used to predict river stage, velocity fields, and bedform as a function of discharge, and to predict velocity and shear stress at multiple points within the river channel. Using well-accepted governing equations, engineers have been able to develop tools to simulate the bulk flow of water at a resolution sufficient to route stage and various conservative and non-conservative constituents. Further refinements allowed engineers to develop tools that could predict the detailed behaviour of flow fields and to approximate dynamic fluvial processes such as channel evolution in response to detailed flow fields. There have been many recent examples of developments in using 1D and 2D hydraulic models for ecological applications, for example, to develop a hydro-biogeochemical model of a large river (Sauvage et al., 2003), to assess the effects of structures on flooding through the riparian zone (Cioffi & Gallerano, 2003), to assess bed-form scenarios for restoring salmon spawning habitat (Pasternack et al., 2004), and to link sand wave motion to bacterial activity (Fischer et al., 2003).

Management needs

Neither the synthetic nor the engineering approaches, by themselves, are adequate to develop science-based management strategies for rivers. At one end of the spectrum of research, advances made by synthesists have been largely conceptual in nature. Their findings are usually presented in a relative sense because the primary goal was to guide the development of research or to develop broad conservation strategies for systems that could not be convincingly reduced to mechanistic models. The synthetic approach is very useful from a heuristic or theoretical standpoint, but cannot be used *a priori* to address many river management issues because it is insufficiently quantitative at the scale at which management decisions must be made in a particular legal or regulatory framework. Relative descriptions of important processes typical of the synthesists cannot be used to support equitable water resources developments when decision-makers expect analysis results as typically provided by incremental, engineering tools. However, the synthetic approach can be used *a posteriori* to guide large-scale restoration schemes (e.g. Poff et al., 1997) or to set general guidelines for conservation action on individual rivers (e.g. Richter et al., 1996, 1997). Implementation of the synthesist approach in the context of site-specific river management is typically viewed as "experimental" (e.g. Stanford et al., 1996; Poff et al., 2003).

The engineering approaches suffer the inverse problem identified for the synthetic approaches. That is, the engineering approaches are generally restricted to simulating physical or chemical processes that can be approximated by the Eulerian-based approach of discretizing complex geometries with a grid (or mesh) and then applying sets of governing equations to each node. The engineering approach works well for simulating certain processes that are easy to aggregate into control volumes, e.g. water flow or water quality, but appears to work less well to describe the response to processes that are not easy to aggregate, e.g. habitat selection by fish and macroinvertebrates.

As long as the techniques and approaches used by engineers and synthesists remain separate, it is unlikely that river managers will have the tools or approaches necessary to manage for sustainability or to develop restoration plans for impacted rivers. Therefore, the logical next step is to promote the development of approaches that will allow the tools of synthesists and engineers to be coupled together.

A way ahead

Three major advances are required to integrate human and environmental water needs in river management and, therefore, to pave the way for implementing ecological model similitude analysis.

Drive to understand variability

There is considerable evidence (Naiman et al., 2002) to support the premise that healthy river ecosystems depend on maintaining the flow variability that characterizes streams in each particular ecoregion. But a fundamental understanding of the ways in which physical and biological processes interact to sustain the ecological integrity of rivers and streams remains to be elucidated. There is an urgent need to determine the variability of key abiotic parameters over a range of spatial scales, to measure and model the effects of these variations upon biota, habitats and ecosystems, to understand the timescales and mechanisms of ecosystem response to hydrological change, and to educate the public about the importance of variability in sustaining riverine ecosystems. There must also be a move to advance models for rivers in 'developed' catchment contexts. This is especially important in regions such as western Europe where most catchments have a long history or urban growth and industrialization. The new models are required to create river reaches with the morphological and ecological configurations compatible with the magnitudes and rates of fluvial processes within the highly developed catchments. They would incorporate the range of special habitats that are typical of natural rivers but lost to most regulated systems, such as: springs, ponds, exposed sediment bars, woody-debris accumulations, and wooded islands and riparian zones. The ultimate approach for improved management of large rivers is commitment to the long-term research designed to better describe processes that characterize large rivers and to convert this knowledge into management tools using the concept of ecological model similitude analysis. Coupled data designed to address abiotic–biotic responses over (at least) decadal timescales are required to underpin the progressive advancement of key models. International collaboration would be beneficial to encompass a range of river and socio-economic settings.

Commitment to advancing knowledge-based approaches to management

This requires more and different data and integrative frameworks that embrace the traditions of both synthesists and holists. Institutional mechanisms (e.g. adaptive management) must be introduced to ensure that knowledge gained through science and project monitoring can be incorporated into management action. Scientists and managers working on large rivers must agree on a common vision that will guide research, tool development, and management concepts. There would be considerable benefits if such a long-term vision could incorporate international collaboration. One element of this long-term vision must be the development of models to evaluate the complex effects of abiotic variations upon habitats, species and communities, and their interactions. The basic science of both engineers and synthesists must be coupled together at a "first principles" level in an hypothesis-testing setting to develop defensible applied scientific tools. Such tools could be used by resource managers to develop management plans for sustainable development of large rivers. Adaptive management would allow a constant feedback between the three integrating levels, particularly between the tool builders and end-users. Clearly, we also need indicators of management performance to advance adaptive management.

Improved communication

Presently, professional specialization limits adequate communications so that basic researchers are partially isolated from applied scientists, but the move toward restoration ecology is bridging the gap. Although the linkage between applied scientists and resource managers is more robust, communication of science to the public, including the evidence-based justification of management decisions, remains a major issue. Consider the problem of conveying to the general public the advantages of natural variability. Ecologically, habitat turnover is important for sustaining a healthy river and we know that even extreme events can be important in resetting communities and rejuvenating the corridor (see Vugteveen et al., 2006). However, the average layperson often perceives a "healthy system" as one that remains constant. Deflections from the status quo are often alarming to the public and result in an outcry for intervention. A major challenge is to educate society of the fact that natural variability is a prerequisite for sustainability and protection of biodiversity.

The EU Water Framework Directive requires Member States to provide information to, and consult with, stakeholders; Member States are also encouraged to actively engage and involve stakeholders in collaborative activities. Across Europe there is more than two decades of effort, with variable success, requiring scientists to move away from their laboratory bench to engage in knowledge and technology transfer with the range of stakeholders and the public. Today, there is recognition that internationally competitive science requires specialist expertise, as does the process of making science accessible. It is one thing to encourage a culture within science that acknowledges and even promotes outreach, it is another to expect all experts to be able to participate in outreach effectively. The roles of skilled scientific translators, facilitators and mediators are now highly regarded. Petts et al. (2005) suggest that although 'hero researchers' able to engage with the policy community can have a dramatic impact, increasingly third party intermediaries who understand but do not do the science are being employed as highly effective translators of science into policy. Experts in such positions serve to help digest and apply academic research to management issues and to promote knowledge transfer to the full spectrum of stakeholders and the public. They also provide research scientists with feedback to identify and prioritize new research challenges and to advance a clearer perspective of critical problems in river basin management.

This paper provides a perspective that spans the 'holistic' and 'incrementalist' schools to establish a framework for advancing the scientific knowledge necessary to develop appropriate tools for water resources management. It highlights themes of multi-scaled variability in lotic systems and the need for better translation of basic science into a management context. Adaptive management involves recursive cycling through 'assessment-planning-action-monitoring-appraisal-new action' and the engagement of stakeholders and the public, ideally, throughout this process. However, an emphasis on participant satisfaction with the *process*, as a criterion of success, must not detract from the outcome or from the need to advance knowledge and improve tools, through basic science. This requires an effective loop between basic scientists, applied scientists and the resource managers who provide the primary interface with stakeholders and the public. The paper raises a number of questions including: Most large rivers have been changed by human actions but are these changes inevitable? Can the key ecological functions be restored in a sustainable way? Can we overcome the impediments to our understanding? We argue for long-term, coupled (biotic–abiotic) data to advance integrated models based upon fundamental principles. But is this just a dream?

Acknowledgements

This paper was inspired by discussions held at a Workshop on Emerging Concepts for Integrating Human and Environmental Water Needs in River Basin Management. This Workshop was held at The University of Birmingham, Edgbaston, UK, from 16th–18th September 2003, and was sponsored by the U.S. Army Corps of Engineers as part of the Water Operations and Technical Support Program (WOTS). Permission to publish this information was granted by the Office of the Chief, U.S. Army Corps of Engineers. We thank all delegates for their contributions: J. Aldrick, A. Baker, M. Baptist, P. Breil, J. Brittain, H. Capra, J. Hall, M.N. Hedger, B. Kleiss, M. Ledger, J. Martin, P.H. Nienhuis, J. O'Neil, J. Sadler, C. Soulsby, E. Tabacchi, K. Tockner and P. Wood. The full Report is available on: http://el.erdc.usace.army.mil/elpubs/pdf/trel05–13.pdf

References

Amon, R. M. W., G. Budeus & B. Meon, 2003. Dissolved organic carbon distribution and origin in the Nordic Seas: exchanges with the Arctic Ocean and the North Atlantic. Journal of Geophysical Research 108/27: 1–17.

Andersson, E., C. Nilsson & M.E. Johansson, 2000. Effects of river fragmentation on plant dispersal and riparian flora. Regulated Rivers: Research and Management 16: 83–89.

Baker, A., 2001. Fluorescence excitation-emission matrix characterisation of some sewage impacted rivers. Environmental Science and Technology 35: 948–953.

Baker, A., 2002. Fluorescence properties of some farm wastes: implications for water quality monitoring. Water Research 36: 189–194.

Boon, P. J., B. R. Davies & G. E. Petts, (eds), 2000. Global Perspectives on River Conservation. Wiley, Chichester.

Cattanéo, F., N. Lamouroux, P. Breil & H. Capra, 2002. The influence of hydrological and biotic processes on brown trout (*Salmo trutta*) population dynamics. Canadian Journal of Fisheries and Aquatic Sciences 59: 12–22.

Capra, H., C. Sabaton, V. Gournaud, Y. Souchon & P. Lim, 2003. A population dynamics model and habitat simulation as a tool to predict brown trout demography in natural and bypassed stream reaches. River Research and Applications 19: 551–568.

Cioffi, F. & F. Gallerano, 2003. A 2D self-adaptive hydrodynamic scheme for the assessment of the effects of structures on flooding phenomena in river basins. River Research and Applications 19: 1–26.

Church, M., 2002. Geomorphic thresholds in riverine landscapes. Freshwater Biology 47: 541–558.

Clark, M. J., 2002. Dealing with uncertainty: adaptive approaches to sustainable river management. Aquatic Conservation 12: 347–363.

Cosgrove, D. & G. E. Petts (eds), 1990. Water, Engineering and Landscape. Belhaven Press, London.

Environment Agency, 2001. Our Vision: Making It Happen. Environment Agency, Bristol.

Fischer, H., A. Sukhodolov, S. Wilczek & C. Engelhardt, 2003. Effects of flow dynamics and sediment movement on microbial activity in a lowland river. River Research and Applications 19: 473–482.

Geerling, G. W., A. M. J. Ragas, R. S. E. W. Leuven, J. H. van den Berg, M. Breedveld, D. Liefhebber & A. J. M. Smits, 2006. Succession and rejuvenation in floodplains along the river Allier (France). Hydrobiologia 565: 71–86.

Goodson, J. M., A. M. Gurnell, P. G. Angold & I. P. Morrissey, 2002. Riparian seed banks along the lower River Dove, UK: their structure and ecological implications. Geomorphology 47: 45–60.

Gouraud, V., J. L. Baglinière, P. Baran, C. Sabaton, P. Lim & D. Ombredane, 2001. Factors regulating brown trout populations in two French rivers: application of a dynamic population model. Regulated Rivers: Research and Management 17: 557–569.

Gurnell, A. M. & G. E. Petts, 2002. Island-dominated landscapes of large floodplain rivers, a European perspective. Freshwater Biology 47: 581–600.

Irvine, K., 2004. Classifying ecological status under the European Water Framework Directive: the need for monitoring to account for natural variability. Aquatic Conservation 14: 107–112.

Junk, W. J., P. B. Bayley & R. E. Sparks, 1989. The flood pulse concept in river-floodplain systems. In Dodge D. P. (ed.), Proceedings International Large River Symposium. Canadian Special Publication Fisheries Aquatic Sciences 106: 110–127.

Latterell, J. J., K. D. Fausch, C. Gowan & S. C. Riley, 1998. Relationship of trout recruitment to snowmelt runoff flows and adult trout abundance in six Colorado Mountain streams. Rivers 6: 240–250.

Leuven, R. S. E. W., A. J. M. Smits & P. H. Nienhuis, 2000. From integrated approaches to sustainable river basin management. In Smits, A. J. M., P. H. Nienhuis & R. S. E. W. Leuven (eds), New Approaches to River Management. Backhuys Publishers, Leiden: 329–347.

Logan, P. & M. Furse, 2002. Preparing for the European Water Framework Directive – making the links between habitat and aquatic biota. Aquatic Conservation 12: 425–437.

Loreau, M., S. Naeem, P. Inchausti, J. Bengtsson, J.P. Grime, A. Hector, D. U. Hooper, M. A. Huston, D. Raffaelli, B. Schmid, D. Tilman & D. A. Wardle, 2001. Ecology – biodiversity and ecosystem functioning: current knowledge and future challenges. Science 294: 804–808.

Milner, A. M., J. E. Brittain & G. E. Petts, 2001. Trends of macroinvertebrate community structure in glacier-fed rivers in relation to environmental conditions: a synthesis. Freshwater Biology 46: 1833–1848.

Naiman, R. J., S. E. Bunn, C. Nilsson, G. E. Petts, G. Pinay & L. Thompson, 2002. Legitimizing fluvial ecosystems as users of water: an overview. Environmental Management 30: 455–467.

Nienhuis, P. H. & R. S. E. W. Leuven, 2001. River restoration and flood protection: controversy or synergism. Hydrobiologia 444: 85–99.

Pasternack, G. B., C. L. Wang & J. E. Merz, 2004. Application of a 2D hydrodynamic model to design of reseach-scale spawning gravel replenishment on the Mokelumne River, California. River Research and Applications 20: 205–225.

Petts, G. E., 1987. Timescales for ecological change in regulated rivers. In Craig, J. & J. B. Kemper (eds), Regulated Streams: Advances in Ecology. Plenum, New York, 257–266.

Petts, G. E., 1996. Water allocation to protect river ecosystems. Regulated Rivers: Research and Management 12: 353–367.

Petts, G. E., 2000. A perspective on the abiotic processes sustaining the ecological integrity of running waters. Hydrobiologia 422/423: 15–27.

Petts, G. E., M. A. Bickerton, C. Crawford, D. N. Lerner & D. Evans, 1999. Flow management to sustain groundwater-dominated stream ecosystems. Hydrological Processes 13: 497–513.

Petts, G. E. & A. M. Gurnell, 2005. Dams and geomorphology: research progress and future directions. Geomorphology 71: 27–47.

Petts, J. I., S. Owens & H. Bulkeley, 2005. Summary of discussions and recommendations. ESRC Transdisciplinary Seminar Series: knowledge and power, exploring the science/society interface in the urban environment context. www.cert.bham.ac.uk/urbanenvironments/index.htm.

Poff, N. L. & J. V. Ward, 1989. Implication of streamflow variability and predictability for lotic community structure: a regional analysis of streamflow patterns. Canadian Journal of Fisheries and Aquatic Sciences 46: 1805–1818.

Poff, N. L., J. D. Allan, M. A. Palmer, D. D. Hart, B. D. Richter, A. H. Arthington, K. H. Rogers, J. L. Meyer & J. A. Stanford, 2003. River flows and water wars? Emerging science for environmental decision-making. Frontiers in Ecology and the Environment 1: 298–306.

Poff, N. L., J. D. Allan, M. B. Bain, J. R. Karr, K. L. Prestegaard, B. Richter, R. Sparks & J. Stromberg, 1997. The natural flow regime: a new paradigm for riverine conservation and restoration. BioScience 47: 769–784.

Poff, N. L. & J. V. Ward, 1990. The physical habitat template of lotic systems: recovery in the context of historical pattern of spatio-temporal heterogeneity. Environmental Management 14: 629–646.

Poff, N. L. & J. D. Allan, 1995. Functional organization of stream fish assemblages in relation to hydrologic variability. Ecology 76: 606–627.

Postel, S. & S. Carpenter, 1997. Freshwater ecosystem services. In Daily, G. C. (ed.), Nature's Services. Island Press. Washington DC: 195–214.

Railsback, S. F. & B. C. Harvey, 2003. An Individual-based Trout Model for Instream Flow Assessment: Validity, Advantages, and Disadvantages International IFIM Users' Workshop. Fort Collins: Colorado USA.

Resh, V. H., A. V. Brown, A. P. Covich, M. E. Gurtz, H. W. Li, G. W. Minshall, S. R. Reice, A. L. Sheldon, J. B. Wallace & R. C. Wissmar, 1988. The role of disturbance in stream ecology. Journal of North American Benthological Society 7: 433–455.

Richards, K., J. Brasington & F. Hughes, 2002. Geomorphic dynamics of floodplains: ecological implications and a potential modelling strategy. Freshwater Biology 47: 559–580.

Richter, B. D., J. V. Baumgartner, J. Powell & P. D. Braun, 1996. A method for assessing hydrologic alteration within ecosystems. Conservation Biology 10: 1163–1174.

Richter, B. D., J. V. Baumgartner, R. Wingington & P. D. Braun, 1997. How much water does a river need? Freshwater Biology 37: 231–249.

Sadler, J. P., D. Bell & A. P. Fowles, 2004. The hydroecological controls and conservation value of beetles on exposed riverine sediments in England and Wales. Biological Conservation 118: 41–56.

Sauvage, S., S. Teissier, P. Vervier, T. Ameziane, F. Garabetian, F. Delmas & B. Caussade, 2003. A numerical tool to integrate biophysical diversity of a large regulated river: hydrobiogeochemical bases. The case of the Garonne River (France). River Research and Applications 19: 181–198.

Sinsaburgh, R. L. & S. E. G. Findley, 2003. Dissolved organic matter: out of the black box and into the mainstream. In: Findlay, S. E. G. & R. L. Sinsabaugh (eds), Aquatic Ecosystems: Interactivity of DOM, Academic Press: 479–488.

Smits, A. J. M., P. H. Nienhuis & H. L. F. Saeijs, 2006. Changing estuaries, changing views. Hydrobiologia 565: 339–355.

Stanford, J. A., J. V. Ward, W. J. Liss, C. A. Frissell, R. N. Williams, J. A. Lichatowich & C. C. Coutant, 1996. A general protocol for restoration of regulated rivers. Regulated Rivers: Research and Management 12: 391–413.

Tabacchi, E., D. L. Correll, R. Hauer, G. Pinay, A. -M. Planty-Tabacchi & R. C. Wissmar, 1998. Development, maintenance and role of riparian vegetation in the river landscape. Freshwater Biology 40: 497–516.

Tockner, K., F. Malard & J. V. Ward, 2000. An extension of the flood pulse concept. Hydrological Processes 14: 2861–2883.

Townsend, C. R., 1989. The patch dynamic concept of stream community ecology. Journal North American Benthological Society 8: 36–50.

Van Stokkom, H. T. C., A. J. M. Smits & R. S. E. W. Leuven, 2005. Flood defense in the Netherlands: a new era, a new approach. Water International 30/1: 76–87.

Vannotte, R. L., G. W. Minshall, K. W. Cummins, J. R. Sedell & C. E. Cushing, 1980. The river continuum concept. Canadian Journal of Fisheries and Aquatic Sciences 37: 130–137.

Vugteveen, P., R. S. E. W. Leuven, M. A. J. Huijbregts & H. J. R. Lenders, 2006. Redefinition and elaboration of river ecosystem health: perspective for river management. Hydrobiologia 565: 289–308.

Walters, C., 1986. Adaptive Management of Renewable Resources. Blackburn Press, New Jersey.

Ward, J. V., K. Tockner, U. Uehlinger & F. Malard, 2001. Understanding natural patterns and processes in river corridors as the basis for effective river restoration. Regulated Rivers: Research and Management 17: 311 323.

Whittaker, D. & B. Shelby, 2000. Managed flow regimes and resources values: traditional versus alternative strategies. Rivers 7: 233–244.

Hydrobiologia (2006) 565:289–308

R.S.E.W. Leuven, A.M.J. Ragas, A.J.M. Smits & G. van der Velde (eds), Living Rivers: Trends and Challenges in Science and Management
DOI 10.1007/s10750-005-1920-8

Redefinition and elaboration of river ecosystem health: perspective for river management

P. Vugteveen*, R.S.E.W. Leuven, M.A.J. Huijbregts & H.J.R. Lenders
Department of Environmental Studies, Institute for Wetland and Water Research, Faculty of Science, Radboud University Nijmegen, P.O. Box 9010, 6500 GL Nijmegen, The Netherlands
*(*Author for correspondence: E-mail: p.vugteveen@science.ru.nl)*

Key words: condition indicators, ecological integrity, ecosystem functioning, ecosystem organization, stressor indicators, sustainability

Abstract

This paper critically reviews developments in the conceptualization and elaboration of the River Ecosystem Health (REH) concept. Analysis of literature shows there is still no consistent meaning of the central concept Ecosystem Health, resulting in models (i.e. elaborations) that have unclear and insufficient conceptual grounds. Furthermore, a diverse terminology is associated with describing REH, resulting in confusion with other concepts. However, if the concept is to have merit and longevity in the field of river research and management, unambiguous definition of the conceptual meaning and operational domain are required. Therefore a redefinition is proposed, based on identified characteristics of health and derived from considering semantic and conceptual definitions. Based on this definition, REH has merit in a broader context of river system health that considers societal functioning next to ecological functioning. Assessment of health needs integration of measures of multiple, complementary attributes and analysis in a synthesized way. An assessment framework is proposed that assesses REH top-down as well as bottom up by combining indicators of system stress responses (i.e. condition) with indicators identifying the causative stress (i.e. stressor). The scope of REH is covered by using indicators of system activity, metabolism (vigour), resilience, structure and interactions between system components (organization). The variety of stress effects that the system may endure are covered by using biotic, chemical as well as physical stressors. Besides having a unique meaning, the REH metaphor has added value to river management by being able to mobilize scientists, practitioners and publics and seeing relationships at the level of values. It places humans at the centre of the river ecosystem, while seeking to ensure the durability of the ecosystem of which they are an integral part. Optimization of the indicator set, development of aggregation and classification methodologies, and implementation of the concept within differing international frames are considered main aims for future research.

Introduction

Rivers serve many societal functions and belong to the most intensively human influenced ecosystems worldwide. Especially the last decades, socio-economic developments have led to their degradation and pollution. Functions of rivers, particularly those that are vital to sustaining the human community have become impaired (Nienhuis & Leuven, 1998). In response, environmental sciences have focused on river condition assessment, system management and rehabilitation measures. Over time, various systemic concepts have emerged in relation to condition assessment, most notably sustainability, ecological integrity and ecosystem health (Callicot et al., 1999).

The ecosystem health concept has emerged as 'river' ecosystem health (REH) or river health in the field of river research and management (Karr, 1999). REH recognizes that water resource problems involve biological, physical and chemical as well as social and economic issues, and is therefore considered a useful concept for directing integrated assessments of river condition (Norris & Thoms, 1999). Furthermore, 'health' is found an appealing term for politicians and water managers (Hart et al., 1999; Rogers & Biggs, 1999) as it is intuitively grasped by stakeholders (Meyer, 1997), making it easy to communicate environmental problems and management measures. As such, bringing back river systems to a 'healthy state' and maintaining this state have become important objectives in national and international water management programs (Karr, 1991; Hart et al., 1999; Rapport et al., 1999). An important legislative framework to mention in this respect is the European Water Framework Directive (European Commission, 2000) that guides developments in European water management today. This directive demands an integrative ecosystem approach, meaning that catchments need to be managed in a holistic way, reflecting the interconnection that exists between the landscape, the water and its uses. This view is also reflected in the concept of ecosystem health, which therefore has good compatibility with the objectives of the Water Framework Directive (Pollard & Huxham, 1998).

Within current elaborations of the REH-concept, three different ways of utilization can be distinguished. Each of them represents a separate dimension of the concept, i.e. meaning, model and metaphor (Pickett & Cadenasso, 2002). The 'meaning' dimension comprises the conceptual definition. The 'model' dimension embodies the specifications (such as elements under study, spatial or temporal limitations) needed to address the actual situations that the definition might apply to. Finally, the 'metaphorical' dimension constitutes the use of REH in common parlance, and in public dialogue. The three dimensions are linked, exemplified by the fact that any application of the model dimension of the REH-concept can only be developed based on a conceptual understanding, i.e. the meaning of the concept. However, use of REH has not always been clear and consistent (Norris & Thoms, 1999). Often it lacks precise definition in conceptual as well as operational elaborations. This can be partly explained by the fact that the concept is interdisciplinary and evolving, which may cause confusion in conceptualization as well as application.

The present paper critically reviews developments of REH and focuses on the 'meaning', 'model' and 'metaphorical' dimensions of the concept. By doing so, it aims to structure and advance the discussion on ecosystem health and assess the significance of the concept for river management. First, the paper proposes a redefinition of REH within a broader context of River System Health after considering existing definitions and differences with related concepts (i.e. meaning dimension). Secondly, it gives insight in the scientific elaboration and assessment framework (i.e. model dimension). Thirdly, this paper briefly addresses the added value to river management (i.e. metaphorical dimension). The paper concludes with a perspective for future research regarding REH applications in integrated assessments and management of river catchments.

Meaningful concept for river functioning

Basic components

For better understanding and insight in the meaning and contents of REH, we will first consider the meaning of its component parts; health, ecosystem and river. This eventuates technical comprehension of the 'ingredients' of the concept and facilitates discussion on the question: what defines REH?

The American Heritage Dictionary (Pickett, 2000) supplies the following definitions of health: '1. The overall condition of an organism at a given time. 2. Soundness, especially of body or mind; freedom from disease or abnormality. 3. A condition of optimal well-being.' The first entry reveals that health describes the overall state of an organism (human being, i.e. a complex system). Taking into account the third entry as well, which defines health as well-being, it appears that health expresses a wholeness perspective, whereby performance (of the organism) cannot be explained by regarding separate parts. From the second entry it can be derived that health requires normative criteria for its definition. Health refers to a state of 'normal functioning' or 'normality' for multiple parts of an organism, free from disease. The standard for being healthy is 'soundness' (i.e. sound functioning) or, based on the last entry, a generalized state of 'optimal well-being'. This shows that health is a flexible notion since what is considered normal, sound or optimal (i.e. healthy) can vary under influence of different geographical and societal constituents, implying that states of reference are required to distinguish unhealthy from healthy (Fig. 1).

The basic definition of an 'ecosystem' by Tansley (1935) encompasses a biotic community or assemblage and its associated physical environment in a specific place. This implicates that the concept of an ecosystem requires a biotic complex, an abiotic complex, interaction between them, and a physical space. This general definition covers an almost unimaginably broad array of instances, as it is neutral in scale and constraint, making it applicable to any case where organisms and physical processes interact in some spatial arena (Pickett & Cadenasso, 2002). Over time, various specifications to the basic concept of ecosystem have emerged, using different foci like energy, nutrients, organisms and the inclusion of human sciences. The first and most broadly accepted definitions of ecosystems aimed to understand what physical environmental processes control and limit the transformation of energy and materials in ecosystems. Odum (1969) focused on ecological succession, whereby an ecosystem was considered a unit in which a flow of energy leads to characteristic trophic structure and material cycles within the system. Others focused on the physical template of ecosystems, resulting in the articulation of ecosystem attributes like resilience (e.g. Holling, 1973). More recent perspectives have widened the ecosystem concept from 'natural' to 'human-inclusive', thereby acknowledging that humans may be regarded as an integral part of ecosystems. This has resulted in ecosystem models that account for economic flows of goods and services (Costanza et al., 1997) and the development of models that incorporate the full range of human institutions (Pickett et al., 1997; Naveh, 2001). Central to all uses of the ecosystem concept is the core requirement that a physical environment and organisms in a specified area are functionally linked.

River systems can be described in five dimensions (Lenders & Knippenberg, 2005). The three physical dimensions (longitudinal, transversal and vertical) are key features of river systems (Ward et al., 2002; Van der Velde et al., 2004). These three physical dimensions have been elaborated in terms of ecological concepts such as the River Continuum Concept (Vannote et al., 1980), the Serial Discontinuity Concept (Ward & Stanford, 1995), the Flood-Pulse Concept (Junk et al., 1989) and the Flow-Pulse Concept (Tockner et al., 2000). The temporal or fourth dimension (Ripl et al., 1994;

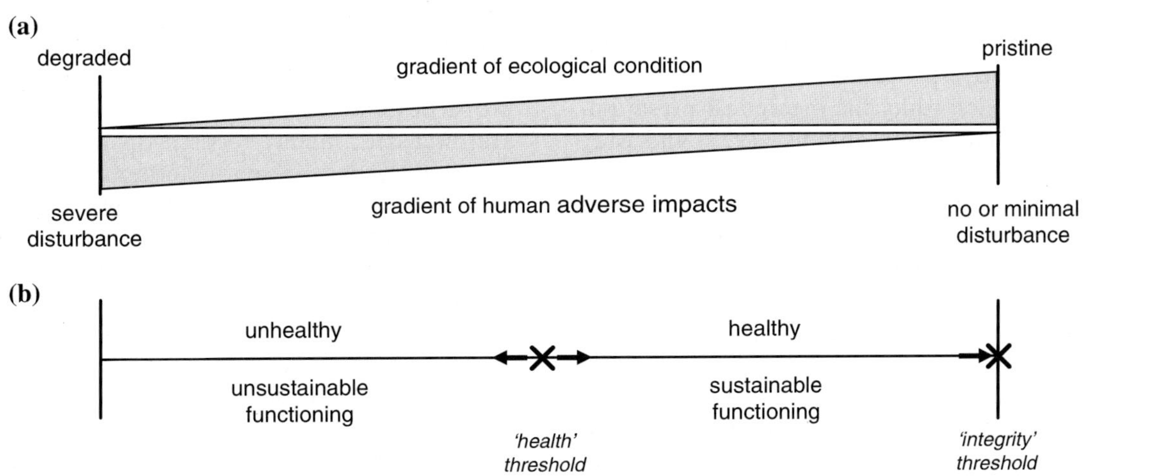

Figure 1. (a) The continuum of human impacts and river condition and (b) the normative valuation of quality in terms of ecosystem health and ecological integrity. Position of thresholds (cross-symbols) is related to valuation of sustainability. Arrows indicate that 'health' threshold is flexible, whereas 'integrity' threshold is rigid. Adapted from Karr (1999).

Boon, 1998; Poudevigne et al., 2002; Lenders & Knippenberg, 2005) represents short- and long-term changes and is usually elaborated in terms of physical river system processes, such as hydro- and morphodynamics, and accompanying phenomena such as succession and rejuvenation. Finally, the social or fifth dimension includes socio-economic activities as well as issues like cultural identity and various positions humans may hold towards nature (Lenders & Knippenberg, 2005).

Key definitions reviewed

Initially, the extension of health to describe ecosystem condition was a response to the accumulating evidence that human-dominated ecosystems became dysfunctional. The health metaphor was used based on the assertion that an ecosystem, like an organism, is built up from the behaviour of its parts (Costanza & Mageau, 1999). The first definitions of ecosystem health focused on the crucial parts of system functioning, the vital signs of a healthy system (Rapport et al., 1985), such as primary productivity and nutrient turnover. This was further elaborated by Costanza et al. (1992) who defined health in terms of activity, organization and resilience. Karr (1991) emphasized the system ability of autonomic functioning, stating that a (biological) system could be considered healthy when its inherent potential is realized, its condition is stable, its capacity for self-repair when perturbed is preserved and minimal external support for management is needed. In these definitions of ecosystem health, stability, resistance and resilience are key properties, portraying an ecosystem model according the theoretical presuppositions of Odum (1969), Holling (1973) and May (1977). This reflects a 'natural' system that is deterministic, homeostatic, and generally in equilibrium. Within the concept, health is defined as freedom from or coping with distress, i.e. in the context of maintaining essential functions. A progression from consideration of how human institutions relate to the biophysical environment ('nature') has led to developments in ecosystem models from 'human exclusive' to 'human inclusive', as articulated in the fifth dimension of river functioning (Lenders & Knippenberg, 2005). The perspective that ecosystems also provide services for humans (e.g. aesthetic pleasure, timber, water purification), has led to definitions of ecosystem health in the context of promotion of well-being and productivity (Calow, 1995), defining it in terms of capacity for achieving reasonable human goals or meeting needs.

The foregoing makes clear that there are divergent meanings given to 'ecosystem health', but the evolution in literature tends to suggest that the full scope of the concept should include ecological criteria as well as (considerations of) human values and uses derived from the system (Boulton, 1999; Fairweather, 1999; Karr, 1999; Rapport et al., 1999). The 'health' concept finds acceptance by an increasing number of researchers (Rapport et al., 1999), but over time there has been scientific debate on whether it is appropriate to use 'health' in an ecological context (Belaoussof & Kevan, 2003) and how to define and apply the concept (Lackey, 2001). Some abandon the health metaphor, arguing that health is not an observable ecological property, lacks validity at levels of organization beyond the individual and is 'value-laden' (Simberloff, 1998; Davis & Slobotkin, 2004).

Table 1 summarizes key definitions of ecosystem health, varying from generalized, systemic definitions to narrow, operational definitions. There is no universal conception of ecosystem health, but the table shows that the broad definitions of ecosystem health generally include reference to stability and sustainability. More confusion arises when health is elaborated for a specific system such as a river. Generally, explicit definition of the meaning of REH is avoided, so it is not always clear what constitutes health. Rather, properties and monitoring criteria of the concept are discussed, mainly focused on the elaboration of the concept in terms of criteria for measures (Boulton, 1999; Bunn et al., 1999; Karr, 1999; Norris & Thoms, 1999; Norris & Hawkins, 2000). Other studies use REH as an umbrella concept for explaining integrated assessments of river condition using specific indicators (Obersdorff et al., 2002) in specific components (Maddock, 1999) or compartments (Maher et al., 1999). Ecological functioning is central in most considerations of REH, but there is general consensus that economic and social functions should be included in the concept (Boulton, 1999). However, economic and social functions are often merely considered as conditional but not as integral parts of the system (see e.g. Fairweather, 1999; Moog & Chovanec, 2000). Economic factors are often stressed as important boundary conditions (e.g. in terms of goods and services to be delivered by the river; e.g. Rapport et al., 1998b), but especially social factors (e.g. sense of belonging, sense of place) are mostly neglected (Kuiper, 1998; Lenders, 2003).

Overall, inconsistency exists in defined meanings of REH, as well as in the extension of its meaning into models (i.e. elaborations). Reason for this may be a disconnect between the academics discussing the concept of ecosystem health and the aquatic scientists deploying methods in the field to assess condition (Norris & Thoms, 1999). Also, a diverse terminology has emerged around REH, due to the extensive scientific and philosophical discussion surrounding its conceptual development (Callicott et al., 1999; Society for Ecological Restoration Science & Policy Working Group (SER), 2004). Table 1 shows that terms like 'sustainable' and 'integrity' are part of the terminology to define health. However, these terms have own conceptual meanings, adding to the confusion in understanding the concept of health. Therefore, further clarification and demarcation of normative concepts related to REH (i.e. sustainability and ecological integrity) are needed in order to ultimately allow a (re)definition of the health concept for river systems.

Integrity, health and sustainability

In environmental management and politics, 'sustainability' appears to be the most comprehensive concept. Though sustainability has been represented as a scientific concept, it is in fact in its broadest sense an ethical precept, being more a concept of prediction instead of being definitional (Costanza & Patten, 1995). In accordance with the Brundtland-commission report 'Our Common Future' (World Commission on Environment and Development, 1987), this concept highlights three fundamental components to sustainable development: environmental protection, economic growth and social equity. These three components should be in balance to 'sustain' them for future generations. Applying the sustainability-concept to river systems implies that river management should set its aims to ecological as well as to economic and social functions (Leuven et al., 2000).

For the ecological subsystem, terms like ecological or biological integrity are often used as either concepts competing with ecosystem health or as synonyms for ecosystem health (Callicot et al., 1999). The common denominator of the integrity and health concepts appears to be the observation that they all bear reference to qualities, i.e. characteristics of the system. Nonetheless, the concepts are distinct in meaning (Mageau et al., 1998; Karr, 1999).

Pickett (2000) defines integrity as '1. Steadfast adherence to a strict moral or ethical code. 2. The state of being unimpaired; soundness. 3. The quality or condition of being whole or undivided; completeness'. In the entries under 2 and 3, integrity within the context of river management requires a reference. Which river condition can be considered as 'unimpaired' and which river state is 'complete'? The first entry also requires a reference but offers the opportunity to apply one's own criteria of moral or artistic (aesthetic) values to be taken into account. The entries 2 and 3 predefine these values as state of non-impairment and state of completeness, respectively. This narrows the meaning of integrity to an absolute quality: a river system is integer or it is not, depending on the answer whether or not the system is unimpaired or complete. In everyday practice the ecological or biological integrity concept also refers often to a pre-disturbance or pristine state (Karr, 1999), defined as '[..] having a species composition, diversity, and functional organization comparable to that of the natural habitat of the region' (Karr, 1991). Apart from the question how to define and to determine this pre-disturbance state, the concept of integrity seems to seek for a maximum exclusion of man and of any influence humans may have (Lenders, 2003; cf. SER, 2004). Furthermore, integrity appears to appeal above all things to the state of organization of a system, emphasizing structure and pattern as important features of the system, while processes are primarily necessary to attain and maintain these features (Callicot et al., 1999; Lenders, 2003).

The above mentioned dictionary entries and conceptual definitions illustrate that health primarily refers to functioning. The acknowledgement that health has been described in terms of

Table 1. Examples of ecosystem health definitions

Study	Ecosystem health definitions	System dimensions	Specification	Approach
Karr (1991)	Healthy when its inherent potential is realized, its condition is stable, its capacity for self-repair when perturbed is preserved and minimal external support for management is needed.	Conceptual ecosystem; Physical and temporal	System characteristics	T
Costanza et al. (1992)	An ecological system is healthy and free from "distress syndrome" if it is stable and sustainable - that is, if it is active and maintains its organization and autonomy over time and is resilient to stress.	Conceptual ecosystem; Physical and temporal	System characteristics	T
Meyer (1997)	A healthy stream is an ecosystem that is sustainable and resilient, maintaining its ecological structure and function over time while continuing to meet societal needs and expectations.	River ecosystem; Physical, temporal and social	System characteristics	T
Harvey (2001)	Healthy ecosystems are characterized by sustainable turnovers of energy, nutrients, organic matter and water, which remain stable over comparatively long periods of time.	Conceptual ecosystem; Physical and temporal	System characteristics	T
SER (2004)	Ecosystem health is the state or condition of an ecosystem in which its dynamic attributes are expressed within 'normal' ranges of activity relative to its ecological stage of development.	Conceptual ecosystem; Physical and temporal	System characteristics	T
Fairweather (1999)	*Not explicitly defined;* river health can be expressed by holistic measures.	River ecosystem; Physical, temporal and social	Indicators of multiple disciplines	T/B
Bunn et al. (1999)	*Describes properties of river health:* Low rates of GPP and R_{24}, net consummation of carbon ($P < R$).	River ecosystem; Longitudinal, lateral	Physical indicators	B

Karr (1999); Oberdorff et al. (2002)	Health is equated to integrity. Evaluation of health through indicator of biological integrity (IBI).	River ecosystem; Longitudinal, temporal	Multimetric index	B
Maddock (1999)	*Not explicitly defined;* measurements of physical habitat features account for health.	River ecosystem; Lateral	Physical measures	B
Maher et al. (1999)	A healthy sediment ecosystem has an acceptable species abundance and diver sity and functions satisfactorily.	River ecosystem; Vertical	Chemical indicators	B
Norris & Thoms (1999)	*Not explicitly defined;* primary needs for a healthy ecosystem are biotic integrity (Karr, 1991) and sustainability.	River ecosystem; Longitudinal, lateral, temporal	Physical indicators in relation to aquatic biota	B
Townsend & Riley (1999)	*Not explicitly defined;* degree of perturbations in physical space account for health.	River ecosystem; Physical and temporal	Multi-scale, multi-temporal measures	B
An et al. (2002)	Health is equated to integrity. Evaluation of biological integrity, habitat conditions and chemical parameters.	River ecosystem; Longitudinal, lateral	Multimetric indices, chemical indicators	B

System dimensions are based on Lenders & Knippenberg (2005): 'conceptual ecosystem': generalized ecosystem, not defined by any spatial scale; 'physical': three dimensions, i.e. longitudinal, lateral and vertical. Approaches can be top-down (T) and/or bottom-up (B). GPP: Gross Primary Production; P: rate of primary production; R: rate of respiration; R_{24}: total respiration over 24 h.

performance and capacity to resist and abate stress and disturbances underlies this statement. Furthermore, health refers to a desired (flexible) condition as opposed to the absolute (rigid) condition that integrity refers to. In addition, health can be regarded more of a relative system quality: there are several levels of health possible, each level being determined by different (ecological) criteria. Utilization of the health concept in river management therefore requires a pre-definition of the desired levels of performance (Costanza & Mageau, 1999; Lenders, 2003). If this desired condition is defined as a pre-disturbance state (unimpaired, complete), as is often the case in river management thinking, health and integrity become almost synonyms (Fig. 1).

When comparing ecosystem health and ecological integrity in relation to their purpose for river management, ecological integrity appears to be rather rigid as a guiding concept for management, referring to an absolute condition and offering few degrees of freedom for other functions (social and economic) within a broader coherent sustainability context. It is therefore a less obvious strategy for densely populated regions of the world where rivers, including their catchment areas and floodplains, have to fulfil a large number of societal functions. We therefore prefer a strategy that aims at ecosystem health as the central concept for sustaining the ecological domain of the river system, whereby the concept of sustainability sets the overarching goals.

Redefinition

Based on the above findings of connotation and scientific meaning, it can be concluded that REH needs to express the ability of the system to function, i.e. to perform and sustain autopoetic processes. Key properties hereby are vigour (throughput or productivity of the ecosystem) and resilience (ability to maintain structure and patterns of behaviour in the face of stress). Self-maintenance of the system depends on system processes in interaction with system structure at various spatial and temporal scales (i.e. organization). Note that health itself is not an ecological property but a societal construct, only having meaning in relation to human beings. The essence of health is an expression of wholeness, self-maintenance and other premises as explained above. However, *qualifications* of health require definition in terms of scientifically-based criteria. Flexibility in defining health status of the ecosystem allows consideration of economic and social functions in a similar fashion as expressed in the concept of sustainability that protects environmental quality within the context of social and economic prosperity. Thus, a healthy status is flexible in definition within the limits of sustainable functioning (Fig. 1) whereby societal values drive the level of ecological quality that is attainable within a river system. Capturing the above-made health propositions, REH is redefined as:

> an expression of a river's ability to sustain its ecological functioning (vigour and resilience) in accordance with its organization while allowing social and economic needs to be met by society.

From a system perspective, the definition acknowledges that besides the ecological domain, the river system also encompasses a social and economic domain, for which ecosystem health is conditional. This fits a broader conceptual context, here referred to as River System Health (RSH), which considers REH to be a component in the overall health status of the river system. As such, RSH is regarded the integration of ecosystem health and the health of the economic and social systems (Fig. 2). RSH expresses that it is not only the ecological component that makes up a sustainable system, but also that ecological qualities should be safeguarded and (re)developed in full accordance with and taking account of social and economic qualities. This means that the three health components are interdependent; the status of an individual health component is conditional for the health of the other two, besides its individual performance. As such, RSH may be considered a holistic representation of people, their activities and their impacts integrated with the ecology and resources of the river system (*sensu* 'coastal health' by Wells, 2003). Though the relation between the health components is clarified as such, elaboration of economic- and social system health is beyond the scope of this paper. Having outlined the above conceptual framework and meaning of REH, the next step is to develop a suitable 'model' that enables assessment of its

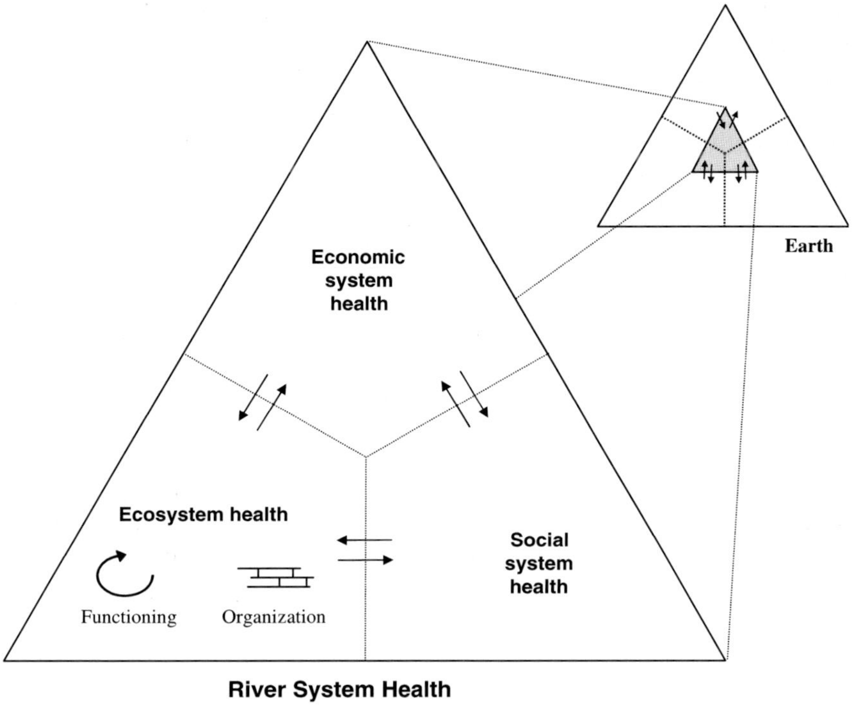

Figure 2. River System Health (RSH) is represented as the overall health status of the ecological, economic and social health components. Ecosystem health is a measure of ecological functioning within the organization of the river system. RSH itself depends on interactions between the river system and the surrounding earth.

status. Construction of such an operational framework will greatly enhance the applicability of the concept in practice.

Assessment framework

REH as an integrative, conceptual notion is not directly measurable or observable, so 'substitute' operational measures (like temperature for human health) are required to enable its assessment. In practice, REH can only be evaluated after ecological endpoints of 'good' health are identified for these measures. The assessment framework is required to measure progress towards these endpoints.

Two complementary approaches have emerged to assess ecosystem health, i.e. the top-down and bottom-up approach. The top-down approach provides a holistic basis for studying river ecosystems focusing on macro-level functional aspects without knowing all the details of the internal structure and processes, but rather knowing the primary responses in system performance under stress (Costanza et al., 1992). This approach removes the necessity of first defining all the elements and their mutual relationships before defining the whole ecosystem (Leuven & Poudevigne, 2002). Stress effects can be detected by assessing response parameters, using so-called condition indicators. However, this necessitates caution when one evaluates REH, as it is difficult to guarantee that all components of whole system performance are considered in an assessment. The bottom-up or reductionist approach emphasizes the structural aspects of natural systems and focuses on identifying ecosystem health on the basis of accumulated data on simple stressor-effect (i.e. causal) relationships. Hereby a stressor is defined as any biological, physical or chemical factor that can induce adverse effects on an ecosystem (Environmental Protection Agency, 1998). Within the context of REH, stressors are mainly understood to arise from human activities and as such pose stress on the natural system. Using the bottom-up approach the current stress status of an area (status assessment) or the progression of river stressor conditions (trend detection) can be assessed. Evaluating REH with this approach involves considerable work to provide information for each spatial and temporal scale, as well as for all the responses of the ecosystem (i.e. changes in structural and functional attributes) to the stressor or set of multiple stressors (Leuven et al., 1998).

Given the restraints of both approaches, a combination of both is suggested to address and link REH status to environmental problems within the river basin (Fig. 3), and offering river managers opportunities to counteract these problems. In practice this necessitates the application and aggregation of a suite of indicators to cover REH, representative of the functioning and organization of the system (condition indicators) as well as the constraints that act upon system functioning (stressor and effect indicators). As such, the combined approach demands various dimensions of river functioning (Lenders & Knippenberg, 2005) to be considered and multiple disciplines to be integrated in the assessment framework (Belaousssoff & Kevan, 2003).

Condition indicators

The system-level attributes vigour, resilience and organization have been traditionally proposed as top-down assessment measures of ecosystem health (Rapport et al., 1998a; Costanza & Mageau, 1999; Holling, 2001). Applied to REH, maintenance of the first two attributes (vigour and resilience) can be considered capacities of sound ecological functioning. Table 2 summarizes available condition indicators that assess system functioning and organization. The table shows that there is a range of condition indicators for ecosystems, but until now relatively few have been developed and tested to assess ecosystem health of river systems. These specific indicators will be shortly described below.

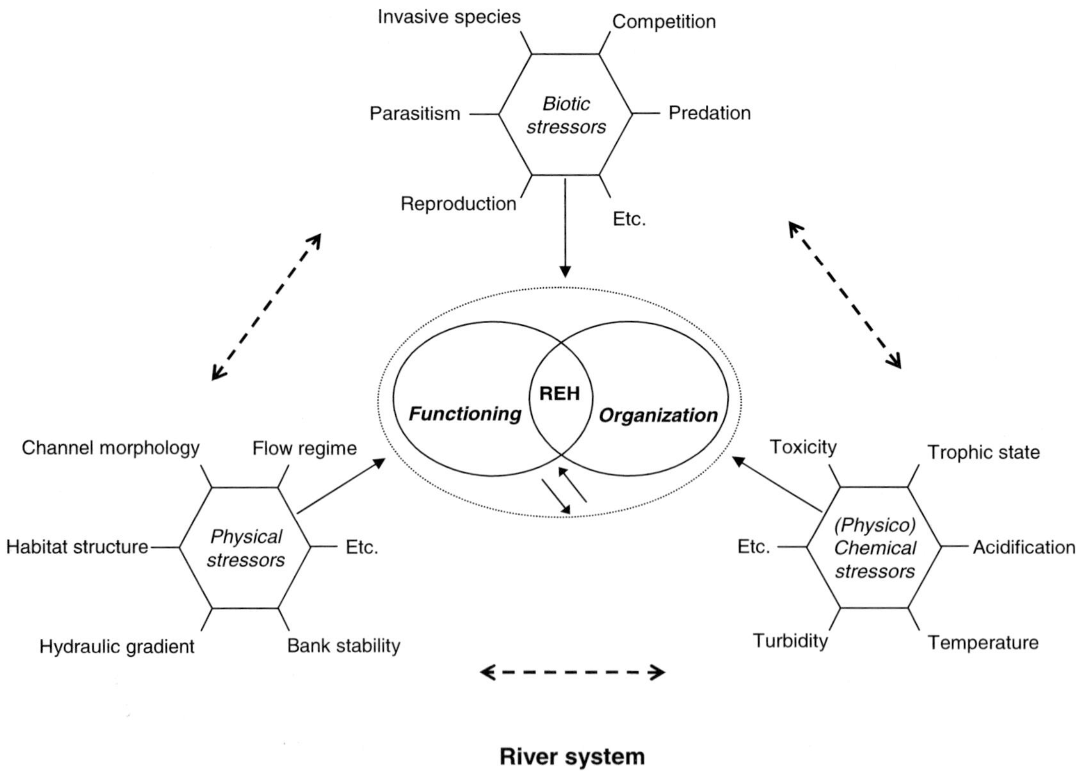

Figure 3. Relation between River Ecosystem Health (REH), condition indicators (functioning and organization) and various stressor indicators. Small opposite arrows signify interaction of river ecosystem with society. Bi-directional broken arrows indicate the interdependence of stressors, i.e. human activities may directly pose either a physical, (physico-)chemical or biotic stress on the river, but most common is a physical change in the system that results in chemical and subsequent biotic stress reactions.

Functioning

The vigour of a system is an attribute of system performance that represents the activity, metabolism or primary productivity of the ecosystem. Available indicators can be measured directly and relatively easy, including gross primary production (GPP) and energy flow measures like resource efficiency, system throughput and cycling (Costanza & Mageau, 1999). The most commonly used empirical measures are GPP, biomass as well as production and respiration ratios (Vannote et al., 1980; Bunn et al., 1999; Xu et al., 2001). The intensity and dynamic of GPP give expression to system vigour (Costanza & Mageau, 1999), by quantifying the magnitude of input (material or energy) available to the system (Bunn et al., 1999). Another measure of system metabolism is the rate of decomposition of terrestrial plant leaves in streams and rivers. It has been suggested for some time as an integrated measure of the effects of human disturbance (Young et al., 2004). Leaf breakdown is potentially an ideal measure because it links the characteristics of riparian vegetation with the activity of invertebrates and microbial organisms, and is affected by natural and human-induced variation in a wide range of environmental factors (Young et al., 2004). Other measures of vigour include resource use efficiency, unit energy flow and system throughput (Ulanowicz, 1986; Mageau et al., 1998; Xu et al., 2001), as well as system cycling (Finns cycling index; Allesina & Ulanowicz, 2004). These indices are part of network analysis, a phenomenological approach that holistically quantifies the structure and function of food webs by evaluating biomasses and energy flows (Ulanowicz, 1986).

Measuring the resilience of a system is difficult because it implies the ability to predict the dynamics of that system under stress (Costanza & Mageau, 1999). Quantifying resilience therefore often includes modelling techniques whereby resilience is expressed in terms of disturbance absorption capacity (Holling, 1987), scope for growth (Bayne, 1987) or population recovery time (Pimm, 1984). A suggested proxy measure is system overhead, which is another network analysis index described by Ulanowicz (1986). It quantifies the number of redundant or alternate pathways of material exchange and may be thought of as a systems ability to absorb stress without dramatic loss of function (Costanza & Mageau, 1999). Ecological buffer capacity is a measure that has been applied to lakes (Xu et al., 1999). It represents the ability of the system to normalise effects by external variables (i.e. pollution input, acidifying precipitation etc.) through changes in internal variables (plankton concentration, phosphorus concentration etc.). It can be expressed as a ratio between external variables that are driving the system and internal variables that determine the system (Xu et al., 1999, 2001).

Organization

Ecosystem organization relates to the complex of interactions between system processes and structure across space and time. Quantifying organization may be more difficult than functioning because quantifying organization involves measuring both the diversity and magnitude of system components (e.g. river sediment and main stream) and the material exchange pathways between them (Costanza & Mageau, 1999). Indicators of organization include the diversity of species and energy flows (i.e. exergy), as well as indirect network analysis measures such as system uncertainty, development capacity, mutual information and predictability (Ulanowicz, 1986; Turner et al., 1989; Mageau et al., 1998). The difficulty of quantifying organization in practice is apparent from Table 2, which shows no indicators that have been elaborated for REH. A suggested indicator is system uncertainty or Shannon diversity of individual flows, which may be easily adaptable and applicable for rivers. This network analysis index represents the total number and diversity of input, output and material flows and is a measure of the total uncertainty embodied in any configuration of flows (Mageau et al., 1998). The Shannon index is also applicable to biodiversity; Xu et al. (2001) measured algal species diversity in a lake ecosystem and showed a low diversity index outcome to be related to ecosystem stress. Based on data of wild bee pollinators, Belaousssoff & Kevan (2003) argue that the degree of deviation of diversity and abundance from log normality can be used as an indicator of ecosystem health. Pollinator communities from fields unaffected by an insecticide showed a log normal distribution of diversity and abundance but those fields affected did not.

Table 2. Set of condition indicators to assess river ecosystem health

Ecosystem attributes	Specification	Indicators	Method	Reference
Functioning	Vigour (activity, metabolisms or productivity)	Gross primary production* (GPP), Standing crop biomass (B), respi ration* (R) and ratios (GPP/R*, GPP/B), carbon assimilation ratio	DM	Odum (1985), Xu et al. (2001); Young et al. (2004); Bunn et al. (1999)
		Leaf litter processing rate*	DM	Young et al. (2004)
	Resilience (counteractive capacity of ecosystems to maintain structure and function)	Systems overhead*	NA	Ulanowicz (1986); Costanza & Mageau (1999)
		Ecological buffer capacity+	DM, SM	Jørgensen (1995); Xu et al. (1999, 2001)
Organization	Interactions between processes and structure across space and time (e.g. diversity).	System uncertainty; Shannon's diversity index+	NA; DM	Ulanowicz (1986); Mageau et al. (1998); Xu et al. (2001)
		Exergy and structural exergy+	SM	Jørgensen (1995); Xu et al. (1999; 2001)
Combinations	Combine both functioning and organization aspects	Functional measures of species richness, abundance & morphology, trophic composition (IBI) & habitat preferences *	DM	Karr (1991); Poff & Allan (1995)
		System ascendancy*	NA	Costanza & Mageau (1999)

*: Applied to river ecosystems (including estuaries); +: Applied to freshwater ecosystems, easy to adapt for river ecosystem application; DM: Direct measurement; IBI: Index of biotic integrity; NA: Network analysis; SM: System modelling.

Another measure of organization is exergy, defined as the amount of work a system can perform when it is brought to thermodynamic equilibrium with its environment. Exergy is expected to increase as ecosystems mature and develop away from the thermodynamic equilibrium. It can be expressed as a function of the biomass in the system and the (genetic) information that the biomass is carrying. Structural exergy can be defined as the ability of the ecosystem to utilize available resources and can be expressed as the exergy relatively to total biomass (Xu et al., 1999).

Combinations

There are also measures that combine both functioning and organization aspects. The Index of Biotic Integrity (sic) incorporates multiple attributes of fish communities to evaluate human influence on a stream and its catchment. It is by far the most used index (in various versions) for assessment of river condition (Karr, 1991). The IBI employs a series of metrics based on assemblage structure and function (fish or invertebrate assemblages) that give reliable signals of river condition to calculate an index score at a site, which is then compared with the score expected in the absence of stress. The multi-metric approach has widely found use (Karr, 1999), for example by Poff & Allan (1995), who added habitat preference measures to measures of trophic composition and fish morphology. The measure of system ascendancy has been articulated by Ulanowicz (1986), who stated that as an ecosystem network develops through time in a stable environment, it becomes more hierarchical and has fewer redundant links. This means that whereas a mature or non-stressed network has few redundant connections, a polluted, stressed, or frequently disturbed network will have many redundant connections (thus low ecosystem ascendancy). Indeed Costanza & Mageau (1999) found lower ascendancy value for polluted estuaries.

Stressor and effect indicators

Biotic, physical and chemical stressors can affect river ecosystems. As outlined before, the proposed assessment framework can be used to address the current stress status of an area (status assessment) or to express the development of river stressor conditions (trend detection). As a first step, we listed a number of indicators related to the different kinds of stress. These indicators can be assessed with methodologies currently in use. Table 3 presents a list that is not exhaustive, but a representative selection of established indicators.

Concerning biotic stressors, there is sufficient evidence that invasive species may negatively affect the occurrence of indigenous species (Bij de Vaate et al., 2002). The number and abundance of invasive species for fish and macro-invertebrates may be considered a good indicator for the stress caused by invasive biota in a river ecosystem. Species richness (Hill, 1973) or a species richness-abundance index, such as the Simpson index (Simpson, 1949) may be used to quantify stress of invasive species in river ecosystems. Another biotic stressor indicator is measurement of size-distribution structure. Studies of aquatic systems show that an increase in stress pressures is accompanied by the decreased dominance of large species and an increased dominance of small species. Quantitative estimates of maximum size attained by fish species can be used to calculate shifts in the size distribution of species (Wichert & Rapport, 1998).

Physical stressors relate to changes in flow regime and habitat structure. Alternations of flow regimes can play a major role in the destruction of river ecosystems. Richter et al. (1997) developed a Range of Variability Approach (RVA) to assess the influence of human activities on the water budget and dynamics of aquatic systems. A suite of 32 hydrological parameters is defined to characterize hydrological variability before and after an aquatic system has been altered by human activities (Richter et al., 1996). A less elaborative method to assess the hydrological functioning of rivers is the Tennant method. A first picture of the hydrological functioning of a river can be obtained by comparing recommended percentages of the historical average annual flow with the actual monthly hydrographs for winter and summer (Tennant, 1976). Apart from water quantity and dynamics, the connectivity of water bodies is of importance for the ecological functioning of river ecosystems, particularly for anadromous fish species. The number and abundance of anadromous fish species may be considered as a good indicator

Table 3. Set of stressor and effect indicators to assess river ecosystem health

Scope	Type	Indicator	Indicator specification	Method	Reference
Biotic	S	Fish, Invertebrates	Invasive species	SR, S (invasive)	Hill (1973); Simpson (1949)
	E		Size-distribution structure	MSD	Wichtert & Rapport (1998)
Physical	S	Flow regime	Quantity and dynamics	RVA, T	Richter et al. (1997); Tennant (1976)
	S	Habitat structure	Depth, width, structure and substrate	QHEI, AQEM	Rankin (1989); Hering et al. (2004)
	E	Connectivity	Fish number and abundance	SR, S (anadromous)	Hill (1973); Simpson (1949)
Chemical	S	Temperature	Temperature change	WQI	Brown et al. (1970); Couillard & Lefebvre (1985)
	S	Trophic state	%DO, BOD, P, NO_3^-, TDS	WQI	Brown et al. (1970); Couillard & Lefebvre (1985)
	S	Turbidity	NTU	WQI	Couillard & Lefebvre (1985)
	S	Acidification	pH	WQI, AQEM	Couillard & Lefebvre (1985); Hering et al (2004)
	E	Toxicity	Toxic stress	msPAF, AQEM	Traas et al. (2002); Klepper et al. (1998); Hering et al. (2004)

Types include S: stressor indicator; E: effect indicator.
Indicator specifications include %DO: percentage dissolved oxygen; BOD: Biological Oxygen Demand; msPAF: multispecies Potentially Affected Fraction of species; NO_3^-: Nitrates; NTU: Nephelometric Turbidity Unit; P: Total Phosphates; TDS: Total Dissolved Solids.
Methods include AQEM: integrated Assessment of the ecological Quality of streams and rivers throughout Europe using benthic Macro-invertebrates; MSD: Maximum Size Distribution; QHEI: Qualitative Habitat Evaluation Index; RVA: Range of Variability Approach; S: Simpson index; SR: Species Richness; T: Tennant method; WQI: Water Quality Index.

for the stress caused by the lack of connectivity in a river. Species richness (Hill, 1973) or a species richness-abundance index, such as the Simpson index (Simpson, 1949), for anadromous fish species may be used to quantify the stress due to lack of continuity along rivers. The Qualitative Habitat Evaluation Index (QHEI) was designed to provide a measure of habitat that generally corresponds to those physical factors that affect fish communities (Rankin, 1989). The QHEI is based on six interrelated metrics: substrate, in-stream cover, channel morphology, riparian zone and bank erosion, pool/glide and riffle/run quality, and gradient. Another way to assess habitat structure destruction is to use information on species occurrences, which are sensitive towards degradation in stream morphology (Hering et al., 2004).

The third group of indicators reflects chemical stressors. Water quality can be assessed in a relatively straightforward way, by measuring a number of key physical attributes and processes. Various methods aim to integrate these measurements to one comprehensive index (BKH, 1994). The Water Quality Index (WQI) of the US National Sanitation Foundation is one of the most widely used of all existing water quality indices, integrating nine water quality parameters, such as pH and Biological Oxygen Demand (Brown et al., 1970; Couillard & Lefebvre, 1985). Although the WQI can be applied in a comprehensive way, it lacks the inclusion of a stress index for toxic pollutants. Species are generally exposed to complex chemical mixtures in the environment. Calculation of the combined ecotoxicological effects of mixtures of chemicals on sets of species can be done according to concentration addition rules of calculus for pollutants with the same mode of action and response additive calculation rules between toxic modes of action (Traas et al., 2002). The toxic stress index reflects the fraction of species expected to be (potentially) affected at a given environmental exposure to a mixture of chemicals (Klepper et al., 1998). Another way to address chemical stress is to use information on species occurrences, which are sensitive towards a specific stressor, such as acidification or organic pollution (Hering et al., 2004).

Tables 2 and 3 present a cross-section of indicators required to assess overall REH status through vital properties of the system (vigour, resilience and organization) and lower-level system parameters that are indicative of (potential) stress causalities impairing REH. The list of condition indicators reveals that a limited number of indicators is yet available to assess comprehensive system properties (e.g. resilience) for freshwater systems. The complexity of the underlying processes seems an obvious factor in this. The presented stressor and effect indicators cover the scope of common stresses, but the set is adaptive to specific local circumstances and policy requirements. More explicit than in the current list, measures may be included of ecosystem services ensuring specific social and economic qualities (stressor measures on harvestable fish species, etc.). Overall, the set of top-down and bottom-up indicators suggests that more integration is required amongst measures to produce practical indices of overall REH. There remains a dilemma in trying to construct a comprehensive evaluation system for REH: on one hand is the desire to ensure that it truly reflects the defining attributes of REH – on the other, the more complex the system the more information is needed, and time or money may not permit its collection (Boon, 2000).

Added value of health metaphor

Next to having a conceptual meaning and being elaborated in models, REH has symbolic and informal use in scientific language, and in public dialogue. This is perceived as the metaphorical dimension of REH (Pickett & Cadenasso, 2002). In river management the health metaphor has added value in two ways. First, it has scientific value as a structural metaphor that perceives ecosystems as organisms. This provides a simple intellectual framework that allows comprehension of the multi-dimensionality and interrelationships that exist in complex systems. As such it has a generative and creative role for developing concepts of ecosystem condition and structuring research questions. Complementary to this is its socio-political role. Within this role the metaphor generally differs from its scientific use as the precision and narrow focus of scientific terms is generally avoided in favour of richness of connotation and in support of societal important values, for example investing in river rehabilitation (Bennett, 2002). As such the metaphor has value in effectively communicating results about the condition of river ecosystems and related environmental problems (Meyer, 1997). Humans have intrinsic comprehension of health and can relate to a physician-like approach that involves diagnosis, prognosis, treatment, and prevention. For this reason, it is now widely used in both popular and academic discussions of environmental problems and has widely found public use in policymaking and management objectives.

The strength of the metaphor lies in its potential to mobilize scientists, practitioners and publics by seeing relationships at the level of values. This way it places human beings at the centre of considerations about development, while seeking to ensure the durability of the ecosystem of which they are an integral part. There can be no sustainable development unless interventions take into account both the well-being of human beings and the survival of the ecosystem (Forget & Lebel, 2001). Therefore it is necessary to include the human institutions that interact with the river and that control its future condition: laws and their enforcers, management agencies, industries etc. (Meyer, 1997). The value of health is recognized by the fact that 'river health' has been adopted in various (inter)national monitoring programs and political objectives, for example in Australia and South Africa, Cambodia, Laos, Thailand and Vietnam (Australian and New Zealand Environment and Conservation Council, 1992; Hohls, 1996; Mekong River Commission, 2003).

Perspective

Central in river ecosystem health is the ability of the system to function, i.e. to perform and sustain (key) processes that are in accordance with system structure at various scales (i.e. organization). A healthy state is flexible in definition within the limits of sustainable functioning (Fig. 1), consequently allowing consideration of economic and social functions for its definition. This fits a broader conceptual context, introduced as River System Health (RSH), which considers REH to be a component in the overall health status of the river system. The framework of RSH extends beyond a separation of a 'natural' and 'societal' river system and aims to fully integrate human attitudes and social institutions that are a part of a rivers' societal catchment, meaning the social and economic structures and institutions that directly influence ecological structure and processes (Meyer, 1997; Fig. 2).

Assessment of ecological health needs integration of measures of multiple, complementary attributes and analysis in a synthesized way. The proposed assessment framework outlines a combined top-down/bottom up approach that combines condition and stressor/effect indicators. For river managers, this poses a framework that is descriptive, i.e. able to evaluate the effects of human interactions on ecological functions, as well as being diagnostic, i.e. indicative of responsible stressors. In order to retrieve an easy-to-use, transparent methodology, efforts need to be directed to define a minimum set of indicators that may reliably represent the scope of REH. The indicators in this paper represent a useful, exemplary selection from a broad range of currently available indicators and are believed to cover the main concept of REH. However, cross-comparisons of indicators are required to optimize the indicator set. Based on findings on the indicative power, mutual relationships and interdependencies of metrics, certain indicators may prove 'redundant' while others may be worth including. For example, An et al. (2002) used a biological assessment (IBI) in combination with habitat (QHEI) and chemical measurements to evaluate REH. Habitat quality showed a strong positive relation to species richness. This suggests that the QHEI can be a predictive tool for changes in biological communities. Another study by Miltner & Rankin (1998) showed a negative correlation between nutrients and IBI, detectable when nutrient concentrations exceeded background concentrations.

Benchmarks need to be set for each indicator that enables distinction between "healthy" and "unhealthy" (i.e. intra-valuation; Norris & Thoms, 1999). These benchmarks need to be based on reference conditions that illustrate the spatial and temporal dynamics of self-maintaining, sustainable functioning river ecosystems. Appropriate river systems of reference can be identified through expert judgment. For some indicators, the benchmark values assigned could and should be determined by existing guidelines, objectives or standards e.g. contaminant levels in sediments.

Attention should be given to time and spatial scales of measured parameters, e.g. regarding seasonal or long-term natural dynamics of parameter values (Innis et al., 2000). Natural dynamics may cause relative impacts of stresses to change across seasons. Considerations of scales are not only necessary for evaluating individual indicators, but also for comparing and integrating the results of multiple indicators.

The set of REH indicators suggested in this paper may together be integrated to construct a REH index. Expressing REH in a single index demands the aggregation of multiple indicators and requires use of suitable aggregation techniques. Managers and/or scientists may value the ecological attributes that these indicators measure differently. A process of weighting is required to differentiate between attributes of differing importance (i.e. inter-valuation) (Wells, 2003). The values of weighing factors need to be defined, based on validated scientific data and expert judgment. This way a 'scoring' or classification system can be developed in which indicators and their metrics are clearly described and the derivation and interpretation of scores can be readily understood. A classification system improves objectivity by ensuring that valuations of health are rigorous, repeatable and transparent (Boon, 2000). Multi-optional visualization and calculation techniques can add to transparency of the weighting, calculation and aggregation process and supply information to managers that is relevant for defined objectives and required information detail. This can provide an effective tool for decision-making that can synthesize knowledge over a range of space and time scales within a nested hierarchy of (sub)systems and be set to multiple levels of assessment intensity, varying from a "superficial" screening to intense diagnostic health assessment.

An index of REH may enable a single judgment of the ecological health status of a river system and evaluation of management objectives. As such, a REH index can support decision-making when a specific health rank is linked with defined policy actions. Such models may be valuable assets in implementation of political frameworks like the Water Framework Directive. In a wider context, the REH concept and its models can provide consistency in ecological assessment approaches, based on flexibility of different scales, hierarchy and information on functioning and organization of the river system. Though the paper has given an assessment framework for managers to work with, practical elaboration will have to be extended on how to relate relevant single effects, values and criteria across fields of impact in a meaningful way and how to make them comparable in order to be able to weight them and trade them off if necessary (Brouwer et al., 2003).

Finally, REH (within the wider context of RSH) has the potential to evolve into a core concept for integrated water management. However, this will require further synchronization with contemporary concepts and methodologies available to achieve the aims set in management, such as restoration, rehabilitation, ecosystem management and adaptive management.

Acknowledgement

Part of this study has been financed by the Interdepartmental Institute Science & Society of the Radboud University Nijmegen (grant W&S 2004-04). This is CWE-publication 427.

References

Allesina, S. & R. E. Ulanowicz, 2004. Cycling in ecological networks: Finn's index revisited. Computational Biology and Chemistry 28: 227–233.

An, K. -G., S. S. Park & J. -Y. Shin, 2002. An evaluation of a river health using the index of biological integrity along with relations to chemical and habitat conditions. Environment International 28: 411–420.

Australian and New Zealand Environment and Conservation Council, 1992. Australian Water Quality Guidelines for Fresh and Marine Waters, National Water Quality Management Strategy. Australian and New Zealand Environment and Conservation Council, Canberra.

Bayne, B. L., 1987. The Effects of Stress and Pollution on Marine Animals. Praeger, New York.

Belaousoff, S. & P. G. Kevan, 2003. Are there ecological foundations for ecosystem health? The Environmentalist 23: 255–263.

Bennett, J., 2002. Investing in river health. Water Science and Technology 45: 85–90.

Bij de Vaate, A., K. Jazdzewski, H. A. M. Ketelaars, S. Gollasch & G. van der Velde, 2002. Geographical patterns in range extension of Ponto-Caspian macroinvertebrate species in Europe. Canadian Journal of Fisheries and Aquatic Sciences 59: 1159–1174.

BKH, 1994. Chemische waterkwaliteitsindices: internationale inventarisatie van technieken en methodieken voor aggregatie en presentatie van chemische waterkwaliteitsgegevens. RO190101/5818L/R5. BKH Adviesbureau, Delft (in Dutch).

Boon, P. J., 1998. River restoration in five dimensions. Aquatic conservation: Marine and Freshwater Ecosystems 8: 257–264.

Boon, P. J., 2000. The development of integrated methods for assessing river condition value. Hydrobiologia 422/423: 413–428.

Boulton, A. J., 1999. An overview of river health assessment: philosophies, practice, problems and diagnosis. Freshwater Biology 41: 469–479.

Brouwer, R., S. Georgiou & R. K. Turner, 2003. Integrated assessment and sustainable water and wetland management. A review of concepts and methods. Integrated Assessment 4: 172–184.

Brown, R. M., N. I. McClelland, R. A. Deininger & R. G. Tozer, 1970. A water quality index – do we dare? Water and Sewage Works 117: 339–343.

Bunn, S. E., P. M. Davies & T. D. Mosch, 1999. Ecosystem measures of river health and their response to riparian and catchment degradation. Freshwater Biology 41: 333–345.

Callicot, J. B., J. B. Crowder & K. Mumford, 1999. Current normative concepts in conservation. Conservation Biology 13: 22–35.

Calow, P., 1995. Ecosystem health – a critical analysis of concepts. In Rapport, D. J., C. Gaudet & P. Calow (eds), Evaluating and Monitoring the Health of Large-scale Ecosystems. Springer-Verlag, Berlin, 33–41.

Costanza, R. & B. C. Patten, 1995. Defining and predicting sustainability. Ecological Economics 15: 193–196.

Costanza, R. & M. Mageau, 1999. What is a healthy ecosystem? Aquatic Ecology 33: 105–115.

Costanza, R., B. Norton & B. J. Haskell, (eds), 1992. Ecosystem Health – New Goals for Environmental Management. Island Press, Washington DC.

Costanza, R., R. d'Arge, R. de Groot, S. Farber, M. Grasso, B. Hannon, K. Limburg, S. Naeem, R. V. O'Neill, J. Paruelo, R. G. Raskin, P. Sutton & M. van den Belt, 1997. The value of the world's ecosystem services and natural capital. Nature 387: 253–260.

Couillard, D. & Y. Lefebvre, 1985. Analysis of water quality indices. Journal of Environmental Management 21: 161–179.

Davis, M. A. & L. B. Slobodkin, 2004. The science and values of restoration ecology. Restoration Ecology 12: 1–3.

European Commission, 2000. Directive 2000/60/EC, Establishing a framework for community action in the field of water policy. European Commission PE-CONS 3639/1/100 Rev 1, Luxembourg.

Environmental Protection Agency, 1998. Guidelines for Ecological Risk Assessment. U.S Environmental Protection Agency EPA/630/R-95/002F, Washington DC.

Fairweather, P. G., 1999. State of environment indicators of 'river health': exploring the metaphor. Freshwater Biology 41: 211–220.

Forget, G. & J. Lebel, 2001. An ecosystem approach to human health. International Journal of Occupational and Environmental Health 7(Supplement): 3–36.

Hart, B. T., B. Maher & I. Lawrence, 1999. New generation water quality guidelines for ecosystem protection. Freshwater Biology 41: 347–359.

Harvey, J., 2001. The natural economy. Nature 413: 463.

Hering, D., O. Moog, L. Sandin & P. F. M. Verdonschot, 2004. Overview and application of the AQEM assessment system. Hydrobiologia 516: 1–20.

Hill, M. O., 1973. Diversity and eveness: a unifying notation and its consequences. Ecology 54: 427–432.

Hohls, D. R., 1996. National Biomonitoring Programme for Riverine Ecosystems: Framework Document for the Programme. NBP Report Series No. 1, Institute for Water Quality Studies, Department of Water Affairs and Forestry, Pretoria.

Holling, C. S., 1973. Resilience and stability of ecological systems. Annual Reviews of Ecology and Systematics 4: 1–23.

Holling, C. S., 1987. Simplifying the complex: the paradigms of ecological function and structure. European Journal of Operational Restoration 30: 139–146.

Holling, C. S., 2001. Understanding the complexity of economic, ecological, and social systems. Ecosystems 4: 390–405.

Innis, S. A., R. J. Naiman & S. R. Elliot, 2000. Indicators and assessment methods for measuring the ecological integrity of semi-aquatic terrestrial environments. Hydrobiologia 422/423: 111–131.

Jørgensen, S. E., 1995. Exergy and ecological buffer capacities as measures of ecosystem health. Ecosystem Health 1: 150–160.

Junk, W. J., P. B. Bailey & R. E. Sparks, 1989. The flood pulse concept in river floodplain systems. In Dodge, D. P. (ed.), Proceedings of the Internationals Large River Symposium (LARS). Canadian Special Publications of Fisheries and Aquatic Sciences 106: 110–127.

Karr, J. R., 1991. Biological integrity: a long-neglected aspect of water resource management. Ecological Applications 1: 66–84.

Karr, J. R., 1999. Defining and measuring river health. Freshwater Biology 41: 221–234.

Klepper, O., J. Bakker, T. P. Traas & D. Van de Meent, 1998. Mapping the Potentially Affected Fraction (PAF) of species as a basis for comparison of ecotoxicological risks between substances and regions. Journal of Hazardous Materials 61: 337–344.

Kuiper, J., 1998. Landscape quality based upon diversity, coherence and continuity. Landscape planning at different planning-levels in the River area of the Netherlands. Landscape and Urban Planning 43: 91–104.

Lackey, R. T., 2001. Values, policy, and ecosystem health. BioScience 51: 437–443.

Lenders, H. J. R., 2003. Environmental rehabilitation of the river landscape in the Netherlands. A blend of five dimensions. Ph.D.-thesis, University of Nijmegen.

Lenders, H. J. R. & L. Knippenberg, 2005. The temporal and social dimensions of river rehabilitation: towards a multidimensional research perspective. Archiv für Hydrobiologie (Large Rivers Supplement) 155/15: 119–131.

Leuven, R. S. E. W. & I. Poudevigne, 2002. Riverine landscape dynamics and ecological risk assessment. Freshwater Biology 47: 845–865.

Leuven, R. S. E. W., J. L. M. Haans, A. J. Hendriks, R. A. C. Lock & S. E. Wendelaar Bonga, 1998. Assessing cumulative impacts of multiple stressors on river systems. In Nienhuis, P. H., R. S. E. W. Leuven & A. M. J. Ragas (eds), New Concepts for Sustainable Management of River Basins. Backhuys Publishers, Leiden, 241–259.

Leuven, R. S. E. W., A. J. M. Smits & P. H. Nienhuis, 2000. From integrated approaches to sustainable river basin management. In Smits, A. J. M., P. H. Nienhuis & R. S. E. W. Leuven (eds), New Approaches to River Management. Backhuys Publishers, Leiden, 329–347.

Maddock, I., 1999. The importance of physical habitat assessment for evaluating river health. Freshwater Biology 41: 373–391.

Mageau, M. T., R. Costanza & R. E. Ulanowicz, 1998. Quantifying the trends expected in developing ecosystems. Ecological Modelling 112: 1–22.

Maher, W., G. E. Batley & I. Lawrence, 1999. Assessing the health of sediment ecosystems: use of chemical measurements. Freshwater Biology 41: 361–372.

May, R. M., 1977. Thresholds and breakpoints in ecosystems with a multiplicity of stable states. Nature 269: 471–477.

Mekong River Commission, 2003. Annual Report. Mekong River Commission, Phnom Penh (available online www.mrcmekong.org).

Meyer, J. L., 1997. Stream health: incorporating the human dimension to advance stream ecology. Journal of the North American Benthological Society 16: 439–447.

Miltner, R. J. & E. T. Rankin, 1998. Primary nutrients and the biotic integrity of rivers and streams. Freshwater Biology 40: 145–158.

Moog, O. & A. Chovanec, 2000. Assessing the ecological integrity of rivers: walking the line among ecological, political and administrative interests. Hydrobiologia 422/423: 99–109.

Naveh, Z., 2001. Ten major premises for a holistic conception of multifunctional landscapes. Landscape and Urban Planning 57: 269–284.

Nienhuis, P. H. & R. S. E. W. Leuven, 1998. Ecological concepts for the sustainable management of lowland river basins: a review. In Nienhuis, P. H., R. S. E. W. Leuven & A. M. J. Ragas (eds), New Concepts for Sustainable Management of River Basins. Backhuys Publishers, Leiden, 7–33.

Norris, R. H. & M. C. Thoms, 1999. What is river health? Freshwater Biology 41: 197–209.

Norris, R. H. & C. P. Hawkins, 2000. Monitoring river health. Hydrobiologia 435: 5–17.

Oberdorff, T., D. Pont, B. Hugueny & J. P. Porcher, 2002. Development and validation of a fish-based index for the assessment of 'river health' in France. Freshwater Biology 47: 1720–1734.

Odum, E. P., 1969. The strategy of ecosystem development. Science 164: 262–270.

Odum, E. P., 1985. Trends expected in stressed ecosystems. Bioscience 35: 419–422.

Pickett, J. P. (ed.) 2000. The American Heritage Dictionary of the English Language, (4th edn). Houghton Mifflin Company, Boston.

Pickett, S. T. A. & L. Cadenasso, 2002. The ecosystem as a multidimensional concept: meaning, model, and metaphor. Ecosystems 5: 1–10.

Pickett, S. T. A, W. R. Burch Jr, T. W. Foresman, J. M. Grove & R. Rowntree, 1997. A conceptual framework for the study of human ecosystems. Urban Ecosystems 1: 185–199.

Pimm, S. L., 1984. The complexity and stability of ecosystems. Nature 307: 321–326.

Poff, N. L. & J. D. Allan, 1995. Functional organization of stream fish assemblages in relation to hydrological variability. Ecology 76: 606–627.

Pollard, P. & M. Huxham, 1998. The European Water Framework Directive: a new era in the management of aquatic ecosystem health? Aquatic Conservation: Marine and Freshwater Ecosystems 8: 773–792.

Poudevigne, I., D. Alard, R. S. E. W. Leuven & P. H. Nienhuis, 2002. A system approach to river restoration: a case study in the lower Seine valley, France. River Research and Applications 18: 239–247.

Rankin, E. T., 1989. The qualitative habitat evaluation index (QHEI): rationale, methods, and application. Division of Water Quality Planning & Assessment, Ecological Assessment Section, Columbus.

Rapport, D. J., H. A. Regier & T. C. Hutchinson, 1985. Ecosystem behavior under stress. American Naturalist 125: 617–640.

Rapport, D. J., R. Costanza & A. J. McMichael, 1998a. Assessing ecosystem health: challenges at the interface of social, natural and health sciences. Trends in Ecology and Evolution 13: 397–402.

Rapport, D. J., C. Gaudet, J. R. Karr, J. S. Baron, C. Bohlen, W. Jackson, B. Jones, R. J. Naiman, B. Norton & M. M. Pollock, 1998b. Evaluating landscape health: interacting societal goals and biophysical process. Journal of Environmental Management 53: 1–15.

Rapport, D. J., G. Böhm, D. Buckingham, J. Cairns Jr., R. Costanza, J. R. Karr, H. A. M. de Kruijf, R. Levins, A. J. McMichael, N. O. Nielsen & W. G. Whitford, 1999. Ecosystem health: the concept, the ISEH, and the important tasks ahead. Ecosystem Health 5: 82–90.

Richter, B. D., J. V. Baumgartner, J. Powell & D. P. Braun, 1996. A method for assessing hydrologic alteration within ecosystems. Conservation Biology 10: 1163–1174.

Richter, B. D., J. V. Baumgartner, R. Wigington & D. P. Braun, 1997. How much water does a river need? Freshwater Biology 37: 231–249.

Ripl, W., J. Pokorný, M. Eiseltová & S. Ridgill, 1994. A holistic approach to the structure and function of wetlands and their degradation. International Waterfowl and Wetlands Research Bureau Publication 32: 16–35.

Rogers, K. & H. Biggs, 1999. Integrating indicators, endpoints and value systems in strategic management of the rivers of the Kruger National Park. Freshwater Biology 41: 439–451.

Simberloff, D., 1998. Flagships, umbrellas, and keystones: is single-species management passé in the landscape era? Biological Conservation 93: 247–257.

Simpson, E. H., 1949. Measurements of diversity. Nature 163: 688.

Society for Ecological Restoration Science & Policy Working Group (SER), 2004. The SER Primer on Ecological Restoration (available online www.ser.org).

Tansley, A. G., 1935. The use and abuse of vegetational concepts and terms. Ecology 16: 284–307.

Tennant, D. L., 1976. Instream flow regimens for fish, wildlife, recreation and related environmental resources. Fisheries 1: 6–10.

Tockner, K., F. Malard & J. V. Ward, 2000. An extension of the flood pulse concept. Hydrological Processes 14: 2861–2883.

Townsend, C. R. & R. H. Riley, 1999. Assessment of river health: accounting for perturbation pathways in physical and ecological space. Freshwater Biology 41: 393–405.

Traas, T. P., D. Van de Meent, L. Posthuma, T. H. M. Hamers, B. J. Kater, D. De Zwart & T. Aldenberg, 2002. Potentially affected fraction as measure of toxic pressure on ecosystems. In Posthuma, L., G. W. Suter II, & T. P. Trass (eds), Species Sensitivity Distributions in Ecotoxicology. Lewis Publishers, Boca Raton, 315–344.

Turner, M. G., R. Costanza & F. H. Sklar, 1989. Methods to compare spatial patterns for landscape modelling and analysis. Ecological Modelling 48: 1–18.

Ulanowicz, R. E., 1986. Growth and development: ecosystems phenomenology. Springer-Verlag, New York.

Van der Velde, G., R. S. E. W. Leuven & I. Nagelkerken, 2004. Types of river ecosystems. 2.7.2.4. In Dooge, J. C. I. (ed.), Fresh Surface Water, Encyclopedia of Life Support Systems (EOLSS). EOLSS Publishers, Oxford (available online www.eolss.net).

Vannote, R. L., G. W. Minshall, K. W. Cummins, J. R. Sedell & C. E. Cushin, 1980. The river continuum concept. Canadian Journal of Fisheries and Aquatic Sciences 37: 130–137.

Ward, J. V. & J. A. Stanford, 1995. The serial discontinuity concept: extending the model to floodplain rivers – regulated rivers. Research and Management 10: 159–168.

Ward, J. V., C. T. Robinson & K. Tockner, 2002. Applicability of ecological theory to riverine ecosystems. Verhandlungen Internationale Vereinigung für Theoretische und Angewandte Limnologie 26: 443–450.

Wells, P. G., 2003. Assessing health of the Bay of Fundy – concepts and framework. Marine Pollution Bulletin 46: 1059–1077.

Wichert, G. A. & D. J. Rapport, 1998. Fish community structure as a measure of degradation and rehabilitation of riparian systems in an agricultural drainage basin. Environmental Management 22: 425–443.

World Commission on Environment and Development, 1987. Our Common Future. Oxford University Press, Oxford.

Xu, F. -L., S. E. Jørgensen & S. Tao, 1999. Ecological indicators for assessing freshwater ecosystem health. Ecological Modelling 116: 77–106.

Xu, F. -L., S. Tao, R. W. Dawson, P. -G. Li & J. Cao, 2001. Lake ecosystem health assessment. Water Research 35: 3157–3167.

Young, R. G., C. R. Townsend & C. D. Matthaei, 2004. Functional indicators of river ecosystem health – an interim guide for use in New Zealand. Report 870, Cawthron Institute, Nelson.

Hydrobiologia (2006) 565:309–316
R.S.E.W. Leuven, A.M.J. Ragas, A.J.M. Smits & G. van der Velde (eds), Living Rivers: Trends and Challenges in Science and Management
DOI 10.1007/s10750-005-1921-7

Opinion Paper

Emergent principles for river management

W.T. de Groot[1,2,*] & H.J.R. Lenders[3]
[1]*Centre of Water and Society, Faculty of Science, Radboud University Nijmegen, P.O. Box 9010, 6500 GL Nijmegen, The Netherlands*
[2]*Institute of Environmental Sciences, Leiden University, P.O. Box 9518, 2300 RA Leiden, The Netherlands*
[3]*Department of Environmental Science, Institute of Wetland and Water Research, Faculty of Science, Radboud University Nijmegen, P.O. Box 9010, 6500 Nijmegen, GL The Netherlands*
*(*Author for correspondence: E-mail: W.deGroot@science.ru.nl)*

Key words: river management, adaptive management, participation, planning approach, co-management, communities, visions of nature

Abstract

Paradigms for land and water management are on the move. New approaches are said to be, or meant to be, more 'participatory', 'integrated', 'adaptive', 'ecosystem-based' and so on. The present paper explores emergent principles for land and water management in ecological management theory, environmental science and the social sciences. These principles comprise adaptive management, opportunity-driven analysis, visions of managers and the public, and co-management that includes local and supra-local rationality. The paper concludes that for river management, these principles largely reinforce each other. This lays a basis for a style of river management in which the river managers may continue to be the guardians of science-based and whole-basin rationality, while at the same time interacting more successfully with society.

Introduction

In 1837, after two centuries of debate, it was decided to drain and impolder the Haarlem Lake (Haarlemmermeer), covering some 100 km^2 between the cities of Leiden and Amsterdam in the Netherlands. In order to organize this great work, a Management Commission was established, composed of high-ranking politicians, administrators and engineers. The Commission had the full responsibility of all aspects of this work – technical, financial, social and spatial. It answered directly to the Secretary of the Interior and to the King, without a need to involve any of the many other departments, local governments and agencies that would nowadays be called 'stakeholders'. On that basis of absolute power, the Commission followed a flexible approach, adapting the technical plan to changing circumstances and rational objections of regional and local organizations. Twenty years later, against all technical and financial odds, the area was dry and parcelled out for agriculture, and was given back to the local government and farmers; testimony of the success of a strictly top-down, autocratic, centralized, rational and yet responsive type of planning and implementation (Jeurgens, 1991).

In present-day views on planning, this elitist 'Haarlem Lake planning style' is regarded as a sure recipe for moral and practical failure. Planning nowadays should always be participatory, i.e. including local stakeholders from the early stages onwards, or even be fully locally based (e.g. Leuven et al., 2000; Lenders, 2003; Morrison, 2003; Wiering & Arts, 2006; Witter et al., 2006). One extreme paradigm of locally based planning is "Endogenous Development" that is grounded solely in indigenous visions that encompass a community's human, natural and spiritual spheres of life, embedded in local identity and the 'bioregional narrative' (Cheney, 1989).

This paper will not review all issues of planning styles and methods that lie between these two (elitist and endogenous) extremes. Rather, we aim to draw some contours of a possible paradigm for our times, geared especially to river and river landscape planning and management. We have tried to do so, moreover, in a non-technical fashion that aims to keep aboard the natural scientist and others for whom the theory and practice of planning and management are only distant (and often rather boring) realities.

Our method will be to first take stock of a number of emergent principles in three relevant disciplinary fields. The first is on ecological management that finds new inspirations in non-equilibrium (or multi-state equilibrium) ecosystem dynamics. The second field is that of interdisciplinary environmental science, focusing especially on the shift from problem-oriented to more opportunity-oriented work. The third is social science, focusing especially on the concept of co-management. It will then be explored to what extent these principles contradict or reinforce each other. Most sections will contain a few examples of contributions of hydrobiology.

Evolution of ecological management concepts

Overall, Western societies have become more conscious of the values of nature and the functions that ecosystems and landscapes have for the quality of human life and community identity. With that, land use planning has in many places become more 'ecosystem-based'. Ecosystem-based planning is even becoming a concept of its own, with an emphasis on involving all actors that are connected to the regional ecosystem in a planning process that takes as its point of departure the functions that the system performs for these stakeholders – instead of first designing what society wants and then try to force nature into this human straightjacket (Imperial, 1999). Ecosystem-based river management means to first take heed of what the river is, how it functions and what it could be in terms of, say, hydro-morphodynamics, biodiversity, connectivity and integrity (Nienhuis & Leuven, 1998), and only then enter into a give-and-take between society and this functioning.

On a more theoretical level, studies on the dynamics of forests, drylands and other ecosystems have lead to new concepts of non-equilibrium ('states-and-transitions') ecology, such as resilience and the adaptive cycle (Holling & Gunderson, 2002) or catastrophic shifts (Scheffer et al., 2001). Although it could be argued that river system components such as wetlands and floodplains are not always the most perfect examples of dominance of adaptive cycles over 'classic' succession (Scholte, 2005), non-equilibrium ecology leads to a new appreciation of rivers as dynamic, self-rejuvenating systems.

Non-equilibrium ecology supports a vision of spatial planning as 'adaptive management', i.e. management that is constantly monitoring and responding to internal and external change of the ecosystem and its social context, rather than management based on static 'ideal' blueprints. See for instance Imperial (1993) on estuarine ecosystems. With Kessler (2003), we may add here that adaptive management requires not only flexibility on the short term, but also a guiding vision of the very long term, a vision that is realistic and yet inspirational, not blocked by all constraints that may exist on the short term (contrary to the concept of 'target image'; Lenders et al., 1998). Such a vision is not only a source of inspiration for stakeholders in the planning process (Mitchell, 2002; Morrison, 2003). It also makes adaptive management less purely reactive and more anticipatory, and it counterbalances the risk that the sum-total of short-term adaptations may later turn out to have been steps on an irreversible and undesired pathway. In the present-day context of room-for-river policies in Europe (Van Stokkom et al., 2005), it is worthy to note that in a book published more than 170 years ago, the historian Bilderdijk (1832) lamented that the Dutch should never have begun to embank their rivers in the first place. If our hydraulic forefathers would have had a truly long-term vision, Bilderdijk writes, we would have left their natural course, floodplains and floods to the rivers, and we would still live above river level.

In a long-term vision as part of an adaptive management strategy, adaptability itself should be a central tenet. Paraphrasing the well-known definition of sustainable development, the results of our adaptive management should not compromise the ability of future generations to do their own adaptive management. The practical value of this principle of reversibility and keeping options open for future generations can easily be imagined when we realize that we ourselves are the future generations of our forefathers. Doing a bit of 'counterfactual history' as a thought experiment with regards to the impoldering of the Haarlem Lake, it could be surmised that present-day society (and hydrobiologists) would be much happier if the 19th-century planning and design of the impoldering would have been guided by a long-term vision of open options for future generations and not, as it happened, by a vision that assumed an eternal value of the two functions (land loss prevention and agriculture) that happened to be the most salient at the time of the decision-making, bypassing all principles of reversibility as well as interests of water management, waterborne (sailing) transport and fisheries. Quite possibly, the lake would have been kept more open, and buildings and infrastructure would have been planned with more reversibility, e.g. by situating them closer to the polder edges. In the course of history then, the lake-cum-polder could have been adapted to the radical change that has taken place in the value of open water versus that of sugar beets. Part of the lake would still be open, other parts could have been reflooded and nature and recreational options could have been developed. Moreover, urban development of the cities of Amsterdam, Haarlem and Leiden around such a lake would have been much more interesting than the faceless suburbs, industries and airport that have sprawled into the polder due to the currently low value of agricultural land.

Present-day Dutch river authorities do have a long-term vision, and one that looks remarkably like Bilderdijk's. The room for river ambitions of the river authorities cannot be explained by political opportunism or cost-benefit analyses using normal discount rates. They are part of a truly long-term vision. Making room for the rivers, contrary to raising ever-higher dikes, is seen as the first step in a long process of re-naturalization of the Dutch water system that is necessary if only for long-term economic reasons. Moreover, room-for-rivers is a particularly good vision for adaptive management, because it leaves room for future generations to adapt the open space to their own needs and visions. A criticism on present-day room-for-river policy could only be that its implementation is not adaptive enough yet. River authorities often seem to try forcing the room for rivers into society in stead of searching for and working with the opportunities at hand (see also Wiering and Arts, 2006).

One type of such opportunities is that, as a by-product of great value for the future, new room for rivers often implies new room for nature and landscape development, too. Due to the dynamic character of the rivers, the specific values of new riverine nature will usually lie in the rareness and naturalness of the processes that build it, rather than in species and pattern diversity *per se*. These dynamic and robust riverine ecosystems can often accommodate a relatively high level of (respectful) human presence in them. Such relatively wild nature that can accommodate human participation will be of increasing social and cultural value in North-Western Europe where other landscapes are rapidly becoming more urbanized and more regulated (De Groot, 2004). This not only enhances the democratic basis of room-for-river policies in general. The satisfaction of urban desires for wild nature also creates new opportunities for economic benefits for riverine communities. Other parts of the widened floodplains could be kept under traditional agricultural management, with the open grasslands that are of great cultural value in Dutch society. This landscape, too, can be made to accommodate much more human participation than it does at present, e.g. with less fences and more footpaths, as in the English countryside.

Evolution of environmental science

Environmental science may be defined as the discipline structured around the need to address the environmental issues of society. Seen this way, environmental science started out in the early 1900s, when biologists and chemical scientists began to study and proclaim the vulnerability of nature and the nastiness of pollution. If solutions happened to be within easy reach (e.g. flushing of city canals or purchasing of natural areas), these efforts of natural scientists often were, as they sometimes still are, sufficient to find and implement solutions for the problems. During the 1970s, when the environmental problems rapidly acquired more urgency and depth, environmental science evolved into a truly interdisciplinary field, with the social sciences added in order to gain more insight into the social causes and solutions of environmental problems, and the humanities added later in order to reach more depth in the ethical and cultural aspects of the environmental problems.

This paradigm of 'interdisciplinarity around the problem core' proved to be productive, as shown not only by the countless applied studies focussing on specific environmental problems but also by the more fundamental development of integrated environmental models and many methodologies of environmental impact assessment, derivation of environmental standards, societal analysis and environmental policy design. Hydrobiologists contributed to this 'classic', problem-oriented environmental science with many studies on the ecology of eutrophication and toxic pollution and the development of assessment systems such as the Saprobic Index (e.g. Zelinka & Marvan, 1961) and species sensitivity distributions (Posthuma et al., 2002); see also Vugteveen et al. (2006) on ecosystem health.

In the course of time, limitations of the problem-oriented paradigm also began to show. Expansions of environmental science are sought, at present, in two directions, both of which aim to make this discipline more fundamental and with that, on the longer run, more broadly effective for society.

The first direction is caused by the limitation that traditional environmental methodologies such as problem in context (De Groot, 1998), even though enabling the researcher to connect environmental problems with their underlying causes in society, do not facilitate to study these underlying structures and processes in their own right. From the environmental problem 'downward', the researcher may tap into the cultural phenomena and the structures of actors causally connected by the mechanisms of power, which he may then use to explain and maybe even solve the environmental problem at hand. Next time he may do this again, and again – but what is the pattern, the 'systemness' of these social causes? Out of the ensuing desire to address these fundamental relationships between environment and society in a more systematic manner grew the study of the society–environment system. Ecological economics and the study of the material metabolism of societies, e.g. in material flows or ecological footprint analyses, are important expressions of this new 'sustainability science'.

Secondly and more importantly for the present exploration, it was increasingly felt that environmental science should not continue seeing the world as only the total sum of environmental problems, ignoring the fact that right next door, as it were, lie the myriad of instances where the environment is not a problem but rather a source of happiness and benefit. The environment is not only a constraint. Nature is not only something vulnerable that should be protected. The environment is also an opportunity for improved quality of life, and nature is also something dynamic that may be further developed. Integration of these insights into the discipline leads to an environmental science that retains its basic normative drive of working for relatively concrete issues of society, but is more broadly future-oriented than its problem-oriented origin. A good example is the book of Mitchell (2002) on environmental management that discusses methods of long-term visioning and participatory opportunity analysis alongside with traditional problem-oriented approaches such as environmental impact assessment (EIA). Another typical product of this happier and more opportunity-driven environmental science are the studies of Van den Born et al. (2001) that aim to elicit the philosophical visions that people in the Netherlands and elsewhere have of the relationship between humans and nature, cast in terms of mastership over nature, stewardship of nature, partnership with nature and spiritual participation in nature. One of the results of these studies is that to a remarkable extent, people in Western Europe express visions that lie far beyond the Cartesian image of mastership over nature, and seek to encounter nature also in its greatness and wildness.

Rivers and river landscapes offer obvious opportunities for this encounter. Put more generally, rivers and other water bodies, besides being objects of pollution, risks and other problems, obviously play many positive roles in the lives of people too. They naturally link up, therefore, with the more opportunity-driven new branch of environmental science and management. Along with that, hydrobiologists find new employ for studies on the restoration and rehabilitation of water systems (e.g. Nienhuis et al., 2002) and, remarkably, also for descriptive studies elucidating the history and intricate beauty of local aquatic ecosystems such as those of Nienhuis (2003) on the Dutch estuaries – studies that used to be associated with the earliest phase of environmental science (see above).

Evolution in social-scientific management concepts

A central distinction within the social sciences is between the 'micro' and 'macro' levels of analysis, denoting, respectively, the level of individual actors and the level of the large social systems ('societies') where actors are anonymous. Within economics, for instance, micro-economics focuses on the rational behaviour of individual people, households and firms and stands besides macro-economics that studies the overall behaviour of large-scale economic systems in terms of aggregate indicators.

Implicitly in this scheme, nothing *between* these two levels is conceptualized as worthy of analysis. A well-known example of this pattern of reasoning and the recommendations it gives rise to is Hardin's (1968) 'Tragedy of the Commons', in which he explained that individual actors (micro level) are always bound to over-exploit and destroy their common good (macro level) even if this is to the detriment of each actor. Obviously then, the recommendation can only be that all common goods, including environmental goods such as forests and water bodies, should either be cut up into privately owned portions or be fully state-owned, so that private actors can be coerced to behave.

Many social scientists became increasingly uncomfortable with this situation. It was found, for instance, that many common goods such as grazing lands, forests and fisheries had been managed sustainably over the ages neither by private actors nor by the state, but by voluntary associations of actors that organised the management of the commons among each other, tuned to the local ecology and culture. If many of such common properties did disappear in the course of history, it was often by usurpation by the state or external private powers rather than by internal incapacity of the communities (Cox, 1985). Observations such as these supported a new interest in community-level institutions, as found, for instance, in the seminal work of Putnam (1993) on social capital and of Ostrom (1990) on environmental common properties. In economics, institutional economists and game theorists try to fill the void between the micro and macro levels.

The new social science supported the ongoing drive towards community-based rather than state-based management of nature and natural resources (Ghimire & Pimbert, 1997). Much of this drive is fuelled by the practical limitations and the moral *défit* of state-based nature conservation in developing countries (e.g. Colchester, 1997) but even in these countries, community-based management is often confronted with strong limitations, too. Communities may be quite capable of small-scale environmental monitoring (e.g. Hunsberger et al., 2005) but may find themselves at a loss when large-scale monitoring of resources is at stake. Fishery communities, for instance, may experience great difficulty in organizing the monitoring of large-scale fish stocks, and a supra-local actor such as the state may then step in to fulfil this function. Other limitations of communities are more fundamental, and especially concern local *motivations* rather than capacities. It is quite difficult for a local community, for instance, to see, let alone be motivated by, the national or global rareness or functions of say, the forest, the species or the floods that are so abundant locally. To a varying but significant degree, therefore, the state, speaking on behalf of supra-local scales and functions, is often quite justified to remain involved in local natural resource management. *Co-management* is the term most used for this pattern of mixed responsibilities and mutual support of communities and supra-local authorities (e.g. Borrini-Feyerabend et al., 2004). Other terms are joint management, collaborative management or cross-scale institutional linkages (Berkes, 2002).

Co-management has a long-standing tradition in fisheries, a sector where strong government motivations tend to clash with strong local communities, so that co-management arrangements arise almost naturally (Wilson et al., 2003). Co-management concepts are now expanding over terrestrial sectors too, e.g. as 'Joint Forest Management' in India. In fact, principles of co-management are applicable in all situations where evident local interests interweave with evident supra-local rationalities, even in centralistic countries such as the Netherlands. One Dutch example, discussed by Brussaard (2004), though without mentioning the term co-management, is a case in the province of Friesland where the government exempted a local farmers group of a generic nitrate control measure, under the condition that the farmers would find their own creative solutions for the nitrate problem.

Rivers, obviously, also belong to the category of candidates for co-management inspirations. Local identities and economic interests are often river-bound but at the same time, rivers are typically large-scale systems that call for a large-scale (state or state-like) management authority. Some river problems such as large-scale pollution may lie beyond the scope of community involvement, but for many other policy issues (e.g. room-for-rivers), river management can develop into directions that seek to include the voice of local communities from the very beginning, in a style of communication and negotiation that is much more fitting, practically and ethically, than traditional 'participation' of communities in essentially state-driven planning.

Could this have implications for (applied) hydrobiology? Could there exist a distinct style and/ or content of applied hydrobiological studies geared towards the co-management of rivers and other water bodies? Brussaard (2004) is optimistic in this respect based on his (terrestrial) case study, when he talks about the role of ecology in the design of locally viable management and monitoring options, and the fact that scientists can learn from this work, but he also mentions the problem that governments have funds for research, while communities do not. Institutions and funding for 'joint science', in our view, are prerequisite for developing an ecology and hydrobiology to underpin co-management of land and water in Western societies.

Confluence or contradiction?

Sciences are connected to societal sectors and interests. To all likelihood, therefore, our focus on ecology, environmental science and social science has ensured that the identified principles are conducive to express the interests of nature, sustainability and communities, respectively. Other disciplines will generate other principles, however, and the fact that we did not focus on, say, economics, logistics, risk analysis or conflict sociology implies that the principles we found may not be equally conducive to express the needs of cost-effectiveness, river-based transport, flood risk management or conflict mediation, e.g. between upstream and downstream interests. Obviously then, the principles discussed in this paper are not the only ones needed for balanced river management. We will not further explore this limitation here, however, and rather focus this concluding section on the degree to which the principles we have identified may contradict or reinforce each other. We will ignore the too obvious cases, e.g. that adaptive management goes well together with opportunity-driven environmental science.

Starting out with the idea of adaptive management that is supposed to go together with a long-term vision, we have seen in the section on ecological concepts that these two do not go together automatically. River authorities can have a strong long-term vision and at the same time be relatively un-adaptive in their mode of operation. On the other hand, nothing seems to stand in the way of change in this respect. We might even say that the better you know where you want to arrive in the long run, the easier it is to make flexible choices along the way.

The principle of ecosystem-based management may appear to contradict the principle of co-management. After all, the idea of co-management is only an idea about how human agents could deal with each other, without the ecosystem being in view. This, however, is nothing specific for co-management, since any type of management is a human affair. In fact, ecosystem-based planning and management may well reinforce the characteristic 'co-' in co-management. In ecosystem-based co-management, the state and the local communities, in stead of addressing each other head on, first listen to the ecosystem as a 'third party' and analyse what the ecosystem means to them. This gives parties a common footing before entering into their more adversary roles, and is comparable to the role of research into the groundwater system in Ostrom's (1990) case study of building institutions for groundwater management in the USA.

The vision of managers in the principle of adaptive-management-with-long-term-vision is a quite different thing from the 'visions of nature' of the public. The content of these visions might therefore well contradict each other, as would be the case, for instance, if the river managers would entertain a long-term vision of an ever more fully regulated river while the public would cherish partnership with nature and participation in a more natural river landscape. It so happens that at least in the Netherlands, this contradiction does not present itself at all, however. Re-naturalization is a shared vision of river managers and society. This does not mean, at the same time, that concrete proposals of river re-naturalization will always be greeted with happiness by the concerned community. People's responses are always highly contextual, and people's ideas on nature vary a lot when going from the 'high visions' down to daily life necessities (De Groot & Van den Born, 2003).

A strong long-term vision of the river managers could easily run counter to the open exchange of views with local communities that is implied in co-management. True motivation for co-management may in fact arise only if the river managers' vision would include not only the re-naturalization of river itself but also the co-managing and adaptive style of getting there. Reaching such a more inclusive vision, to a depth equal to the present-day vision that focuses on the river only, requires a rethinking of present-day paradigms and restyling of the way that river managers feel and communicate. But this is certainly not a mission impossible because the rethinking and restyling does not contradict the most basic motivations of river managers. Also in settings of adaptive co-management, river managers can continue to be the guardians of technical, science-based and supra-local rationality, and they can continue the pursuit of re-naturalization of the great rivers – only, possibly, more successfully.

References

Berkes, F., 2002. Cross-scale institutional linkages: perspectives from the bottom up. In Ostrom, E., T. Dietz, N. Dolšak, P. C. Stern, S. Stonich & E. U. Weber (eds), The Drama of the Commons. National Academy Press, Washington, 293–321.

Bilderdijk, W., 1832. Geschiedenis des Vaderlands I, Amsterdam (in Dutch).

Borrini-Feyerabend, G., M. Pimbert, M. T. Farvar, A. Kothari & Y. Renard, 2004. Sharing Power. Learning by doing in co-management of natural resources throughout the world. IIED and IUCN/ CEESP/ CMWG, Cenesta, Tehran. (http://www.iucn.org/themes/ceesp/Publications/sharingpower.htm#download).

Brussaard, L., 2004. De liefde tot het land. In van der Zandt, A.N., Tussen Onland en Verwonderland. Nijmegen University Press, Nijmegen: 49–61 (in Dutch).

Cheney, J., 1989. Postmodern environmental ethics: ethics as bioregional narrative. Environmental Ethics 11: 117–134.

Colchester, M., 1997. Salvaging nature: indigenous peoples and protected areas. In Ghimire, K. B. & M. P. Pimbert (eds), Social Change and Conservation. Earthscan, London: 97–130.

Cox, S. J., 1985. No tragedy in the commons. Environmental Ethics 7: 49–61.

De Groot, W. T., 1998. Problem-in-Context: a framework for the analysis, explanation and solution of environmental problems. In Nath, B., L. Hens, P. Compton & D. Devuyst (eds), Environmental Management in Practice, Volume I. Routledge Publishers, London: 22–43.

De Groot, W. T., 2004. Eens worden de rivieren onze contramal. Landschap 2: 5 (in Dutch).

De Groot, W. T. & R. J. G. van den Born, 2003. Visions of nature and landscape type preferences. Landscape and Urban Planning 63: 127–138.

Ghimire, K. B. & M. P. Pimbert (ed.), 1997. Social Change and Conservation. Earthscan, London.

Hardin, G., 1968. The tragedy of the commons. Science 162: 1243–1248.

Holling, C. S. & L. H. Gunderson, 2002. Resilience and Adaptive Cycles. In Gunderson, L. H. & C. S. Holling (eds), Panarchy. Island Press, Washington: 25–62.

Hunsberger, C. A., R. B. Gibson & S. K. Wismer, 2005. Citizen involvement in sustainability-centred environmental assessment follow-up. Environmental Impact Assessment Review 25: 609–627.

Imperial, M. T., 1993. The evolution of adaptive management for estuarine ecosystems. Ocean and Coastal Management 20: 147–180.

Imperial, M. T., 1999. Institutional analysis and ecosystem-based management. Environmental Management 24: 449–465.

Jeurgens, J., 1991. De Haarlemmermeer; een studie in planning en beleid 1836–1858. PhD thesis, Leiden University, NEHA, Amsterdam (in Dutch).

Kessler, J. -J., 2003. Working Towards SEAN-ERA. Tropical Resource Management Papers 43. Wageningen University, Wageningen.

Lenders, H. J. R., 2003. Environmental rehabilitation of the river landscape in the Netherlands. A blend of five dimensions. PhD thesis, Radboud University, Nijmegen.

Lenders, H. J. R., B. G. W. Aarts, H. Strijbosch & G. van der Velde, 1998. The role of reference and target images in ecological recovery of river systems: lines of thought in the

Netherlands. In Nienhuis, P. H., R. S. E. W. Leuven & A. M. J. Ragas (eds) New Concepts for Sustainable Management of River Basins. Backhuys Publishers, Leiden: 35–52.

Leuven, R. S. E. W., A. J. M. Smits & P. H. Nienhuis, 2000. From integrated approaches to sustainable river basin management. In Nienhuis, P. H., R. S. E. W. Leuven & A. M. J. Ragas (eds) New Concepts for Sustainable Management of River Basins. Backhuys Publishers, Leiden: 329–347.

Mitchell, B., 2002. Resource and Environmental Management. Pearson Education, New York.

Morrison, K., 2003. Stakeholder involvement in water management: necessity or luxury? Water Science and Technology 47: 43–51.

Nienhuis, P. H., 2003. Het rammelend kunstgebit in de Oosterschelde. Bionieuws 13(19): 7.

Nienhuis, P. H. & R. S. E. W. Leuven, 1998. Ecological concepts for the sustainable management of lowland river basins: a review. In Nienhuis, P. H., R. S. E. W. Leuven & A. M. J. Ragas (eds), New Concepts for Sustainable Management of River Basins. Backhuys Publishers, Leiden: 7–33.

Nienhuis, P. H., A. D. Buijse, R. S. E. W. Leuven, A. J. M. Smits, R. J. W. de Nooij & E. M. Samborska, 2002. Ecological rehabilitation of the lowland basin of the river Rhine (NW Europe). Hydrobiologia 478: 53–72.

Ostrom, E., 1990. Governing the Commons . Cambridge University Press, New York.

Posthuma, L., G. W. Suter II & T. P. Traas, 2002. Species Sensitivity Distributions in Ecotoxicology. Lewis Publishers, New York.

Putnam, R. D., 1993. Making Democracy Work: Civic Traditions in Modern Italy. Princeton University Press, Princeton.

Scheffer, M., S. Carpenter, J. A. Foley, C. Folke & B. Walker, 2001. Catastrophic shifts in ecosystems. Nature 413: 591–596.

Scholte, P., 2005. Floodplain rehabilitation and the future of conservation and development: towards adaptive management of success in Waza-Logone, Cameroon. PhD thesis, Leiden University, Leiden.

Van den Born, R. J. G., H. J. R. Lenders, W. T. de Groot & E. Huijsman, 2001. The new biophilia: an exploration of visions of nature in Western countries. Environmental Conservation 23: 65–75.

Van Stokkom, H. T. C., A. J. M. Smits & R. S. E. W. Leuven, 2005. Flood defense in the Netherlands: A new era, a new approach. Water International 30: 76–87.

Vugteveen, P., R. S. E. W. Leuven, M. A. J. Huijbregts & H. J. R. Lenders, 2006. Redefinition and elaboration of river ecosystem health: perspective for river management. Hydrobiologia 565: 289–308.

Wilson, D. C., R. J. Nielsen & P. Degnbol, 2003. The Fisheries Co-management Experience. Kluwer Academic Publishers, Dordrecht.

Wiering, M. A. & B. J. M. Arts, 2006. Discursive shifts in Dutch river management: 'deep' institutional change or adaptation strategy? Hydrobiologia 565: 327–338.

Witter, J. V., H. T. C. van Stokkom & G. Hendriksen, 2006. From river management to river basin management: a water manager's perspective. Hydrobiologia 565: 317–325.

Zelinka, M. & P. Marvan, 1961. Zur Präzisierung der biologischen Klassification der Reinheit fliessender Gewässer. Archiv für Hydrobiologie 57: 389–407.

Hydrobiologia (2006) 565:317–325 © Springer 2006
R.S.E.W. Leuven, A.M.J. Ragas, A.J.M. Smits & G. van der Velde (eds), Living Rivers: Trends and Challenges in Science and Management
DOI 10.1007/s10750-005-1922-6

Opinion Paper

From river management to river basin management: a water manager's perspective

J.V. Witter[1,*], H.T.C. van Stokkom[1] & G. Hendriksen[2]
[1]*Brabantse Delta Water Board, Bergschot 69-1, 4817 PA, Breda The Netherlands*
[2]*Ministry of Transport, Public Works and Water Management, P. O. Box 9070, 6800 AD Arnhem, The Netherlands*
(**Author for correspondence: E-mail: j.witter@brabantsedelta.nl*)

Key words: network authority, sustainable water systems, stakeholder analysis, environmental policy, joint planning

Abstract

Because of climate change and the need for sustainable water systems, water management has changed considerably in recent years, from river management to basin management. This change is illustrated for both the main Dutch river system and a small, representative regional water system in the Netherlands. This change results in four major challenges, (1) creating an effective regional administration which functions as a network authority, (2) increasing collaboration with other water managers, (3) increasing efficiency and achieving goals at minimum cost, and (4) communicating in an open manner with stakeholders. The article outlines the way in which water managers try to address these challenges. Special attention is given to the need for increased coordination between environmental policies and water policies.

Introduction

Water management in the Netherlands is the responsibility of both water boards and the state. Water boards manage the smaller, regional water systems, while the large rivers are managed by the national government.

Water management has changed profoundly in recent years, partly because the side-effects of traditional water management in the Netherlands have become apparent. In many cases, the ecological and hydrological resilience of water systems has been severely reduced or has disappeared entirely. This loss of resilience is even more alarming because climatic change and the resulting sea level rise are imposing even greater demands on the water systems. Therefore, emphasis in water management has been shifted towards sustainability, which has necessarily led to a change from river management to river basin management (Leuven et al., 2000). River basin management, however, is not a task that the government and the water boards can perform on their own. They need the co-operation of other stakeholders such as other governmental organisations, citizens, land users and the business community.

The change to river basin management is accompanied by two other developments. Firstly, citizens have become increasingly well-informed and are aware of threats and opportunities, pursuing individual interests and critically evaluating government actions. This has reinforced the above-mentioned need for participation by citizens, land-users, etc. The second development was the merging of water boards. This process has been going on for a considerable time now and has resulted in the number of water boards decreasing from 2500 in 1946 to 129 in 1990 (Huisman et al., 1998) and to a mere 25 in 2005. These two developments have interacted in a complex way with the above changes in water management.

This paper illustrates the change to river basin management for a small regional river basin and

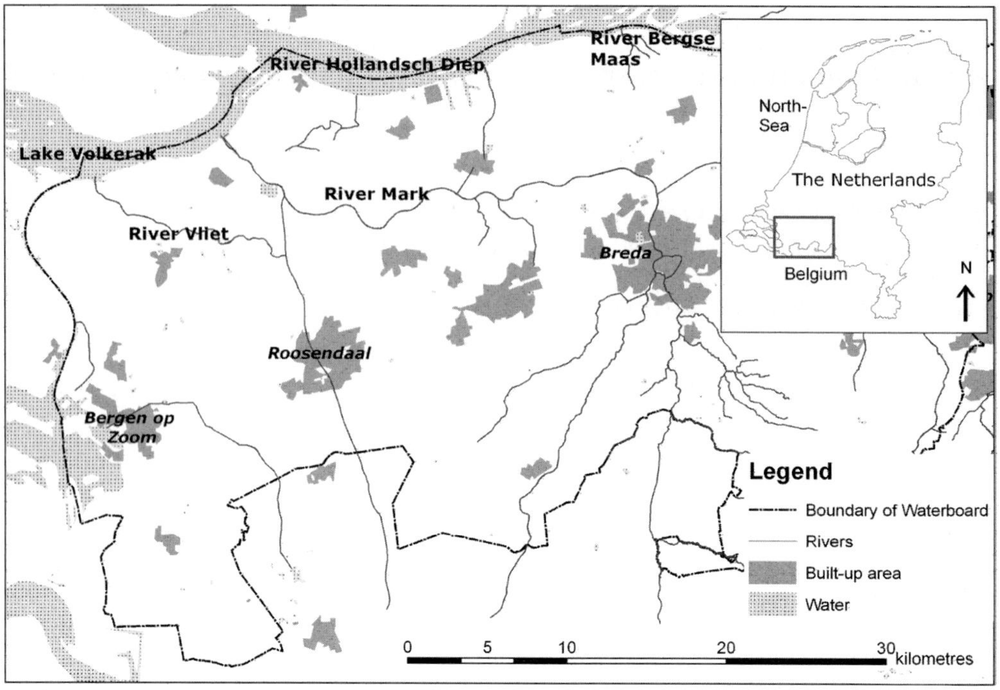

Figure 1. Map of the Mark basin.

for the main Dutch river system, presented from a water manager's perspective. Starting from an analysis of stakeholders of water management, it also presents the four main challenges to water management, and outlines the way in which water managers try to address these challenges.

Changes in water management

Regional water systems: the case of the Mark river basin

The basin of the river Mark (Fig. 1) is situated in the south of the Netherlands, near the Belgian border. It is a transboundary basin, with 440 km^2 situated in Belgium and 970 km^2 situated in the Netherlands. The basin consists of two parts. The southern part is a gently sloping, sandy, Pleistocene area of about 1000 km^2. At the Dutch/Belgian border, the area is situated about 20–25 m above sea level, while the towns of Breda, Roosendaal and Bergen op Zoom (Fig. 1) are situated more or less at sea level. The northern, Holocene, downstream part of the basin (about 400 km^2) consists of clay polders situated at or slightly below (−1 m) sea level. The river drains into Lake Volkerak (Fig. 1), which is part of the Meuse system.

With an area of about 1410 km^2, the Mark basin is a small basin. Water discharges can vary strongly: within 24 h, water levels can rise from normal to critical levels, in particular in the city of Breda, which is located at the transition from the sloping Pleistocene part to the flat Holocene part. Recent years in particular have seen frequent discharge peaks and annual water level maxima have shown a distinct upward trend over the last 15 years.

As early as 1800, there were recurrent complaints about inundations upstream of Breda. The navigability had also deteriorated: the construction of dikes had considerably reduced the tidal volume of the Mark, which caused the rapid silting up of the river. Due to these problems, the 'Hoogheemraadschap van de Mark en Dintel' water board was founded in 1804. The water board took care of the annual dredging of the Mark and started the construction of sluices at the outlets of the rivers Mark and Dintel.

Since 1820, a long series of measures have been taken that have affected the drainage conditions of the basin (Anonymous, 1936). The stream flow area of the Mark increased more than threefold as a result of the river being deepened and widened. The river was shortened by 7.5% and its hydraulic resistance decreased because a number of sharp bends were removed. In other words, the river became far more efficient in terms of runoff capacity.

The driving forces behind these changes were the need for more space and better drainage conditions, because of economic growth, population growth and intensification of land use – in particular the modernization of agriculture after World War II. The side-effects on the river system were (Witter & Raats, 2001):

- an increase in peak discharges by about 40%;
- a shift in the water balance resulting in more runoff and less storage;
- a decrease in mean surface water and groundwater levels in the upstream part of the basin by about 20 cm;
- a decrease in riverbank storage by about 20 km^2 as floodplains have been converted into polders.

The improvements in the water discharge capacity of the river Mark led to a significant decrease in maximum water levels (Witter, 2003), despite an upward trend in the 10-day rainfall preceding these maxima. Recent research has shown that this upward trend in rainfall is mainly caused by an increase in the number of days with large amounts of rainfall (Bruin, 2002).

Large transnational rivers

Located in the delta formed by the Rhine and Meuse rivers, the Netherlands has a long history of adapting its natural water and river systems to user functions such as housing, agriculture and shipping. Large parts of the Netherlands are still subsiding, while sea levels are rising. Approximately 25% of the country is situated below mean sea level. Without the dikes and dunes along the coast, 65% of the most densely populated parts of the Netherlands would be flooded every day (Huisman et al., 1998).

Since 1798, the overall responsibility and coordination of water management in the Netherlands has been a task of the national government (Van de Ven, 2004). In the period 1800–1900, most flooding events in the large rivers Rhine and Meuse were caused by ice dams. As for open water, there was, in the words of the national River Commission in 1821 'only rarely any danger to be feared'. Van der Ham (2004) concluded in his overview of river management over the past three centuries that measures to improve the river bed, such as normalisation, improvements to the river mouth, straightening of bends, the construction of groins and the conversion of floodplains and river branches into polders, together with the improvement of dikes, have proved very effective in the past. These measures can be summarized as a combination of 'cleaning up' and 'constructing impediments', and have led to a significant decrease in inundation frequencies. During the past century, hardly any significant flooding events have occurred along the Dutch part of the Rhine.

However, the large-scale reclamation of wetlands and the regulation and harnessing of rivulets have reduced the hydromorphological resilience of the large rivers. Water and sediment discharge patterns have been affected in such a way that periods of high or low precipitation rates are now immediately reflected by high or low water levels in the river. The large-scale draining of agricultural land and the expansion of urbanized areas, consisting almost completely of impervious materials, have caused rapid run-off of rainwater and high water discharge peaks in the rivers. River regulation schemes and the embanking of floodplains have augmented this problem (Dister et al., 1990).

The need for a new flood defence policy

This existing flood defence policy of continually increasing the water discharge capacity of large and regional rivers has reached a dead end. In its Fourth Memorandum on Water Management, the Dutch government stated that new, sustainable measures are preferred to meet the desired level of flood protection (Anonymous, 1999). These measures were inspired by the near floods of 1993 and 1995. In 1995, some 250,000 people were evacuated from their homes for some days due to the questionable stability of dikes that had been exposed to protracted flooding and had become saturated with water. The estimated damage to agriculture, industrial activities and private enterprises amounted to more than € 1 billion. The government immediately decided to initiate the so-called Large Rivers Delta Plan, which stipulated that all river dikes had to be adapted to meet current design water levels. In 2001, the design water levels were reviewed because of the ongoing climate change. This resulted in increased design water levels, and the policy that was announced in the Fourth Memorandum on Water Management was aimed at implementing measures that would prevent a new round of raising and reinforcing dikes, despite the increased design water levels. The policy of 'impediment' had reached a dead end and had to make way for a policy of river restoration, in which expanding the floodplain of a river by moving dikes further inland is preferred to raising the dikes (Nienhuis et al., 2002; Buise et al., 2005; Van Stokkom et al., 2005). In view of the effects of climate change – increased rainfall and rising sea levels – merely raising the dikes is pointless in the long run. Moreover, periodically reinforced and raised dikes may give local authorities a false sense of security (Van Stokkom et al., 2005).

One should also realise that social and economic development is a continuous process. The economic investments and concurrent developments in the Netherlands imply that the effects of flooding and, consequently, the risks increase by a factor of 10–100 every hundred years. As a consequence, flooding risks should be examined periodically, taking into account the potential loss of human lives, damage, social and economic disruption and economic value within the potentially affected regions. At present, such an evaluation is being undertaken in the Netherlands (RIVM, 2004).

The new flood defence policy provides an answer to both flood defence problems and ecological and water quality problems. Flood defence is increasingly formulated in an international context, agreements being concluded for the large transnational rivers as regards design flood discharges on either side of the borders. At the same time, water quality policy is also increasingly being formulated within a European context. The European Water Framework Directive constitutes an important step towards integrated river basin management. In 2009, the first transboundary river catchment plans will come into force. The most prominent effect will be much stricter water quality standards and rehabilitation of hydromorphological characteristics of rivers (Buise et al., 2005).

Stakeholder analysis and challenges to water management

The water boards as well as the regional departments of the Ministry of Transport, Public Works and Water Management are responsible for flood protection and the management of water quantity and quality within their territory. The Ministry of Transport, Public Works and Water Management sets goals for flood protection and water management, supervises the implementation of water policy by the water boards and provincial authorities and is also responsible for the management of state-owned water bodies and water infrastructure. The Ministry of Housing, Spatial Planning and the Environment is responsible for spatial planning, water quality and emission standards, soil and groundwater protection and guidelines for drinking water and sewerage. The 12 provincial authorities in the Netherlands are responsible for defining the regional water management policies that the water boards are supposed to implement. They also define and supervise the responsibilities and activities related to flood protection, with the exception of state-managed waters and infrastructure. Municipal authorities are responsible for spatial planning and thus act as partners to the water boards by facilitating changes in land-use and the identification of areas for the storage of excess water. Since they are also responsible for the collection of sewage water, they are also partners to the water boards with respect to the purification of sewage water.

As a consequence of the complex administrative structure, involving an international context and several national governmental levels, decision-making in water management is becoming ever more complex. Water managers no longer dominate the decision-making process. Negotiated agreement, interactive planning and the involvement of the public opinion and various stakeholders are promising ways to reconcile conflicting interests. This changing perception of water management is discussed elsewhere in this issue by Wiering & Arts (2006). Water management should be seen as a process and not as a series of problems to be solved (Morrison, 2003). Water managers are therefore increasingly involved in stakeholder analysis (Hunsberger et al., 2005).

The Ministry of Transport, Public Works and Water Management distinguishes three different domains ('own operations', 'shared operations', 'operations by third parties') for its three main activities: water quality, water quantity and shipping management. For each of these activities, a relational network is built and increasing effort is being invested in maintaining these relations. With respect to stakeholders, the main focus is on public interests like the safety and sustainability of water systems, which are seen as interests of the majority of stakeholders.

Water boards are also actively involved in stakeholder analysis. Table 1 shows the stakeholder analysis for the Brabantse Delta water board. This table clearly reflects the complexity of the external context of water management.

As water managers need to broaden their scope to include basin management and as they no longer dominate the decision-making process, they need to invest in these relations. This leads to the first and second challenges for water management:

1. Water management must become a regional administration, which is effective, credible, strongly rooted within the region and functioning like a 'network authority'.
2. Since the physical water system itself is also a network and cannot be transformed into a sustainable water system in isolation, increased collaboration among water managers is needed.

These challenges must be addressed in a world that has become a network world, in a context of critical citizens and customers for the services supplied by water management, and within the context of a critical national and European administration. This leads to the third challenge:

3. Water management must become an efficient regional authority that meets its goals and commitments at minimum cost.

Modern society is an open society, with critical and well-informed citizens, pursuing individual interests and critically evaluating government actions – the same being true for other stakeholders like businesses, farmers and the interests of nature – which leads to the fourth challenge:

4. Water management must become a modern and professional regional authority that communicates in an open manner with its stakeholders and makes choices in order to meet conflicting interests.

Challenges to water management and how to cope with them

Network authority

Water managers are currently very much seeking collaboration with partners, not only while implementing projects, but particularly in the earlier goal-setting phase. The trend is shifting from making their own water management plans (sometimes formulated after some form of consultation with other parties) towards joint planning. To give some examples:

- In recent years, there has been a number of outbreaks of animal diseases in the Netherlands, affecting meat production on farms. In solving these veterinarian problems, an approach was adopted aiming at a general rethinking of land use in the rural environment. This resulted in bottom-up planning processes being set up regionally. The Mark basin covers three of these regions. In these bottom-up planning processes, the local and regional authorities participate, as well as the various stakeholders. The resulting views on a sustainable rural environment include the selection of locations for joint projects for water storage, ecological restoration, water quality improvement, etc.
- In 2005 and 2006, water boards and municipal authorities will formulate joint plans to solve nuisance by inundations, high water tables and other water-related problems in built-up areas. The projects to solve these problems should be completed by 2007.
- In 2009, the water management plans for the basins of the Meuse, Rhine and Scheldt will come into force. These plans will be formulated within an international context and will be translated into more detailed plans for each of the sub-basins.

Table 1. Stakeholder analysis for the Brabantse Delta water board

Stakeholders	Principal interests	Ways they exert their influence
European Union, international River Commissions	Standards for water quality	Regulations, fines
National Government	Standards for water quality and quantity	Regulations, negotiation, permits and fines
Regional departments of the Ministry of Transport, Public Works and Water Management	Fellow water managers; joint interests but also possibly conflicting interests at 'water transfer points'	Regulations, permits and fines
Provincial authorities	Setting goals for water management, given European and national standards	Regulations, negotiation, permits and fines
Municipal authorities	Partners in sewage handling Partners in river restoration projects Responsible for spatial planning, which means that their consent is essential in finding locations for temporary water storage	Regulations, negotiation, permits and fines
Other water boards	Fellow water managers; joint interests but also possibly conflicting interests at 'water transfer points'	Regulations, negotiation
Citizens	Individual, not necessarily collective interests, Low local taxes	Complaints Campaigns, public debate Tax collection problems
Industry/commercial firms	Low local taxes, Flexible and timely solutions	Complaints Campaigns, public debate Tax collection problems Competition by private purification plants
Farmers	Low local taxes, Flexible and timely solutions	Complaints Campaigns, public debate Tax collection problems
Nature conservation interests/environmental groups	Restoration of nature and environmental rehabilitation	Complaints Campaigns, public debate Tax collection problems
Research institutes	Innovation with respect to solutions (costs, sustainability, flexibility) Innovation aimed at stakeholder participation Education of future water professionals	Joint projects Joint projects Young professional programs
Mass media	Looking for stories, either favourable or unfavourable	Campaigns, public debate

As the water boards and the regional departments of the Ministry of Transport, Public Works and Water Management operate regionally, they are better equipped to understand the regional and local needs. They also have the technical skills to design, implement and maintain a sustainable water management strategy. The present annual expenditures on water management by the national government and the water boards are approximately €1 billion and €2 billion, respectively. The greater part of this money is being invested in measures like retention areas, riverbank restoration, water quality improvement and restoration of water-related habitats, the consequences of which are felt locally. These measures need to fit in with the spatial planning schemes designed by municipal and provincial authorities, as well as with national and provincial environmental policies. The water boards, which have their own boards of governors, are particularly suited to form the local coalitions needed to implement such complex projects, with often conflicting interests and many partners. The government benefits from the regional roots of the water boards by increasingly collaborating with them in the implementation of its flood management projects for the large rivers. This is resulting in an organisational structure involving many water boards, which have front offices at the local level and pay much attention to the relations with the regional municipal authorities. Our operating processes emphasise communication, consultation and direct participation by stakeholders in projects.

Nevertheless, more changes will be needed for a truly effective and credible system. There is an almost continuous public debate on the question whether water management should be carried out by some kind of dedicated department (yet to be created) of either the national or provincial authorities. Given the complexity of the network involved in water management, an effective organisation has to be able to set its own priorities, to bear direct responsibility for its actions and to have its own financial resources. It would have to be able to generate a substantial part of the funds needed for water management. This would help to maintain a clear direction, even in times when its views on water management are not very welcome, for instance in times of economic stagnation or when the consequences of water management are unevenly distributed over different groups of stakeholders. As operation and maintenance have low and usually inferior profiles compared to the construction of new works, there is a constant need to 'put water on the agenda'.

This is why water boards cherish their position as independent institutions for local governance, which enables them to collect their own taxes and set their own priorities. They therefore need to invest in building public confidence in this institution. This is being done in a variety of ways: by scaling up the organisation and emphasising professionalism, by investing much effort in public participation in the elections for its board of governors and by formulating clear views on what has to be achieved during each board's term of office.

Increased collaboration between water managers

The joint plans referred to above lead to a multitude of joint projects by the water boards and the Ministry of Transport, Public Works and Water Management. For instance, the new flood defence policy has resulted in plans to move dikes further inland and in plans for 'green rivers', that is, rivers which are only used in conditions of imminent flooding (Van Stokkom et al., 2005; Anonymous, 2005). Both options involve finding solutions for the main river system – which is operated by the state – within the regional water systems – which are operated by water boards.

Other examples of joint planning by the water boards and the Ministry of Transport, Public Works and Water Management include a national drought study and a study on the ecological restoration of the delta of the rivers Rhine, Meuse and Scheldt. The national drought study is a joint project of the Ministry of Transport, Public Works and Water Management, provincial authorities and water boards to assess the effects of climate change on water availability. It is envisaged that future summers will be drier and that summer discharges of the Rhine will fall as the glaciers in the Swiss Alps become smaller. The study aims to identify future water shortages and possible solutions. The study on the ecological restoration of the delta of the rivers Rhine, Meuse and Scheldt is being undertaken in response to water quality problems caused by the compartmentalisation of the delta area (Smits et al., 2006). This compartmentalisation is due to the building of dams in the so-called Delta Project, which have changed the former estuarine delta into an artificial, stagnant freshwater system, rich in nutrients. This has led to a number of problems, including algal blooms and other water quality problems (De Jonge & De Jong, 2002; Smits et al., 2006). Possible answers to these problems include restoration of the former estuarine conditions, periodically flushing the compartments with excess fresh water in order to wash away the algae, and/or a drastic cleaning up of pollution sources. All these solutions have their pros and cons and involve the regional water systems within the delta region. A joint study involving the national and provincial authorities and the relevant water boards has been started, to identify the best solution.

Collaboration between water managers is also being fostered by legal and administrative instruments such as water agreements and what are called 'blue nodal points'. The latter are locations within a waterway where water is transferred from one basin to the other or – in an administrative sense – from one water board to another. The general idea is that it is not allowed to pass on problems from one basin or one administration to another. To this end, rules and instruments are being developed. Water agreements include a set of agreements between two water managing administrations to ensure that problems are not passed on. Examples of such agreements are those about maximum water levels, maximum discharges, ways to communicate and solve disputes, etc. Agreements on water quality have so far not been concluded in this context, but the trend is towards incorporating this aspect as well.

Goals and commitments achieved at minimum cost

A recent goal of water managers is process improvement to achieve targets at minimum cost. The changing climate and European water legislation are imposing major demands on the water system. Meeting these demands requires large investments and optimised use of scarce resources. Water managers are trying to cope with this in a number of ways. Firstly, it is important for them to prioritise scarce financial and staffing resources. The 'Brabantse Delta' water board has recently started to prioritise its expenditures on the basis of four criteria:

- achievement of goals;
- return (in terms of goal achievement) on investments;
- degree of co-operation by third parties;
- risks (participation, financial, etc.).

Secondly, there is a constant need to improve operating processes to achieve maximum efficiency. The Ministry of Transport, Public Works and Water Management and the water boards are both actively involved in programmes to improve their operating processes.

Communicating in an open manner with stakeholders and make choices to reconcile conflicting interests

This is still very much a learning process. It is, however, evident that water boards are definitely prepared to make this change. This is evidenced in the first place by the various joint planning activities. In addition, however, the style of communication is also changing, one new phenomenon being the introduction of policy statements by the new board of governors of the 'Brabantse Delta' water board in which they explicitly state the goals they want to achieve within their term of office. Another example is the emphasis water boards and the national government are putting on communication with their partners and stakeholders through the Internet. This involves paying attention to the three relations an administrative body has with its clients:

- developing a co-operative relation, trying to achieve goals through collaboration rather than through unilateral action;
- optimising a service-providing relation, offering services like safety, water level control and purification of sewage water;
- reshaping an administrative relation which is under pressure, in the sense that the water managers reduce the administrative burden on citizens and modernize the relationship with them.

Closing remarks

Water management, both by the national government and by water boards, is changing rapidly, and is beginning to be implemented by professional administrative bodies that communicate in an open manner with stakeholders and work closely together with colleagues. This also implies that water management is more than simply 'minding the shop'; it needs to look ahead. In order to do this effectively, water management needs to establish a business agenda in order to actively pursue the solution of problems featuring on this agenda. Further, as water quality is increasingly becoming a top priority, partly because of European regulations, environmental and water policies need far stronger co-ordination. Just as the need for water storage locations induced by climate change and the need for sustainable water systems have in recent years resulted in an increased co-ordination between spatial planning and water management, the coming years will witness a similar process with respect to environmental policy.

References

Anonymous, 1936. Verslag inzake onderzoek verbetering afwatering Mark en Dintel en verbinding Mark-Vliet. Province of Noord-Brabant, Den Bosch (in Dutch).

Anonymous, 1999. Fourth national policy document on water management. Dutch Ministry of Transport, Public Works and Water Management, The Hague.

Anonymous, 2005. Ontwerp planologische kernbeslissing Ruimte voor de Rivier. Dutch Ministry of Transport, Public Works and Water Management and Ministry of Housing, Spatial Planning and Environment, The Hague (in Dutch).

Bruin, A. T. H., 2002. Verandering in neerslagkarakteristieken in Nederland gedurende de periode 1901–2001. Royal Dutch Meteorological Institute, De Bilt (in Dutch).

Buijse, A., F. Klijn, R. S. E. W. Leuven, H. Middelkoop, F. Schiemer, J. Thorp & H. Wolfert, 2005. Rehabilitation of large rivers: references, achievements and integration into river management. Archiv für Hydrobiologie Supplement 155 (1–4) Large Rivers 15 (1–4): 715–738.

De Jonge, V. N. & D. J. de Jong, 2002. Ecological restoration in coastal areas in the Netherlands, concepts, dilemmas and some examples. Hydrobiologia 478: 7–28.

Dister, E., P. Obrdlik & E. Schneider, 1990. Water management and ecological perspectives of the Upper Rhine's floodplains. Regulated Rivers 5: 1–15.

Huisman, P., W. Cramer, G. van Ee, J. C. Hooghart, H. Salz & F. C. Zuidema, 1998. Water in the Netherlands. Netherlands Hydrological Society, Delft.

Hunsberger, C. A., R. B. Gibson & S. K. Wismer, 2005. Citizen involvement in sustainability-centred environmental assessment follow-up. Environmental Impact Assessment Review 25: 609–627.

Leuven, R. S. E. W., A. J. M. Smits & P. H. Nienhuis, 2000. From integrated approaches to sustainable river basin management. In Smits, A. J. M., P. H. Nienhuis & R. S. E. W. Leuven (eds), New Approaches to River Management. Backhuys Publishers, Leiden: 329–347.

Morrison, K., 2003. Stakeholder involvement in water management: necessity or luxury? Water Science and Technology 47: 43–51.

Nienhuis, P. H., A. D. Buise, R. S. E. W. Leuven, A. J. M. Smits, R. J. W. de Nooij & E. M. Samborska, 2002. Ecological rehabilitation of the lowland basin of the river Rhine (NW Europe). Hydrobiologia 478: 53–72.

RIVM, 2004. Risico's in bedijkte termen: een evaluatie van het beleid inzake de veiligheid tegen overstromen. Dutch Institute for Public Health and Environment, Bilthoven, Report no 500799002. (in Dutch).

Smits, A. J. M., P. H. Nienhuis & H. L. F. Saeijs, 2006. Changing estuaries, changing views. Hydrobiologia 565: 339–355.

Van der Ham, J., 2004. Afleiden of opruimen? De strijd om de beste aanpak tegen rivierbederf. Dutch Ministry of Transport, Public Works and Water Management (RIZA), Lelystad. (in Dutch).

Van de Ven, G. P, 2004. Man-made Lowlands. Uitgeverij Matrijs, Utrecht.

Van Stokkom, H. T. C., A. J. M. Smits & R. S. E. W. Leuven, 2005. Flood defense in The Netherlands: a new era, a new approach. Water International 30: 76–87.

Wiering, M. A. & B. J. M. Arts, 2006. Discursive shifts in Dutch river management: 'deep' institutional change or adaptation strategy? Hydrobiologia 565: 327–338.

Witter, V. J. & T. Raats, 2001. Hydrological criteria for durable water systems. In Schumann, A. H., M. C. Acreman, R. Davis, M. A. Marino, D. Rosbjerg & J. Xia (eds), Regional Management of Water Resources. IAHS Publication 268: 231–238.

Witter, V. J., 2003. Small hydrological basins in delta regions, past and future (climatic) changes: the Mark-Vliet basin. Water Resources Systems – Water availability and Global Change. IAHS Publication 280: 3–11.

Hydrobiologia (2006) 565:327–338
R.S.E.W. Leuven, A.M.J. Ragas, A.J.M. Smits & G. van der Velde (eds), Living Rivers: Trends and Challenges in Science and Management
DOI 10.1007/s10750-005-5923-2

Review Paper

Discursive shifts in Dutch river management: 'deep' institutional change or adaptation strategy?

M.A. Wiering* & B.J.M. Arts
Political Sciences of the Environment, Nijmegen School of Management, Radboud University Nijmegen, P.O. Box 9108, 6500 HK Nijmegen, The Netherlands
*(*Author for correspondence: E-mail: m.wiering@fm.ru.nl)*

Key words: discourse analysis, institutional change, new modes of governance, river basin approach, policy arrangement approach, water management

Abstract

This paper argues that a discursive shift is taking place in Dutch water policy, from 'a battle against water' to 'living with water' or 'accommodating water'. Yet we ask ourselves whether this shift is just an adaptation strategy of the existing elite group of water managers, who pay lip-service to new management approaches in order to maintain their vested interests, as some authors claim, or whether it implies 'deep' institutional change, e.g. in terms of the emergence of new water institutions, power relations and procedures. While investigating this question, we make use of the 'policy arrangement approach', which pays attention to institutional and discursive aspects of policy making alike. Our conclusion is that we are currently observing institutional changes beyond 'policy talk', particularly in terms of new legislation and procedures. However, it is too early to speak of 'deep' institutional change in Dutch water management, because the former water institutions are still maintaining their power positions, despite the availability of additional resources for policy and research as well as the emergence of several new modes of governance.

Introduction

Nowadays environmental scientists, hydrological engineers and policy makers are using a new lexicon when it comes to water issues. Water management can be expressed in such terms as 'room for the river', 'resilient water systems' or 'dynamic enforcement of the coastal zone' (Van Stokkom et al., 2005). In general, these policy concepts have been partly lent from older ecosystem based perspectives on water systems, which seem to have found themselves amidst a new political and societal momentum, because of the public attention aimed at water management. In this article we focus on river policies in the Netherlands and attempt to evaluate developments from an institutionalisation perspective. Is this new lexicon, or this 'discursive shift', a predecessor of fundamental change in the institutions of water management? Or should it be seen as a strategy used by water managers and scientists, consciously or unconsciously, in order to adapt to a changing environment, while the underlying institutions remain the same?

Van Hemert (1999) has a rather cynical explanation for the reason why we should create 'room for the river': it is meant to create more room for the new projects of The Directorate-General of the Ministry of Transport, Public Works and Water Management, in order to continue the engineering of rivers. She claims that the changes proposed are only 'discursive' instead of 'factual'. She describes 'room for the river' as an adaptation strategy of the Directorate-General for Transport and Water Management in order to give 'room to the engineer'. In terms of institutional dynamics, it is a pattern in which certain 'near core' discourse elements are changed without changing the 'deep core beliefs' so as to ensure that the 'old' organisational structures – organisations, interaction rules, resources – are perceived as being necessary and that they therefore should be maintained (cf. Sabatier & Jenkins Smith, 1993).

We are not certain about this explanation and that is why we must search for an answer stemming from our own theoretical perspective and empirical viewpoints. In order to do this, we must introduce another lexicon, that of institutional dynamics and policy arrangements. Next, we will give a brief history of past developments in Dutch river management, concerning mainly the contrast between 'the battle against water' and 'accommodating water'. Subsequently, we will apply our theoretical framework and then discuss the nature and magnitude of institutional changes in river management.

Policy arrangements and institutional change

For a theoretical framework which combines insights on both discursive shifts and institutional change, we decided to choose the 'Policy Arrangement Approach' (PAA) (Arts & Leroy, 2003; Boonstra, 2004; Van Tatenhove et al., 2000). This approach builds upon multi-actor policy network models. However, it pays more attention than these models to: (1) institutional contexts in which policy actors must operate, (2) the substance of policy making and (3) the power relations between the policy actors involved (for policy network models see: Marsh & Rhodes, 1992; Kickert et al., 1997).

We can define a policy arrangement as the way in which a certain policy domain – such as water management – is shaped in terms of organisation and substance (Van Tatenhove et al., 2000). In addition, we conceive a policy arrangement as consisting of four analytical dimensions: 'discourses', 'rules', 'actors' and 'resources' (*Ibidem*). Actors, resources and (some) rules refer to organisational aspects, whereas discourses and (various other) rules refer to substantive aspects. Furthermore, we have been able to provide a number of 'indicators' for each dimension on the basis of which change over time can be assessed (Table 1). However, the notion of 'indicator' is placed between brackets, as we are aware that – in a strict methodological sense – we are not dealing with 'true' indicators here, namely empirical assets which can be immediately observed. Yet these 'indicators' can help us to analyse institutional change empirically and more thoroughly. In order

Table 1. Operationalisation of the Policy Arrangement Approach (PAA)

Concept	Aspects	Dimensions	Change indicators
Policy arrangement	Substance	Discourses	Change in:
			*Paradigms
			*Utopias
			*Policy programmes
		Rules[a]	*Legislation
	Organisation	Actors	*Procedures
			*Political culture
			*Actor constellation
			*Interaction patterns
			*Coalitions and oppositions
		Resources	*Resource constellation
			*Power relations
			*Political influence

[a]Rules can be both substantive and organisational in nature.

to assess the 'depth' of change, we have made an analytical distinction between 'shallow' and 'deep' institutional change. The former relates to changes in policy discourses only, the latter to a change in the entire policy arrangement.

In our view, a discourse refers to a set of ideas, concepts, buzzwords and stories which combined give meaning to a certain phenomenon in the real world (Hajer, 1995). An example is the sustainability discourse, which brings together notions such as economic, ecological and social sustainability, sustainable development, a belief in the possibility to integrate economy and ecology, examples of win–win situations, etc. This language gives meaning to a world characterised by poverty and ecological degradation, but also to a world which has the potential to become sustainable after all. Theoretically, a discourse consists of three 'layers': ontological, normative and strategic (Therborn, 1980). Discourses can be essentially ontological in nature, related to questions such as: How do we see reality? How do we define problems? What do we think is taking place? Can we be certain about our risk management strategies? Here we deal with the 'paradigms' or 'world views' of policy actors. Whether these (fundamentally) change over time, or not, is our first discursive 'indicator'. At the same time, discursive space is filled with normative expressions, concerning the values at stake and the goals that are set. Here we deal with the 'utopias' or 'ideals' of policy actors. Whether these (fundamentally) change over time, or not, is our second discursive 'indicator'. The third layer of discourse consists of the route or 'road map' from what we see as 'real' to what we conceive of as 'desirable', from problem to solution. Here we deal with the 'policy programmes' of policy actors. Whether these (fundamentally) change over time, or not, is our third discursive 'indicator'.

The next dimension, rules, consists of 'legislation', 'procedures' and 'political culture' (Giddens, 1984; Rittberger, 1993). Legislation refers to the formalisation and transposition of policy discourses into binding law. Therefore, an important aspect of 'deep' institutional change is the extent to which changes in discourse are reflected in changes in legislation. Nevertheless, rules are not only 'substantive' in nature, but also 'organisational' (Giddens, 1984). This latter aspect refers to procedures, to how political participation and decision-making processes are codified in 'the rules of the game'. One might wonder whether discursive shifts in a policy domain also imply that there are new rules to the game, e.g. in terms of the participation of new actors in decision-making. For example, does a cross-border river basin approach (discourse!) indeed lead to the formal participation (rules!) of German policy actors in Dutch water management? Besides the distinction between substantive and organisational rules, one can distinguish between formal and informal ones. The former refers to legislation and procedures, previously dealt with, whereas the latter refers to 'political culture'. For example, the Dutch 'polder model' generally causes other types of policy processes to occur rather than the German 'formal-legalistic model' (Haverland, 1999). Such national 'policy cultures' colour the way in which policies are shaped. Nevertheless, it should be noted that these informal rules do not change that easily.

The third policy arrangement dimension of policy actors is analysed on the basis of 'actor constellation', 'interaction patterns' and 'coalitions & oppositions' (based on: Marsh & Rhodes, 1992; Godfroij & Nelissen, 1993; Kickert et al., 1997). The first 'indicator' relates to the set of (key) policy actors in a given policy domain (such as water management). The question is 'Who is involved in agenda-setting, decision-making and policy implementation, both formally as well as informally?' And do we see changes over time within this constellation of actors? Secondly, we can perhaps observe changes in the way these players interact, quantitatively and qualitatively. For example: does interaction increase or decrease? Do we see more co-operation or, in contrast, more conflict? As a consequence of these changing interaction patterns, we might find the emergence of new coalitions and oppositions, making this our third 'indicator'. However, it should be noted that a change of interaction patterns does not necessarily imply that new coalitions or oppositions have been established. This is just a possibility and, for that reason, we distinguish between this second and third 'indicator'.

Finally, the dimension of resources is elaborated upon in the 'indicators' resource constellation, power relations and political influence (Huberts & Kleinnijenhuis, 1994; Arts, 1998). The first relates to assets which policy actors have or can mobilise on the basis of which they can exercise power, e.g. authority, money, knowledge or technology. The relevance of these resources and their usefulness may vary depending on the setting and the time span. Generally, these assets are not equally divided among policy actors, which leads to a situation in which not all of the actors share similar capacities to achieve (political) outcomes. Here, unequal power relations between policy actors exist, although one should acknowledge that these relations are dynamic (in time and space). A further question to be asked is whether policy actors do indeed make use of their capacities to achieve the outcomes they wanted. In other words, power should be 'operation' into influence. Although intrinsically related, there is no one-to-one relationship between power and influence, as policy actors may decide not to make use of their resources and/or fail to achieve the outcomes they wanted. Although it should be noted that power (in terms of resources and capacities) is relatively easy to assess, whereas it is very hard to measure political influence. Therefore, in the context of this article, we will limit our analysis to resources and power relations.

Discursive shifts in river management

The traditional discourse in Dutch water management reflects the history of the Netherlands: fighting against the sea, storms and frequent flooding, losing land, building dikes, conquering land from the sea, embanking and cultivating it. Similarly, the Dutch streamlined their rivers, minimised the river basins, closed creeks and small streams and replaced them with canals. Water was mostly viewed as a 'threat' and had to be regulated and controlled. (Van de Ven, 2004; Van Steen & Pellenbarg, 2004) This 'battle against the water' discourse led to a river management that was, up until very recently, focused on building dikes in order to keep peoples feet dry (Wiering & Driessen, 2001). This perspective was widely spread amongst the civil engineers and (other) policy makers of the Ministry for Transport, Public Works and Water Management and its Directorate-General, as well as the regional water boards. From the beginning of the 1960s up until the mid 1980s river management was politically overshadowed by the coastal works (the famous Delta works) and subsequently upon finishing these major projects, river flooding management was discussed, but had low priority on the Dutch political agenda. A long lasting discussion on water safety norms for river flooding was characterised by Van Eeten (1999) as a 'dialogue of the deaf'. Dike enhancement was even more slowed down by protest and litigation from river landscape protectors and environmental agencies, who expressed a 'counter-movement' distrust in Dutch water authorities as well as a Not In My Backyard (NIMBY)-effect among civilians. Yet, in general and among the public at large, feelings of trust in Dutch water management could be sensed.

From the mid-1980s river management gradually incorporated the upcoming issues of water quality, environmental concern and nature conservation, and evolved towards the concept of integrated water management. In 1985, a memorandum of the Ministry (Dealing with Water) promoted a system-oriented and integrated view on water management, thereby initiating a fundamental discussion during the following years. Gradually appeared a new, 'system ecology' discourse on river management, in addition to the – still hegemonic – 'battle against water' discourse of the hydraulic engineering water manager (Van Hemert, 1999; Disco, 2002).

The relatively low priority placed on the (river-) flooding management in the Netherlands changed rapidly in 1995. In January of that year, the Dutch faced a near national disaster, as the water rose to extreme levels in the country's major rivers – the river Rhine (and its branches, the Waal, the Lower Rhine and the IJssel) and the river Meuse. Almost 250,000 people, and a large number of livestock belonging to farmers, had to be evacuated. The authorities feared that the dikes would not hold. The economic damage and evacuation costs were great. In the end, the dikes along the Rhine and its branches did hold, but only just. Large stretches of the river Meuse have only quays and natural embankments, and the people living in the southern part of the Meuse floodplain suffered the most material damage. In 1995 the near flood disaster can be considered as a real *shock event* felt in Dutch society.

Initially, this strengthened the traditional discourse relation to the division of water and land use. In terms of the different discursive elements we mentioned earlier, we could say that ontologically, water proved to be a threat to society (once again) and that there was a strong feeling of urgency in regard to 'defending ourselves against water'. According to tradition, the Dutch named the policy programme 'Major Rivers Delta Plan' and 'Major Rivers Delta Act' as a blatant reference to the world famous 'Delta Works' which were constructed as a defence against the North Sea. Discursively, this was an important step. The normative discursive space was dominated by 'direct safety first' and the strategy was focussed on strengthening the dikes. Accompanied by a strong sense of urgency and by new legislation, the Dutch diking operation between 1996 and 2000 was, in itself, an example of successful project management (Wiering & Driessen, 2001).

But this was only a short-term response to the shock event and the near-flooding disaster had also a strong impact on the ideas and plans for *future* water management. Policy makers and scientists began to realise that the former policy of dividing water and land – and marking borderlines with dikes – was not sufficient in order to meet the goals of safety and the reduction of risks in the long run. Some even used the term 'control paradox' (Remmelzwaal & Vroon, 2000; Wiering & Immink, 2003): by building and strengthening dikes an idea of safety is created, giving way to more social and economical activities behind the dikes. This, in turn, could lead to an increase in the detrimental social effects and the economic damage suffered when occasional flooding does happen and to an eventual increase in the feelings of *insecurity*. Gradually water management seemed to be moving away from merely building higher dikes towards adjusting and extending the flood plains and giving 'room to the river' (Van Stokkom et al., 2005; Smits et al., 2000). After the first policy guidelines in this direction were established, more radical policy plans were published. We could witness a discursive turn towards a new discourse of 'accommodating water'.

This idea of 'accommodating water' was not new. Environmental scientists, some of the hydrological engineers and even planning agencies, already described ways of water management that took natural water systems into account for many years, especially since the 'Dealing with Water' Memorandum of 1985. Disco (2002) stresses the growing importance of the ecological conceptualisation of water systems in the domain of central water agencies and institutions. He called this the 'ecological turn' of Dutch water management in the 1980s and 1990s, an evolutionary development that is also visible in the different planning reports on the national 'water household'.

How, then, can we judge these processes by way of our discourse-indicators, changes in ontological, normative and strategic discourse? We predominantly witness changes in policy aims and goals (normative discourse) and route-changes (strategic discourse): 'room for the river', 'space for water' and a 'good water quality' are the aims of the water managers. These are to be reached by new strategies: a more system-oriented approach towards water issues, implementing regional views on river basin management, new tools for integrated water management, etc. This is also promoted by communicating with society, by raising the public awareness of water related policy problems and by visualising water as a possibility, and not merely as a threat (Smits et al., 2000).

Interpreting the ontological indicator of discourse is, however, more problematic. We see the traditional 'hydraulic engineering' -perspective of the old 'battle against water' -discourse eroding, but the new discourse 'accommodating water' is to be found somewhere in between the old 'battle against water' and the more radical 'system ecology' discourses. Moreover, concepts such as 'room for the river' and 'space for water' are open to interpretation and can inhabit both natural ecosystem-based or more traditional hydraulic 'room for the engineer'-based views of water systems and similar solutions. We can conclude, though, that the traditional fixation on dike enhancement in river basins has made way for a variety of options to combine flooding management with land use and nature conservation (Van Stokkom et al., 2005).

After having given giving, in general terms, the different discourse-elements that involve fighting over hegemony in water management, we can now turn to the other aspects of the policy arrangement. We will focus on a few major processes of change in the three remaining institutional dimensions: rules of the game, power/resources and actors/coalitions. We will discuss these in more detail, in order to give insight into the nature of these changes, and to see if these discursive shifts are actually being institutionalised in the current water management.

Changing rules

Legislation

The first organisational dimension in which changes have become visible is the 'rules of the game'- dimension. We shall first, very briefly, characterise this dimension. Present-day water management has differentiated into a set of rules, which have been laid down in, for example, the Water Management Act, the Groundwater Act, the Embankment Act, the Pollution of Surface Waters Act and the Pollution of Sea Water Act. Besides these laws, there are formalised or informal rules concerning the jurisdiction and competences of the water policy agents. The water legislation is looked upon as rather complex and difficult for outsiders to penetrate. Because of its complexity and fragmentation, the relevant authorities often rely on informal agreements concerning the division of tasks. Thus, when it comes down to new rules in general, one of the first questions to answer is whether the Dutch water legislation is on the brink of being redesigned.

In 2002 the Vice-Minister of Transport, Public Works and Water Management asked a special Committee to give its advice whether the legal design had to be revised and the course of possible institutional changes. The Committee argued that there were several reasons for changing the legislation. First of all, there is a lack of internal coherence and transparency in the complex set of rules regarding water management. Secondly, there is an indistinct relationship between the rules on general water policy (e.g. 'water household' planning, norms, standards, general procedures) and the specific rules of water management in practise, i.e. dike enhancement and maintenance, dams and embankments. The Committee concluded that a new, Integrated Water Management Act should be created, which would include most existing water legislation, as mentioned above, but would exclude the rules concerning 'Water Chain Management' and the (constitutional) legislation regarding competencies of organisations. It was only quite recently that the Dutch government confirmed that it is indeed necessary to redesign and create such an integrated Water Management Act (Memorandum, 2004). Some of the arguments refer to the overall policy intended to streamline rules and to reduce rule density. More importantly, the arguments (of both the Committee and the Cabinet) are connected to the following two policy developments: (1) to anticipate the implementation of the European Water Framework Directive (WFD) and (2) to (better) embed the central concept of integrated water management (referred to as 'water system management') – and to improve its instrumentation. In other words, the proposed legislative changes are though to be an important step in switching from sector-based water management to integrated 'water system management'. These changes are stimulated by obligations, rules and concepts stemming from the European policy arena (WFD).

Change in procedures

Another sign of institutional change can be found in the instrumentation of water management in relation to other policy domains. An exponent of new procedures is the development of the process instrument of the so-called 'Water Assessment'. This procedure is to be considered as a form of 'water impact assessment' (partly resembling the environmental impact assessment) as a result of which water management will change its interrelations with spatial planning in the Netherlands. According to the Water Management in the 21st Century Advisory Committee – in the following referred to as the WB21-Committee – the Cabinet paid insufficient attention to safety and water-related problems in the past. As a result, a great deal of space was gradually reclaimed from the water management system. 'New spatial planning decisions may not exacerbate the challenges to safety and leave water-related problems unnoticed' (Ministry of Public Works, Transport and Water Management, 2000). The water test explicitly addresses water-related aspects in all relevant spatial plans and must result in a separate section in the explanatory policy document. It considers both water quantity issues (impacts on retention and storage capacity, risks of flooding, drought, and groundwater level) and water quality issues (surface- and groundwater pollution, water sanitation, biodiversity). It must be applied at an early stage in the spatial planning process and water/related problems can not be passed from one watershed to another. If decisions are taken that have negative effects on the water system, adequate measures must be identified to compensate or mitigate these effects. Seen as a new procedure, the water test is expected to influence both the position of water managers and the responsibilities of spatial planners. Water managers will gradually shift from a 're-active' attitude in the realm of spatial planning to a 'pro-active' position, in which water managers are supposed to be involved in an early phase of policy formation of spatial plans (location, impact on water systems, etc.). Spatial planners are expected to take water aspects into account when making decisions concerning spatial plans.

Change in political culture

By using the term 'political culture' we point at the general patterns of the (mainly) informal and implicit rules of the game which can be associated with certain policy domains or which can characterise national politics (see section 2). The political culture of water management can be typified by three features, which are momentarily in a state of flux. First of all, the governmental authorities are the ultimate locus of authoritative power in water management. Both market and civil society are relatively weak regulation mechanisms in this field. A centralised water management planning system exists and water agencies exhibit a rather hierarchical organisational culture. The policy arrangement involved here can thus be called 'state-oriented' or 'etatist' (cf. Van Tatenhove et al., 2000). However, this situation was not problematic during the past decades. Water quality and water safety have always been considered to be important public goods in Dutch society; goods which should be provided for by the state.

Secondly, because of the past history of strong sector-based politics, the water managers are used to operating in a rather autonomous and *isolated* policy field. There has hardly been any public support or protest from social groups, citizens or the business community, except, perhaps, for the traditional strong participation of farmers on Dutch water boards (Wiering & Immink, 2003). Thirdly, water management is *technocratic* in nature. This is the result of the relatively closed policy domain in combination with specific functional governmental tasks and a specific epistemic community focusing on 'hydraulic engineering'.

In the wake of near-flooding disasters, excessive rain fall, drought problems and expected climate change, the political and societal attention paid to water problems is increasing. Because of this, the etatist, closed and technocratic features of the Dutch water policy arrangement have been contested in recent years. A 'socialising' of water management has emerged, including an aspiration of the domain itself to act less hierarchically and to decentralise decisions (Van Leussen, 2002). Moreover, the water boards are being subjected to a democratisation-process. As a consequence, water management has gradually been forced to 'open up', and to become more transparent to its citizens by abandoning its isolated, expert-based and technocratic policy style. Signs of this transition, from 'government-to-governance', can be found in the explicit need for stakeholders to become more involved. However, one could (again) question the 'depth' of this institutional change, because in truth the basic administrative structures have still remained unchanged thus far, as we will see in the next section.

Policy actors

Actor constellation

Dutch constitution consists of three general administrative levels, the municipal and provincial authorities and the national administration. Only the issue of water management has an additional fourth layer: the water boards. The (public) authority of the water boards is geared towards the management of regional water quantity and water quality and those of the province towards groundwater management and the planning aspects of regional water management. The national General-Directorate is responsible for the main water infrastructure of large rivers and canals, and, finally, the local authorities deal with urban water and sewerage.

Two central policy developments arise, in searching for changes in the actor constellation. In the context of the European Water Framework Directive, the river basin approach asks for co-operation within the four river basins of which the Netherlands forms a part – Meuse, Rhine, Scheldt and Eems – and thus, it asks for cross-border water management. But there is no prescription on *how* to co-operate. Up until now, it has not led to any new arrangements in the Netherlands, but instead the co-operation between the existing water authorities, nationally and internationally (such as the International Rhine Commission) continues to predominate (Backes, 1999). When it comes to water quantity issues, the successive discussions on the WB21-Committee -proposals eventually led to a so-called 'National Administrative Agreement on Water', between the national administration and the representatives of the municipalities, provinces and water boards. In short, in both water quality (the WFD) and water quantity (the WB21-policies) the Dutch authorities have clung to the existing organisational order and division of tasks and competencies. The Vice-Minister of Transport, Public Works and Water Management illustrated this nicely by saying that she did not want 'an institutional discussion' concerning the administrative organisation of Dutch water management.

Interaction patterns

Although there are no crucial changes in the actor constellation, there is a change in interaction patterns, namely a move towards decentralising responsibilities and empowering the regional authorities. This can be illustrated by the process architecture of the 'Room for the River' operation. The provinces have obtained an essential voice in this policy process by giving a so-called 'weighty advice', and because of their strong representation in the advisory steering committees. Both the WB21-policies and the WFD have stimulated territorial shifts towards the region and sub-river basins. We can conclude that the interactions between the different governmental layers are being intensified and that this is heading towards more co-operative and horizontal interrelations, accompanied by the empowerment of provinces and – to a lesser extent – the water boards and the municipalities.

Finally, we should pay attention to changes in the way the water management agencies and the public interact. Here we refer to the ways in which communication has changed since the discourse of 'living with water' and the 'disclosure' of water agencies. It is again difficult to draw conclusions on 'real' institutional change. On the one hand, a change has definitely taken place in policy style, through communication campaigns directed at the public, increasing information flow, stakeholder involvement, etc. On the other hand, when it comes to taking critical steps towards formulating and preparing policy, such as the first stages of the implementation of the WFD or the selection of the location of so-called emergency flooding areas (these are strongly contested in some of the preserved areas), stakeholder participation is much less appreciated and the policy style of the Ministry and the Directorate-General for Public Works and Water Management is, once again, regarded as being technocratic.

Coalitions and oppositions

Are we witnessing new coalitions between water management and other policy actors in the field? Here we notice that, since the upcoming ecosystem-based discourse in river management, agencies that were traditionally primarily engaged with nature conservation, biodiversity and forest policies are entering the policy arena of water management. For example, the National Forest Service has presented interesting ideas on how to combine water safety issues with nature development and the ecological management of river basins (e.g. Ministry of Agriculture, Nature Conservation & Food Quality, 2003). Even though today's water policy seems to have been broadened towards issues of environment, landscape and nature conservation, and even though 'non-water' authorities are sometimes key players in *ad hoc* co-operation in floodplain projects, we still cannot conclude that this has led to new actor coalitions in the water policy arrangement as a whole. At some points the National Forest Service, with a strong nature conservation perspective, has in fact competed with the Directorate-General for Transport, Public

Works and Water Management, which is mainly focussed on safety issues in river reconstruction (Van de Bilt, 2004).

Shifts in power

The two indicators 'resource constellation' and 'power relations' point at the possibility of new and/or more resources in a policy domain on the one hand, and the (re)division of these resources over the different key players, potentially implying new power relations, on the other. With regard to the former, we can observe two important changes: (1) there are more financial resources for water management at our disposal; and (2) there are more resources to further develop the knowledge infrastructure in particular. However, with regard to the second indicator, we can not see structural changes in power balance. As was previously stated, the organisational structure of Dutch water management has remained rather stable so far. We can observe the same key players and (more or less) the same power relations, as the new resources have strengthened those who were already 'in power', neither the smaller parties nor the newcomers. Below we will elaborate on these observations.

Resource constellation

Rudely awakened by the (near-)floods in the 1990s, and triggered by the concern over the effects of climate change, new resources have become available for water management, both internationally and nationally. First of all, extra public money was spent on strengthening the river dikes as quickly as possible (Wiering & Driessen, 2001). Secondly, money was set aside in order to design the *new* policies for water management, in line with the results of the WB21-Committee. In the National Administrative Agreement on Water is was stated that, from 2003 to 2015, an amount of 8 billion euros should be reserved; for the period preceding 2050 about 16 billion euros. Over the short term this entails an investment of 1.3 billion euros until 2007 (Ministry of Transport, Public Works and Water Management, 2003).

Besides these overall financial resources belonging to the policy domain, another essential resource is its knowledge infrastructure. There are several important power resources in water management. For instance, the supreme technical knowledge possessed by water managers, and – to a large extent – the public trust that has traditionally been given to this functional layer – combined with important legislative powers (e.g. risk norm-setting and water-related taxes). The knowledge system of Dutch water management is dominated by specialised governmental services, knowledge institutes and universities. The most important agency is the Directorate-General of the Ministry itself and within this central agency there are specialised services such as the Royal Institute for the Coastal Zone and Sea (RIKZ) or the Institute for Inland Water Management and Waste Water Treatment (RIZA), which are now operating on a more independent basis. The external knowledge infrastructure of Dutch water management is also extensive; especially WL Delft Hydraulics has to be mentioned here (Delft University is traditionally the cradle of Dutch 'hydraulic engineering') but also other institutes are active in the field: the Netherlands Organisation for Applied Scientific Research (TNO), the Meteorological Institute (KNMI) as well as the universities of Wageningen, Nijmegen, Twente and Utrecht. The differentiated knowledge infrastructure (hydraulic engineering, hydrobiology, ecology, policy analysis, rural and urban areas) of these various institutes is gathered together in the Netherlands Centre for River Studies (NCR). Furthermore, there is intensive co-operation between the specialised water management services, the environment and health research institute (RIVM) and Alterra, the research institute for the green living environment. The knowledge infrastructure on water is strong, but it is also narrowly focused on technical issues. A background report on the knowledge for integrated water management (Wisserhof, 2000) claimed that the financial impulses for multi-disciplinary projects did not, as a rule, stem from the water sector, but from other ministries such as the Ministry for Housing, Spatial Planning and Environment or the Ministry for Agriculture, Nature Conservation and Food Quality. But, according to Wisserhof (2000), the broadening of the knowledge system and co-operation is 'in statu nacendi'.

What about the new resources that have become available for research on water management, both in the Netherlands and Europe? Indeed, there are new research programmes, for instance the NWO Research Council for Earth and Life Sciences (ALW) and the Netherlands Foundation for the Advancement of Tropical Research (WOTRO) have launched a research programme for fundamental research regarding fresh water; there is a new ICES/KISS programme 'living with water' (in total 45.7 million euros, with co-financing) and there are many new initiatives which concern climate change, water and space, water and society, etc. In general, more financial means have been allocated to water management and research on the topic.

Power relations

What does these new resources for water policy and -research signify when it comes to the reallocation of resources and power relations? With regard to the supplementary public funds made available for the traditional policy of strengthening the dikes (after the 1995 near-flooding disaster), the financial resources went to the core players of Dutch water management: the Directorate-General, the provinces and the water boards. With that, the position of the existing authorities, services and institutes was in fact strengthened. Moreover, in the new 'room for the river' -policy and other WB21-policy measures, as well as the Water Framework Directive, the existing agencies continue to dominate policy formulation and policy implementation, although there is more involvement of other policy domains and more sensibility towards other political arenas and knowledge resources. Considering the content of various research programmes, the focus has (partially) shifted and a gradual broadening of the knowledge infrastructure has occured, with more attention being paid to ecological research, social science research, policy analysis and socio-economical research. At the moment, it is impossible to identify the changing allocation of budgets and their impact on individual research agencies exactly, but we can sense several new initiatives that indicate the general trends. First of all, the Netherlands Centre for River Studies (NCR) is an attempt to integrate the various knowledge sources. The NCR managed a large research umbrella project in the framework of the EU Structure fund (the IRMA- Interregional Rhine/Meuse Action – Programme). We can also witness several new clusters geared towards technical knowledge, i.e. hydraulic engineering (i.e. Delft cluster) and new portals that show co-operation between existing knowledge institutes (e.g. Coordinated Programme on Water and Climate). Apparently, the existing well vested research institutes on water, nature conservation, environment, climate and health are searching for new joint ventures in the water domain. All in all, we can observe an increase in public resources for water policy and a gradually broadening of water research, although the division of these resources over the different key players in the water policy field have roughly remained the same. Granted, additional money went to newcomers, e.g. in policy analysis, and ecologists seem to have been strengthened by the broadening and division of resources, but these are minor shifts, compared to the budgets of the 'big players' in the field.

Conclusion

In conclusion, we cannot share Van Hemert's rather 'cynical' conclusion that the new policy discourse 'more room for the river' in the Netherlands is only an adaptation strategy of the Dutch Directorate-General of Transport and Water Management in order to maintain 'room for the engineer' (Van Hemert, 1999), now that there is ever more talk of integrated river basin management, new water partners, 'wet' nature development, biodiversity, etc. After all, we can observe institutional changes beyond discursive shifts, particularly in terms of new legislation and procedures. The Dutch government is preparing an integrated Water Management Act, partly as a consequence of the EU Water Framework Directive, and has already implemented the so-called 'water test' for spatial plans. Furthermore, the traditionally closed, technocratic and 'etatist' political culture of the Dutch water management has become more open-minded and new coalitions on nature development in river basins are emerging in flood plain projects. Hence, besides 'government', we can also observe the first signs of an emergence of 'new modes of governance' in the water policy domain. In addition, more research funds are being spent on ecological and socio-economic issues related to water quality and quantity.

However, it is too early to speak of a 'deep' institutional change in Dutch water management, particularly when its administrative organisation and power structure has been taken into account. Here we do not see much renewal. Newcomers (at least, in the heart of water management) such as the National Forest Service, environmental NGOs and ecologists still hold a weak position in the Dutch water policy arrangement and, as far as new resources for policy making and research becoming available, these have mainly strengthened the positions of the traditional power holders. All in all, from the institutional perspective of the four dimensions of a policy arrangement, we do see a substantial change in terms of policy discourses and rules in the water policy domain, but hardly any change in terms of (key) policy actors, new coalitions and power balance. With that, current renewal in the Dutch water management sector should be positioned somewhere in the 'grey zone' between shallow and deep institutional change. Consequently, the chances that the innovative water policy discourses will indeed institutionalise in new, open and stabilised structures of new agencies and coalitions, because of shifts in resources, are still rather bleak.

This is even more so, since the discourse of 'accommodating water' is in fact ambiguous at the ontological level. It may seem attractive as an 'ecosystem-based' story line – in order to meet new standards of water safety and to design new policy options beyond the traditional strategy of building higher dikes - but part of the policy routes for 'room for the river' is still to build higher dikes. Moreover, on an epistemological level, most water system-engineers keep on believing in 'the engineering of society', and they, normatively, still prioritise water safety over other river values and functions (e.g. landscape and biodiversity). The safety norms and related policy measures in river basins of the hegemonic water agencies are not easily and openly discussed. In other words, 'room for the river' could also be interpreted as combining safety and nature conservation in an open discussion on human and ecological risks. Nevertheless, in less than a decade since the near-flood disaster of 1995, Dutch water management has now set course for a truly institutionalised form of integrated water management. 'Accommodating water' is one of the story lines that is contributing to this development. Although discursive shifts may sometimes seem to be 'a lot of talk', they actually have a function in the long-term process of institutionalisation, which can eventually bring about structural changes in policy.

References

Arts, B. J. M., 1998. The Political Influence of Global NGOs. Case Studies on the Climate and Biodiversity Conventions. International Books, Utrecht.

Arts, B. & P. Leroy (eds), 2003. Verandering van politiek, vernieuwing van milieubeleid. Klassieke en post-moderne arrangementen. Nijmegen University Press, Nijmegen (in Dutch).

Backes, Ch. W., 1999. Towards international water boards and cross-border spatial plans? In Van Hall, A., G. Th. Drupsteen, & H. J. M. Havekes (eds), The State of Water. Koninklijke Vermande, Lelystad: 179–194 (in Dutch).

Boonstra, F., 2004. Laveren tussen regio's en regels. Verankering van beleidsarrangementen rond plattelandsontwikkeling in Noordwest Friesland, de Graafschap en Zuidwest Salland. Van Gorcum, Assen (in Dutch).

Disco, C., 2002. Remaking nature: the ecological turn in Dutch water management. Science, Technology and Human Values 27: 206–235.

Giddens, A., 1984. The Constitution of Society. Polity Press, Cambridge.

Godfroij, A. J. A. & N. J. M. Nelissen (eds), 1993. Verschuivingen in de Besturing van de Samenleving. Coutinho, Bussum (in Dutch).

Hajer, M. A., 1995. The Politics of Environmental Discourse. Ecological Modernization and the Policy Process. Oxford University Press, Oxford.

Haverland, M., 1999. National Autonomy, European Integration and the Politics of Packaging Waste. Thela-thesis, Amsterdam.

Huberts, L. & J. Kleinnijenhuis (eds), 1994. Methoden van Invloedsanalyse. Boom, Meppel (in Dutch).

Kickert, W., E. Klijn & J. Koppejan (eds), 1997. Managing Complex Networks. Strategies for the Public Sector. SAGE, London.

Marsh, D. & R. A. W. Rhodes (eds), 1992. Policy networks in British government. Clarendon Press, Oxford.

Memorandum, 2004. Integratie van Waterwetgeving. Parliamentary Paper 29694, No. 1. Tweede Kamer der Staten-Generaal, Den Haag (in Dutch).

Ministry of Agriculture, Nature Conservation & Food Quality, 2003. Lonkend Rivierenland. National Forest Service, Driebergen (in Dutch).

Ministry of Transport, Public Works and Water Management, 2000. A Different Approach to Water, Water Management Policy in the 21st Century. [English version 2001] http://www.verkeerenwaterstaat.nl.

Ministry of Transport, Public Works and Water Management, 2003. Nationaal Bestuursakkoord Water. Ministerie van Verkeer en Waterstaat, Den Haag (in Dutch).

Remmelzwaal, A. & J. Vroon 2000. Werken met water, veerkracht als strategie. Report 2000.021. RIZA, Lelystad (in Dutch).

Rittberger, V. (eds), 1993. Regime Theory and International Relations. Clarendon Press, Oxford.

Sabatier, P. A. & H. C. Jenkins Smith (eds), 1993. Policy Change and Learning, an Advocacy Coalition Approach. West View Press Inc., Boulder, Colorado.

Smits, A. J. M., P. H. Nienhuis & R. S. E. W. Leuven (eds), 2000. New Approaches to River Management. Backhuys Publishers, Leiden.

Therborn, G., 1980. The Ideology of Power and the Power of Ideology. Verso NLB, London.

Van de Bilt, S., 2004. Eén terrein, waar twee beheren. MSc-thesis Political Sciences of the Environment. Radboud University, Nijmegen (in Dutch).

Van Eeten, M., 1999. Dialogues of the Deaf, Defining New Agendas for Environmental Deadlocks. Eburon, Delft.

Van Hemert, M., 1999. Ruimte voor de ingenieur, rivierbeheer in Nederland eind jaren negentig. Kennis en Methode, Tijdschrift voor Empirische Filosofie 23: 361–387.

Van Leussen, W., 2002. Leven met water, vermaatschappelijking van het waterbeheer – consequenties voor de civiel ingenieur en voor waterbeheerorganisaties. Inaugural address Universiteit Twente, Enschede (in Dutch).

Van Steen, P. J. M. & P. H. Pellenbarg, 2004. Water management challenges in the Netherlands. Tijdschrift voor Economische en Sociale Geografie 95: 590–598.

Van Stokkom, H. T. C., A. J. M. Smits & R. S. E. W. Leuven, 2005. Flood defense in the Netherlands: a new era, a new approach. Water International 30: 76–87.

Van Tatenhove, J., B. Arts & P. Leroy (eds), 2000. Political Modernization and the Environment. The Renewal of Environmental Policy Arrangements. Kluwer Academic Publishers, Dordrecht.

Van de Ven, G. P., 2004. Man-made Lowlands. History of Water Management and Land Reclamation in the Netherlands. Matrijs, Utrecht.

Wiering, M. A. & P. P. J. Driessen, 2001. Beyond the art of diking. Interactive policy on river management in the Netherlands. Water Policy 3: 283–296.

Wiering, M. A. & I. F. G. Immink, 2003. Nieuwe beleidsarrangementen voor waterbeheer en ruimtelijke ordening? In Hidding, M. & M. Van der Vlist (eds), Water en Ruimte. SDU Uitgevers bv, The Hague (in Dutch): 183–196.

Wisserhof, J., 2000. Inventarisatie kennisinfrastructuur. Report NRLO/2000/4. Nationale Raad voor Landbouwkundig Onderzoek, Adviesraad voor het Wetenschaps- en Technologiebeleid, Raad voor het Milieu- en Natuuronderzoek. The Hague (in Dutch).

Hydrobiologia (2006) 565:339–355
R.S.E.W. Leuven, A.M.J. Ragas, A.J.M. Smits & G. van der Velde (eds), Living Rivers: Trends and Challenges in Science and Management
DOI 10.1007/s10750-005-1924-4

Opinion Paper

Changing estuaries, changing views

A.J.M. Smits[1,3,*], P.H. Nienhuis[2] & H.L.F. Saeijs[3]

[1]*Centre for Water and Society, Radboud University, P.O. Box 9010, 6500 GL Nijmegen, The Netherlands*
[2]*Institute for Wetland and Water Research, Department of Environmental Science, Radboud University, P.O. Box 9010, 6500 GL Nijmegen, The Netherlands*
[3]*Erasmus Centre for Environmental Studies, Erasmus University, Burg. Oudlaan 50, 3062 PA Rotterdam, The Netherlands*
(**Author for correspondence: E-mail: a.smits@science.ru.nl*)

Key words: closure dams, estuaries, flooding, offensive water management, potential risks, return tidal dynamics, SW Netherlands

Abstract

During the design and the execution of the Delta project, carried out after the storm flood of 1953 in the SW Netherlands, the importance of the long-term effects of morphological and ecological developments driven by tides and currents, have been underestimated. Due to these processes the height of the dams will have to be increased for centuries to come, because the land behind the levees cannot grow in elevation anymore with the rising of the sea. Maintenance of the civil-engineering structures, and mitigating their unpredictable impacts on ecosystems, involve very high recurrent costs. The chance of flooding is reduced, but the potential damage after a storm flood is enlarged: seawalls and dykes provide a false sense of safety against flooding. Changes in the role of agricultural use in the European context, offer an opportunity to abandon arable fields and to retrocede them to the sea in order to absorb tidal energy and to allow the land to rise concomitant with the sea. A cost-benefit analysis of this approach should assess the direct and indirect economic values, as well as the non-use (intrinsic) values, whereby public engagement in management questions, facilitates decision-making processes. Reversible and resilient economic measures within the limits of the natural processes are preferable. A future, speculative perspective is an urbanised landscape, where people and investments are located in safe places, e.g. on floating, or sea-encircled artificial dwelling-mounds, surrounded by a landscape that is ruled by the forces of nature. New approaches such as developed in the Westerschelde offer flexible solutions to flooding problems, and are worth a broader evaluation. A worldwide platform of experts should be organised to study the future management of estuaries and deltas, and to develop and exchange new ideas and techniques.

Introduction

The history of the SW Netherlands is marked by a continuous struggle between man and the sea. Since the year 1000 man reclaimed salt-marsh areas and transformed those into agricultural land. But irregularly occurring storm floods broke the man-built seawalls and recaptured parts of the gained land. Figure 1 illustrates the typical geomorphology of the delta system in the SW Netherlands between 1900 and 1950. The area of roughly 10,000 km^2 was characterised by a considerable number of small and larger islands and peninsulas, deep and shallow tidal channels, extensive intertidal sand and mudflats reaching up to 20 km off the coast, vegetated coastal plains, salt marshes and brackish marshes above mean high water. The most land inwards parts of the estuaries, where the rivers Rhine, Meuse and Scheldt enter the delta, were characterised by freshwater tidal marshes and willow coppice.

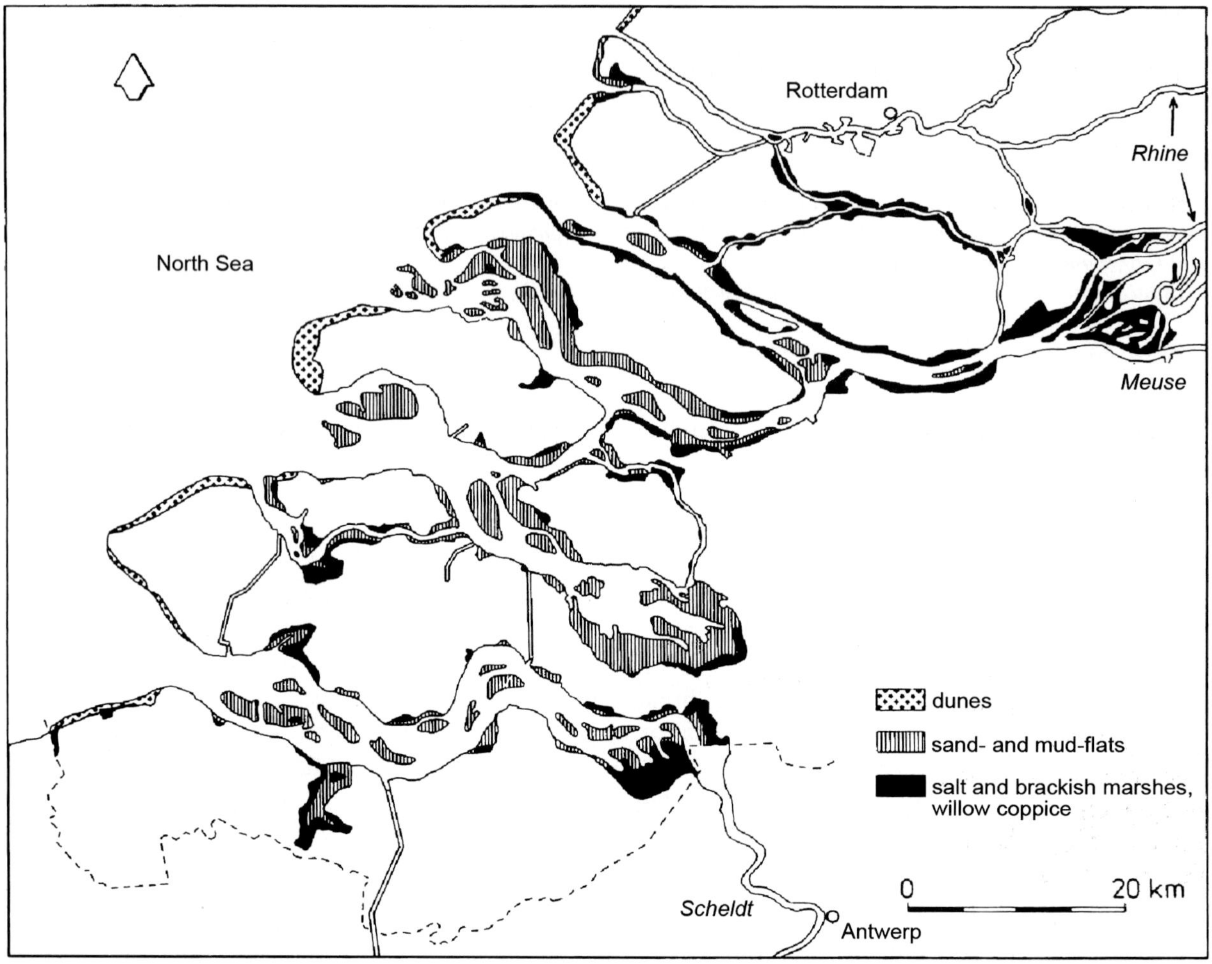

Figure 1. The Delta of the rivers Rhine, Meuse and Scheldt between 1900 and 1950; an extensive estuarine and coastal geo-morphological system (adapted from Nienhuis, 1982).

On February 1, 1953, a north-westerly storm induced tides to 3.4 m above normal levels, breached approximately 180 km of coastal-defence dikes and flooded 160,000 ha of polder-land in the SW Netherlands. 1835 people lost their lives in this large storm flood, more than 46,000 farms and buildings were destroyed or damaged, and approximately 200,000 farm animals drowned (Slager, 2003). The Delta project, formalized in 1957 by an act of the Dutch parliament, was conceived as an answer to the continuous risk of flooding, which threatens lives and property in this low-lying region. Because of the low mean elevation and premium on space in the Netherlands, the Dutch have a long tradition of coastal-defence construction and land reclamation. The need for continuous coastal construction has intensified over the years as a result of population growth, land subsidence and rising sea level. The potential threat of storm surges from the North Sea had already led to the closure of the Brielse Meer in 1950. The core of the Delta Project, to maintain a safe coastline as short as possible, called for the closure of the main tidal estuaries and inlets in the SW Netherlands, except for the Westerschelde and the Nieuwe Waterweg (Fig. 2). Along the Westerschelde the existing dikes have been raised, for reasons of continued international shipping access to Antwerp. In the Nieuwe Waterweg, the shipping route to the mainport of Rotterdam, the construction of the 'Maesland kering', a barrier protecting Rotterdam from storm surges, was finished in 1992. This enterprise was considered to be the final phase of the Delta project.

It is recognized that the decision, following the flood of 1953, to build a large, solid and inflexible 'wall against the sea', when placed in the cultural context of the time, was understandable. The aim of this paper is to show that this rigid project was not the best solution, in the light of modern standards and ways of thinking about sustainable water management. Focused on the Dutch delta, new insights in combination with advanced technological developments, call for a different strategy, comprising more flexible solutions. This paper will start with a short survey of the Delta project, and the underlying safety standards in the past and in the future, with regard to the risks of being hit by a severe storm flood. The lessons learnt from the Delta project are reviewed, ongoing solutions are discussed and new ideas are proposed, in order to mitigate and solve morphological and ecological problems. To stimulate the international discussion, alternative approaches and speculations about a society in balance with nature, coping with the tidal dynamics, will be given. The paper ends with a number of recommendations and views on future perspectives.

The Dutch Delta project

The Delta project formally started in 1957. A prerequisite for the construction of the primary sea-walls in the mouths of the estuaries was the need to reduce tidal-current velocities in the estuaries, before the construction of the primary barriers could be undertaken. Tidal velocities were lowered by constructing secondary compartmental barriers (Zankreekdam, Grevelingendam and Volkerakdam; Fig. 2) to reduce the extent of the Delta area subject to tidal influence. This resulted, in turn, in a reduced tidal volume and, therefore, lower current velocities through the main estuaries. The former (semi-)estuaries Veersche Gat and Grevelingen were closed off from the North Sea by high sea-walls in 1961 and 1971, respectively, and turned into non-tidal lakes or lagoons filled with brackish or saline water, whereas the Haringvliet was closed in 1970 by the construction of large sluices, meant to function as an outlet for the rivers Rhine and Meuse (Fig. 2).

The original plan for the Oosterschelde estuary called for a dam across the mouth of the estuary, a distance of 9 km, to be finished in 1978. The tidal basin would than have been changed into a stagnant lake filled with – polluted – water from the river Rhine. But the final form of the present barrier differs drastically from the simple dam that has been envisaged originally. Through the 1960's and early 1970's, conservationists and fishermen provoked an awareness in many people of the need, to protect the area's outstanding natural resources and its unique tidal habitat, including an extensive shellfish (oyster) industry, the only one in the Netherlands. The Dutch government decided to change the design of the dam in 1974. After several years of desk studies the Dutch parliament accepted in 1976 a compromise solution: a storm-surge barrier. On the one hand the barrier allows the reduced tides to enter the estuary freely, thus safeguarding the tidal ecosystem, including the plant and animal communities. On the other hand the barrier guarantees safety for the human population and for the properties of the inhabitants when storm floods threaten the area. This barrier design marked a turning point in the Dutch political decision-making process with regard to the natural environment. The storm-surge barrier was constructed between 1979 and 1986 in the western inlet of the estuary (Fig. 2; Nienhuis & Smaal, 1994).

The positive state of mind about the chosen solution for the Oosterschelde gradually tempered in the course of the years. The disturbed hydrodynamic tidal balance in the estuary enhanced the erosion of the tidal flats: all flats, managed as precious nature reserves, will gradually disappear under water. In fact, the closing of the four main branches of the Rhine–Meuse estuary brought the natural transitions between fresh, brackish and salt water to an end. The complicated interplay between deposition and erosion of marine and river sediments in all four estuarine areas was ceased, and large uncontrolled changes in long-term hydrodynamic and geomorphologic processes were set in motion. The original natural habitats disappeared, and were replaced by man-made habitats. This is reflected in the changes in biodiversity: characteristic estuarine species disappeared, as was the case with migratory species (e.g. fish species), used to travel between the rivers

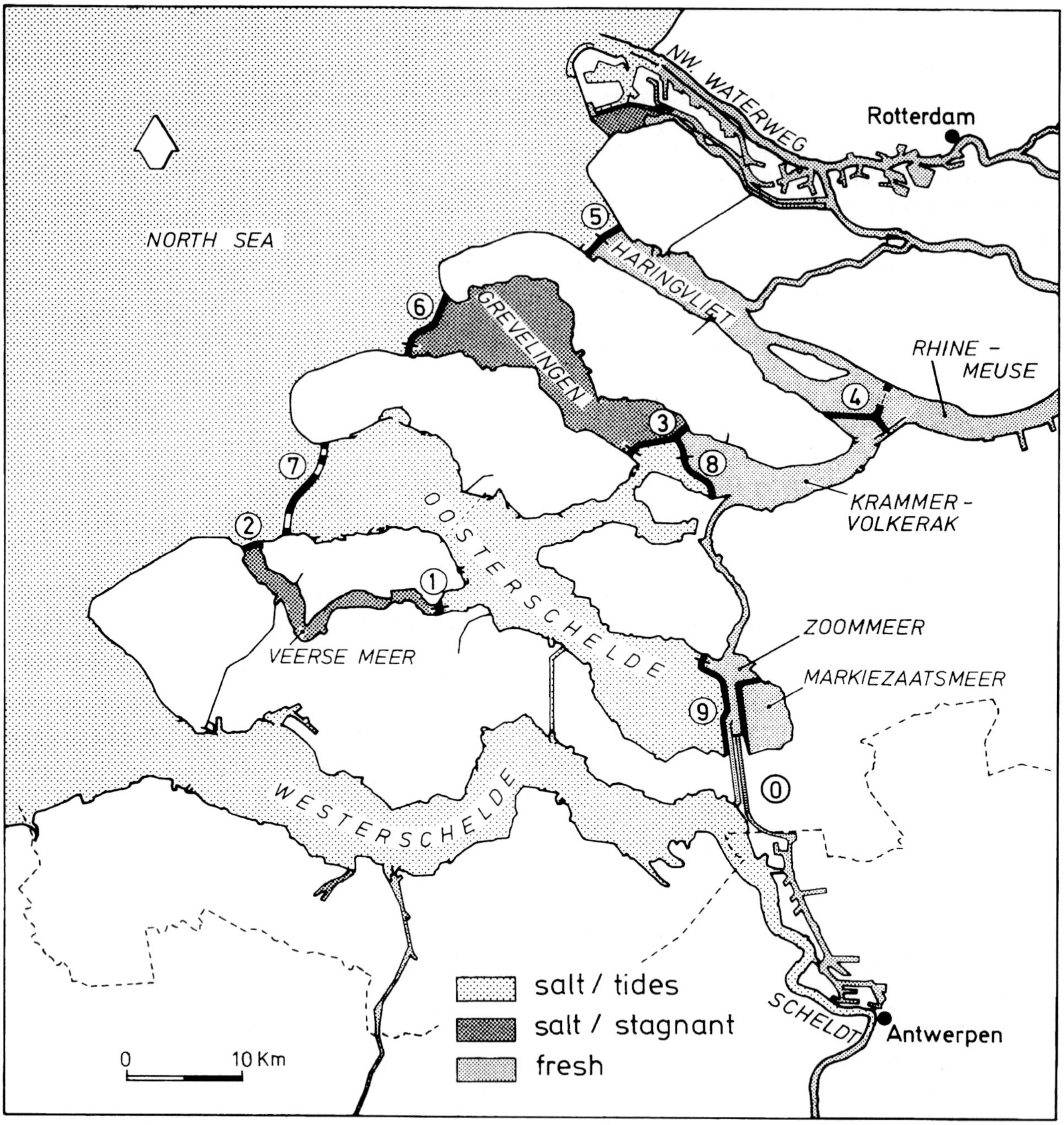

Figure 2. Map of the Delta area of the rivers Rhine, Meuse and Scheldt in the SW Netherlands, with various water bodies as resulting from the Delta project-engineering scheme. 1 = Zandkreekdam, 1960; 2 = Veersegatdam, 1961; 3 = Grevelingendam, 1964; 4 = Volkerakdam, 1969; 5 = Haringvlietdam, 1970; 6 = Brouwersdam, 1971; 7 = Oosterschelde storm surge barrier, 1986; 8 = Philipsdam, 1987; 9 = Oesterdam, 1986. Markiezaatsmeer has been closed off from Zoommeer by Markiezaatsdam in 1983. The connection between Oosterschelde and Westerschelde was already closed in 1867 (0 = Kreekrakdam). Projects preceding the Delta act of 1957, such as the closure of the Brielse Maas (1950), the Braakman (1952) and the construction of the flood barrier in the Hollandse IJssel (1958) near Rotterdam, have not been indicated (compare Figure 4; Nienhuis & Smaal, 1994).

and the sea. However, an increasing number of exotics have established themselves, covering large subtidal and intertidal areas (e.g. the Japweed, *Sargassum muticum* and the Japanese oyster, *Crassostrea gigas*; De Jonge & De Jong, 2002).

The Delta project is considered as the culmination of a long tradition of land reclamation and defence against the sea. Almost the entire area of the land around the estuaries was reclaimed from the sea in a trial-and-error process, over more than a millennium of constructing and repairing dykes in the muddy salt marshes and the former peat bogs. The combination of the rising sea level and subsidence of the reclaimed land (particularly the peat areas) dramatically changed the difference in surface levels between sea and land. Most polderland now lies far below the level of the sea (Fig. 3).

The flood disaster of 1953 has not been followed by an evaluation of the practice of traditional water and land management. Instead, the event worked as a catalyst for the decision to persist with large-scale measures in the existing tradition: to build larger and more rigid dams. There was a strong conviction that technology would always remain to be able to control the energy of the sea. The execution of the Delta project brought Dutch water engineers world fame. The skills and experience gained, became a significant export product of the Netherlands: Dutch engineering firms were asked to plan and execute similar large water projects in other parts of the world. For many countries the Dutch approach became the model for water management technology (www.rikz.nl).

From a socio-economic point of view, the impression of safety bestowed by the massive dykes, invited people to invest money behind them. Towns and villages prospered and tended to grow. Although the frequency of a potential disaster has diminished, the potential damage to lives and goods has increased: the impression of complete safety is therefore false. It is, in fact, the strong believe in technological solutions that made the Dutch population blind to the real risks. Particularly during periods of poor maintenance of the dykes (war, recession) that became only too obvious. The answer to a devastating flood has always been: build higher and stronger dykes. The effects have always been: more investments behind the dykes, but the repetitive consequences were: larger damage during a subsequent catastrophe. Concerning the Delta project, a cost-benefit analysis was carried out only after (and not before) the main decision to close off the estuaries was made (Tinbergen, 1959; Van Dantzig, 1959). Surprisingly, the analysis did not compare the different solutions; it simply calculated whether the costs of the chosen solution were in equilibrium with the expected benefits.

The chosen solution, to cut off the estuaries from the sea by large dams, will partly be irreversible in practice. The costs were so high that a

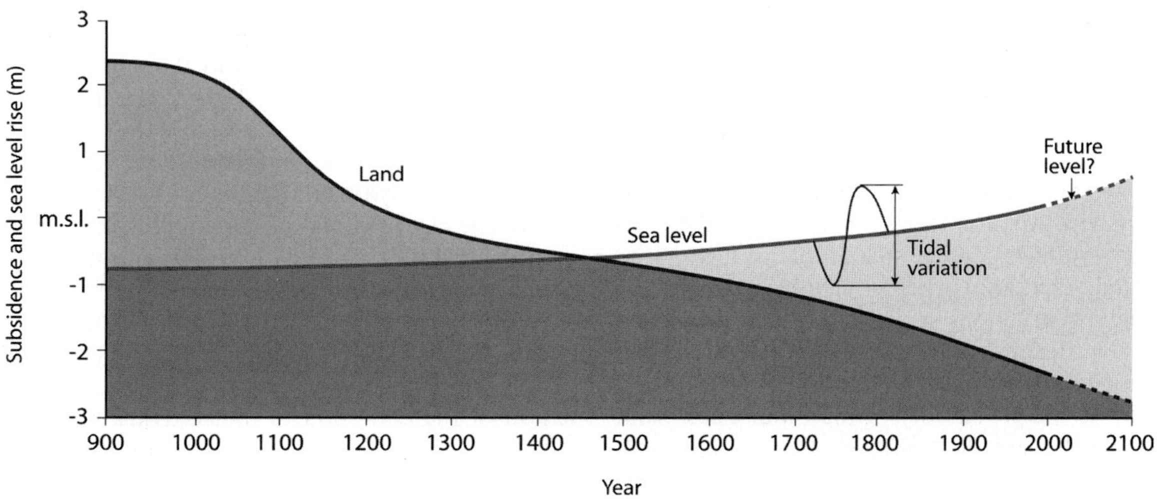

Figure 3. Scheme of the subsidence of the land in the western part of the Netherlands, and the simultaneous sea level rise over time (derived from Huisman et al., 1998).

reversal would mean a tremendous write-off of the investment: you cannot do the same job twice. Secondly, changes in the use of the lakes that originated behind the dams were far-reaching. New and strong economic interests developed, profiting from freshwater for agriculture and extensive tide-free shipping routes. These interests resist attempts to change the situation. The conclusion is that the decision to design and build large civil-engineering structures such as the Delta-dams, determines the pathway for the future. Rising sea levels will be countered by heightening the dams, thus increasing the potential damage.

Accepted risks and safety standards

A striking observation is that as soon as a dyke or dam is constructed, the inhabitants' relation with the natural dynamics of the system weakens and finally disappears. Shortly after construction, an embanked area will be used more intensively for living and working, usually to a degree which is out of proportion to whatever natural dynamics still occur in the area. Every improvement in 'safety' against the dangers of the natural environment is followed by new investments, and thus the risk is enlarged to, or beyond, the former level. The long history of transformation of the Dutch delta shows that spatial changes, very often comprise a series of small, apparently unimportant adaptations, such as the draining of polders and the expansion of a city near the river mouth. These processes take place gradually, and their impact on the environment is generally not perceived. At a certain moment, the development can no longer be stopped. At that moment, it can be postulated that economic growth can only be realised by more extensive and more radical alterations to the natural system. Apparently, it is extremely difficult to change strategies, once spatial plans have been executed, especially when it involves large engineering constructions such as the Delta project.

When dealing with a flood-prone area, the concept of risk is a central notion. Risk is often defined as the probability of the occurrence of an (unwanted) event multiplied by the consequences of that event: Risk = Probability × Effect. When we try to make this risk calculation for the year 1953, and roughly fifty years later for 2005, we find that the probability of a disaster has declined, but the potential effect has increased dramatically: more people and infrastructure will now be damaged by a major storm-flood than in 1953. However, it has to be recognised, that the effect of a flood disaster nowadays might be less severe than in 1953 as a result of improved communication and evacuation plans. Nevertheless, the rise of the water level in an inundated polder would occur faster than it did 50 years ago: the land has subsided and the dykes have been raised, which means that the polders would fill up more rapidly with seawater. While everybody feels safe in the shelter of the large constructions, it can be hypothesised that the risk has actually increased, not decreased. From this point of view, the Delta project was just a step forwards in a process, that had lasted for over 1000 years. The constant factor is that every measure to improve the safety of the area is followed by more investment and an increasing human population: greater technical safety (lower probability of flooding) is always cancelled out by the risk of larger numbers of deaths and more costly damage (RIVM, 2004).

In the Netherlands safety standards with regard to the risk of being hit by a flood are set by the Act of Defence against Flooding of 1996 (www.wetten.overheid.nl). Under this law the standards set by the execution of the Delta project, have been maintained (1/4000 to 1/10,000 per year). Three sets of arguments have shaped these standards. Firstly, arguments dealing with the predicted water levels should be mentioned, viz. the expected maximum impact of super-storms, including the tidal phase and the wind direction, the sea-level rise, and lowering of the land (due to man-induced subsidence as well as the autonomous sinking of part of the continental shelf). The quality of the dams, dykes and sluices constitute the second set. The expected loss of human life, and the value of the investments behind the dykes, is the third set of arguments. Although several of these factors can be calculated or standardised, others cannot. In the end, the standards are a political choice. The law says that every five years there must be an evaluation of the condition of dams, dykes and sluices. Whenever a shortcoming is found it has to be repaired. Whenever calculations derived from the first set, e.g. the rising sea level, make it necessary, dykes and dams must be adapted. There is, however, no obligation to reconsider the standard of safety against the arguments of the third set, the value of goods and human life.

Although legal obligations to maintain the agreed levels of flood protection exist, it is our opinion that this strategy cannot be sustained forever. Because of the subsidence of the western part of the Netherlands, the rising sea level and the more extreme fluctuations in river water discharges, due to climate change, we will end up living behind enormous dykes. Large pumps will have to be built to remove the continuous input of seepage water. The difference in surface level between the sea and the hinterland will further increase, enhancing the effects of a potential flood to catastrophic proportions. This situation can hardly be referred to as 'safe'. It has to be concluded that, seemingly, there is no way back: from the moment the first dyke was built by the monks, the endless spiral of 'fighting the waters' had begun. To our opinion, at some stage, the current approach of raising higher and higher dykes and building larger and larger dams will no longer be an option.

Lessons learnt and solutions proposed

Changing views with regard to technological 'solutions'

Decisions about managing the Dutch delta were based on contemporary knowledge, available in the 1950s and 1960s. This applies to the significance of the morphological and ecological attributes of the ecosystems, but also to the technical solutions chosen. In the 1950s it was clear: the struggle against the violent forces of the sea had to be won using physical barriers. Raising the existing dykes following the 1953 disaster, obviously, was no longer a convincing solution, considering the arrears of maintenance; it would take decades to reach the desired safety level. A statement had to be made, drastically shortening the coastline: the Delta Plan with its huge dams was designed with 'safety forever' in mind.

During extensive enterprises such as the Delta project (that took more than 30 years), it is common use that new ideas or concepts frequently displace those underlying the original project design (e.g. saline, stagnant Lake Grevelingen and the Oosterschelde with reduced tidal amplitude; Nienhuis, 1982). Over the past ten years, an increasing number of questions have been raised about the desirability and usefulness of rigid technological approaches. Are more and higher dams really the best solution to fight storm-floods? Are these measures not neglecting the rising level of the sea? Is an approach that opposes the natural behaviour of an estuary the best idea, or is it possible to envisage solutions that match with the forces of the sea and the characteristics of the estuary, rather than measures that counteract natural processes?

In 1953, the larger part of the affected area in the SW Netherlands had a rural destination, and agriculture was the dominant economic activity. The chosen solution (Delta project) was the best under the prevailing circumstances and within the cultural tradition. Agriculture, however, has since lost its pre-eminence in the now-urbanised Dutch society. Agriculture itself has been industrialised. Capital-intensive trades such as meat, vegetables and flower production need more and more investments, and less farmland. In the Netherlands, the future of types of agriculture that use large areas of land is very doubtful. Some economists predict an extreme decrease in land use for agricultural purposes, because the opening of the world market and the enlargement of the European Union to eastern European states makes it cheaper to produce elsewhere. Others predict a development towards more extensive land use: that is, fewer but larger farms that use more land and less intensive techniques. These two developments could occur at the same time. The outcome of a cost-benefit analysis depends to a considerable degree on the forecasts concerning the use of 'space'. It can be postulated that once the drive for more (agricultural) land has declined, a process opposite to that of the last 1000 years becomes possible: giving land back to the sea (Schuijt, 2001).

The maintenance costs of the Delta project are very high, much higher than estimated in the cost-benefit analysis (CBA) of 1959. Tens of millions of euros are spent each year to keep the civil-engineering constructions in good condition. The maintenance of the storm-surge barrier in the Oosterschelde alone, for instance, costs 15 million euros annually. Other costs were not foreseen (or accepted) at all, for example the projects to locally restore disturbed nature, and to control the water quality problems. To mitigate large-scale erosion of the foreshore of the former estuaries, bank protection over several hundreds of kilometres was put into place, at a cost of one million euro per kilometre. Moreover, considering the worldwide degradation of ecological quality, the economic value of ecosystems is being more widely recognized (Costanza et al., 1997). This added value was neglected in the past, and this concerns also the CBA framework used by Tinbergen (1959) and Van Dantzig (1959). Attempts are now being made to include the economic value of (aquatic) ecosystems in the decision-making process, when interventions are planned. Applying the economic values attached to an estuary, e.g. with regard to fisheries interests, it has been shown repeatedly that human intervention results in huge economic losses. Calculations along these lines, show that the economic value of the Dutch south-western delta has declined by 40% between 1900 and 2000 (Bouma & Saeijs, 2000).

In the end it is impossible to attach monetary values to all that nature offers. The ethical discussion should be added because intrinsic values are complementary to monetary values. What is it worth to be able to walk along a natural beach and leave problems and stress behind you? How much do we want to pay to let our children enjoy the sea and the wetlands without worrying about pollution problems? It is to society to decide what is the price to be paid to enjoy a healthy ecosystem.

The values of a natural delta: bring back dynamics

The execution of the Delta project, which followed centuries of smaller interventions, triggered several (unexpected) environmental problems. The building of the delta dams rigorously cut off the hydrologic and ecologic river continuum, both at the seaside as well as at the side of the rivers. The annihilation of the dynamic tidal gradient was foreseen by ecologists, but their voice was not heard in the 1960s when the provision of safety after the 1953 disaster was the main societal issue, and ecological arguments hardly played any role (Nienhuis, 2006). A primary function of the former estuaries was the discharge of river water to the North Sea. The closure of the Haringvliet and the Volkerak is obviously hindering the unrestrained discharge of river water during peak floods. The 1995 river flood has opened the eyes of the river managers for that problem, and measures are taken now to mitigate that problem, both along the rivers proper (core decision of physical planning 'Room for the River'; www.ruimtevoorderivier.nl; Van Stokkom et al., 2005) as well as in the delta region. A problem may arise when an extreme river flood coincides with a north-westerly storm flood at the North Sea, which requires the storm-surge barriers to be closed.

Sealing off the estuaries in the Dutch delta has led to the accumulation of polluted sludge in the northernmost river branches (Fig. 4). Although an increase in sedimentation was expected as a result of the closure of the Haringvliet in 1970, the large quantity of polluted sediments that settled in subsequent years in the Dutch delta overran all predictions. Biesbosch–Hollands Diep and Haringvliet became the downstream chemical depot for the Rhine and Meuse rivers. More than 150 million m^3 of highly polluted sludge have settled here. Fortunately, the quantity of pollutants in river effluents has steeply decreased in recent decades, and the toxic sediments are now being covered with relatively clean sediments, but the underlying potential negative effects are still available. It is likely that, even if the original sources of pollution should be removed, contaminated sediments would continue to deliver emissions over many decades (Smit et al., 1997).

The lack of tidal currents in the Delta compartments is the cause of many of the environmental problems that recently emerged. However, it is technically possible to bring tidal rhythms back into the area. If the Haringvliet sluice is to be turned into a storm-surge barrier (Fig. 4), and if both Lake Oostvoorne and Lake Brielle could be opened up again in a controlled, reduced way, (Fig. 4), it can be hypothesised that the natural dynamics of the northern part of the coastal delta, comprising tides, and a salinity and sedimentation gradient, could be restored. The opening of the two lakes mentioned, would nevertheless be difficult because of the intensive urban infrastructure and harbour developments in the area. A connection with the developing sandy islands in the coastal delta, where a new shallow sea is emerging, could upscale the natural values by a substantial

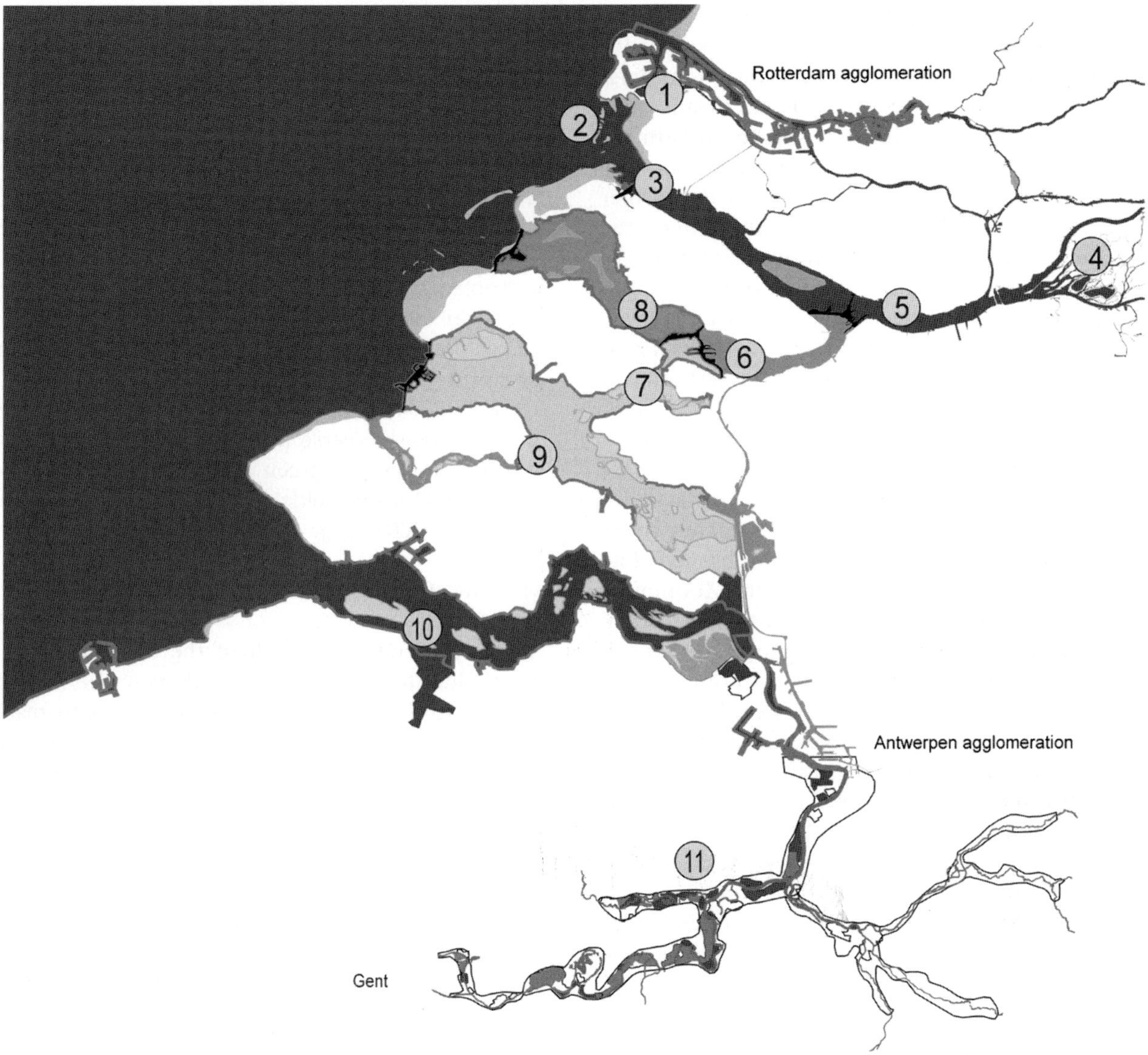

Figure 4. Changing estuaries, changing views: the return of the (lost) tidal dynamics. (l) Reconnecting Lake Oostvoorne and Lake Brielle to the sea and the tides. (2) Voordelta: undisturbed development of sandbanks alongside the coast, a new outlet for the rivers Rhine and Meuse. (3) Opening of the Haringvliet Sluices to restore the tides and the brackish gradient from freshwater to salt water. (4) Removing dykes and restoring the freshwater tidal area in the Biesbosch. (5) Sedimentation zone in the Hollands Diep branch for clean sandy sediments from the river Rhine. (6) Flushing the eutrophic Krammer–Volkerak with cleaner water from the river Rhine. (7) Connecting the Oosterschelde to the rivers Rhine and Meuse. (8) Connecting the Grevelingen both to the sea and to the river. (9) Connection between the stagnant Veerse Meer and the tidal Oosterschelde. (10) Returning the reclaimed Braakman area to the sea, for energy dissipation. (11) Controlled Inundation Areas, controlled reduced tides along the tidal river Scheldt (Saeijs et al., 2004).

amount (Fig. 4). If combined with the development of large natural tidal zones along the borders of Lake Oostvoorne and Lake Brielle, it would result in an extremely scarce and highly valued ecosystem, in the neighbourhood of one of the largest industrial concentrations of Western Europe (Stikvoort et al., 2002; www.rikz.nl; www.ecologisch-herstel.nl).

Experiments to re-introduce reduced tidal movements in the Haringvliet have been carried out and were successful. If the Haringvliet estuary is to be restored, this should be done as soon as possible, since irreversible geomorphological processes, combined with the extinction of migratory species (particularly fish species) are continuing. Opening of the Haringvliet sluice means also the (partial) restoration of tidal movements in the Biesbosch (Fig. 4), an area created during the giant storm-flood of 1421. In that year the former polder, the Grote Waard, was swallowed by the sea and turned into a shallow freshwater tidal area, and it took the river Rhine four centuries to fill up most of the area with sediments to above sea level. During this period, the Biesbosch was a vast and unique freshwater tidal area, the largest one in Europe (Kuijpers, 1995; Kerkhofs et al., 2005).

During the most severe period of pollution from the Rhine and Meuse, the Krammer–Volkerak (Fig. 4) was closed off from the Haringvliet in 1969, and this prevented the contaminated riverwater to enter the adjacent estuarine branch. The enclosed water mass, including the Zoommeer (Fig. 4) was conceived as a freshwater system, almost exclusively fed by the discharge from a few small rivers in Noord-Brabant, although some input of Rhine and Meuse water had to be accepted. Over the years, however, it became clear that nutrient accumulation, causing mass blooming of blue-green algae, nevertheless occurred, fed by agricultural run off, mainly brought by the smaller rivers. A recent survey of possible solutions to the eutrophication problem was carried out. One suggestion was to flush the artificial lakes with enough freshwater, in order to decrease the residence time of the eutrophicated water and hence to prevent the development of algal blooms. The problem is that this measure cannot be applied in dry summer periods when little river water is available, but algal blooms are at their peak. Creating a saltwater lake or a semi-tidal estuarine area are the suggested directions for a sustainable solution. The problem here is that the adjoining agriculture needs the freshwater supply (Tosserams et al., 2002; www.rikz.nl; www.ecologisch-herstel.nl).

In 1986 the building of the storm surge barrier in the Oosterschelde was finished. Owing to this enormous technical and financial effort, two thirds of the tidal movements have been maintained in this estuary. As this estuarine branch was cut off from the river, the connection with incoming nutrients, and the transition zone between salt and freshwater, were lost. The deterioration of the natural system, i.e, the irreversible erosion of the tidal flats, is continuing. Partial restoration of the estuarine gradient is a feasible option, however, by re-introducing a quantity of freshwater from the river Rhine via the Krammer–Volkerak (Fig. 4). The connection between the stagnant non-tidal saltwater Lake Grevelingen (Fig. 4) and the sea has already been restored in a restricted way. The flushing of the lagoon, however, can be enhanced by expanding the capacity of the already existing siphon in the eastern dam, connecting Grevelingen with the Oosterschelde (RIKZ, 2004). The connection between the stagnant, brackish Veerse Meer – suffering from massive blooms of the green alga *Ulva lactuca* – and the tidal Oosterschelde has already been accomplished in 2005 (Fig. 4; www.rikz.nl).

A useful approach: the Flemish view

The southernmost branch of the delta, the Westerschelde, has remained in open connection with the North Sea, to allow sea going vessels to reach Antwerp harbour. Artificial deepening of the navigation channel for ever bigger ships enhanced the man-induced tendency of the Westerschelde to enlarge its tidal volume. The potential danger of this ongoing process was particularly felt in the narrowing, upstream part of the estuary, and it was modelled that the city of Antwerp experienced an ever greater risk of being flooded. The construction of a storm surge barrier in the Westerschelde (compare with the Oosterschelde dam) was financially and politically out of the question, and in the meantime the thoughts about the values of natural systems were evolving. After thorough studies, leading to better understanding of the tidal system, Dutch water managers responsible for the Westerschelde came up with a completely new approach within the limits of the natural characteristics of the estuary (Saeijs et al., 1993). During a centuries-long history, the medieval river upstream of Antwerp gradually changed into a tidal river, and large parts of the floodplain were reclaimed in the course of time. By constructing a new dyke more land inwards, and by subsequent removal of the old dyke, the original floodplain upstream of Antwerp could be given back to the river ('ontpolderen' in Dutch). These measures create more space for the river: the tides will come in twice a day, occupying the new floodplain, and hence changing the river forelands into freshwater tidal marshes, and increasing the tidal volume once again. The advantages are obvious: lower water levels would give more safety; more water exchange would lead to less dredging measures, and better quality of nature.

Flemish engineers, morphologists and ecologists adopted these ideas and came up with an adapted approach (Van den Bergh et al., 1999, 2003). To prevent flooding, the Flemish designed a system of Controlled Inundation Areas (CIA) along this estuarine river: giving back the floodplain to the tidal system in a controlled way (Fig. 5). Simply removing dykes in places where (uninhabited) floodplain were still intact would not have been effective. Instead, the dykes have been lowered in such a way that, during severe storms when Antwerp is in danger, the dykes start to overflow into selected polders (Fig. 6). Newly built, higher dykes in the hinterland protect the adjoining villages and cities. The top of the flood is thus removed in an 'elastic' way, playing with the natural forces instead of opposing them, within the limits of the natural system.

This system has some more advantages. Firstly, up to eight times a year the dykes of the controlled overflow polders may be overtopped, which means that the process of the sedimentation of silt during extreme tides may continue. In this way the area will slowly be elevated, following the rise of the sea level. A second advantage is that by using the in- and outlet devices of the polders, the opportunity will be opened to restore the natural conditions of a (internationally extremely rare) freshwater tidal system. At the moment, several areas covering a total of about 2000 ha are under construction (Existing Controlled Inundation Areas in Fig. 5). There are plans to enlarge the area to up to 4500 ha

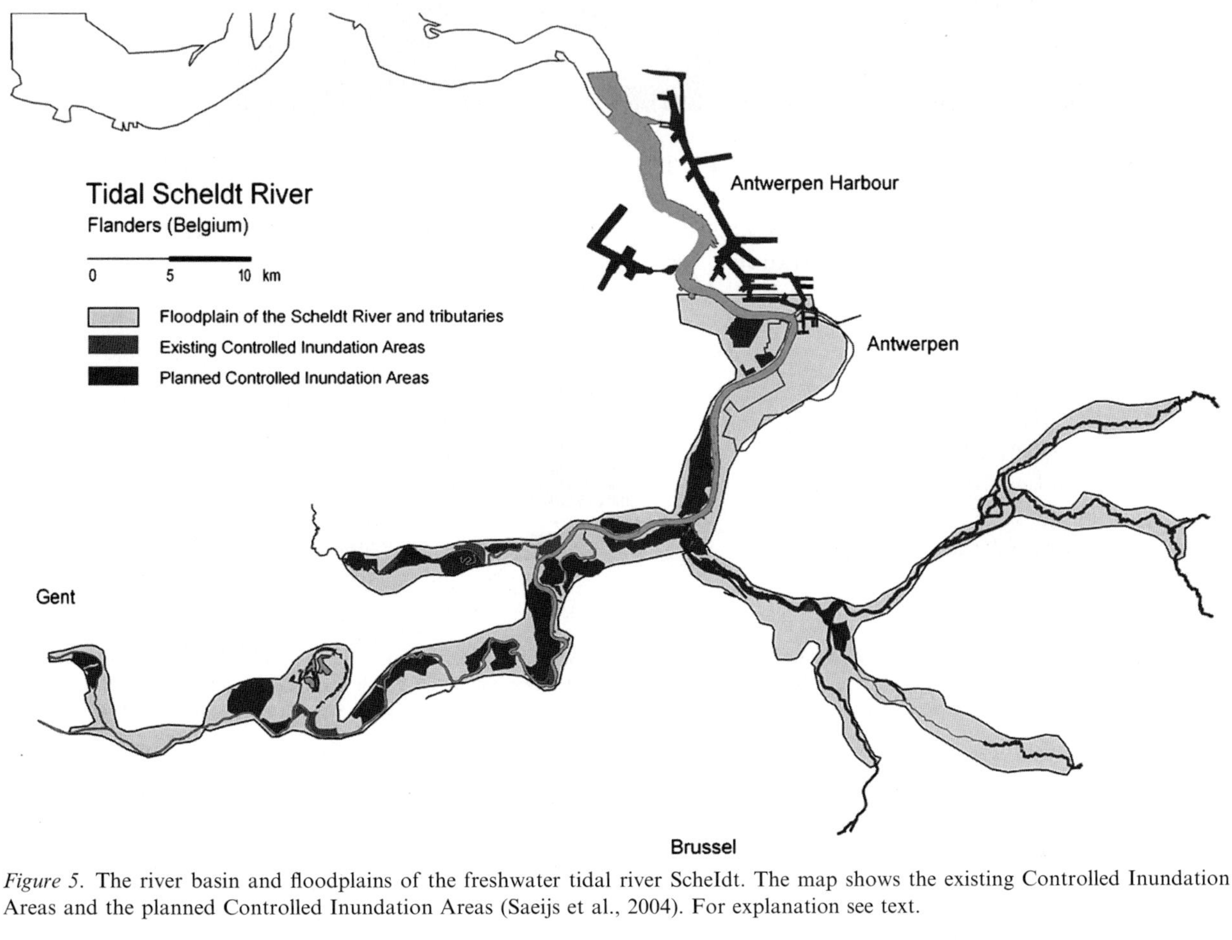

Figure 5. The river basin and floodplains of the freshwater tidal river Scheldt. The map shows the existing Controlled Inundation Areas and the planned Controlled Inundation Areas (Saeijs et al., 2004). For explanation see text.

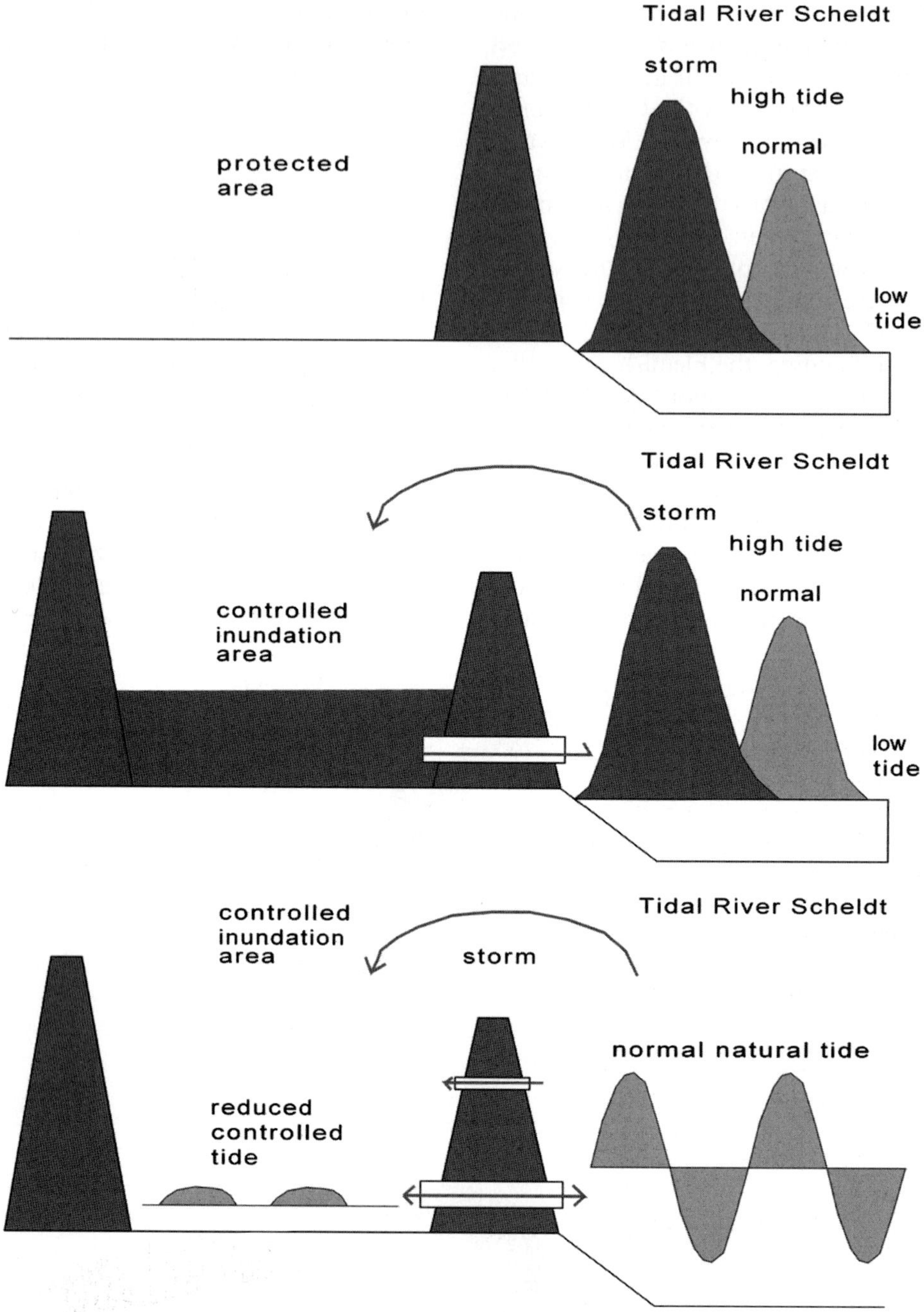

Figure 6. Controlled Inundation Areas along the freshwater tidal river Scheldt. Upper panel: Scheme of the tidal river Scheldt, existing situation, danger of uncontrolled flooding. Middle panel: Controlled Inundation Area along the river Scheldt provides protection against flooding. Lower panel: Reduced controlled tides: freshwater tidal areas along the river Scheldt (Saeijs et al., 2004).

(Planned Controlled Inundation Areas in Fig. 5). Apart from the improved safety conditions, and the possibility to keep pace with the sea level rise, a natural tidal freshwater river will reclaim its tidal plains. In the very densely urbanised area of Flanders, including Antwerp, Ghent and Brussels, a beautiful natural park will be created with unprecedented ecological and recreational potentials.

In 2003 Flemish and Dutch experts formulated an outline of a common sketch for sustainable development of the Scheldt estuary over its full length of about 160 km, based on an earlier accepted long term vision for the Scheldt-estuary. The Flemish Controlled Inundation Area solution was presented through an international body (ProSes), and similar solutions for the Dutch part of the estuary were presented as well. A new idea was formulated by the ProSes study (Van den Bergh et al., 2003), viz. dissipation of tidal energy. According to this idea the energy of the tidal wave is absorbed by its contact with the soil of the permanent water bodies and of the (periodically) inundated areas. In wide and shallow systems the energy of the tides is absorbed to a much higher degree than in waters with limited contact with the streambed and banks of the channel. In terms of energy dissipation, the 1000-year technological development of building dykes, cutting of shallow branches from the sea, and deepening the navigation channel, have diminished the energy absorption capacity greatly, allowing the floods to develop more force and cause more damage. Serious drawbacks of dredging measures in tidal navigation channels on the erosion–sedimentation cycle (in the Ems estuary) were discussed by De Jonge (1983).

Dykes are not entirely safe, and decision makers in the harbours and industries of Westerschelde area had no intention to rely on the 1:10,000 or 1:4000 per year safety standards. The huge investment in heavy industry and, even more crucial, the existence of three nuclear plants (two in Flanders and one in the Netherlands), put the value of potential damage very high indeed, and called for additional solutions. In order to survive extremely high tides, artificial dwelling mounds were in use in the estuary already in the early Middle Ages (Bloemers & Van Dorp, 1991). Even the highest floods obey the rhythm of the tides so that on an artificial mound, the worst that could happen was a shallow covering of water for a few hours. Along the estuary most places are situated below Mean Sea Level, sometimes several metres. Industrial managers made a partial return to the safest possible strategy for flood prevention: large-scale artificial mounds. In fact, many of the large harbour facilities and industrial estates were built on mounds. Even in the worst case, i.e. a flood exceeding the 1:10,000 level, the mounds will remain dry or only suffer from shallow flooding during a short period. Curiously enough, urban planning did not follow this strategy (Saeijs et al., 2003).

Alternative approaches: a society in balance with nature

For centuries the primary economic drivers in the Dutch delta were agricultural interests. Since globalisation has enlarged the trade markets, the rationale behind the fight against the sea, to maintain reclaimed land for regional food supply, has lost much of its value. Keeping in mind agricultural over-production in Europe, it is easy to conclude that there is even too much land used by farmers. Resource-use strategies should be looked for, which are compatible with the characteristics of the geographic region. It may be postulated that an economy based on marine and estuarine fisheries and aquaculture is in better harmony with the resources of the delta than forced agricultural practice.

The delta area could become even more important for recreation than it is already nowadays. Leisure time and spending for amenity reasons have grown during the last decades, and the millions of people living in the densely populated western part of the Netherlands provide a ready market. The recreation industry could be expanded by changing land use; for instance by returning the ancient agricultural lands to nature. These lands have settled over hundreds of years of use, lowering and compacting the ground. Here, freshwater could stagnate, leading to interesting freshwater nature reserves. As these lands are situated near picturesque old towns, a perfect combination of culture and nature could be created. Recently, a computer model has become available which simulates varying conditions related to land and water management and climate change (*SimDelta: a spatial model for coastal morphology on mega-time scale*; www.rikz.nl). Application of the model allows an exploration of possible morphological changes in the south-western delta area under specified conditions. Within the context of this paper, although speculative, two scenarios are intriguing.

1. No dykes, open estuaries, and no forced drainage of the dry land. If the inhabitants of the Dutch delta had not been focused on land reclamation but on aquaculture and fishery, the south-western estuary would be much larger than it is today. In this scenario only the villages and cities would be embanked or otherwise protected from, or accommodated to, high water levels. The importance of the wetlands and open estuaries as nurseries for fish and molluscs, and breeding places and feeding grounds for waterfowl would have been recognised and preserved. The number of flood events would have been larger, but the impact – both socially and economically – is assumed to have been lower. A disaster on the scale of the 1953 flood would probably not have occurred because estuarine sedimentation processes, and the growth of peat in isolated river-fed areas would have matched the irreversibly rising sea level, caused by geophysical and climate changes. The model assumes that the population density could have been approximately at the same level as it is nowadays. All the costs linked to dyke construction, closing of the estuaries, water pollution and nature restoration would have been saved.
2. Embanked islands, open estuaries, and no overstrained drainage of the dry land. If the inhabitants had decided to embank the islands without applying intensive drainage programs, the total area of land above mean sea level would have been larger than it is now, but smaller than in scenario 1. In scenario 2, the inhabitants would have been focused on trading, tourism and, to a smaller extent, on aquaculture and fishery. The spatial planning of the islands would have catered for the storage, handling and transport of goods. After embankment, sedimentation processes would have ceased but peat growth would have prevented subsidence of the soil. The weakness of the soil structure would, however, have required specifically adapted construction methods for housing, buildings and infrastructure. Flooding frequency would have been higher than in scenario 1 because the embankments would have raised the water levels in the river branches. However, the impact of incidental flooding would have been relatively low in the absence of soil subsidence. Hypothetically, it is unlikely that the damage of a flood of the dimensions seen in 1953 would have happened. All the costs linked to closing the estuary, water pollution and nature restoration would have been saved.

Again, although it is speculative, it can be concluded from the above model exercises that a delta without dykes is safer than a delta with dykes, because natural processes will weaken the effects of extreme storm floods. A dynamic delta is more sustainable and robust, while it is flexible enough to adapt to changing situations. A static delta, on the contrary, is not safe and increasingly vulnerable to severe flooding. Projected into the future, this insight means that, in the Dutch case, if the importance of agriculture diminishes (as expected), new choices will open up. A modern, sea-oriented society could be developed, with (large) harbours, sea-oriented industries, shipping and fisheries and aquaculture trades. Housing on islands would be safe against flooding and surrounded by a shallow sea. We have every opportunity to do this moderately and wisely, by keeping the land we need and giving back to the sea what is necessary, to turn the environment into a more safe natural system.

This line of thinking opens a new perspective for different approaches to spatial planning in areas that are at risk of flooding. The delta of the rivers Rhine, Meuse and Scheldt is an example of an area where people could live and work in places that are sometimes inundated. Dykes could be displaced, enabling (temporary) inundation and sedimentation processes to take place again. The surface level of the land will be raised by natural sedimentation processes, providing a durable alternative for the unreliable dykes. When the natural dynamics are used in an optimal way, the effects of sea level rise, climate change and land subsidence can be partially counteracted. Maintenance will always remain necessary, however, because unpredicted and unwanted erosion and sedimentation processes have to be managed and guided. At sheltered localities houses could be built in the floodplain, constructed in such a way that the water cannot affect them, for example floating during floods or built on small artificial mounds as was practised in past centuries. While the 'mound-strategy' is already used in the Dutch delta by large investors, the private inhabitants should be offered the same advantage. Risks to life would be reduced compared to the situation in which an unexpected breach of a dyke could take place.

Implementing the measures described above would introduce a sustainable way of rehabilitating nature and developing the specific characteristics of the estuaries, so that ecological productivity and biodiversity will be optimised. Economically, this type of development needs less maintenance, leads to lower costs for flood defence, fulfils the precautionary principle and decreases flood risks. Next to the people economically bound to the region, the environment would be very attractive for retired people: Zeeland could become the "Florida" of the Netherlands, providing unique living conditions in harmony with the natural water environment. It could grow into an example of spatial planning in accordance with the natural system instead of a defensive planning against the basic characteristics of the ecosystems (cf. Van den Born et al., 2001).

As we have seen in the section '*A useful approach: the Flemish view*', along the Westerschelde, proposals have been launched to give land back to the sea in order to dissipate tidal energy, and to enhance the quality of the natural environment. Another successful example of changed land use can be found in England (European FRAME project). Dykes that were raised on the Alkborough flats along the east coast after the 1953 storm, are being removed. As a result, part of the Humber estuary (440 ha) will be changed from agricultural land into natural salt marsh and other estuarine habitats. An additional advantage is that the intertidal flats will break the incoming waves. The saving in maintenance and building costs of the surrounding dykes is estimated at 18 million euros (www.frameproject.org).

Recommendations

Drawing on contemporary experience and knowledge, it would be easy to point the finger at past mistakes and wrong decisions. We cannot change history, but we can adjust our own future and share with other countries the Dutch experience and changing views, related to the exploitation of estuaries. In summary, the lessons we have learned from the design, execution and follow-up of the Delta Works in the SW Netherlands have led us to the following recommendations and visions:

1. The Delta Works approach has underestimated the importance of long-term hydromorphological and ecological processes and changes: the height of the dams will have to be increased for centuries to come, and the land behind the levees cannot grow anymore with the rising of the sea.
2. Realise that on-going costs for maintenance and mitigation of side effects are ultimately higher than those for the original building of the civil engineering constructions. Maintenance costs for technical constructions and mitigation costs of hydrological and ecological negatively valued moves are a perpetual expense. In this context the Dutch taxpayers are charged for several tens of millions of euros each year, and these costs will not decline but will only rise, due to the progressing deterioration of the engineering works.
3. 'Look before you leap': safety declines! There is only a toilsome way back after land reclamation and dyke construction, and it is therefore better not to start this process. The 1953 flood was a disaster waiting to happen, and it might happen again. The dependence on dykes and other infrastructure will intensify over time, which means that opting for short-term safety is inevitably connected to increasing long-term vulnerability. Moreover, once a sea-dyke is built, people will use the new land to the maximum: new investments are made, so the economic damage of a flood disaster increases steadily.
4. Better to be safe than sorry. Dykes can never guarantee full safety; they provide a false sense of safety against flooding. The huge dams may be technical masterpieces for control of the dynamics of the sea, but society fails to control the socio-economic processes the dykes unleash during a flood disaster, and their existence is to a large extent irreversible. The chance of flooding has been reduced, but the potential damage is enlarged, so the net

economic result is probably negative. Risk assessment should focus not only on minimising the risk of a prevailing flood disaster, but also on minimising the impact by not allowing large investments at vulnerable locations.

5. There is a call for reversible and flexible solutions. When modifications of the natural system seem to be inevitable, try to use reversible measures. As knowledge develops, other solutions might be found which could than be applied. The hydro-morphological and ecological changes that will occur when an estuary is modified are still poorly understood, and predictions have to deal with a great amount of uncertainty. Changes in the importance of agricultural land in the European context, offer an opportunity to give arable fields back to the sea in order to absorb tidal energy and to allow the land to rise concomitant with the sea.
6. Look for suitable economic drivers. Each (semi-natural) landscape has its own characteristics and, considering the fact that the values of the existing ecosystems are recognised and integrated into economic development strategies, extreme disturbance of the environment should be avoided. Try to develop economic drivers that are compatible with the conditions of the natural environment, in the delta e.g. fishery, aquaculture, salt-water crops, and marine-wetland oriented recreation and trades. Reversible and resilient economic measures within the limits of the natural processes are preferable.
7. A future perspective is a new urbanised landscape, where people and investments are located in safe places, surrounded by a landscape that is ruled by the forces of nature. Industry took refuge on artificial mounds, the most ancient and the most modern way to survive storm-floods. Floating, or sea-encircled artificial dwelling-mounds deserve full attention as a long-term strategy for safe building in the lowest parts of the Netherlands and Flanders. New approaches such as developed in the Westerschelde offer flexible solutions to flooding problems, and are worth a broader evaluation.
8 Try to make a complete environmental cost-benefit analysis, including the direct and indirect economic use values, and the non-use (intrinsic) values. The use and non-use values of (unaffected) ecosystems, and the long-term value of the safety that nature provides if we do not interfere, should be included in cost-benefit analyses
9. Create public awareness related to ecosystem functioning and safety for people and their goods. Public participation may result in better understanding of the structure and functioning of ecosystems. When the short- and long-term consequences of, for example, obstructing the connectivity of the river-estuary–sea continuum is explained, people will better understand the importance of an intact ecosystem. The result may be greater goodwill for sustainable developments, even if unpopular measures have to be taken such as giving land back to the sea.
10. Consult a world-wide group of estuarine experts, from the natural as well as from the social sciences. As shown in this paper, lessons can be learned from countries that have experience with interference in estuaries. As water management practices are traditionally exported by Dutch experts, the Netherlands and Flanders (Belgium) could take the lead in such an initiative.

Acknowledgements

This paper is a shortened an updated version of the report ‘Changing estuaries, changing views’ (Saeijs et al., 2004) comprising the farewell speech of professor Saeijs and the inaugural speech of professor Smits, held on 16 June 2004 at the Erasmus University Rotterdam. Researchers of the Erasmus University Rotterdam (the Erasmus Centre for Sustainability and Management) in collaboration with the Radboud University of Nijmegen (Centre for Water and Society), commissioned by the Worldwide Fund for Nature (WWF) have compiled evidence for this paper, sustained by experts from the Ministry of Transport, Public Works and Water Management, and Dutch and Flemish Institutions. The constructive remarks of three anonymous reviewers are gratefully acknowledged.

References

Bloemers, J. H. F. & T. van Dorp, 1991. Pre- and protohistorie van de lage landen. Open Universiteit, Unieboek, Houten (in Dutch).

Bouma, J. J. & H. L. F. Saeijs, 2000. Eco-centric cost-benefit analysis for hydraulic engineering in river basins. In Smits, A. J. M., P. H. Nienhuis & R. S. E. W. Leuven (eds), New Approaches to River Management. Backhuys Publishers, Leiden: 167–178.

Costanza, R., R. d'Arge, R. de Groot, S. Farber, M. Grasso, B. Hannon, S. Naeem, K. Limburg, J. Paruelo, R. V. O'Neill, R. Raskin, P. Sutton & M. van den Belt, 1997. The value of the world's ecosystem services and natural capital. Nature 387: 253–260.

De Jonge, V. N., 1983. Relations between annual dredging activities, suspended matter concentrations, and the development of the tidal regime in the Ems estuary. Canadian Journal of Fisheries and Aquatic Sciences 40 (suppl. 1): 289–300.

De Jonge, V. N. & D. J. de Jong, 2002. Ecological restoration in coastal areas in the Netherlands, concepts, dilemmas and some examples. Hydrobiologia 478: 7–28.

Huisman, P., W. Cramer, G. van Ee, J. C. Hooghart, H. Salz & E. C. Zuidema (eds), 1998. Water in the Netherlands. Netherlands Hydrological Society, Delft: 1–186.

Kerkhofs, M. J. J., T. Tiebosch, J. A. van der Velden & J. W. M. Kuijpers, 2005. Alternative management of the Haringvliet sluices: the first step towards major rehabilitation of the Rhine-Meuse estuary. Archiv für Hydrobiologie Supplement 155, Large Rivers 15: 569–578.

Kuijpers, J. W. M., 1995. Ecological restoration of the Rhine-Maas estuary. Water Science and Technology 31: 187–195.

Nienhuis P. H., 1982. De oecologische consequenties van de Deltawerken. In Wolff, W. J. (ed.), Wadden, Duinen, Delta. Biologische Raad Reeks, PUDOC, Wageningen: 101–132 (in Dutch).

Nienhuis, P. H., 2006. Water and values: ecological research as the basis for water management and nature management. Hydrobiologia 565: 261–275.

Nienhuis, P. H. & A. C. Smaal, 1994. The Oosterschelde estuary (the Netherlands), a case-study of a changing ecosystem. Hydrobiologia 282/283: 1–597.

RIKZ, 2004. Verlopend getij. Oosterschelde, een veranderend natuurmonument. RIKZ Rapport (2004).028. Rijksinstituut voor Kust en Zee, Middelburg: 1–80 (in Dutch).

RIVM, 2004. Risico's in bedijkte termen. Rijksinstituut voor Volksgezondheid en Milieu, Milieu- en Natuurplanbureau – RIVM rapport 500799002, RIVM, Bilthoven: 1–242 (in Dutch).

Saeijs, H. L. F., Pieters T. & Turkstra E., 1993. Van graven met de hand tot beheren met verstand. Visie op de toekomst van het Schelde-estuarium. Symposium volume Hogeschool Zeeland Gazet 6 (8/9): 5–9 (in Dutch).

Saeijs, H. L. F., A. J. M. Smits, W. Overmars & D. Willems, 2004. Changing Estuaries, Changing Views. Erasmus University Rotterdam, Radboud University Nijmegen, XXL Press, Nijmegen.

Schuijt, K., 2001. The economic value of lost natural functions of the Rhine river basin; costs of human development of the Rhine river basin ecosystem. Erasmus University Rotterdam, Publication Series 36: 1–42.

Slager, K., 2003. De Ramp; een reconstructie van de watersnood van 1953. Atlas, Amsterdam/Antwerpen (in Dutch).

Smit, H., G. van der Velde, R. G. Smits & H. Coops, 1997. Ecosystem developments in the Rhine-Meuse Delta during two decades after enclosure and steps towards estuary restoration. Estuaries 20: 504–520.

Stikvoort, E., J. Graveland & R. Eertman, 2002. Leve(n)de Noordrand. Pragmatische toekomstvisie voor het ecologische herstel van het Rotterdamse havengebied. Rapport RIKZ, Den Haag/2002.032 (in Dutch).

Tinbergen, J., 1959. Socio-economic aspects of the Delta Plan. Report Delta Commission, Contribution VI, Research with importance to the design of dykes and dams. Rijkswaterstaat, Den Haag: 66–74 (in Dutch).

Tosserams, M., E. H. R. R. Lammens & M. Platteeuw, 2002. Het Volkerak–Zoommeer; de ecologische ontwikkeling van een afgesloten zeearm. RIZA Rapport 2000.024. Ministerie van Verkeer en Waterstaat, Lelystad (in Dutch).

Van Dantzig, D., 1959. The economic decision problem related to flood management in the Netherlands. Report Delta Commission, Contribution II,2, Research with importance to the design of dykes and dams. Rijkswaterstaat, Den Haag, 58–110 (in Dutch).

Van den Bergh, E., S. van Damme, J. Graveland, D. J. de Jong, I. Baten & P. Meire, 2003. Voorstel voor natuurontwikkelingsmaatregelen ten behoeve van de Ontwikkelingsschets 2010 voor het Schelde estuarium. Werkdocument RIKZ/OS/2003.825x. RIKZ, Middelburg (in Dutch).

Van den Bergh, E., P. Meire, M. Hoffmann & T. Ysebaert, 1999. Natuurherstelplan Zeeschelde: drie mogelijke inrichtingsvarianten. Rapport IN 99/18. Instituut voor Natuurbehoud, Brussels (in Dutch).

Van den Born, J. G., H. J. R. Lenders, W. T. De Groot & E. Huijsman, 2001. The new biophilia: an exploration of visions of nature in Western countries. Environmental Conservation 28: 65–75.

Van Stokkom, H. T. C., A. J. M. Smits & R. S. E. W. Leuven, 2005. Flood defence in the Netherlands: a new era, a new approach. Water International 30: 76–87.

Part Four
Synthesis

Hydrobiologia (2006) 565:359–367
R.S.E.W. Leuven, A.M.J. Ragas, A.J.M. Smits & G. van der Velde (eds), Living Rivers: Trends and Challenges in Science and Management
DOI 10.1007/s10750-005-1925-3

Living rivers: trends and challenges in science and management

G. van der Velde[1,5,*], R.S.E.W. Leuven[2,5], A.M.J. Ragas[2,5] & A.J.M. Smits[3,4,5]
[1]*Department of Animal Ecology and Ecophysiology, Institute for Wetland and Water Research, Radboud University Nijmegen, P.O. Box 9010, 6500 GL Nijmegen, The Netherlands*
[2]*Department of Environmental Science, Institute for Wetland and Water Research, Radboud University Nijmegen, P.O. Box 9010, 6500 GL Nijmegen, The Netherlands*
[3]*Centre for Water and Society, Radboud University Nijmegen, P.O. Box 9010, 6500 GL Nijmegen, The Netherlands*
[4]*Erasmus University Rotterdam, Erasmus Centre for Sustainability and Management, P.O. Box 1738, 3000 DR Rotterdam, The Netherlands*
[5]*Member of the Netherlands Centre for River Studies (NCR), P.O. Box 177, 2600 MH Delft, The Netherlands*
*(*Author for correspondence: E-mail: G.vandervelde@science.ru.nl)*

Key words: environmental rehabilitation, ecological rehabilitation, ecosystem-based water management, river functioning, sustainability

Abstract

This paper discusses and summarizes the major conclusions of the special issue entitled 'Living rivers: trends and challenges in science and management' of *Hydrobiologia/Developments in Hydrobiology*. It is structured along three lines: (1) Functioning of river systems; (2) Ecological rehabilitation of river systems; (3) Challenges and obstacles to sustainable management. Sustainable river basin management requires a thorough understanding of natural, physical, chemical and ecological processes as well as socio-economic functioning. Environmental rehabilitation of river systems inevitably requires an ecosystem-based approach involving solutions supported by the public. The challenge is to combine ecological and socio-economic sustainability. Rethinking our efforts in both directions must lead to more innovative solutions than those available at present.

Introduction

Rivers can be regarded as the arteries of water catchments (Wetzel, 2001). They have always attracted human settlement, because of their useful functions in society, in particular transport by shipping, so cities and villages have developed along the elevated river banks. Human impact has increased over the centuries, leaving few rivers or only parts of rivers in a pristine state (Nienhuis et al., 1998, 2002; Smits et al., 2000; Buijse et al., 2005). The rivers came to serve more and more functions, including water transport, drinking and irrigation water extraction, fisheries, discharge of pollutants, recreational purposes, sand and gravel mining, agriculture in the floodplains, hydropower generation, cooling water for power stations and industries, freshwater reservoirs and prevention of salt intrusion in coastal areas (Leuven et al., 2000).

Since many lowland river floodplains have been embanked by dikes to protect the hinterland against floods, the natural floodplains, with their wide range of habitats, have been narrowed to so-called river forelands (Van Stokkom et al., 2005; Bij de Vaate et al., 2006). The connections between rivers and the former riverine habitats, like moorland pools in river dunes and peat bogs, have become completely blocked. As a result, our view of riverine systems has also become narrowed to systems immediately adjoining the river, and nowadays former riverine systems are even regarded as independent systems, which need to be preserved (cf. Van Kleef et al., 2006). Because processes to generate or rejuvenate such systems have also become limited or stopped completely, conservation is a difficult task, requiring permanent human intervention in the ecological succession that would have been reset in a pristine system by fluvial and aeolic dynamics. The course of rivers became fixed by constructions, dams changed the river's hydrology and connectivity, and newly constructed canals linked the formerly isolated river basins. Flooding disasters always led to more measures to keep water out of inhabited areas and to reduce safety risks, and protection dams or storm barriers were constructed against floods from the sea, changing the highly dynamic estuaries into freshwater or saline lakes and changing sedimentation and sediment transport processes (Smits et al., 2006).

Nowadays it is recognized that all these drastic human interventions have resulted in a loss of biodiversity, a decline in the populations of migratory species including commercial fish, and reduced landscape connectivity and ecological functioning (Nienhuis et al., 1998; Smits et al., 2000). Attempts are being made to restore or rehabilitate these ecosystems as much as possible within the constraints put on river systems by society. This is a great challenge to water management, and can only be achieved by an interdisciplinary approach to river management, implemented by well-informed managers who combine engineering insights and skills with knowledge of ecological processes and who are sensitive to society's demands. This means that ecosystem processes should form the basis of our thinking on ways to use river systems. Although these processes and their dynamics may seem to have been tamed by human intervention, recent disasters (e.g. the severe floods along the rivers Rhine, Elbe and Odra in Europe and hurricane Katrina in the United States of America) have brought us down to earth. Each disaster causes us to rethink the way we have dealt with the problems in the past. However, as in wars, the solutions chosen have often been those that should have been chosen to overcome the previous disaster. It is time now to restart our thinking on river management from the viewpoint of natural processes and their dynamics. However, the main question is whether we can reconcile ecological processes with the increasing societal demands on river systems or whether this is an unsolvable dilemma. This was the key question debated at the symposium called 'Living rivers: trends and challenges in science and management' that was organized on 31 October 2003 at Radboud University Nijmegen in honour of Piet Nienhuis on his retirement (Leuven et al., 2006a). This special issue of *Hydrobiologia/Developments in Hydrobiology* presents the proceedings of this symposium, together with a number of complementary invited papers (Leuven et al., 2006b). The present paper primarily summarizes the results reported in this special issue. Its structure reflects the three sections of this special issue: (1) Functioning of river systems; (2) Ecological rehabilitation of river systems; (3) Challenges and obstacles to sustainable management. The synthesis concludes with a discussion of trends and challenges in river science and management.

Functioning of river systems

With the increasing need to restore or rehabilitate river systems, fundamental questions have arisen on the functioning of pristine rivers. This has led to the development of several ecological river concepts, such as the River Continuum Concept, the River Productivity Concept, the Flood Pulse Concept and the Intermediate Disturbance Concept (Nienhuis & Leuven, 1998; Van der Velde et al., 2004). These concepts are holistic views, thoughts and models which show the river as a dynamic ecotone (gradient or cline). It is very clear from these concepts, which are tools to understand natural processes and their impacts on ecosystem functioning, that processes upstream influence processes downstream and vice versa. Calculations with the Basin Box multimedia fate model show that spatial differences between upstream, midstream and downstream areas of large river basins may have a considerable impact on the environmental concentrations of chemicals (Hollander et al., 2006). Other examples are migration of fish species and natural damming, e.g. by beavers (Johnston & Naiman, 1990). Terrestrial allochthonous input from the riparian vegetation can be equally important as autochthonous primary production for the ecological functioning of river systems. Connectivity is another key word with respect to the natural functioning of riverine systems. Across the width of the floodplain, the aquatic, semi-aquatic and terrestrial biotopes normally show several degrees of connectivity, leading to a gradient of habitats and resulting in increased biodiversity. Conservation or rehabilitation of such gradients is essential to maintain the characteristic biodiversity, as demonstrated by De Nooij et al. (2006) for protected and endangered plant and animal species. Tockner et al. (2006) demonstrate this for amphibians. Open connections between tributaries and the main channel are essential to stock the river with fish when spawning places in the river channel itself are lost (Pollux et al., 2006). Tockner et al. (2006) emphasize that in an active floodplain, vegetated islands and large woody debris are important in maintaining both a variety of habitats and enhancing amphibian diversity and density. Maintaining a wide floodplain creates opportunities for biogeochemical gradients, increasing plant and other biodiversity (Lamers et al., 2006). River reconstruction and management should therefore aim at re-establishing the entire hydrodynamic gradient, increasing spatial heterogeneity and improving hydrodynamic conditions (De Nooij et al., 2006). Heterogeneity resulting from different landscape features is also important for the relatively slow process of recolonization by animals, for example small mammals, after floods (Wijnhoven et al., 2005, 2006).

Flooding can also cause a zonation of plant species, especially through the influence of occasional summer floods. It proved possible to relate this to the different flooding tolerance responses demonstrated by experiments (Van Eck et al., 2006).

The roughness of the three-dimensional floodplain vegetation structure, especially for the forest vegetation, can nowadays be measured by airborne laser scanning. Such information is useful for ecological studies as well as for the hydrodynamic modelling of rivers (Straatsma & Middelkoop, 2006). Geerling et al. (2006) studied aerial images to establish the spatio-temporal heterogeneity in a meandering part of the Allier river, analysing ecotope composition and dynamics. A characteristic senescence and rejuvenation rate was established and the authors conclude that the combination of hydro-morphological processes and ecological succession resulted in a temporal diversity of the riparian area. Ecotope diversity remained stable at a minimum river stretch of about 1.5 meander lengths for several years (a type of dynamic equilibrium or meta-stability state). The framework and approach of linking hydrological and ecological models allowed predictions and comparisons of the consequences of alternative naturalization scenarios of moist-soil plant species for the Illinois floodplain–river system (Ahn et al., 2006) and of vegetation composition for the Sava River (Baptist et al., 2006).

Ecological fieldwork remains the basis for management strategies (Nienhuis, 2006). Petts et al. (2006) argue for advancing science to underpin water resource and river management, to establish a unifying framework that will lead to new, appropriate tools for water resource management. Such developments have been necessitated by the European Union's Water Framework Directive. In the authors' opinion, there is a lack of data on the ecological significance of flow variability for habitat variability, biological population responses and the biological interactions between these populations. They propose novel sets of indicators of structural integrity and ecosystem health within fluvial hydrosystems, which could be linked to process and function for large floodplain rivers (Petts et al., 2006; Vugteveen et al., 2006).

Ecological rehabilitation of river systems

Ecological rehabilitation or restoration is required for the return and conservation of characteristic flora and fauna. A precondition for such ecological rehabilitation is good water quality, and in this context, the BasinBox model can be a useful tool to predict the fate of chemicals from various sources in river catchments (Hollander et al., 2006). Although water quality can be improved and can be regarded as one of the success stories of river rehabilitation (Admiraal et al., 1993; Bij de Vaate et al., 2006), the return of indigenous flora and fauna has been a slow and often unpredictable process (Buijse et al., 2005; Bij de Vaate et al., 2006).

Lenders & Knippenberg (2005) and Bij de Vaate et al. (2006) state that many alterations in the Rhine Delta are irreversible, so opportunities for the development of riverine biotopes are primarily to be found in the river forelands and not in the main river channels. Rivers that are heavily influenced by human activities suffer from a series of constraints, which hamper a complete return to natural values. These constraints have often also included transboundary problems, such as navigation, diffuse pollution, eutrophication and loss of habitats. Improving water quality while these constraints remained in place has often led to increased colonization by alien invaders (Van der Velde et al., 2002; Bij de Vaate et al., 2006). This tendency has been accelerated by new canals connecting river catchments (such as the Rhine–Main–Donau link). These canals have allowed endemic faunas from rivers to mix, changing communities, occupying the food web and forming a new constraint on the return of the original fauna (Van Riel et al., 2006). Lamers et al. (2006) comment that the present-day river forelands are too narrow for the development of interesting gradients in water quality that might enhance vegetation biodiversity. Furthermore, biogeochemical constraints result from letting in chemically changed river water elsewhere in the river catchment, and from the unnatural water level regimes in rivers caused by the influx of this water during dry periods. The influx of river water leads to eutrophication not only through the nutrients present in the river water itself but also by a process called internal eutrophication, which accelerates the decay of organic matter present in these waters. These changes caused by river water influx have occurred over more than 60% of the surface area of the Netherlands. However, some small reserves remain which are not strongly eutrophied and alkaline.

Ecological restoration measures have also been necessary in soft-water pools and raised bogs, which used to be part of the pristine river wetland (Van Duinen et al., 2006; Van Kleef et al., 2006). Formerly, natural large rivers featured many gradients from eutrophic, alkaline to oligotrophic, soft-water bodies. Measures taken to restore the original vegetation were, however, drastic and damaged macroinvertebrate biodiversity. Microinvertebrate biodiversity does not seem to change very much after rewetting of peat bog remnants, in contrast to that of macroinvertebrates. Van Kleef et al. (2006) and Van Duinen et al. (2006) show that biological traits and dispersal rates can predict recolonization patterns after restoration measures in these wetland environments.

Towards sustainable management

Since the industrial revolution, there have been various stages of single-interest river management. Since the end of the 1970s, this has evolved into more integrated approaches, focusing on the multifunctional use of the coherent whole of water systems (i.e. water, sediment, bed, bank, technical infrastructure and biological components). It has gradually evolved into integrated river basin management, which focuses on the entire river catchment (integrated catchment management). Integrative approaches mainly refer to technical, organizational and institutional resources and planning processes in river management. Sustainable river basin management refers to the management goal, i.e., sustainable development (Leuven et al., 2000). This requires a process of change in which the exploitation of resources, the allocation of investments, the orientation of technological developments and institutional arrangements must be in harmony and increase the present, as well as the future, opportunities to accommodate human needs (Brundtland Commission, 1987). Indispensable elements of sustainable river basin management are (Leuven et al., 2000):

- acceptance of the right to safe water resources as one of the basic human rights;
- attention to the long-term impacts of economic development on river systems and future generations;
- interdependence between environmental quality (e.g., healthy river and riverine ecosystems) and socio-economic development (e.g., improving the livelihoods of the basin's inhabitants);
- inter-basin water diversion and utilization with special attention for the improvement of in-stream uses of water, including fish production, protection of endangered species and recreational opportunities;

- use of water resources without degrading their quality, reducing their quantity or exceeding the carrying capacity of the river system, and inclusion of inevitable environmental damage costs in cost-benefit analyses;
- strategic planning with clear visions, goals and strategies for entire river basins (i.e., integrated or comprehensive catchment management planning);
- transparency and public participation via open planning processes and free access to credible and reliable data, information and knowledge about the quality status and impacts of management alternatives for river systems;
- mutual understanding and respect between riparian nations, and international co-operation, in particular to solve transboundary problems concerning water quality and quantity, and to guarantee security or solve conflicts over water resource diversion;
- development and application of sophisticated technical capabilities, e.g., integrated monitoring, modelling, decision support systems, user-friendly knowledge and information systems supporting stakeholder participation and ecologically sound hydraulic infrastructure, navigation and land use concepts.

The papers included in the section 'Challenges and obstacles to sustainable management' of this special issue pay attention to several aspects listed above. Wiering & Arts (2006) explored whether sustainable water management is only an adaptation for survival of the institutions or really constitutes institutional change. There are many arguments for the former view, because of the many hard economic functions of the present-day rivers, such as safety, shipping, drinking water supply and recreation, which make ecological sustainability a difficult goal to attain. De Groot & Lenders (2006) demonstrate that river management has been dominated mostly by scientific arguments for too long, but that sound management must also involve a cultural and historical dimension to be acceptable to society. Smits et al. (2006) present a provocative story with the aim of drastically changing our infrastructural thinking in such way that the river is working for society; in other words, they advocate ecosystem-based water management. We should not waste all of our energy taming natural processes, with inevitable loss of money and biodiversity. This, however, requires a totally different way of thinking: not battling against water but living with water or accommodating water (Wiering & Arts, 2006). In fact, whereas this new thinking requires the founding of new institutions, traditional water institutions still maintain their position of power, hampering such changes. However, there is a tendency to create effective regional administrations which function as network authorities (Witter et al., 2006). In addition, several tools have been developed that can help underpin sustainable river management, such as the River Ecosystem Health (REH) concept. The advantage of this concept is that it mobilizes scientists, practitioners and the public all alike (Vugteveen et al., 2006).

Trends and challenges

Based on his own vast experience, Nienhuis (2006) provides an historical overview of the way ecological research has changed water and nature management. It is very clear that over the last fifty years, firm concepts, sophisticated laboratory and field experiments and simulation models have produced a much better understanding of the processes in large riverine ecosystems. However, ecological insights have to be presented in such a way as to be accessible to the public, and have to be transformed into appropriate measures.

Another aspect is the state of affairs as regards the development of river science. International thinking in river research and management remains a distant prospect in most cases, in spite of the availability of more and more concepts and a rapidly growing body of literature. It seems that each group in each region has its own school (e.g. the Anglo-American, continental European and Australian approaches; cf. Ryder & Boulton, 2005). Most problems are local but we are still far from a general framework for river management, in the sense that a synthesis in which the whole is more than the sum of the parts is lacking.

River managers may continue to be guardians of science-based and whole-basin rationality, while at the same time interacting more successfully with society (Ryder & Boulton, 2005; De Groot &

Lenders, 2006). This forms the basis of the latest trend and challenge emerging in western European countries. National governmental agencies are becoming aware that they have the overall responsibility to protect their river basins against floods and/or environmental disasters. At the same time, however, they also realize that top-down policies and legislation will have the intended effect only in a few cases. For example, the policies formulated by the International Commission for the Protection of the Rhine (ICPR) in the 1970s proved to be rather effective in controlling unbridled emissions of pollutants in the river Rhine and its tributaries. However, when it comes to putting the concept of alternative land use into practice, so that society learns once again to live with natural dynamics instead of fighting them, things appear to be much more difficult. If the people involved fail to see the necessity of this new approach, it is very difficult to support these policies and people will do their utmost to find ways around them.

Furthermore, our thinking about river management seems to be driven by local and regional disasters. It may be questioned how much we learn from each other. We seem to remain ignorant about lessons learned elsewhere and discuss solutions only on a national scale. There are also large differences of opinion about what flooding risks can be considered acceptable in different countries and different situations. For example, hurricane Katrina recently breached the dikes of Lake Pontchartrain and caused extensive floods in and around New Orleans (Bohannon & Enserink, 2005). Before the disaster, a flooding risk of once every 200 years has been considered acceptable. Scientists at Louisiana State University had already predicted in 2001 that a disaster was waiting to happen and that thousands of people could die (Fischetti, 2001; Brouwer, 2003). The whole situation surprised engineers in the Netherlands (Knip, 2005). According to Dutch legislation, flooding risks must not exceed once every 10,000 years for the coastal areas and once every 1250 years for the central river district (Van Stokkom et al., 2005). We also know that people's memories are short and that disasters are soon forgotten, after which all measures taken are gradually neglected, and research funding may dwindle after a period of 5–10 years.

Long-term protection against floods and droughts and the creation of healthy river ecosystems require that rivers are given more space in their floodplains. Alternative land use, which can also provide interesting economic perspectives while simultaneously giving the river and its dynamics more opportunities, may be an option. Debate on these options is still ongoing, however, because negative effects also have to be considered (Van Stokkom et al., 2005). This grass-roots approach is the focal point of the European Interreg IIIB project entitled 'Freude am Fluss' (enjoy the river) which is being jointly carried out by French, German and Dutch organizations and centres of expertise (Anonymous, 2003). The natural dynamics of rivers and their water discharge patterns are no longer being depicted as potential threats, but as opportunities for making a fresh start and reducing flood damage risks. Within the next 5 years, two regional land-use plans and three local land-use plans for the river basins of the Rhine and Loire will be modified in accordance with the 'Freude am Fluss' concept. Moreover, this project will also present a show-case example of how natural dynamics and vegetation succession in floodplains can be managed to achieve a good balance between ecology, economy and safety (Peters et al., 2005; Geerling et al., 2006).

Conclusions

The major conclusions of the symposium and special issue 'Living rivers: trends and challenges in science and management' are:

- Example calculations with the Basin Box multimedia fate model show that spatial differences between upstream, midstream and downstream areas of large river basins may have a considerable impact on the environmental concentrations of chemicals.
- Moist soil plants are good indicators for water level regimes, but not for river–floodplain connectivity, for which fish species are the best indicators. Summer floods largely determine the long-term spatial gradient of grassland plants differing in flooding tolerance. This relation has also been experimentally confirmed, giving the observations greater predictive power.

- Monitoring vegetation structures is important for the modelling of habitat suitability, hydraulic roughness and erosion–sedimentation patterns in river floodplains. Modern remote sensing techniques, such as airborne laser scanning, can provide synoptic information on vegetation structure over large areas. Measuring senescence and rejuvenation processes in river-floodplain ecosystems by remote sensing techniques reveals high hydromorphological dynamics al local scale but also show meta-stability at landscape scale.
- Open connections between lowland tributaries and the main channels of regulated rivers are necessary to maintain phytophilic fish populations, as these lowland tributaries provide vegetation and shallow water, thus functioning as nurseries.
- In embanked floodplains along regulated rivers, terrestrial fauna recolonization after flooding events is a relatively slow process, which starts from elevated areas, indicating the importance of these areas for floodplain biodiversity. One can distinguish various types of species-specific recolonization of heterogeneous river floodplains by small mammals after recurrent flooding, depending on the connectivity between suitable landscape types.
- Protected and endangered species depend on the entire gradient of biotopes provided by the floodplain. Amphibians are also able to use the whole natural floodplain gradient, even for reproduction. Numbers of species, however, are generally higher in the riparian forest pools.
- Restoration of large rivers within current human constraints will lead to dominance of exotic invasive species and limited return of native species. Spatial heterogeneity, together with further water quality improvement, is necessary to restore native biodiversity. Restoration should include natural intra-basin connectivity in all dimensions, but also the prevention of further inter-basin connectivity between rivers in various ecoregions.
- In contrast to macro-invertebrates, micro-invertebrate assemblages in remnant peat bog areas can be restored with relatively simple rewetting measures.
- Rigorous measures to restore soft-water vegetation have led to a decreased diversity of macro-invertebrates with certain biological traits. Phasing rehabilitation measures over time and compartmentalizing the measures can help overcome these problems.
- Letting in river water in surrounding wetlands to safeguard agriculture leads to unnatural hydrological regimes but also to internal eutrophication of fen areas induced by high alkalinity and sulphate input. Maintaining biogeochemical diversity to ensure a biodiverse aquatic vegetation requires wider floodplains.
- Ecological insights need to be made more accessible to the public and to be transformed into appropriate measures. Our infrastructural thinking about river management should be drastically changed, in such way that the river is working for society (ecosystem-based water management). This requires a totally different way of thinking: not battling against the water but living with it or accommodating it. Greater emphasis on social and economic aspects and on community involvement is required to gain public support for river rehabilitation and sustainable river basin management. There is currently a tendency to create effective regional water administrations which function as network authorities. In addition, several novel tools for sustainable river basin management have been developed, such as the River Ecosystem Health concept. River management based on this concept mobilizes scientists, practitioners and the public alike.

Acknowledgements

The synthesis of this special issue was made possible by the contributions of all authors. The authors thank Mr. J. Klerkx for linguistic improvements to the manuscript. This is CWE publication No. 430.

References

Admiraal, W., G. van der Velde, H. Smit & W. G. Cazemier, 1993. The rivers Rhine and Meuse in the Netherlands: present state and signs of ecological recovery. Hydrobiologia 265: 97–128.

Ahn, C., D. M. Johnston, R. E. Sparks & D. C. White, 2006. Analysis of naturalization alternatives for the recovery of moist-soil plants in the floodplain of the Illinois River. Hydrobiologia 565: 217–228.

Anonymous, 2003. Freude am Fluss project, Interreg IIIB North-West Europe, Application Form Radboud University Nijmegen. Interreg IIIB NWE Secretariat ENO, Lille.

Baptist, M. J., M. Haasnoot, P. Cornelissen, J. Icke, G. van der Wedden, H. J. de Vriend & G. Gugić, 2006. Flood detention, nature development and water quality along the lowland river Sava, Croatia. Hydrobiologia 565: 243–257.

Bij de Vaate, A., R. Breukel & G. van der Velde, 2006. Long-term developments in ecological rehabilitation of the main distributaries in the Rhine delta: fish and macroinvertebrates. Hydrobiologia 565: 229–242.

Bohannon, J. & M. Enserink, 2005. Hurricane Katrina. Scientists weigh options for rebuilding New Orleans. Science 309: 1808–1809.

Brouwer, G., 2003. The creeping storm. Civil Engineering – ASCE 73: 46–55.

Brundtland Commission, 1987. Our Common Future. World Commission on Environment and Development of the United Nations. Oxford University Press, Oxford.

Buijse, A., F. Klijn, R. S. E. W. Leuven, H. Middelkoop, F. Schiemer, J. Thorp & H. Wolfert, 2005. Rehabilitation of large rivers: references, achievements and integration into river management. Archiv für Hydrobiologie Supplement 155(1–4) Large Rivers 15(1–4): 715–738.

De Groot, W. T. & H. J. R. Lenders, 2006. Emergent principles for river management. Hydrobiologia 565: 309–316.

De Nooij, R. J. W., W. C. E. P. Verberk, H. J. R. Lenders, R. S. E. W. Leuven & P. H. Nienhuis, 2006. The importance of hydrodynamics for protected and endangered biodiversity of lowland rivers. Hydrobiologia 565: 153–162.

Fischetti, M., 2001. Drowing New Orleans. Scientific American 285: 76–85.

Geerling, G. W., A. M. J. Ragas, R. S. E. W. Leuven, J. H. van den Berg, M. Breedveld, D. Liefhebber & A. J. M. Smits, 2006. Succession and rejuvenation in floodplains along the River Allier (France). Hydrobiologia 565: 71–86.

Hollander, A., M. A. J. Huijbregts, A. M. J. Ragas & D. van de Meent, 2006. BasinBox: a generic multimedia fate model for the predicting of the fate of chemicals in river catchments. Hydrobiologia 565: 21–38.

Johnston, C. A. & R. J. Naiman, 1990. Browse selection by beaver: effects on riparian forest composition. Canadian Journal of Forest Research 20: 1036–1043.

Knip, K. 2005. Overstroming New Orleans was zoals voorspeld. NRC Handelsblad 7 September 2005 (in Dutch).

Lamers, L. P. M., R. Loeb, A. M. Antheunissen, M. Miletto, E. C. H. E. T. Lucassen, A. W. Boxman, A. J. P. Smolders & J. G. M. Roelofs, 2006. Biogeochemical constraints on the ecological rehabilitation of wetland vegetation in river floodplains. Hydrobiologia 565: 165–186.

Lenders, H. J. R. & L. Knippenberg, 2005. The temporal and social dimensions of river rehabilitation: towards a multi-dimensional research perspective. Archiv für Hydrobiologie Supplement 155 (1–4) Large Rivers 15 (1–4): 119–131.

Leuven, R. S. E. W., A. J. M. Smits & P. H. Nienhuis, 2000. From integrated approaches to sustainable river basin management. In Smits, A. J. M., P. H. Nienhuis, & R. S. E. W. Leuven (eds), New Approaches to River Management. Backhuys Publishers, Leiden, 329–347.

Leuven, R. S. E. W., P. J. M. van den Heuvel, M. M. van Katwijk, P. M. J. Herman, G. van der Velde & A. M. J. Ragas, 2006a. Pieter Hendrik Nienhuis: aquatic ecologist and environmental scientist. Hydrobiologia 565: 1–18.

Leuven, R. S. E. W., A. M. J. Ragas, A. J. M. Smits & G. van der Velde (eds), 2006b. Living rivers: trends and challenges in science and management. Hydrobiologia 565: 1–367.

Nienhuis, P. H., R. S. E. W. Leuven & A. M. J. Ragas (eds), 1998. New Concepts for Sustainable Management of River Basins. Backhuys Publishers, Leiden.

Nienhuis, P. H. & R. S. E. W. Leuven, 1998. Ecological concepts for the sustainable management of lowland river basins: a review. In Nienhuis, P. H., R. S. E. W. Leuven & A. M. J. Ragas (eds), New Concepts for Sustainable Management of River Basins. Backhuys Publishers, Leiden, 7–34.

Nienhuis, P. H., A. D. Buijse, R. S. E. W. Leuven, A. J. M. Smits, R. J. W. de Nooij & E. M. Samborska, 2002. Ecological rehabilitation of the lowland basin of the river Rhine (NW Europe). Hydrobiologia 478: 53–72.

Nienhuis, P. H., 2006. Water and values: Ecological research as the basis for water management and nature management. Hydrobiologia 565: 261–275.

Peters, B., E. Kater & G. W. Geerling, 2005. Handboek Cyclisch Beheer – Omgaan met Spontane Natuur en Hydraulische Opstuwing. Staatsbosbeheer, Rijkswaterstaat, Stichting Ark & Radboud Universiteit, Nijmegen (in Dutch).

Petts, G.E., J. Nestler & R. Kennedy, 2006. Advancing science for water resources management. Hydrobiologia 565: 277–288.

Pollux, B. J. A., A. Korosi, W. C. E. P. Verberk, P. M. J. Pollux & G. van der Velde, 2006. Reproduction, growth and migration of fishes in a regulated lowland tributary: Potential recruitment to the river Meuse. Hydrobiologia 565: 105–120.

Ryder, D. & A. J. Boulton, 2005. Redressing the limnological imbalance: trends in aquatic ecology, management and conservation in Australia. Hydrobiologia 552: 159–166.

Smits, A. J. M., P. H. Nienhuis & R. S. E. W. Leuven (eds), 2000. New Approaches to River Management. Backhuys Publishers, Leiden.

Smits, A. J. M., P. H. Nienhuis & H. L. F. Saeijs, 2006. Changing estuaries, changing views. Hydrobiologia 565: 339–355.

Straatsma, M. W. & H. Middelkoop, 2006. Airborne laser scanning as a tool for lowland floodplain vegetation monitoring. Hydrobiologia 565: 87–103.

Tockner, K., I. Klaus, C. Baumgartner & J. V. Ward, 2006. Amphibian diversity and nestedness in a dynamic floodplain river (Tagliamento, NE-Italy). Hydrobiologia 565: 121–133.

Van der Velde, G., I. Nagelkerken, S. Rajagopal & A. bij de Vaate, 2002. Invasions by alien species in inland freshwater bodies in western Europe: the Rhine delta. In Leppäkoski, E., S. Gollasch, & S. Olenin (eds), Invasive

Aquatic Species of Europe. Distribution, Impacts and Management. Kluwer Academic Publishers, Dordrecht: 360–372.

Van der Velde, G., R. S. E. W. Leuven & I. Nagelkerken, 2004. Types of river ecosystems. In: Fresh surface water. Water Sciences, Engineering and Technology Resources. Encyclopedia of life support systems (EOLSS). UNESCO, EOLSS Publishers, Oxford, 1–29 (http://www.eolss.net/).

Van Duinen, G. A., Y. Zhuge, W. C. E. P. Verberk, A. M. T. Brock, H. H. van Kleef, R. S. E. W. Leuven, G. van der Velde & H. Esselink, 2006. Effects of rewetting measures in Dutch raised bog remnants on assemblages of aquatic Rotifera and microcrustaceans. Hydrobiologia 565: 187–200.

Van Eck, W. H. J. M., J. P. M. Lenssen, H. M. van de Steeg, C. W. P. M. Blom & H. de Kroon, 2006. Seasonal dependent effects of flooding on plant species survival and zonation: a comparative study of ten grassland species. Hydrobiologia 565: 59–69.

Van Kleef, H. H., W. C. E. P. Verberk, R. S. E. W. Leuven, H. Esselink, G. van der Velde & G. A. van Duinen, 2006. Biological traits successfully predict the effects of restoration management on macroinvertebrates in shallow softwater lakes. Hydrobiologia 565: 201–216.

Van Riel, M. C., G. van der Velde, S. Rajagopal, S. Marguillier, F. Dehairs & A. de bij Vaate, 2006. Trophic relationships in the lower Rhine food web during invasion and after establishment of the Ponto-Caspian invader *Dikerogammarus villosus*. Hydrobiologia 565: 39–58.

Van Stokkom, H. T. C., A. J. M. Smits & R. S. E. W. Leuven, 2005. Flood defense in the Netherlands: a new era, a new approach. Water International 30: 76–87.

Vugteveen, P., R. S. E. W. Leuven, M. A. J. Huijbregts & H. J. R. Lenders, 2006. Redefinition and elaboration of river ecosystem health: perspective for river management. Hydrobiologia 565: 289–308.

Wetzel, A. G., 2001. Limnology: Lake and River Ecosystems (3rd edn.). Academic Press, San Diego.

Wiering, M. A. & B. J. M. Arts, 2006. Discursive shifts in Dutch river management: 'deep' institutional change or adaptation strategy? Hydrobiologia 565: 327–338.

Wijnhoven, S., G. van der Velde, R. S. E. W. Leuven & A. J. M. Smits, 2005. Flooding ecology of voles, mice and shrews: the importance of geomorphological and vegetational heterogeneity in river floodplains. Acta Theriologica 50: 453–473.

Wijnhoven, S., G. van der Velde, R. S. E. W. Leuven & A. J. M. Smits, 2006. Modelling recolonisation of heterogeneous river floodplains by small mammals. Hydrobiologia 565: 135–152.

Witter, J. V., H. T. C. van Stokkom & G. Hendriksen, 2006. From river management to river basin management: a water manager's perspective. Hydrobiologia 565: 317–325.

Hydrobiologia (2006) 565:369–371
R.S.E.W. Leuven, A.M.J. Ragas, A.J.M. Smits & G. van der Velde (eds), Living Rivers: Trends and Challenges in Science and Management
DOI 10.1007/s10750-006-0229-6

Subject index

H

I

J

L

M

N

O

P

R